M-Health

Emerging Mobile Health Systems

TOPICS IN BIOMEDICAL ENGINEERING
INTERNATIONAL BOOK SERIES

Series Editor: Evangelia Micheli-Tzanakou
Rutgers University
Piscataway, New Jersey

Signals and Systems in Biomedical Engineering:
 Signal Processing and Physiological Systems Modeling
Suresh R. Devasaahayam

Models of the Visual System
Edited by George K. Hung and Kenneth J. Ciuffreda

PDE and Level Sets: Algorithmic Approaches to Static and Motion Imagery
Edited by Jasjit S. Suri and Swamy Laxminarayan

Frontiers in Biomedical Engineering:
 Proceedings of the World Congress for Chinese Biomedical Engineers
Edited by Ned H.C. Hwang and Savio L-Y. Woo

Handbook of Biomedical Image Analysis:
 Volume I: Segmentation Models Part A
Edited by Jasjit S. Suri, David L. Wilson, and Swamy Laxminarayan

Handbook of Biomedical Image Analysis:
 Volume II: Segmentation Models Part B
Edited by Jasjit S. Suri, David L. Wilson, and Swamy Laxminarayan

Handbook of Biomedical Image Analysis:
 Volume III: Registration Models
Edited by Jasjit S. Suri, David L. Wilson, and Swamy Laxminarayan

M-Health: Emerging Mobile Health Systems
Edited by Robert S.H. Istepanian, Swamy Laxminarayan, and
Constantinos S. Pattichis

Robert S.H. Istepanian
Swamy Laxminarayan
Constantinos S. Pattichis
(Editors)

M-Health

Emerging Mobile Health Systems

With 182 Illustrations

 Springer

Robert S.H. Istepanian
Kingston University
London
UK

Swamy Laxminarayan
Idaho State University Biomedical Research
   Institute and the Institute of Rural Health
Pocatello, Idaho
USA

Constantinos S. Pattichis
University of Cyprus
Nicosia
Cyprus

ISBN 978-1-4419-3892-3   ISBN 978-0-387-26559-9 (eBook)

Printed on acid-free paper.

9 8 7 6 5 4 3 2 1

springeronline.com

"Ask, and it shall be given you; seek, and

you shall find; Knock, and it shall be opened

to you. For everyone who asks, receives;

and he who seeks, finds; and to him who

knocks, it shall be opened."

Matthew 7:7-8.

Dedication

To the God Almighty for His Blessings and also
to my family (my Wife Helen and to my two Daughters, Carolyn and Sarah).

Robert S.H. Istepanian

To the tireless work and contributions of all my students all around the world and
to my family (my wife Marijke, my son Vinod and my daughter Malini).

Swamy Laxminarayan

To the memory of my cousin Andreas Procopiou who, despite physical disability,
lived a fruitful life, offering happiness to the people around him.
He believed in innovative technology for making the life of the disabled better.

Constantinos S. Pattichis

Contributors

Editors

Robert S.H. Istepanian
Mobile Information and Network Technologies Research Centre (MINT),
School of Computing and Information Systems,
Kingston University, London
UK
r.istepanian@kingston.ac.uk
http://technology.kingston.ac.uk/MINT

Swamy Laxminarayan
ISU Biomedical Research Institute and the
Institute of Rural Health, Pocatello, Idaho
USA
s.n.laxminarayan@ieee.org

Constantinos S. Pattichis
Department of Computer Science
University of Cyprus
Cyprus
pattichi@ucy.ac.cy
http://www.cs.ucy.ac.cy/People/Profiles/

Section Editors

Lymberis Andreas
European Commission, Directorate General Information Society, Av. De Beaulieu 31,
1160 Brussels
Andreas.lymberis@cec.eu.int
www.cordis.lu/ist/directorate_c/ehealth/index.html

Nugent Chris D.
University of Ulster, Jordanstown, Northern Ireland, BT37 0QB.
cd.nugent@ulster.ac.uk;

Pattichis Marios S.
Dept of Electrical and Computer Engineering, and Computer Engineering
The University of New Mexico
ECE Building, The University of New Mexico, The University of New Mexico, Albuquerque,
NM 87131-1356.
pattichis@ece.unm.edu

Pavlopoulos Sotiris A.
National Technical University of Athens Biomedical Engineering Laboratory, Department
of Electrical and Computer Engineering, National Technical University of Athens, 9 Iroon
Polytecniou str., Zografou Campus, 15773, Athens, Greece
spav@biomed.ntua.gr
www.biomed.ntua.gr

Vieyres Pierre
Université d'Orléans, Laboratoire Vision et Robotique, 63 Ave de Lattre de Tassigny,
18020 Bourges Cedex, France
Pierre.Vieyres@bourges.univ-orleans.fr

Authors

Abdallah Rony
The George Washington University Hospital, The George Washington University Hospital,
2150 Pennsylvania Avenue NW 6A, Washington DC 20037
ronya@gwu.edu

Alesanco Álvaro
University of Zaragoza, Communication Technologies Group (GTC), Aragón Institute of
Engineering Research (I3A), University of Zaragoza, Zaragoza, SPAIN
alesanco@unizar.es;

Altieri Roberto
Area Manager, Nergal S.r.l. Viale B. Bardanzellu, 8 - 00155 Rome, Italy
altieri@nergal.it
www.nergal.it

Andreou Panayiotis
Department of Computer Science, University of Cyprus, University of Cyprus,
P.O. Box 20537, Nicosia, CY 1678, Cyprus
cs98ap1@cy.ac.cy

Arbeille Philippe
Université de Tours, Centre Hospitalier Universitaire Trousseau, 37032 Tours Cedex,
France
Arbeille@med.univ-tours.fr

Bali Rajeev K.
Knowledge Management for Healthcare (KMH) subgroup, Biomedical Computing
Research Group (BIOCORE), Coventry University, Priory Street, Coventry, CV1 5FB,
United Kingdom
r.bali@ieee.org
www.mis.coventry.ac.uk/biocore/kmh

Beglinger Christoph Prof. Dr. med
University Hospital Basel, Department of Gastroenterology, Petersgraben 4,
CH-4031 Basel, Switzerland
beglinger@ tmr.ch

Ben Shaphrut Oded
Hebrew University Jerusalem and Interuniversity Institute, Eilat, Israel,
P.O.B 469, Eilat 88103, Israel
odediver@pob.huji.ac.il

Black Norman D.
University of Ulster, Jordanstown, Northern Ireland, BT37 0QB, UK
Nd.black@ulster.ac.uk

Bonis Julio
IMIM. IMAS. Fundació IMIM, c/ Doctor Aiguader, 80, 8003 Barcelona
jbonis@imim.es

Bults Richard
University of Twente, Faculty of Electrical Engineering, Mathematics and Computer
Science, University of Twente, PO Box 217, 7500 AE, Enschede, The Netherlands
r.g.a.bults@ewi.utwente.nl

Christodoulou Eleni
Department of Computer Science, University of Cyprus,
P.O. Box 20537, Nicosia, CY 1678, Cyprus
Eleni_Christodoulou@cytanet.com.cy

Clamp Susan
Clinical Information Science Unit, University of Leeds, 26 Clarendon Road,
Leeds LS2 9NZ, United Kingdom
s.clamp@leeds.ac.uk

Cornelis Jan
Department of Electronics and Information Processing (ETRO),
Vrije Universiteit Brussel, Pleinlaan 2, B-1050 Brussels, Belgium
jpcornel@etro.vub.ac.be
www.etro.vub.ac.be/Members/cornelis.jan/personal_private.htm

Courrèges Fabien
Université d'Orléans, Laboratoire Vision et Robotique, 63 Ave de Lattre de Tassigny,
18020 Bourges Cedex, France
Fabien.Courreges@bourges.univ-orleans.fr

de Toledo Paula
Grupo de Bioingeniería y Telemedicina, Universidad Politécnica de Madrid,
E.T.S. Ingenieros de Telecomunicación. Ciudad Universitaria. 28040 Madrid
paula@gbt.tfo.upm.es

del Pozo Francisco
Grupo de Bioingeniería y Telemedicina, Universidad Politécnica de Madrid,
GBT ETSI Telecomunicación
Ciudad Universitaria sn. 28040 MADRID, SPAIN
fpozo@gbt.tfo.upm.es

Dhaen Christoffel
Language and Computing, Maaltecenter Blok A, Derbystraat 79,
9051 Sint-Denijs-Westrem, Belgium
christoffel@landc.be

Dikaiakos Marios
Department of Computer Science, University of Cyprus,
P.O. Box 20537, Nicosia, CY 1678, Cyprus
mdd@ucy.ac.cy

Doarn Charles R, Executive Director
Center for Surgical Innovation, Research Associate Professor, Department of Surgery,
University of Cincinnati, SRU 1466, Cincinnati, OH 45267-0558
charles.doarn@uc.edu

Dokovsky Nikolai
Faculty of Electrical Engineering, Mathematics and Computer Science, University of
Twente, PO Box 217, 7500 AE, Enschede   The Netherlands
n.t.dokovski@ewi.utwente.nl

Dos Santos Mariana Casella
Language and Computing, Maaltecenter Blok A, Derbystraat 79,
9051 Sint-Denijs-Westrem, Belgium
mariana@landc.be
www.landc.be

Drion Benoit
Airial, 3 rue Bellini, 92800 Puteaux, France

Dunbar Angela
IMIM. IMAS. Fundació IMIM, c/ Doctor Aiguader, 80, 8003 Barcelona
adunbar@imim.es

Dwivedi Ashish N.
Knowledge Management for Healthcare (KMH) subgroup, Biomedical Computing
Research Group (BIOCORE), Coventry University, Priory Street, Coventry, CV1 5FB
United Kingdom
ashishdwivedi@ntlworld.com
www.mis.coventry.ac.uk/biocore/kmh/

Dyson Anthony
Eddabbeh Najia
BFC, 196 rue Houdan, F-92330 Sceaux (France)
najia.eddabbeh@business-flow.com

Eich Hans-Peter
Heinrich-Heine-University, Dusseldorf, Moorenstr. 5, D - 40225 Düsseldorf, Germany
eich@uni-duesseldorf.de

Falas Tasos
Department of Computer Science, University of Cyprus,
P.O. Box 20537, Nicosia, CY 1678, Cyprus
tfalas@ucy.ac.cy

Finlay Dewar D.
University of Ulster, Jordanstown, Northern Ireland, BT37 0QB.
d.finlay@ulster.ac.uk

Fischer Hans Rudolf, PhD
University Hospital Basel, Department of Gastroenterology, Petersgraben 4,
CH-4031 Basel, Switzerland
hrfischer@eblcom.ch
www.uhb-moebius.ch

García José
Communication Technologies Group (GTC), Aragón Institute of Engineering Research
(I3A), University of Zaragoza, Zaragoza, SPAIN
jogarmo@unizar.es

García-Olaya
Angel Grupo de Bioingeniería y Telemedicina, Universidad Politécnica de Madrid, GBT,
ETSI Telecomunicación, Ciudad Universitaria sn., 28040 MADRID, SPAIN
agarcia@gbt.tfo.upm.es

Georgiadis Dimosthenis
Department of Computer Science, University of Cyprus,
P.O. Box 20537, Nicosia, CY 1678, Cyprus
cspggd@ucy.ac.cy

Giovas Periklis, MD
Cardiology Laboratory, 1st Department of Propaedeutic Medicine,
Laikon Hospital, Athens University, Athens, Greece
periklisgiovas@ath.forthnet.gr

Goens Beth M.
Pediatric Cardiology, University of New Mexico, MSC10 5590,
1Albuquerque, NM 87131-0001
mbgoens@salud.unm.edu

Gómez Enrique J.
Grupo de Bioingeniería y Telemedicina, Universidad Politécnica de Madrid,
GBT, ETSI Telecomunicación, Ciudad Universitaria sn., 28040 MADRID, SPAIN
egomez@gbt.tfo.upm.es

Gutzwiller Jean-Pierre
University Hospital, CH-4031 Basel, Switzerland.

Hadjileontiadis Leontios J.
Dept. of Electrical and Computer Engineering, School of Engineering,
Aristotle University of Thessaloniki, GR-54124 Thessaloniki, Greece
leontios@auth.gr

Hernando M. Elena
Grupo de Bioingeniería y Telemedicina, Universidad Politécnica de Madrid,
GBT, ETSI Telecomunicación, Ciudad Universitaria sn., 28040 MADRID, SPAIN
elena@gbt.tfo.upm.es

Herzog Rainer
Ericsson GmbH, Ericsson GmbH, Maximilianstrasse 36/RG,
D-80539 Munich, Germany
rainer.herzog@ericsson.com

Incardona Francesca
Informa s.r.l. - Arakne s.r.l., V. dei Magazzini Generali,
31, 00154 Roma, Italy
f.incardona@informacro.info
http://www.informacro.info

Istepanian Robert S.H.
Mobile Information and Network Technologies Research Centre (MINT),
School of Computing and Information Systems, Kingston University,
Kingston upon Thames, London, UK.
r.istepanian@kingston.ac.uk
http://technology.kingston.ac.uk/MINT

Jimenez Silvia
Grupo de Bioingeniería y TelemedicinaUniversidad Politécnica de Madrid,
E.T.S. Ingenieros de Telecomunicación., Ciudad Universitaria. 28040 Madrid
sjimene@gbt.tfo.upm.es z

Jones Val
Faculty of Electrical Engineering, Mathematics and Computer Science,
University of Twente, PO Box 217, 7500 AE, Enschede, The Netherlands
V.M.Jones@ewi.utwente.nl
aps.cs.utwente.nl/

Jovanov Emil
Electrical and Computer Engineering Dept.,
The University of Alabama in Huntsville, Huntsville, AL, 35899, USA
jovanov@ece.uah.edu
www.ece.uah.edu/~jovanov

Kirke Chris
Clinical Information Science Unit, University of Leeds, 26 Clarendon Road,
Leeds, LS2 9NZ, United Kingdom

Kirkilis Harris, Sales Manager
Relational Technology, 13 Posidonos Ave., "AEGEAN" Building, 174 55 Alimos,
Athens, Greece
hkirki@relational.gr
www.relational.gr

Konstantas Dimitri
University of Geneva, Centre Universitaire D'Informatique, Rue General-Dufour 24,
CH-1211 Geneve 4, Switzerland
Dimitri.konstantas@cui.unige.ch

Kontaxakis Georgios
Universidad Politécnica de Madrid   ETSI Telecomunicación, Dpto. Ing. Electrónica,
Ciudad Universitaria s/n, 28040 Madrid, Spain
gkont@die.upm.es

Koprinkov George
Faculty of Electrical Engineering, Mathematics and Computer Science, University of
Twente, PO Box 217, 7500 AE, Enschede, The Netherlands
g.t.koprinkov@ewi.utwente.nl

Koutsouris Dimitris
Head, Biomedical Engineering Laboratory, Department of Electrical and Computer
Engineering, National Technical University of Athens, 9 Iroon Polytecniou str., Zografou
Campus, 15773, Athens, Greece
dkoutsou@biomed.ntua.gr
www.biomed.ntua.gr

Kovačević Branko
University of Belgrade, Faculty of Electrical Engineering, Bulevar Kralja Aleksandra 73,
11000, Belgrade, Serbia and Montenegro

Kyriacou Efthyvoulos
Department of Computer Science, University of Cyprus, 75 Kallipoleos str,
P.O.Box 20537, 1678, Nicosia, Cyprus
ekyriac@ucy.ac.cy
www.medinfo.cs.ucy.ac.cy

Lauzan José Enrique
SCHULERBERGSEMA, Albarracín, 25 28037 Madrid, Spain

Lavigne Kevin
Vermont Arctic Education Program
losepia@yahoo.com

Litos George
Informatics and Telematics Institute, 1st Km Thermi-Panorama Road, Thermi,
Thessaloniki GR-57001, Greece

Lugg Desmond J
Director, Division of Extreme Environments, Office of the Chief Health and medical Office,
NASA HQ. Washington, DC, 20546, USA
desmond.e.lugg@nasa.gov

Marković Milan, Dr
Project Manager, IT Department, Delta banka a.d., 7b Milentija Popovića,
11070 Belgrade, Serbia and Montenegro
milan.markovic@deltabanka..co.yu
www.deltabanka.co.yu

Merrell Ronald C., MD
Medical Informatics and Technology Applications Consortium,
Virginia Commonwealth University, 1101 E Marshall Street,
PO Box 980480, Richmond, VA 23298-0480
rcmerrel@mail2.vcu.edu

Montyne Frank
Language and Computing, Maaltecenter Blok A, Derbystraat 79,
9051 Sint-Denijs-Westrem, Belgium
frank@landc.be

Munteanu Adrian
Department of Electronics and Information Processing (ETRO),
Vrije Universiteit Brussel, Pleinlaan 2, B-1050 Brussels, Belgium
acmuntea@etro.vub.ac.be
www.etro.vub.ac.be/Members/MUNTEANU.Adrian/personal_private.htm

Naguib Raouf N.G.
Biomedical Computing Research Group (BIOCORE), Coventry University,
Priory Street, Coventry, CV1 5FB, United Kingdom
r.naguib@coventry.ac.uk
www.mis.coventry.ac.uk/biocore

Nassar Nahy S
Biotech Associates Limited, PO Box 3156, COVENTRY, West Midlands CV8 3YU,
England
n.nassar@ biotechassociates.co.uk
www.biotechassociates.co.uk

Nicogossian Arnauld E., MD
School of Public Policy, George Mason University, 440 University Drive,
MS 3C6 Finley Building
anicogos@gmu.edu

Nikolakis George
Informatics and Telematics Institute, 1st Km Thermi-Panorama Road,
Thermi-Thessaloniki GR-57001, Greece

Novales Cyril
Université d'Orléans, Laboratoire Vision et Robotique, 63 Ave de Lattre de Tassigny,
18020 Bourges Cedex, France
Cyril.Novales@bourges.univ-orleans.fr

Ohmann Christian
Heinrich-Heine-University, Dusseldorf Moorenstr. 5,
D - 40225 Düsseldorf, Germany
ohmannch@uni-duesseldorf.de

Olmos Salvador
Communication Technologies Group (GTC), Aragón Institute of Engineering Research
(I3A), University of Zaragoza, Zaragoza, SPAIN
olmos@unizar.es

Orphanoudakis Stelios C.
FORTH and University of Crete   Director, Foundation for Research and Technology
Hellas (FORTH), PO Box 1385, GR 711 10, Heraklion, Crete, Greece
orphanou@ics.forth.gr
www.ics.forth.gr/cmi-hta/orphanoudakis.html

Owens Frank J.
University of Ulster, Jordanstown, Northern Ireland, BT37 0QB.
Fj.owens@ulster.ac.uk

Papachristou Petros
ATKO SOFT, 3 Romanou Melodou, Str. Marousi 151 25, Athens, Greece
ppap@atkosoft.com

Papadopoulos Constantinos
Department of Computer Science, University of Cyprus,
P.O. Box 20537, Nicosia, CY 1678, Cyprus
cspgcp3@cs.ucy.ac.cy

Papadopoulos George A.
Department of Computer Science, University of Cyprus,
P.O. Box 20537, Nicosia, CY 1678, Cyprus
george@cs.ucy.ac.cy

Papadoyannis Demetrios
Cardiology Laboratory, 1st Department of Propaedeutic Medicine, Laikon Hospital, Athens
University, Athens, Greece

Papazachou Ourania, MD
Cardiology Laboratory, 1st Department of Propaedeutic Medicine,
Laikon Hospital, Athens University, Athens, Greece
rpapazachou@hotmail.com

Peuscher Jan
Twente Medical Systems International, TMS International BV, P.O. Box 1123,
7500 BC Enschede, The Netherlands
jan.peuscher@tmsi.com

Pitsillides Andreas
Department of Computer Science, University of Cyprus,
P.O. Box 20537, Nicosia, CY 1678, Cyprus
Andreas.Pitsillides@ucy.ac.cy
www.ditis.ucy.ac.cy

Pitsillides Barbara
PASYKAF Larnaca, Kariders Court, Larnaca
cspitsil@cytanet.com.cy

Poisson Gérard
Université d'Orléans, Laboratoire Vision et Robotique, 63 Ave de Lattre de Tassigny,
18020 Bourges Cedex, France
Gerard.Poisson@bourges.univ-orleans.fr

Priddy Brent
Electrical and Computer Engineering Dept.,
The University of Alabama in Huntsville, Huntsville, AL, 35899, USA
toopriddy@gmail.com

Raskovic Dejan
Dept. of Electrical and Computer Engineering,
University of Alaska Fairbanks, Fairbanks, AK, 99775-5915, USA
raskovic@computer.org
www.uaf.edu/ece/raskovic.htm

Reichlin Serge
Division of Endocrinology, Diabetes, Metabolism and Molecular Medicine,
Tufts University School of Medicine, New England Medical Center, Boston, USA.
reichlin@dakotacom.net

Reid Innes
Clinical Information Science Unit, University of Leeds, 26 Clarendon Road,
Leeds, LS2 9NZ, United Kingdom

Ricci Roberto
Research Project Coordinator, Informa srl, via dei Magazzini Generali,
31 – 00154, Rome, Italy
r.ricci@informacro.info
www.informacro

Rienks Rienk
Heart Lung Centre and Central Military Hospital, Utrecht,
The Netherlands,
Heidelberlaan 100, 3584 CX Utrecht
r.rienks@chello.nl

Rodriguez Sergio
IMIM. IMAS, Fundació IMIM, c/ Doctor Aiguader, 80, 8003, Barcelona
srodriguez@imim.es

Rodriguez Paul V.
Dept of Electrical and Computer Engineering, ECE Building, The University of New
Mexico, Albuquerque, NM 87131-1356
prodrig@ece.unm.edu

Sakas Georgios
Dept. Cognitive Computing and Medical Imaging,
Fraunhofer Institute for Computer Graphics (IGD), Fraunhoferstr. 5,
Darmstadt, Germany D-64283
gsakas@igd.fhg.de

Samaras George
Department of Computer Science, University of Cyprus,
P.O. Box 20537, Nicosia, CY 1678, Cyprus
cssamara@ucy.ac.cy

Sancho Juan J.
HOSPITAL DEL MAR. IMIM. IMAS., Servei de Cirurgia General i Digestiva,
Passeig Maritim, 25-29, 08003 Barcelona, Spain
sancho@imim.es

Savić Zoran
SmartIS Group, NetSeT, Karađorđeva 65, 11000 Belgrade.

Schelkens Peter
Department of Electronics and Information Processing (ETRO),
Vrije Universiteit Brussel, Pleinlaan 2, B-1050 Brussels, Belgium
Peter.Schelkens@vub.ac.be
www.etro.vub.ac.be/Members/SCHELKENS.Peter/personal_private.htm

Shashar Nadav
Hebrew University Jerusalem and Interuniversity Institute, Eilat, Israel,
P.O.B 469, Eilat 88103, Israel
nadavs@cc.huji.ac.il

Shimizu Koichi
Laboratory of Biomedical Engineering,
Department of Bioengineering and Bioinformatics,
Graduate School of Information Science and Technology,
Hokkaido University, Sapporo, Japan
shimizu@bme.eng.hokudai.ac.jp

Skodras Athanassios N.
Department of Computer Science, School of Science and Technology,
Hellenic Open University, GR-262 22, Patras, Greece
skodras@eap.gr
http://dsmc.eap.gr

Smith-Guerin Natalie
Université d'Orléans, Laboratoire Vision et Robotique,
63 Ave de Lattre de Tassigny, 18020 Bourges Cedex, France
Natalie.Smith@bourges.univ-orleans.fr

Spanakis Manolis
FORTH and University of Crete   Institute of Computer Science (ICS),
Foundation for Research and Technology – Hellas (FORTH),
PO Box 1385, GR 711 10,
Heraklion, Crete, Greece
spanakis@ics.forth.gr
www.ics.forth.gr/cmi-hta/spanakis.html

Strintzis Michael
ITI CERTH   Informatics and Telematics Institute, 1st Km Thermi-Panorama Road,
Thermi-Thessaloniki GR-57001, Greece
strintzi@eng.auth.gr

Thierry Jean Pierre
SYMBION, 109 Rue des Cotes, 78600 Maisons-Laffitte, France
symbion@club-internet.fr

Thomakos Demetrios , General Manager
Proton Labs Ltd, Athens, Greece
protonlabs@ieee.org

Thomos Nikolaos
ITI CERTH   Informatics and Telematics Institute, 1st Km Thermi-Panorama Road,
Thermi-Thessaloniki GR-57001, Greece
nthomos@iti.gr

Torralba Verónica
Grupo de Bioingeniería y Telemedicina, Universidad Politécnica de Madrid,
GBT, ETSI Telecomunicación, Ciudad Universitaria sn., 28040 MADRID, SPAIN
torralba@gbt.tfo.upm.es

Traganitis Apostolos
FORTH and University of Crete   Institute of Computer Science (ICS),
Foundation for Research and Technology – Hellas (FORTH),
PO Box 1385, GR 711 10, Heraklion, Crete, Greece
tragani@ics.forth.gr
www.ics.forth.gr/cmi-hta/tragani.html

Triantafyllidis George
ITI CERTH   Informatics and Telematics Institute, 1st Km Thermi-Panorama Road,
Thermi-Thessaloniki GR-57001, Greece
gatrian@iti.gr

Tristram Clive
Ets TRISTRAM Clive (Futur Dessin), Les Rives, 86460 Availles Limouzine, FRANCE
Clive@free [clive.tristram@free.fr]
www.med-mobile.org/MEMO3/index.php

Tsiknakis Manolis
FORTH Institute of Computer Science (ICS), Foundation for Research and Technology –
Hellas (FORTH), PO Box 1385, GR 711 10, Heraklion, Crete, Greece
tsiknaki@ics.forth.gr
www.ics.forth.gr/cmi-hta/tsiknakis.html

Tzovaras Dimitris
Informatics and Telematics Institute, Center for Research and Technology Hellas,
57001 Thessaloniki, Greece
Dimitrios.Tzovaras@iti.gr

van Halteren Aart
Faculty of Electrical Engineering, Mathematics and Computer Science, University of
Twente, PO Box 217, 7500 AE, Enschede   The Netherlands
a.t.vanhalteren@ewi.utwente.nl

Vierhout Pieter
Faculty of Business, Public Administration and Technology, University of Twente,
P.O.Box 1123, 7500 BC Enschede, the Netherlands
p.vierhout@orde.nl

Virtuoso Salvatore
TXT e-Solutions, Via Frigia 27, 20126 Milano, Italy
Salvatore.virtuoso@txt.it

Walter Stefan
MedCom Gesellschaft für medizinische Bildverarbeitung mbH,
Rundeturmstr. 12, 64283 Darmstadt, Germany
swalter@medcom-online.de

Wang Haiying
University of Ulster, Jordanstown, Northern Ireland, BT37 0QB.
Haiying@infj.ulst.ac.uk

Widya Ing
Faculty of Electrical Engineering, Mathematics and Computer Science,
University of Twente, PO Box 217, 7500 AE, Enschede, The Netherlands
i.a.widya@ewi.utwente.nl

Preface

M-health can be defined as the 'emerging mobile communications and network technologies for healthcare systems'. This definition can also infer that m-health is the result of evolution of the e-health systems and the 'addition' of emerging information and computing technologies in biomedicine to the modern advances in wireless and nomadic communication systems. The recent years have witnessed a major revolution in the technological advances of the next generation of wireless and network technologies paving the way towards the 4G wireless systems. It is clear from these advances and in particular from the anticipated convergence between the future data rates of nomadic and wireless systems within the next decade or so, that such developments will have a challenging and profound impact on future e-health systems.

The recent research relevant to m-health such as advances in nano-technologies, compact biosensors, wearable, pervasive and ubiquitous computing systems will all lead the successful launch of next generation m-health systems within the next decade. They will encompass all these technologies for future healthcare delivery services with the vision of 'empowered healthcare on the move'.

This book paves the path toward understanding the future of m-health technologies and services and also introducing the impact of mobility on existing e-health and commercial telemedical systems.

The book also presents a new and forward looking source of information that explores the present and future trends in the applications of current and emerging wireless communication and network technologies for different healthcare situation. It also provides a discovery path on the synergies between the 2.5G and 3G systems and other relevant computing and information technologies and how they prescribe the way for the next generation of m-health services.

The book contains 47 chapters, arranged in five thematic sections: Introduction to Mobile M-Health Systems, Smart Mobile Applications for Health Professionals, Signal,

Image, and Video Compression for M-Health Applications, Emergency Health Care Systems and Services, Echography Systems and Services, and Remote and Home Monitoring. Each section begins with a section overview, and ends with a chapter on future challenges and recommendations.

This book is intended for all those working in the field of information technologies in biomedicine, as well as for people working in future applications of wireless communications and wireless telemedical systems. It provides different levels of material to researchers, computing engineers, and medical practitioners interested in emerging e-health systems.

The book serves as the basis for understanding the future of m-health technologies and services, exemplifying the impact of mobility on existing m-health and commercial telemedical systems.

We wish to thank all the section editors for their valuable time and efforts in putting together the section chapters, the authors for their hard work and for sharing their experiences so readily, and the numerous reviewers for their valuable comments in enhancing the content of this book. Furthermore we would like to express our sincere thanks to Elena Polycarpou for her excellent secretarial work in communicating with the section editors, authors, and reviewers and for putting together this book. We thank also Dr Henry Wang from the MINT center, and the EU for their funding of most of the work in the MINT centre. Last but not least, we would like to thank, Aaron Johnson, Krista Zimmer and the rest of the staff at Springer for their understanding, patience and support in materializing this project.

We hope that this book will be a useful reference for all the readers in this important and growing field of research and to contribute to the roadmap of future m-health systems and improved and effective healthcare delivery systems.

Robert S. H. Istepanian
Swamy Laxminarayan
Constantinos S. Pattichis

Contents

II. SMART MOBILE APPLICATIONS FOR HEALTH
PROFESSIONALS
Andreas Lymberis, Section Editor

Juan J. Sancho, Susan Clamp, Christian Ohmann, José.E.Lauzàn, Petros Papachristou, Jean P.Thierry, Angela Dunbar, Julio Bonis, Sergio Rodríguez, Chris Kirke, Innes Reid, Hans-Peter Eich, and Clive Tristram

III. SIGNAL, IMAGE, AND VIDEO COMPRESSION FOR M-HEALTH APPLICATIONS
Marios S. Pattichis, Section Editor

I. INTRODUCTION TO MOBILE M-HEALTH SYSTEMS

Constantinos S. Pattichis
Robert S.H. Istepanian
Swamy Laxminarayan

Section Editors

UBIQUITOUS M-HEALTH SYSTEMS AND THE CONVERGENCE TOWARDS 4G MOBILE TECHNOLOGIES

Robert S. H. Istepanian[*], Costantinos S. Pattichis, and Swamy Laxminarayan

1. INTRODUCTION

The increased availability, miniaturization, performance and enhanced data rates of future mobile communication systems will have an increasing impact and accelerate the deployment of m-health systems and services within the next decade. M-Health can be defined as 'emerging mobile communications and network technologies for healthcare.' This emerging concept represents the evolution of e-health systems from traditional desktop 'telemedicine' platforms to wireless and mobile configurations. The expected convergence of future wireless communication, wireless sensor networks and ubiquitous computing technologies will enable the proliferation of such technologies around healthcare services with both cost-effective, flexible and efficient ways. These advances will have a powerful impact on some of the existing healthcare services and will reshape some of the mechanisms of existing healthcare delivery routes.

It is expected that the developments in this area will evolve towards two complementary tracks. The first will focus on the emerging advances in Third Generation (3G) and beyond 3G (4G) technologies whilst, the second will track the recent developments in biosensors and their networks connectivity issues combined with the pervasive, wearable and ubiquitous computing m-health systems. The current (3G) mobile systems represent a distinctive leap from the 2.5G mobile systems supporting higher data rates and better Quality of Services (QoS) that are more suitable for demanding healthcare applications. It is well known that 3G mobile systems will be able to facilitate the demanding m-health applications with 'always on healthcare services'. These can now support a variety of medical and well being services within the operational bit rates from

[*] Robert S.H. Istepanian, Mobile Information and Network Technologies Research Centre (MINT), School of Computing and Information Systems, Kingston University, London KT22 , UK, e-mail: robert.istepanian@kingston.ac.uk

144 Kbps in mobile conditions and up to 2 Mbps in indoor scenarios.

This chapter outlines some of the current m-health systems and presents an overview of some successful applications. It also presents some of the future challenges and research issues dealing with the migration of these systems towards 4G mobile technologies and their convergence with such systems from the view of the two complementary tracks outlined above. The chapter also outlines the contributions presented in the introduction to mobile-health section of the book that represent some of these recent developments and illustrate the multidisciplinary nature of this important and emerging area. This will provide an informative reading for the next chapters of the book.

2. AN OVERVIEW OF CURRENT AND EMERGING M-HEALTH SYSTEMS

Recent years have seen the convergence of both computing and wireless communication systems towards next generation m-health systems. The recent announcement from the European Commission for plans to adapt common e-health policies and timeframe to implement European health information networks by 2008 is a clear example of the future trends in these critical areas of future healthcare systems (Europe Adopts e-health Timetables, 2004). The advances in the mobile technologies will soon prove to have a powerful impact on the way different healthcare delivery organizations will reshape the future of these services globally by utilizing such emerging technologies for the enhancement of future healthcare services. M-health is already redefining the original definition and concept of telemedicine as 'medicine practiced at a distance' to include the new mobility and 'invisible communication technologies' to reshape the future structure of global healthcare systems.

The increased availability, miniaturization, performance, enhanced data rates and the expected convergence of future wireless communication and network technologies around mobile health systems will accelerate the deployment of m-health systems and services within the next decade. These will have a powerful impact on some of the existing healthcare services and will reshape some of the mechanisms of existing healthcare delivery routes. For example, development of smart intelligent sensors and drug delivery devices, some of them implanted, will allow communication with a personal server in complete mobility. The personal server provides global connectivity to the telemedical server using a Wireless Personal Area Network (WPAN), Wireless Local Area Network (WLAN), or wireless Wide Area Network (WAN). The evolving mass markets mainly drive developments in these areas for cell phones and portable computing devices and represent an evolution of the previous generation of telemedical systems. A comprehensive overview of the recent advances in m-health systems can be cited in (Pattichis et al, 2002; Istepanian et al, 2004a; Istepanian et al, 2004b). However, a brief discussion on these technologies will also be presented here for completeness.

In general m-health systems can be classified into the following two categories that are briefly presented here: (i) Current 2.5G and 3G m-health systems; and (ii) Beyond 3G m-health and evolving systems. These are based on the synergy and evolution of the 'enabling technologies' in the areas of wireless communications, mobile networks, biosensors and computing technologies.

2.1 Current 2.5G and 3G M-Health Systems

Traditionally 'wireless telemedicine' is associated with the 'biomonitoring' concept. These have been used extensively in the last two decades to perform different data acquisition tasks mostly, without timely integration of data into the medical record; thus no immediate action occurs if abnormalities are detected. Typical examples are Holter monitors that are routinely used for electrocardiogram (ECG) and electroencephalogram (EEG) monitoring (Istepanian et al, 2000; Istepanian et al, 2004b).

Historically, 'wireless monitoring' includes physiological monitoring of parameters such as heart rate, blood pressure, blood oximetry and other physiological signals. Other areas include physical activity monitoring of parameters such as monitoring of movement, fall detection, location tracking, gastrointestinal telemetry and other physical activities (Pattichis et al, 2002). The benefits of the wireless technology have been illustrated in a number of different examples and applications (Istepanian et al, 2003).

Today with wireless technology, healthcare professionals from any given location could access patient records by connection to the institution's information system. Physicians' access to patient history, laboratory results, pharmaceutical data, insurance information, and medical resources would be enhanced by mobile technology, thereby improving the quality of patient care. Handheld devices can also be used in home health care, for example, to fight diabetes through effective monitoring and mobile management (Li et al, 2003).

In recent years there was a surge in commercial m-health systems that are based in General Packet Radio Service (GPRS) technologies and on their deployment strategies in healthcare systems (Istepanian et al, 2004b). The next few years will also see the deployment of similar systems based on the existing 3G wireless technologies.

The major drivers for mobility in healthcare can be summarized in the following factors:

(i)   Improved access to medical information and data.
(ii) Improved patient care and healthcare services.
(iii) Better specialist care and enhanced medical productivity.
(iv) To provide cost savings and reduced costs.
(v)   'Automation' of the medical data collection and processing mechanisms.

However, there are some limitations on the existing wireless technologies especially the existing commercial 'm-health services'. Some of these drawbacks can be summarised as follows (Istepanian et al, 2004b):

(i)   The lack of an existing flexible and integrated 'm-health-on-demand' linkage of different global mobile communication options and standards for these services. This lack of linkage and compatibility for existing systems is due to the difficulty of achieving operational compatibility issues for the operators and terminal providers and to the mobile devices standards 'm-health protocols'.

(ii) The high cost of communication links, especially between satellites and global mobile devices and the limitation of the existing wireless data rates especially for the globally available 2.5G and 3G services for some e-health services. This is also combined with the availability of secure mobile internet connectivity and information access especially for e-health systems.

(iii) Existing healthcare is a very complex industry and difficult to change and organizational changes are very often required for healthcare institutions to benefit from e-health and m-health services.

(iv)   The short term and long-term economic consequences and working conditions for physicians and healthcare experts using these technologies are not yet fully understood or properly investigated.

(v)   The methods of payment and re-imbursement issues for such e-health and m-health services are not yet fully developed and standardized.

(vi)   There is a lack of integration between existing e-health services and other information systems e.g. referral and ordering systems, medical records etc.

(vii) There are still no clear economic benefits from the ongoing demonstration projects that will show that m-health services show real savings and cost effective potential.

The above represent some of the factors that have hindered the wider applications of m-health technologies thus far across healthcare systems. However, it is hoped that the current deployment of 3G based systems with global operational modalities will alleviate some of these issues and will provide a better and more effective platform for mobile healthcare services.

2.2 Beyond 3G M-Health and Evolving Systems

It is expected that the 4G mobile systems will focus on seamlessly integrating the existing wireless technologies including GPRS, 3G, wireless LAN, Bluetooth, and other newly developed wireless systems into ' IP-based Core network of Heterogeneous access networks'. Some key features of 4G networks can be highlighted as follows (Istepanian et al, 2004a):

(i)   High usability. 4G networks are all IP based heterogeneous networks that allow users to use any system at anytime and anywhere. Users carrying an integrated terminal can use a wide range of applications provided by multiple wireless networks.

(ii)   Support for multimedia services at low transmission cost. 4G systems provide multimedia services with high data rate, good reliability and at low per-bit transmission cost.

(iii) This new-generation network will provide personalized services, in order to meet the demands of different users for different services.

(iv)   4G systems also provide facilities for integrating services. Users can use multiple services from any service provider at the same time.

These advances will reflect the implementation of the main objectives of the future wireless generation beyond 3G systems. These will not simply be able to provide higher data rates, more capacity and more spectrum efficiency, but more importantly to provide a successful framework of integrating all wired, mobile access, nomadic and wireless access systems into a cohesive, integrated and cooperating IP-based architecture interworking with all these existing and future networks.

Ongoing research in these areas should target overcoming the following challenges: applicable to m-health systems and services:

(i)   Challenges in user and device planes- the future architecture(s) will focus on enhancing user interaction capabilities especially with 'context-aware' applications. This issue from the m-health perspective will be challenging as it requires the clear definition of the users (patients, healthcare providers, roaming health-worried and other anticipated mobile users) of such systems. This will also require to define the exact m-health terminal requirements and the future 'personalized healthcare services' that these architecture(s) will bring together.

(ii) The security challenges- The m-health services are considered one of the main services for future mobile systems (i.e. 4G) that will require much research into the provision of secure and reliable access and exchange of medical data and information.

(iii) Challenges in the network and services planes- These will identify and provide independently the variety of 'asymmetric and symmetric' future medical data traffic requirements and hence allow the seamless delivery of the 'on-demand m-health services' that will adapt to both the capability of the underlying network from different healthcare operational modalities and environments.

(iv) The challenges from the socio-economic and business perspective- These will also need future research to focus on the integration of both the technological advances with the future demand of future m-health driven market and services that will evolve in the next few years.

(v) Global connectivity and policy issue and challenges- This will be a new paradigm that will present itself with the wider and global spread of mobile users and services that will provide a new focus on policy makers and the development of new standards for global healthcare services.

In summary, it is expected that 4G advances will provide both mobile patients and normal working end users the choices that will fit their lifestyle and make it easier for them to interactively get the medical attention and advice they need, when and where is required and how they want it regardless of any geographical barriers or mobility constraints. The concept of including high-speed data and other services integrated with voice services is emerging as one of the main points of the future telecommunication and multimedia priorities with the relevant benefits to citizen centered healthcare systems. These creative methodologies will support the development of new and effective medical care delivery systems into the 21st century. The new wireless technologies will allow both physicians and patients to roam freely, while maintaining access to critical medical data and information exchange.

2.3 The Future Scenarios and 4G M-Health Applications

Forthcoming 4G m-health systems could possibly focus their research in the following areas:

(i) The 'Personalized Predictive Healthcare'. The future mobile systems will be able to provide healthcare services that are based on the personal healthcare profile and history of the users. These are particularly important in areas such as in 'chronic diseases' populations that are estimated to be the major target group of such future mobile systems. These groups will make use of the future availability of technologies such as Body Area Networks (BAN) and their wireless connectivity with emerging cellular and nomadic systems.

(ii) 'Mobile On–Demand Home Health Care' services. These services form the major trend that will integrate future advances in pervasive computing and wearable technologies with 4G wireless systems. These will enable patients that are either home bound or with limited mobility to be monitored and managed interactively and virtually with their specialist nurse, healthcare providers and centers of care.

(iii) Virtual Mobile Hospitals and Specialised M-Health Centres. These will provide a direct wireless connectivity of the medical experts with their patients together with their In-Hospital medical data and information in secure and reliable fashion. Hence, a virtual environment to interact with their patients and consult with the non-experts and hospital residents from their remote locations and the provision of high quality medical multimedia access will be provided. These centers can also be very useful in cases of natural disasters.

3. AN ADVANCED APPLICATION OF A 3G M-HEALTH SYSTEM

In this section we briefly present an advanced m-health system as an example for the provision of futuristic m-health services. In particular we will present some of the critical issues on the performance of a compact Robotic Tele-Echography System over 2.5G and 3G mobile communication links. The performance analysis of 3G wireless communication links of OTELO (mObile Tele-Echography using an ultra Light rObot) end-to-end tele-echography m-health system is also presented. This system is also presented in greater detail in chapters 35 and 36.

3.1 The Mobile Tele-Echography Robotic System OTELO

It is well know that teleradiology and especially remote ultrasound echography is considered a quite mature application in telemedicine. This specific service can also be considered as one of the most applicable examples for future m-health systems (see section IV on Echography Systems and Services). The reasons for this is the well known fact that tele-ultrasound can provide specialised healthcare in remote medical settings and can also be used to evaluate the degree of emergency care for patients in remote areas or in urgent cases. Technically, these systems prove to be challenging due to the high data rate requirements of the relevant medical traffic and medical video streams of the system. The main drawback of current portable and mobile tele-ultrasound systems is that the quality of the examination depends highly on the operator's specialised skills, skills that are lacking in most health centres, especially in many small medical centres, isolated sites, or rescue vehicles. In some cases, for example no appropriate well-trained sonographer can be found to perform the first ultrasound echography examination from which the emergency level can be evaluated. Hence, the need for an advanced robotic system that can make available the specialist in a virtual setting to scan the remote subjects in such scenarios. In this section we present the OTELO system which was developed to provide a fully integrated end-to-end mobile tele-echography system for population groups that are not served locally, either temporarily or permanently, by medical ultrasound experts. It comprises of a fully portable tele-operated robot allowing a specialist sonographer to perform a real-time robotised tele-echography to remote patients (see chapters 35 and 36). This tele-echography system is composed of three main parts as shown in Fig. 1 in case of 3G wireless connectivity.

(i)   An « Expert » site where the medical expert interacts with a dedicated patented pseudo-haptic fictive probe instrumented to control the positioning of the remote robot and emulates an ultrasound probe that medical experts are used to handle, thus providing better ergonomic features.

(ii)   The communication options of the system. The OTELO system has developed an embedded communication software based upon IP protocol to adapt to the different communication options. These include the wired (ISDN, ADSL) based links, terrestrial and mobile communication links (see Fig. 1).

(iii) A « Patient » site made up of the 6 degrees of freedom light weight robotic system and its control unit (see Fig. 2).

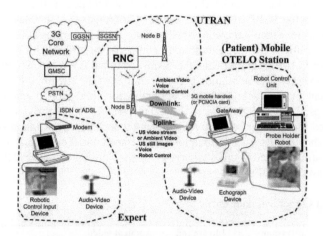

Figure 1. OTELO 3G wireless connectivity and telemedical configuration

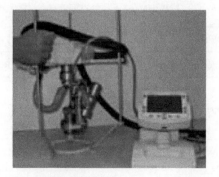

Figure 2. OTELO robot with the tele-echography unit

The functionality and telemedical set-up of the system in the case of 3G wireless connectivity is shown in Fig. 1. This illustrates the virtual operation of the system between the specialist and remote mobile robotic stations. The details of these are presented in Courreges et al, 2004a and Courreges et al, 2004b. The functional modalities of the OTELO system are summarized in Table 1. These include medical video streaming and ambient (video conferencing data) and also voice and robotic data communications between the specialist and remote sites (see also chapters 35 and 36).

3.2 OTELO 3G Performance and Evaluation

In this study we analyse the asymmetrical medical data transmission of the system with the relevant QoS requirements. No sensitive delay data transmission is based on

FTP/TCP/IP (e.g. still ultrasound images) and delay sensitive data transmission is based on UDP, RTP (Real Time Protocol), (e.g. robot control, stream of ultrasound images and ambient video). Hence we made different approaches by emulations and real network tests, to measure the asymmetrical data rates, real-time interaction, delay, network jitter and the average throughput for all types of data for real-time and no real-time performance.

The OTELO traffic is tabulated in Table 1. It is mapped to the three major traffic classes defined by the 3GPP UMTS QoS Classes (Courreges et al, 2004c). The best-suited UMTS QoS class for video streaming is 'Streaming class, streaming RT' which preserves time relation (variation) between information entities of the stream. But for medical image sequence with real-time requirements, the 'Conversational RT' class would be necessary. In addition to preserving time relation between entities of the stream, it has conversational pattern (stringent and low delay) which is preferable for a real-time interaction. For the 3G experimental tests, the selected uplink throughput was 64 Kbps provided by Vodafone, UK. Ultrasound video scans were streamed using a commercial program (Netmeeting™ software). Further details on the compression methods for the medical images and the medical video streams used in this study are given in (Courreges et al, 2004b; Courreges et al, 2004c; Garawi et al, 2004).

Table 1. OTELO medical data traffic

	Ultrasound video stream	Ultrasound still images	Ambient video stream	Voice	Robot control data
Flow direction	Simplex: Patient to Expert	Simplex: Patient to Expert	Duplex	Duplex	Duplex
Trans-port protocol	RTP/ UDP/IP	TCP/IP	RTP/ UDP/IP	RTP/UDP/IP	UDP/IP
Speed requirement	Real-time	Non Real-time	Real-time	Real-time	Real-time
Payload data rate requirement over the air-interface	15 frames/s @ 210 Kbps Uplink	1 frame/ 10s Uplink	1 to 15 frame/s Symmetrical	9.0 Kbps Symmetrical	0.3 Kbps Symmetrical

3.2.1 Throughput performance

Figure 3 shows the performance of the real-time image transmission throughput performance received at the expert station of the 3G OTELO system within the 64 Kpbs data rate. The results of this test, indicate the efficient requirements of this rate to deliver medically acceptable ultrasound images from the robotic system.

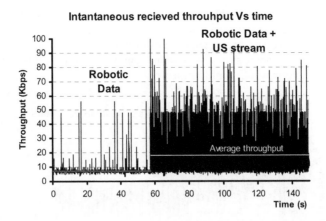

Figure 3. Real-time received ultrasound (US) images throughput by the expert station @ UMTS-64 Kbps

Figure 4 shows the real-time performance of the 3G system for the medical video streaming case. Figures 3 and 4 also show that when the robotic data are sent simultaneously with the ultrasound stream, the average throughput occupied by the ultrasound video stream is 10.6 Kbps (Fig. 4) while the average throughput occupied by the robotic data is still 6.8 Kbps as when no video is sent. During this period the average total throughput is 18.7 Kbps with a standard deviation of 12.7 Kbps and the range is as high as 158 Kbps. The minimum rate encountered during the testing phase was 4 Kbps.

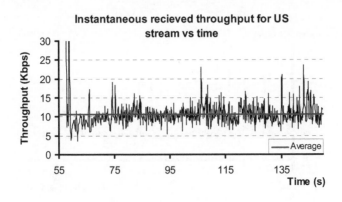

Figure 4. Real-time received throughput of the ultrasound video (US) streams @UMTS-64Kbps

3.2.2 Delay and End-to-End Latency Performance

The other important QoS metric requirements tested for the system are: the received packets performance, delay jitters encountered by the robotic data and the end-to end latency of the system. The estimated time as the robotic data packets are sent periodically is every 70 ms. Figure 5 shows the received ultrasound medical video data packets of the OTELO system in the tested 3G network system. It has also to be noted that the jitter performance was evaluated as the standard deviation between these time intervals. The jitters of the video packet transmission are encountered due to the adaptive nature of the H263 protocol and it's adaptive working mechanism to the image changes and in providing the relevant motion compensations. A baseline removal algorithm is necessary (low pass filtering with a Butterworth fifth order filter).

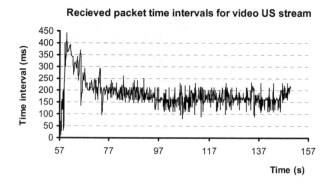

Figure 5. Performance of the ultrasound (US) video data packets received at the expert station

Figure 6 shows the relevant latency performance for the robotic data of the system. The test has shown that the average delay is approximately 80 ms and remains stable during the test duration.

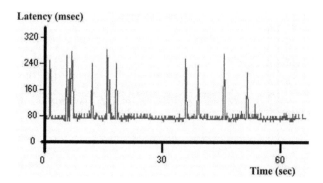

Figure 6. OTELO- 3G latency performance of the robotic control from the expert site to the patient site

In summary, the results of these experimental tests of the OTELO robotic m-health system show that the system performed satisfactorily. These results also show the potential of 3G wireless systems as an important mobility platform for future m-health services with demanding data rate requirements. Further work is ongoing to test and evaluate the medical video streams in real-medical settings and to provide a subjective medical analysis of the system from the medical experts in UK and other European hospital centers.

4.   A BRIEF OVERVIEW OF THE INTRODUCTION TO THE MOBILE M-HEALTH SYSTEMS SECTION

The aim of this section of the book is to provide an introduction to the wider spectrum of the recent advances in m-health technologies and the role of the emerging mobile and network technologies in m-health systems and future healthcare applications.

The efficacy of the m-health paradigm is presented in the second chapter by *Dwivedi et al.* This chapter complements this introductory chapter and provides an insight into the development of the m-health systems and their evolution from the desktop telemedical systems. It also provides an overview of the economic and business models that can be applied in this area.

The third and fourth chapters provide complementary issues to the m-health systems by *Jovanov et al.* and *Priddy et al.* The evolution of the wireless sensors and intelligent wireless networks for m-health systems are discussed in these two chapters. The details of the development of some prototype biosensor models that can be used in future m-health applications especially from the emerging and future 'wireless monitoring' strand are discussed in detail. The biosensors integration is considered an important element in the m-health systems development especially in the 'wearables' area.

One of the most important barriers for the larger deployment of these systems will be the security issues in the future wireless and m-health systems and healthcare services. This important topic that will witness further research and challenges in the near future from the m-health perspective is described in chapters 5 and 6 of this section by *Tsiknakis et al.* and *Markovic et al.*

The last chapter of the section by *Samaras et al*, presents the computational models most appropriate for the wireless/mobile environment, examines wireless networks and their limitations and characteristics, application architectures and models, the collaboration concepts for healthcare virtual teams (and virtual teams in general) and proposes a computational model most suitable for the collaborative healthcare environment.

The topics presented in this section will provide the basis for the more detailed technical sections in the book that describe the relevant challenges and current and future developments within the different research disciplines of mobile health systems. More specifically, the following sections will cover the topics of: Introduction to Mobile M-Health Systems, Smart Mobile Applications for Health Professionals, Signal, Image, and Video Compression for M-Health Applications, Echography Systems and Services, and Remote and Home Monitoring.

5.   CONCLUSIONS

In this chapter we introduced some current and future-challenging issues that relate to the forthcoming advances in m-health technologies. We have discussed some of the fundamental issues and future scenarios regarding the next generation of mobile telemedicine systems and their migration towards 4G mobile systems. It is conclusive that some of the current and successful m-health systems will be more geared towards the emerging wireless solutions that are not feasible within the current 2.5G/3G generation of wireless systems. The chapter also discusses the performance analysis of a compact mobile robotic tele-echography system. The 2.5G and 3G performance of such an advanced telemedical system was also validated and discussed. The next decade will emerge as the most prolific in terms of mobile users globally and hence the wider use of m-health services and applications.

6. ACKNOWLEDGEMENT

The authors would like to thank Dr. H. Wang, Dr. F. Courreges, Mr. S. Garawi, and Mr. Nilesh Prag from the MINT centre for providing their valuable input and comments to this work.

7. REFERENCES

Courreges, F., Vieyres, P. and Istepanian, R.S.H., 'Advances in Robotic Tele-Echography Services- The OTELO System', *Proceedings of the 26th. Annual International Conference of the IEEE Engineering in Medicine and Biology*, 1-5 September 2004, San Francisco, California, 2004a.

Courreges, F., Istepanian, R.S.H. and Garawi, A.S., ' Wireless Challenges for Mobile Robotic Tele-Echography System, Proceedings of the 2nd International Conference on Smart home and health Telematics-ICOST04, Sept 2004, Singapore, In *'Toward a Human –Friendly Assistive Environment,'* Zhang, D. and Mokhtari M. (Eds.), IOS Press, pp.15-22, 2004b.

Courreges, F., Vieyres, P. and Istepanian, R.S.H., 'Advances in Robotic TeleEchography Services . The OTELO System', *Proceedings of the 26th Annual International Conference of the IEEE-EMBS*, San Francisco, CA, USA • September 1-5, pp. 5371- 5374, 2004c.

Europe Adopts e-health Timetables, In 'www.healthdatamanagement.com- Health Data Management', Accessed in May 2004.

Garawi, S. A., Courreges, F., Istepanian, R S. H., Zisimopoulus, H., and Gosset, P., 'Performance Analysis of a Compact Robotic Tele-Echography E-Health System over Terrestrial and Mobile Communication Links', Proceeding of the 5th. IEE International Conference on 3G Mobile Communication Technologies- 3G 2004, London, 18-20, October, pp.118-122, 2004.

Istepanian, R.S.H., and Laxminaryan, S., 'UNWIRED, The next generation of Wireless and Internetable Telemedicine Systems- Editorial Paper', *IEEE Trans. Information Technology in Biomedicine*, Vol.4, 3,pp.189-194, Sept. 2000.

Istepanian, R. S. H. and Wang, H., 'Telemedicine in UK', in *European Telemdicine Glossary of Concepts, Standards, Technologies and Users*, Luciano Beolchi,(Ed.), 5th. Edition, pp. 1159-1165, European Commission- Information Society Directorate –General, Brussels, 2003.

Istepanian, R S. H., Philip, N., Wang, X.H. and Laxminaryan, S., 'Non-Telephone HealthCare: The Role of 4G and Emerging Mobile Systems for Future m-Health Systems', *Medical and Care Compunetics-1*, Lodewijk, B. Laxminarayan, S. and Marsh, A. (Eds.) , IOS Press , Vol. 103, pp. 465-470, 2004a.

Istepanian, R.S.H., Jovanov, E. and Zhang, Y.T., 'm-health: Beyond Seamless Mobility and Global Wireless Healthcare Connectivity', *IEEE Trans. Information Technology in Biomedicine,* Vol.8, 4, Dec. 2004b (In Press).

Li, M. and Istepanian, R S. H., '3G Network Oriented Mobile Agents for Intgelligent Diabetes Management: A Conceptual Model', *Proceedings of the 4th. International IEEE_EMBS Special Topic Conference on Information Technology Applications in Biomedicine*, Birmingham, UK, pp. 31-34, 24-26 April, 2003.

Pattichis, C. S., Kyriacou E., Voskarides S., and Istepanian, R.S.H., Wireless Telemedicine Systems: An Overview, *IEEE Antennas and Propagation*, Vol.44, 2, pp.143-153, 2002.

Tachakra, S., X. H. Wang, Robert S. H. Istepanian, Y.H. Song, 'Mobile E-Health: the Unwired Evolution of Telemedicine', *Telemedicine Journal & E-Health*, Vol.9, No.3, pp.247-257, 2003.

THE EFFICACY OF THE M-HEALTH PARADIGM: INCORPORATING TECHNOLOGICAL, ORGANISATIONAL AND MANAGERIAL PERSPECTIVES

Ashish N. Dwivedi, Rajeev K. Bali, Raouf N.G. Naguib, and Nahy S. Nassar[*]

1. THE ROLE OF TECHNOLOGY IN TELEHEALTH

Technology strongly influences the way we work and is creating opportunities and new demands for a range of different approaches to telehealth (Feldman and Gainey, 1997). Telecommunications have evolved and have been accompanied by an evolution in attitudes to information and communications technologies (Stanworth, 1998). Previously, only companies owned computers and it was the IT specialists, rather than ordinary users, who determined their use and application. Today's response to technological change is profoundly different. On average, around 1 in 4 European households already owns a personal computer; in some countries this rises to more than 50% and in some local communities it is even higher.

A recent study confirms this trend and predicts that, in two years time, it is expected that the use of information communication technology will increase markedly (Marien, 1989). The ease with which we use them and the take-up of remote working in the European Union continues at a rapid pace. Recent estimates (European Telework Organization, 1999) show that approximately 6.7 million Europeans (4.5% of the workforce) were practising remote working in one form or another at the beginning of 1999.

Social, cultural, economic and regulatory factors determine how we organise our business, our work and, hence, our lives (Stanworth, 1998). Technology-led change opens up opportunities for new working methods in three main ways: allowing existing activities to be carried out more rapidly, with more consistency and at a lower cost than could previously be achieved.

---

[*] Rajeev K. Bali, BIOCORE, School of Mathematical and Information Sciences, Coventry University, UK.

Today, the explosive growth of the Internet has promoted the trend for investment in information and communication devices and the healthcare industry is an active participant in this trend (Kazman and Westerheim, 1999). It would be fair to state that advances in communications technology are dramatically changing the delivery of healthcare services (Schooley, 1998).

2. THE HEALTHCARE SECTOR

Modern healthcare is the largest sector of the US economy (Kazman and Westerheim, 1999). However, IT expenditure in healthcare organisations, as a portion of revenues, is in the region of 2%, far below the 7-10% mark in other information-intensive industries (Moran, 1998). HIs are demanding healthcare applications that offer a number of utilities whilst they themselves allocate only a very small component of the total funds at their disposal (see table 1) (Morrissey, 2000, 2001 and 2002).

Analysts are confident that the above situation is about to undergo a sea change. Investor confidence in technology growth in healthcare is so strong that, between 1992 to 1996, there was a quintuple leap in the number of publicly traded health information technology companies. In 1998, the top 35 companies had market capitalization of over $25 billion (Moran, 1998). In 1999, about 43% of US-based Internet users used the Web to locate healthcare related information (Kazman and Westerheim, 1999). This clearly indicates that e-healthcare and its applications (such as m-health) are here to stay.

2.1 Budgeted Expenditure on IT in Healthcare Institutions and Healthcare Trends

An analysis (see Table 1) of the Budgeted IT expenditure (as a ratio of the information systems budget to the total operating expenses) shows that some of the major technologies that have showed a lot of promise are Workflow Management Systems, Mobile Computing technologies such as Personal Digital Assistant (PDA), Wireless Local Area Network (WLAN) and Object Oriented Technology (Dwivedi et al, 2001a, 2001b; 2002).

2.2 The Role of Telecommunications Technology in Health

The changes in information technology, particularly in the telecommunications technology have brought about fundamental changes throughout the healthcare process (Applebaum and Wohl, 2000). The change process undergone is confirmed by research which states that in the period 1997-2000, 85% of healthcare organizations have undergone some sort of transformation (Sherer, 1995).

One of the most important technological changes in electronics has been the ability to convert signals from an analogue to a digital medium (images or signals are converted into digital code by using an analogue-to-digital conversion device) – a process referred to as Digitization (Wallace, 1997). Digitization in healthcare has meant that it is now possible to take healthcare related information in different formats (audio, video, sound) and deliver the same at high speeds in the same basic format.

Table 1. Modern Healthcare's annual survey of information system trends in the healthcare industry
Adapted from (Morrissey, 2000, 2001 and 2002)

Year	2000	2001	2002
No of healthcare organizations surveyed	224 healthcare organizations	212 healthcare organizations	255 healthcare organizations
Budgeted IT expenditure (Ratio of the information systems budget to the total operating expenses)	Average IT expenditure at 2.6% Over 50% put it between 2% and 4%	Average IT expenditure at 2.53% 25% spent about 1.5% on IT whilst 60% are spent around 2.5%	Average IT expenditure at 2.56% More than 50% spent between 2% and 3.5%
Information systems (IS) priorities	Internet and intranets - priority Clinical information systems (CIS) tops the IS priority General-accounting software was the predominant application that was being used on the Intranet	65% considered CIS as the most important IS priority 73% were not willing to outsourcing clinical information systems and services	71% considered CIS as the most important IS priority Need felt to improve IS for better clinical communication for physicians 80% - No interest in outsourcing clinical information systems and services
Clinical Use of Web technology (Intranets)	60% - felt that could IT could facilitate data exchange among caregivers i.e. physician ordering of tests and access to test results	Low interest in maintaining a patient's personal health record accessible via the WWW and matching patients with clinical research. However there is renewed importance of addressing changes in this area due to regulatory obligations	Despite acknowledging that medication interaction and dosing alerts are possible within most IS - implementation has not commenced The few organizations who had made big investments in different HIS (EPR and pharmacy) are reporting substantial returns
General Uses of Web and Intranet technology	Limited use as shown by the following 15% - to share clinical guidelines13% - to access multiple databases simultaneously 33% - as a bridge to other information systems 40% - for network wide communication of any kind	Some early success from linking "billing and insurance-query operations to payers via the Web" "Significant interest …in using the Web to improve data exchange with physicians and their office staff" About 50% indicated that they had no plans to try anything Web-related in the care-management area	33% - Using existing clinical and financial information sources to construct data repositories so as to that help spot trends and improve decision-making Further 22% are working to implement such practices whilst about 13% plan to start implementation of similar activities within a year

Simultaneously there has been a change in technology (from simple copper wires to optical fibres) via which information is transmitted. This in turn has exponentially increased the bandwidth (Wallace, 1997). The phenomenon of Multimedia has made possible the exchange of information – that can be combined from different formats (sound, video, animation, text and graphics) and presented in an interactive manner. This is fast making multimedia technology the "technology of choice" for the delivery of information (Wallace, 1997).

3. M-HEALTH : THE TECHNOLOGY FOR HEALTHCARE DELIVERY IN THE FUTURE

In the healthcare sector, different information technology applications such as clinical information systems, electronic patient records and telemedicine have been used successfully thereby demonstrating their potential to greatly improve the standard of medical care and healthcare administration (Rao, 2001).

Advances in information technology applications have resulted in an "accelerated pace of innovation" (Johns, 1997). Such innovation has resulted in the creation of new opportunities and healthcare concepts such as healthcare information – a term indicating the combined synergistic application of "a science of information, technology, and knowledge…to 'health care" (Johns, 1997). All this has led industry experts to predict that, in the near future, healthcare technologies (and, in particular, technologies such as m-health) will be computerized to a considerable extent (Crompton, 2001).

4. TELEMEDICINE AND M-HEALTH : ORIGINS AND SCOPE

Telemedicine has been derived from two Greek and Latin words. "Tele" in greek translates as distance while "Mederi" in latin means to heal (Rao, 2001). In a modern context, telemedicine can be stated to be a method of healthcare delivery where advanced video communications technologies are used to bridge the geographical gap that exists between the licensed caregivers and/or the care receiver, so as to provide medical diagnosis and treatment (European Health Telematics Observatory, 1999; Charles, 2000; Nairn, 2001; Garets and Hanna, 1998).

Telehealth has a much broader scope in comparison to telemedicine, as telehealth relates to the bigger issues in healthcare administration and regulation, whilst telemedicine is concerned with the clinical aspects of healthcare delivery (European Health Telematics Observatory, 1999 and Johnson, 2000). Some authors (Nairn, 2001; Noring, 2000) have further delineated between the two by positing that, in telemedicine, healthcare providers fall into the category of physicians whilst, in telehealth, the category of healthcare providers can be extended to include non-physicians, as telehealth encompasses health promotion and disease prevention.

The earliest documented use of telemedicine can be traced to the 1920s when radio was employed to link up physicians located on land with ships at sea who were facing medical emergencies. The next leap in telemedicine took place in the 1960s when the US-based National Aeronautics and Space Administration (NASA) pioneered the use telemedicine in space (astronauts had their pulse rates and blood pressure monitored remotely). By the 1970s, telemedicine additionally took advantage of new emerging satellite technologies (Rao, 2001).

UK, Canadian and Malaysian governments have seized the opportunity to make substantial efforts to link electronically different healthcare centres. In the UK, the Government has committed about USD $1.4 billion to bring about a transformation in all facets of healthcare. A significant component of this funding is being used in developing a nationwide electronic platform (Crompton, 2001). This high volume of investment worldwide in healthcare technologies has brought about the emergence of several m-health schemes all over the world (Crompton, 2001; Collins, 2001;The Economist, 1997).

4.1. Objective of M-Health Applications

The aim behind any m-health application is to transfer the expertise of the caregiver from one location to another (Johns, 1997). One of the most widely used applications of m-health is teleradiology (use of "image acquisition, storage, display, processing and transport") from one geographical location to another location for diagnosis (Johns, 1997). With advances in technologies, such as telecommunications, multimedia and IT healthcare applications, m-health has the potential to transform the delivery of healthcare permanently.

4.2. Current M-health Technologies

The cost of setting up an average m-health station works about to $50,000 and incorporates "a computer workstation with 21-inch monitor, electronic stethoscope, ear, nose and throat scope, and an exam camera through which the patients and doctors can see each other"(Tieman, 2000).

There are two main models in m-health: (1) interactive video and (2) store-and-forward (Kazman and Westerheim, 1999; Nairn, 2001). The main difference between them is that interactive video allows real-time patient care, whilst the store-and-forward technology is asynchronous (there is a gap between transmission of data and patient care diagnosis). Today, store-and-forward technology applications in m-health include telepathology and teleradiology (Nairn, 2001). The use of email to transmit medical prescriptions by physicians to their patients is fast becoming another major application of store-and-forward technology in m-health (Convey, 2000).

Since store-and-forward technology is asynchronous (communication over telephone lines linking two computers or other peripheral devices using start and stop bits), applications based on this type of technology are being more widely used in comparison to interactive video applications, as they can easily be transmitted over a network. Modern store-and-forward technology applications, in conjunction with advances in telecommunications technologies (such as digital imaging, WAP and fibre optics) are resulting in the creation of a much larger m-health market (Johnson, 2000; Beavan and Frederick, 2000).

Technologies that offer healthcare videoconferencing as a substrate are still evolving. It is possible to send large amounts of clinical multimedia data (compressed audio and video images) on high speed lines such as broadband technologies over the internet (Nairn, 2001). Given the current pace of advances in internet and videoconferencing technologies, interactive m-health applications will feature heavily in futuristic healthcare systems.

The advantages of m-health include enabling direct links between the caregivers and/or care receivers thereby enabling effective medical care especially to rural populations, saving time and money for caregivers and faster diagnosis and treatment for care receivers (Kazman and Westerheim, 1999; Schooley, 1998; Charles, 2000; Huston and Huston, 2000; Fishman, 1997). Whilst it is clear that m-health is more viable compared to traditional telephonic consultations (Sandberg, 2001), in a normal patient care scenario it enables patients to have faster access to alternative specialists and, more importantly, to have access to information about their sickness (Blair, 2001).

One of the major bottlenecks affecting the uptake of m-health in the US is the fact that insurance coverage of m-health is generally limited to teleradiology and a few cardiac monitoring procedures (Health Care Strategic Management, 2000). Furthermore in the US, legislation affecting m-health is different in each state.

Another major limitation in m-health is that there is no adequate regulatory structure which addresses such issues as licensure, credentialing, intellectual property and MediCare payments (in the US) (Schooley, 1998; Edelstein, 1999). Governments have started to address these issues with the US Congress taking a pioneering stand in this regard. In 1999, it introduced 22 pieces of legislation relating to m-health (Edelstein, 1999).

5. THE INTERNET AND M-HEALTH

An American Medical Association study in 1999 (Swartz, 2000) indicated that 37% of all physicians in the US were using the Web and by 2000 this figure had risen to about 50%. It has been pointed out that more caregivers have adopted modern ICT applications such as wireless phones and PDAs which allow them to be in contact with patients and, in certain circumstances, to save lives. The use of e-mail by physicians as a method of keeping in touch with patients tripled in less than one year -10% of all physicians now use e-mail on a daily or weekly basis to be in contact with their patients (Swartz, 2000).

Initial web-based multimedia patient record systems have been developed which give remote access to telecare providers (Nairn, 2001). We believe that, in the future, web-based multimedia patient administration systems will become the norm for m-health. A similar concept has been put forward by the National Health Service (NHS) in the UK, where healthcare institutions are being asked to adopt an Electronic Patient Record (EPR) system at six varying levels of implementation (NHS, 1998).

One of the biggest indicators that portends the rise of m-health has been the ruling by the Federal Communications Commission (FCC) in the US who had recently cast "...an historic vote dedicating a portion of the radio spectrum for wireless medical telemetry devices such as heart, blood pressure and respiratory monitors" (Health Management Technology, 2000, p12).

5.1. Mobile Computing and Workflow

In the near future, third generation technologies such as Code Division Multiple Access (CDMA) 2000 are expected to raise the transmission standard to about 2.0 Mbps (Cancela, 2001). When this happens, it is quite likely that the vision of an "integrated voice, data, video technology" (Cancela, 2001) will become a reality which is likely to have a significant impact on the use of m-health.

The success of second-generation wireless networks has led to an explosion in the use of wireless applications to transfer voice and data services. This has raised the possibility that future wireless networks might support cost-effective broadband multimedia services (Pinto and Rocha, 1999). Technologies such as Wireless Application Protocol (WAP) have enabled patients and doctors to remain in closer contact. There are successful WAP-based products (such as LifeChart.com) through which doctors can monitor online their patient's condition, and take care of their healthcare needs (Purton, 2000).

Another example of the use of WAP is WirelessMed, through which UK Doctors have wireless access to clinical data on Medline, the largest US government database comprising more than 12 million medical references (which supports download speeds of up to 400-words in a few seconds). Another example of WAP-enabled healthcare products is MedicinePlanet which aims to bring local health information (health news, current health alerts, details on prevailing healthcare systems) to travellers using mobile phones (Purton, 2000). WAP technology is facing strong competition from other medical wireless systems based around PDA (Personal Digital Assistant) platforms which support downloads of a standard patient image in 10-15 seconds (Parkes, 2001).

The main disadvantages of mobile computing (limited battery and processor power) should diminish as new technologies, allowing higher bandwidths, are introduced onto the market (Satyanarayanan, 1996). Workflow-based applications, in conjunction with mobile computing technologies such as WAP and PDA, have the potential to transform the delivery of healthcare information.

Modern day IT applications in healthcare use protocols centered on Mobile Computing and the Internet. Some of them such as Wireless Local Area Network (WLAN) have already demonstrated their potential and financial viability. WLAN-based mobile computing allow healthcare workers to interact in real-time with the hospital's host computer system to enter, update and access patient data and associated treatment from all clinical departments. All this is possible not only from the patients' bedside, but also from a number of geographical locations within the hospital where the WLAN is installed.

The fact that a WLAN takes about one hour to be made operational has been trumpeted as one of the biggest advantages in comparison to a more traditional network, installation of which would take significantly longer (McCormick, 1999). It has been found that the average pay-back period for the initial costs of WLAN installations is 8.9 months. In a survey of WLAN healthcare installations, 97% of customers indicated that "WLANs met or exceeded their expectation to provide their company a competitive advantage" and that "if the productivity benefits are measured as a percentage return on the total investment … the return works out to be 48%" (McCormick, 1999, p13).

The use of Personal Digital Assistant (PDA) by physicians has witnessed rapid acceptance in recent times.   About 40% of all physicians currently use PDAs (Serb, 2002).   However, the majority of physicians are using PDAs to perform static functions. Most of them use PDAs to collect reference material with ePocrates - a drug reference application that can assist physicians in looking up drugs by name or diagnoses, cross-reference analogous medications or generic alternatives and obtain alerts on interactions. This application has been categorised by the US based Journal of the American Medical Association to be indispensable (Serb, 2002).

A few pioneering physicians have started to use PDAs in an interactive way i.e. to write prescriptions, to keep a record of all daily clinical patient interactions and for bedside charting.   The financial viability of PDAs have been demonstrated by a pilot study which has shown that for every US $1 that was invested in PDAs, the return in the form of lower administrative costs was over $4 (Serb, 2002).

6. FINANCIAL ESTIMATES FOR M-HEALTH

Estimates for worldwide m-health services suggests that the market is valued at $2.5 billion a year (Surry, 2001). Other estimates for 2002 project the US market alone to be around $3 billion – a big leap from $65 million in 1997 (Industrial Robot, 1998).

A study by Waterford Telemedicine Partners Inc in Feburary 2000 has indicated that m-health is projected to have an annual growth rate around 40% in the first decade of the 21st century. The study predicts that, by the year 2010, m-health will account for at least 15% of healthcare expenditure (Health Care Strategic Management, 2000).

M-health technologies, in conjunction with state-of-the-art Electronic Health Record (EHR) systems, are changing the face of healthcare.   M-health technologies have the potential to replace 5% of hospital stays, 5% of nursing home care, and 20% of home health visits (Fishman, 1997), resulting in savings of time and money for both patients and doctors.   Additionally, caregivers have more time to devote to clinical activities such as medical diagnosis and treatment.   A study on the medical reimbursement of m-health applications indicates that "telemedicine for radiology, prisoner health care, psychiatry, and home health care are the most cost effective applications ...that are paid for by insurers" (Charles, 2001, p66).

7. WORLDWIDE APPLICATIONS OF M-HEALTH

One of the most widely used applications of m-health is teleradiology (Rao, 2001). The use of m-health applications to monitor patients has been recommended in asthma, congestive heart failure and for diabetes (Friedewald Jr, 2001).

A study carried out by the Agency for Healthcare Research and Quality in the US identified 455 telemedicine programs worldwide out of which 362 are being used in the US (Trembly, 2001). The study has indicates that m-health is most commonly used for: 1] Consultations or second opinions (performed in 290 programs), 2] Diagnostic test interpretation (169), 3] Chronic disease management (130), 4] Post-hospitalization or post-operative follow-up (102), 5] Emergency room triage (95), 6] Visits by a specialist (78) and that about 50 programs provide services in patients' homes.

Countries such as Malaysia have already integrated m-health with the electronic health record concept and there is a national telemedicine strategy in place. Teleconsultations are being carried out on a regular basis in Malaysia. In Sweden, m-health is being used to reduce the stay of children in hospital. This is being achieved by using local telecommunications to connect to the health monitoring equipment (for heart rate, rhythm and blood pressure) and is installed at the residence of the patient (Collins, 2001). In the UK, at the West Surrey Health Authority, patients who are regarded as potential heart failures are monitored electronically for 24 hours a day at their residences (Crompton, 2001).

A report by the US based Association of Telehealth Service Providers has indicated that in 1996 there were about 22,000 telemedicine consultations, which rose to 42,000 in 1997 and by 1998 to about 58,000 (Tieman, 2000).

7.1. True Life Examples that Validate M-Health Applications

In the UK, the North Manchester primary care group has used telemedicine applications to reduce the "average waiting time for a dermatology appointment from 18 months to 17 days" (Cross, 2000). In the US, m-health is said to have had tremendous benefits in reducing hospital intensive care costs (ICU). Each day in ICU costs on average $2,500 to the insurance company, which can be reduced to $35 by a routine teleconsultation (Cross, 2000).

The use of m-health can significantly aid patients in the battle to combat diabetes whilst reducing the associated expenditure. It has been estimated that the use of m-health can save the US government, about "$247 million per year through early intervention and nearly double that if telemedicine can extend the reach of treatment" (Blair, 2001, p4).

In Cornwall (UK) a pilot telemedicine project for teledermatology was undertaken. General Practitioners (GPs) from three surgeries (medical centers) used to send pictures of skin conditions from Cornwall to the county's two consultant dermatologists. They would then make an assessment as to whether the patient was required to visit the dermatologists for treatment or whether the GP could treat them in Cornwall itself. This m-health application found that "one in four patients did not need to see a specialist and could be treated by their family doctor" and reduced the workload of the two overworked dermatologists (The Guardian, 2001).

In 2001, a telemedicine application was used for the first time to carry out telesurgery. Doctors via computer from New York operated on a 68-year-old woman in Strasbourg, France to remove her gall bladder (Alpert, 2001; Johnson, 2002). The patient was released from the hospital 48 hours after surgery and recommended regular activities the following week (Johnson, 2002).

Telemedicine has been used very successfully in the US state of Arizona. Due to the geographical nature of Arizona, it is not possible for its residents which number in thousands to have immediate access to leading edge healthcare, as closet centres are exist 150 miles away or more (Health Management Technology, 2001). The Arizona Telemedicine Program today has more then 96 telephysicians representing 60 medical sub-specialities such as teleradiology, teledermatology, telepsychiatry, telecardiology, teleorthopedics, teleneurology and telerheumatology. These telephysicians have seen more than 11,000 patient cases and have provided primary diagnoses, expert consultation and provision of second opinions. Another pioneering advantage of telemedicine in Arizona has been to decrease risky travel, especially for patients with unstable medical conditions (Health Management Technology, 2001).

The Arizona Telemedicine Program has resulted in significant cost saving for healthcare delivery as the average cost "for a rural patient's visit to an urban health center has dropped from \$520 to \$105" and there exists significant potential in lowering "the average cost for routine home health visits … from \$140 to as low as \$55 per visit". One of the more significant achievements of the Arizona Telemedicine Program was to reduce healthcare expenditures in prisons, with the average cost for a prison inmate's healthcare visit falling from \$415 to \$140 (Health Management Technology, 2001, p47).

One of the most important technological enablers that is likely to affect transmission of healthcare information and healthcare delivery is the connectivity via satellite technology. It was estimated that there were about 150 satellites in orbit in 2001 (Rao, 2001). More significant is the fact that about "1,700 commercial satellites are scheduled for launch in the next decade worldwide" (Rao, 2001, p227). This could be the technological catalyst that brings about a transformation in the manner how healthcare delivery takes place, particularly in those areas where there is a geographical gap between the caregivers and carereceivers or where there is a substantial time constraints on specialized caregivers. Tele-consultations between caregivers and carereceivers and between themselves could see a likely increase (Rao, 2001).

8. IMPORTANT INNOVATIONS THAT HOLD GREAT POTENTIAL TO BE REVOLUTIONARY CHANGE CATALYSTS FOR THE HEALTHCARE SECTOR

8.1. Biomedical Knowledge: The Untapped Potential in Healthcare

Advances in technology have been the hallmark of the healthcare sector, particularly with regard to advances in biomedical sciences. Today there are "10,000 known diseases, 3,000 drugs, 1,100 lab tests, 300 radiology procedures, 1,000 new drugs and biotechnology medicines in development and 2,000 individual risk factors" (Pavia, 2001, p12). This has had an enormous impact on healthcare and, in particular, has rendered the concept of an expert in a particular domain in healthcare irrelevant as shown above it is not possible for one human being to have all the relevant knowledge in their domain of speciality (Pavia, 2001; Rockefeller, 1999).

For example today, "Organ and tissue scanning speed is doubling every 26 months, making tests both faster and cheaper…Image resolution is doubling every 12 months…the increase in computer power (four-fold over the next 20 years) and the availability of inexpensive bandwidth" (Pavia, 2001, p12). All of the above is likely to change our own perception of what information is available and even possible for one human being to acquire.

8.2. The Genetic Engineering (un)revolution

Advances in modern day genetic sciences have increased the number of potential drug compositions from a mere 400 to over 4,000 (Pavia, 2001). This has occurred despite the fact that the rate of adoption of computer applications in healthcare is slower in comparison to other industries (Johns, 1997). The pace of discovery of new drugs may well undergo an exponential leap when the above observation is seen in context of the forecasted increase in computing power as discussed previously in this chapter.

Perhaps the biggest tragedy in the history of modern science was the fact that the announcement regarding the completion of the Human Genome project did not create any ripples in the mindset of healthcare decision-makers and academics or propel a new wave of healthcare discoveries (Jones, 2001). We hypothesize that this situation is not likely to prevail for much longer. The impact of the completion of the project will profoundly change the concept of healthcare itself within the next 25 years (Jones, 2001).

Unfortunately, the contemporary focus in m-health is only on how best to disseminate the information, which could be fatal for the future of telehealth and m-health. Rather than creating or disseminating contextual knowledge, m-health applications are being used to disseminate data and/or information. Futuristic m-health schemes would need to support the transfer of information with context (i.e., such schemes would have to become dynamic in nature). One of the big drawbacks of m-health is that most systems force the caregiver specialist to look at medical issues in isolation, whereas more detailed information (such as the patient's medical history) might help in arriving at a better informed medical diagnosis (Nairn, 2001).

9. DISCUSSION

This section summarises how m-health will affect healthcare delivery. It is argued that m-health will alter healthcare delivery in the following ways:

(a) It would be important to consider having a definition of what is meant by the term "healthcare". This is by no means an easy task. We believe that m-health will reduce both critical and non-critical healthcare treatments. However, in the immediate future m-health is likely to reduce the cost of routine consultations. Advances in technology will help patients carry out routine medical tests, reducing the number of visits to the physician, thereby reducing costs for routine consultations. However there could be an increase for non–routine expenditure for complex treatments.

(b) m-health applications such as teleradiology will increase. In addition, m-health will be used increasingly for the following purposes: consultations or second opinions, diagnostic test interpretation, post-hospitalization or postoperative follow-up, emergency room triage and televisits. The use of m-health applications to monitor patients will see an increased use (eg. for asthma, congestive heart failure, diabetes).

9.1. Need for Incorporating Organisational Perspectives in M-Health Processes

The last two decades of the twentieth century has witnessed a shift in the concept of healthcare information. This shift has been marked by the coming of age of paradigms such as m-health, bioinformatics, biomedical and genetic engineering resulting in exponential advances in healthcare knowledge (Dwivedi et al, 2002). One of the most significant implications of this shift has been the realization that the applications of IT advances in the healthcare sector have caused an information explosion. Individual healthcare stakeholders are not going to be in position to adapt to the above with ease. Any solution would call for significant integrated technological support in the human healthcare decision-making process (Dwivedi et al, 2001a; 2001b; 2002).

It has been proposed that holistic health will emerge as an alternative to complement traditional medicine (Church et. al, 1996; Dervitsiotis, 1998). As patients' homes become the homebase for delivering more and better types of care, people will expect "King Quality, Queen Value" (Nelson, Batalden and Mohr, 1998, p3). Healthcare organisations need to be fully aware of the organisational implications of telehealth initiatives.

The technology associated with m-health schemes transcends geographical, institutional and disciplinary boundaries. M-health redefines organisational roles and responsibilities and by disseminating knowledge and information, it allows healthcare professionals and patients to relate to each other. The astonishing rate of change makes strategic planning extremely difficult.

Appreciating the role of management and how it controls and monitors resource requirements needs is crucial. Having identified suitable individuals and jobs, it must be emphasized that m-health is an additional health-delivery avenue and no healthcare provider should be forced to use the new technology. M-health delivery may be better suited to people who tend to exhibit such traits as a greater ability to structure their workday, more efficiently separate work and family life, or those whose jobs are more independent and proactive.

9.2. Knowledge Management

Knowledge management (KM) is considered as a source of great competitive advantage (Nonaka, 1991; Wiig, 1994). Knowledge can either be tacit or explicit (Beijerse, 1999; Gupta , Iyer and Aronson, 2000; Hansen, Nohria, and Tierney, 1999). Explicit knowledge typically takes the form of company documents and is easily available, whilst tacit knowledge is subjective and cognitive. The ultimate objective of KM is to transform tacit knowledge into explicit knowledge to allow effective dissemination (Gupta , Iyer and Aronson, 2000).

Knowledge Management and Information and Communication Technologies (ICT) as disciplines do not have commonly accepted or de-facto definitions. However some common ground has been established which covers the following points.

KM is a multi-disciplinary paradigm (see Figure 1) which uses technology to support the acquisition, generation, codification and transfer of knowledge in the context of specific organisational processes. ICT refers to the recent advances in applications of communication technologies that have enabled access to large amounts of data and information when seeking to identify problems or solutions to specific issues (Dwivedi et al, 2001a).

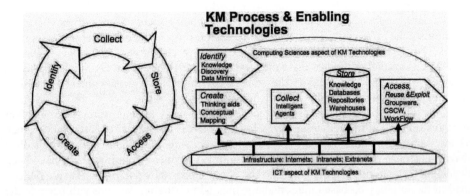

Figure 1. KM Process and Enabling Technologies. Adapted from (Skyrme, 1999)

9.3. Impact of KM on M-health

There are many factors which may influence m-health which can be reduced to three fundamental areas: technological, business and social (Bali, 2000b). These areas all consider the current pace of technological change. The commercial environment is undergoing a period of accelerated information technology change, which some would argue is a revolution. Developments in technology, social considerations, government fiscal policy and business aims and objectives need to be fully understood in order to fully exploit the social and economic benefits that are emerging as a result of m-health (Bali and Naguib, 2001). Healthcare organisations are in a constant state of change and m-health is both a key manifestation and enabler of this change. However, researchers and practitioners need to appreciate the implications and ramifications of such a change.

The multi-disciplinary aspect of KM research has resulted in a multitude of models and approaches, all of which look at KM from perspectives similar to m-health. KM is viewed as a methodology involving the interaction between people, culture, information technology, and organisations. A different perspective discusses KM's relationships between culture, content, process, and infrastructure. Another approach reflects that a successful KM programme must bring together political, organisational, technical and cultural organisational aspects (Bali, 2000a; Puccinelli, 1998; Chait, 1999; Havens and Knapp, 1999)

We would like to establish the interrelationship between KM and m-health by stating that both have been brought about by the ICT revolution, and that both are bringing about fundamental changes which are redefining the work place of contemporary healthcare organisations. Another common point is that both KM and m-health are concerned with dissemination of information in a manner which ensures that information is available when required.

We believe that the difference between KM applications and m-health applications lies in the application of ICT. As compared to KM, ICT in telehealth and telemedicine is in its relative infancy. Unfortunately, the contemporary focus is only on how best to disseminate the information - which could be fatal for the future of m-health (i.e., current use is static). Rather than creating or disseminating contextual knowledge, m-health applications are being used to disseminate data and or information. Futuristic m-health schemes would have to support the transfer of information with context (i.e., such schemes would have to become dynamic in nature).

One of the big drawbacks of m-health is that most systems force the caregiver specialist to look at medical issues in isolation, whereas more detailed information (such as the patient's medical history) might help in arriving at a better informed medical diagnosis (Nairn, 2001). Initial web-based multimedia patient records systems have been developed, which give remote access to the telecare providers (Nairn, 2001). We believe that, in the future, web-based multimedia patient administration systems will become the norm for m-health. A similar concept has been put forward by the NHS (National Health Service) in the UK, where healthcare institutions are being asked to adopt an Electronic Patient Record (EPR) system at six varying levels of implementation (NHS, 1998).

Healthcare institutions require a framework which would help assess how best to identify and create knowledge from internal and external organisational experiences and how best to disseminate these on an organisation-wide basis. This would call for the contextual recycling of knowledge which has been acquired from the adoption of m-health trials. KM can assist m-health to become viable by giving healthcare information context, so that other healthcare providers can use m-health to extract knowledge and not information.

For this to happen, futuristic m-health systems would have to shift their emphasis to deal with the intangibles of knowledge, institutions and culture and that the KM paradigm is aptly suited for this role. This is due to the fact that one of the important reasons behind the emergence of the KM concept is that, even though our access to data and information has increased exponentially, our capability to acquire knowledge (by giving the information context) has not become an industry-wide reality. This also holds true for the healthcare industry.

10. THE NEED FOR A KM FRAMEWORK FOR M-HEALTH

Healthcare managers are being forced to examine the costs associated with healthcare and are under increasing pressure to discover approaches that would help to carry out activities better, faster and cheaper (Dwivedi et al, 2001b; Latamore, 1999). For this to happen, the m-health sector needs to shift it's emphasis to deal with the intangibles of knowledge and culture (Dwivedi et al, 2001a). Healthcare institutions (HIs) adopting m-health applications would require a KM framework which, in light of their ICT implementation level, would assist in the discovery and creation of new knowledge (see Figure 2).

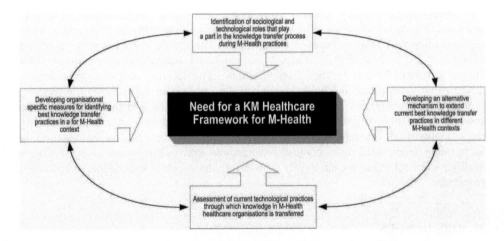

Figure 2. Need for a KM Framework for the m-health sector

HIs need to identify key sociological and technological roles affecting the knowledge-transfer process and to develop organizational-specific measures for identifying best knowledge transfer practices (see Figure 2).

Any KM strategy that is being proposed should extend best knowledge transfer practices on an organization-wide basis. Our contention above has been confirmed by the Canadian Department of Health (Health Canada, 2001). KM can assist healthcare institutions to become viable by giving healthcare information context, so that other healthcare providers can extract knowledge and not information (Dwivedi et al, 2001a). The cornerstone of this chapter is that the KM paradigm can enable the healthcare sector to successfully overcome the information and knowledge overload in healthcare (Dwivedi et al, 2002).

11. CONCLUSIONS

We have discussed important technologies that are driving forces in healthcare and have considered the implications of their advances on healthcare in general. We contend that if the impact of these healthcare technologies are seen together, then the conclusion from a healthcare informatics perspective is clear. In the future, m-health systems would have an increased interest in knowledge recycling of the collaborative learning process acquired from previous m-health practices.

Twenty first century clinical practitioners need to acquire proficiency in understanding and interpreting clinical information so as to attain knowledge and wisdom whilst dealing with large amounts of clinical data. This would be dynamic in nature and would call for the ability to interpret context-based healthcare information. This challenge cannot be met by an IT led solution. The solution needs to come from a domain that supports all the three integral healthcare system components (i.e. people, processes and technology) of the future. We believe that the KM paradigm can offer solutions to healthcare institutions, allowing them to face the challenge of transforming large amounts of medical data into relevant clinical information by integrating information using workflow, context management and collaboration tools, and give healthcare a mechanism for effectively transferring the acquired knowledge, as and when required.

12. REFERENCES

Alpert, M. (2001), "Surgeons without borders," *Scientific American*, volume.285, issue.5, pp.28.

Applebaum S.H. and Wohl, L. (2000), "Transformation or change: some prescriptions for health care organizations," *Managing Service Quality*, volume.10, no.5, pp. 279-298.

Bali, R.K. (2000a), "Towards a qualitative informed model for EPR implementation: considering organizational culture," *Proc of the IEEE Int Conf on Information Technology Applications in Biomedicine (ITAB-ITIS)* Arlington, USA, pp.353-358.

Bali, R.K. (2000b), "Organizational and social impacts of telehealth: a cause for concern," *Proc of the IEEE Int Conf on Information Technology Applications in Biomedicine (ITAB-ITIS)*, Arlington, USA, pp.54-59.

Bali, R.K. and Naguib, R.N.G. (2001), "Towards gestalt telehealth: considering social, ethical and cultural issues," [CD-ROM] *IEEE Canadian Conference on Electrical and Computer Engineering*, Toronto, Canada.

Beavan, B. and Frederick, J. (2000), "Telemedicine raises some points to ponder," *Defense Counsel Journal*, volume.67, issue.3, pp. 400-403.

Beijerse, P.uit.R. (1999), "Questions in knowledge management: defining and conceptualising a phenomenon", *Journal of Knowledge Management*, volume. 3, issue. 2, pp. 94-109.

Blair, R. (2001), "Slice of life," *Health Management Technology*, volume.22, issue. 2, pp. 4.

Cancela, T.A. (2001), "Promises to keep," *"Health Management Technology,"* volume.22, issue.4, pp. 38-40.

Chait, L.P. (1999), "Creating A Successful Knowledge Management System." *The Journal Of Business Strategy*, volume. 20, issue. 2, pp. 23-26.

Charles, B.L. (2000), "Telemedicine can lower costs and improve access," *Journal of the Healthcare Financial Management Association*, volume.54, issue. 4, pp. 66-69.

Church, A.H., Siegal, W., Javitch, M., Waclawski, J., and Burke, W.W. (1996), "Managing organizational change: what you don't know might hurt you," *Career Development International*, volume.1, no.2, pp. 25-30.

Collins, J. (2001), "So far and yet so near with telemedicine," [1DD Edition], *The Times*, London (UK), Jan 9, pp. 2.

Convey, M.C. (2000), "Telemedicine concerns healthcare risk mgrs" *National Underwriter*, volume.104, issue. 46, pp.8.

Crompton, S. (2001), "Virtual hospital speeds recovery," [Final 1 Edition]. The Times, London (UK), Jun 5., pp.2.

Cross, M. (2000), "Society: "Lives online: Tony Blair has committed the government to an electronic revolution of its care programme by 2005. Is this realistic? Can IT really change the way we deliver public services?," *The Guardian*, Manchester (UK), Nov 15, pp. 2.

Dervitsiotis, N.K. (1998), "The challenge of managing organizational change: exploring the relationship of re-engineering, developing learning organizations and total quality management," *Total Quality Management*, volume.9, no.1, pp.109-22.

Dwivedi, A., Bali, R.K., James, A.E. and Naguib, R.N.G. (2001a), "Telehealth Systems: Considering Knowledge Management and ICT Issues", *Proc. of the IEEE-EMBC 23rd Annual International Conference of the IEEE Engineering in Medicine and Biology Society (EMBS)*, Istanbul, Turkey, pp. 3614-3617.

Dwivedi, A., Bali, R.K., James, A.E. and Naguib, R.N.G. (2001b), "Workflow Management Systems: the Healthcare Technology of the Future?", *Proc. of the IEEE EMBC-2001 23rd Annual International Conference of the IEEE Engineering in Medicine and Biology Society (EMBS)*, Istanbul, Turkey, pp. 3887-3890.

Dwivedi, A., Bali, R.K., James, A.E., Naguib, R.N.G. and Johnston, D. (2002), "Merger of Knowledge Management and Information Technology in Healthcare: Opportunities and Challenges", *Proc. of the IEEE Canadian Conference on Electrical and Computer Engineering (CCECE) 2002*. Winnipeg, Canada, volume.2, pp.1194-1199.

Edelstein, S.A. (1999), "Careful telemedicine planning limits costly liability exposure," *Journal of the Healthcare Financial Management Association*, volume.53, issue. 12, pp. 63-69.

European Health Telematics Observatory, (1999), "Draft International Convention on Telemedicine and Telehealth," Medicine and Law committee of the International Bar Association, http //www.ehto.org/ legal /draftconvention.doc.

European Telework Organization, (1999), "Status report on European telework," *New Methods of Work*.

Feldman, D. and Gainey. T. (1997), "Patterns of telecommuting and their consequences: framing the research agenda", *Human Resource Management Review*, volume.7, No.4, pp.369-388.

Fishman, D.J. (1997), "Telemedicine: bringing the specialist to the patient," *Nursing Management*, volume.28, issue.7, pp.30-32.

Friedewald Jr, V.E. (2001), "Returning home," *Health Management Technology*, volume.22, issue.9, pp. 22-27.

Garets, D. and Hanna, D. (1998), "Emerging managed care technologies," *Health Management Technology*, volume.19, issue.11, pp. 28-32,

Gupta, B., Iyer, L.S. and Aronson, J.E. (2000), "Knowledge management: practices and challenges", *Industrial Management & Data Systems*, volume. 100, issue. 1, pp. 17-21.

Hansen, M., Nohria, N. and Tierney, T. (1999), "What's your strategy for managing knowledge?," *Harvard Business Review*, volume.77, issue. 2, pp. 106-116.

Havens, C. and Knapp, E. (1999), "Easing Into Knowledge Management", *Strategy & Leadership*, volume. 27, issue. 2, pp. 4-9.

Health Canada, (1998), "Vision and Strategy for Knowledge Management and IM/IT for Health Canada" (WWW document), Available from: http://www.hc-sc.gc.ca/iacb-dgiac/km-gs/english/vsmenu_e.htm. [Accessed March 20, 2001].

Health Care Strategic Management. (2000), "Telemedicine will grow 40% annually, help health systems create 'centers of excellence'," *Health Care Strategic Management*, volume. 18, issue.2, pp. 5.

Health Management Technology. (2000), "Advice from the American Hospital Association," *"Health Management Technology,"* volume.21, issue.10, pp. 12.

Health Management Technology. (2001), "Powering the Arizona telemedicine program," volume.22, issue.6, pp. 46-47.

Huston, T.L. and Huston, J.L. (2000), "Is telemedicine a practical reality?. Association for Computing Machinery", *Communications of the ACM*, volume. 43, No. 6, pp.91-95.

Industrial Robot. (1998), "Clinical telemedicine to expand by 2002," *Industrial Robot*, volume. 25 no. 3, pp. 15.

Johns, P.M. (1997), "Integrating information systems and health care", *Logistics Information Management*, volume.10, No.4, pp.140–145, viewed 17 October 2002, ProQuest database ABI/INFORM, 86066571.

Johnson, D. (2002), "The telesurgery revolution," *The Futurist*, volume.36, issue.1, pp. 6-7.

Johnson, D.E.L. (2000), "Telehealth expands unrealized dream," *Health Care Strategic Management*, volume.18, issue.6, pp. 2-3.

Jones, W.J. (2001), "Genetics: Year zero", *Health Forum Journal*, volume. 44, issue. 6, pp. 14-18.

Kazman, W. and Westerheim, A.A. (1999), "Telemedicine leverages power of clinical information", *Health Management Technology*, volume. 20, No. 9, pp. 8-10.

Latamore, G.B. (1999), "Workflow tools cut costs for high quality care", *Health Management Technology*, volume. 20, issue. 4, pp. 32-33.

Marien, M. (1989), "IT: you ain't seen nothin yet" in T Forester ed. Computers in the Human Context, Basil Blackwell, Oxford, pp. 7-41.

McCormick, J. (1999), "Wireless hospitals: New wave in healthcare technology," *Health Management Technology*. volume.20, issue.6, pp. 12-13.

Moran, D.W. (1998), "Health information policy: on preparing for the next war," *Health Affair*, volume.17, issue. 6, pp. 9-22.

Morrissey, J. (2000), "Internet dominates providers' line of sight," *Modern Healthcare*, volume.30, issue. 15, pp. 72-92.

Morrissey, J. (2001), "Wanting more from information technology," *Modern Healthcare*, volume.31, issue.6, pp. 66-84.

Morrissey, J. (2002), "High on tech, low on budget," *Modern Healthcare*, volume.32, issue.4, pp. 57-72.

Nairn, G. (2001), "Technology pulls together medicine's diagnostic tools: telemedicine," *Financial Times*, London, Feb 21, pp. 9.

Nelson, C.E., Batalden, P.B. and Mohr, J.J., (1998), "Building a quality future," *Frontiers of Health Services Management*, volume.15, no.1, pp 3.

NHS. (1998), "Overview of benefits from electronic patient records," http // www.nhsia.nhs.uk/ strategy/ full/2.htm.

Nonaka, I. (1991), "The Knowledge-Creating Company", *Harvard Business Review*, volume. 69, issue. 6, pp. 96-104.

Noring, S. (2000), "Telemedicine and telehealth: principles, policies, performance, and pitfalls," *American Journal of Public Health*, volume. 90, issue. 8, pp. 1322.

Parkes, S. (2001), "Medicine may be one of Wap technology's 'killer' applications: mobile services for doctors," *Financial Times*, London, Feb 21, pp. 04.

Pavia, L. (2001), "The era of knowledge in health care", *Health Care Strategic Management*, volume. 19, issue.2, pp. 12-13.

Pinto, P. and Rocha, R. (1999), "Trends on wireless systems radio and satellite," *European Health Telematics Observatory*, http // www. ehto.org / projects/ trends.html.

Puccinelli, B. (1998), "Strategies for sharing knowledge", *Inform*, volume. 12, issue. 9, pp. 40-41.

Purton, P. (2000), "Good prognosis for mobile health services: medical applications," *Financial Times*, London, Sep 20, pp.24.

Rao, S.S. (2001), "Integrated health care and telemedicine," *Work Study*, volume.50, no.6 pp. 222 – 228.

Rockefeller, R. (1999), "Informed shared decision making: Is this the future of health care?" *Health Forum Journal*, volume. 42, issue. 3, pp. 54-56.

Sandberg, L.A. (2001), "The pediatric promise," *Health Management Technology*, volume.22, issue.2, pp. 46-47.

Satyanarayanan, M. (1996), "Mobile information access," *IEEE Personal Communications*, volume.3, no.1, pp 26-33.

Schooley, A.K, (1998), "Allowing FDA regulation of communications software used in telemedicine: a potentially fatal misdiagnosis", *Federal Communications Law Journal*, volume. 50, issue. 3, pp. 731-751.

Serb, C. (2002), "Health care at your fingertips?," *Hospitals & Health Networks*, volume.76, issue.1, pp. 44-46.

Sherer, J. (1995), "The human side of change," *Healthcare Executive*, volume.12, no.4, pp.8-14.

Skyrme, D.J. (1999), *Knowledge Networking: creating the collaborative enterprise*, Butterworth-Heinemann, Oxford.

Stanworth, C. (1998), "Telework and the information age", *New Technology, Work and Employment*, volume. 13, No.1, Mar 1998, pp.51-62.

Surry, M. (2001), "The cyber medics," *Asian Business*, volume. 37, issue.10, pp. 46-47.

Swartz, N. (2000), "Wireless house calls," *Wireless Review*, volume.17, issue.24, pp. 48-53.

The Economist. (1997), "Big sister is watching you: telemedicine,", volume.342, issue.7999, pp. 27, Jan 11.

The Guardian. (2001), Society: frontline: "How skin specialists became west country pioneers," *The Guardian*, Manchester (UK), Aug 1, pp. 10.

Tieman, J. (2000), "Monitoring a good opportunity," *Modern Healthcare*, volume.30, issue.43, pp. 75-80, viewed 26 September 2002, ProQuest database ABI/INFORM, 62701984.

Trembly, A.C. (2001), "Federal study backs use of telemedicine," *National Underwriter*, volume.105, issue.19, pp. 13.

Wallace, S. (1997), "Health information in the new millennium and beyond: the role of computers and the Internet," *Health Education*, no. 3, pp. 88–95.

Wiig, K.M. (1994), "Knowledge management: the central focus for intelligent-acting organizations," *Schema Press*.

WIRELESS INTELLIGENT SENSORS

Emil Jovanov[*] and Dejan Raskovic

1. INTRODUCTION

Traditionally, personal medical monitors used to perform only data acquisition (Advanced Medical, 2004; Agilent, 2004; Cleveland Medical Devices, 2004; Protocol, 2004). Typical examples are holter monitors that are routinely used for ECG and EEG monitoring. Recent development of wireless and mobile technology (Agrawal and Zeng, 2003; Delin and Jackson, 2000; Hsu at al., 1998; Priddy and Jovanov, 2004) laid a foundation for the new generation of wireless intelligent sensors (DARPA, 2004; NIST, 2004). The same technology made feasible implementation of intelligent medical monitors that can provide real-time feedback to the patient, either as a warning of impending medical emergency or as a monitoring aid during exercise. Intelligent medical monitors can significantly decrease the number of hospitalizations and nursing visits (Heidenreich, 1999), acting as a personal "guardian angel" as proposed by Gordon Bell (Morse, 1997). Increased device intelligence will change a whole paradigm of personal monitoring in M-Health systems (Hoyt at al., 2002; Jones at al., 2004; Jovanov et al., 2000; Satava, 2001; Sarcos, 2004; Sontra, 2004; Widya et al., 2004).

A wearable health-monitoring device using a Personal Area Network (PAN) or Body Area Network (BAN) usually requires multiple wires connecting sensors with the processing unit, which can be integrated into user's clothes (GATECH, 2004; Gorlick, 1999). This system organization is unsuitable for longer and continuous monitoring, particularly during the normal activity. For example, monitoring of athletes and computer assisted rehabilitation commonly involve unwieldy wires to arms and legs that restrain normal activity.

Therefore, we expect the new generation of medical monitors to integrate intelligent sensors into wireless body area network as a part of telemedical monitoring system. In this paper we present a survey of existing research and commercially available systems and outline current trends in the development of wireless intelligent sensors.

---

[*] Emil Jovanov, Electrical and Computer Engineering Dept., The University of Alabama in Huntsville, Huntsville, AL, 35899, USA, Phone: (256) 824 5094, Fax: (256) 824 6803, e-mail: jovanov@ece.uah.edu.

2. WIRELESS PATIENT MONITORING

Convenient monitoring during normal activity could be achieved using a wireless PAN (WPAN) of intelligent sensors. In addition to wireless communication of the monitoring unit with the hospital/health information system, we propose a small-range wireless communication between individual physiological sensors on the body. Wireless communication is very convenient for the user because of lack of unwieldy wires between the sensor and the personal server/monitor. It improves user's mobility and hides sensors, protecting user's privacy.

Historically, the development of patient monitors was going toward larger patient mobility and physiological sensor independence in the following order:

1. Static patient connected to the monitoring equipment in the Hospital Information System (HIS) (support for data acquisition and archiving).
2. HIS connected to the Internet. This step allowed remote access to the Personal Medical Record (PMR) stored on a remote telemedical server.
3. Wireless Local Area Network (LAN) connectivity of the personal monitor to the HIS. That allowed patient mobility within the hospital using static gateways networked into HIS. Individual sensors are wired to the personal monitor, but the monitor is wirelessly connected to the HIS using wireless LAN standards, such as 802.11, Bluetooth, or IR.
4. Wireless Wide Area Network (WAN) connectivity of Personal Server (PS) to the telemedical server using satellite and cell phone links.
5. Wireless BAN connectivity between individual intelligent sensors and PS. The system features extremely low power consumption, limited range, and low bandwidth. Ultimate examples include implanted sensors, where the battery recharging/replacement is very limited or impossible. Therefore, it is necessary to enable multi-year functionality or externally powered sensors.

Intelligent wireless sensors perform data acquisition and processing. Individual sensors monitor specific physiological signals (such as EEG, ECG, Galvanic Skin Response (GSR), etc.) and communicate with each other and the personal server as represented in Figure 1.

Personal server integrates information from different sensors and communicates with the upper layer of the hierarchically organized M-Health system (Raskovic et al., 2001). The communication could be either direct using WAN connectivity directly from the personal server, or indirect, using a low-power LAN connection to the gateway (Priddy and Jovanov, 2004).

In this paper we present our prototype implementation of Wireless Intelligent SEnsor (WISE) based on a very low power consumption microcontroller and a PDA based personal server. PDA is a standard mobile platform that could be used as a graphical user interface (GUI) for medical monitoring device. PDA could also be used to access patient's medical record on the telemedical server (Kilman and Forslund, 2004).

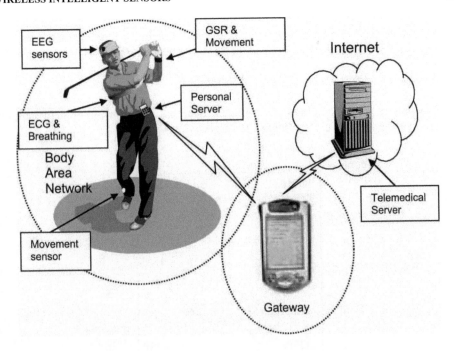

Figure 1. Wireless patient monitoring in the telemedical environment

3. WIRELESS INTELLIGENT MONITORS

Technological advances in low power microprocessors/microcontrollers, application specific integrated circuits (ASICs), battery capacity, and wireless technology made possible increased intelligence of personal health monitors.

Wireless connectivity of the "guardian angel" can integrate the monitor into tele-medical system via a wireless network, so that 911 or a specialized medical response service could be contacted in the event of medical emergencies. Moreover, telemedi-cal monitor can periodically upload and archive episodes that are not life-threatening but still require physician's attention. The major obstacles to wider use of intelligent wearable medical monitors include size and weight of the monitor, sensor implementation and connectivity (Martin et al., 2000). In the current generation, data holters are typically the size of a Walkman, which is larger and heavier than desirable for continuous monitoring during normal activity. Additional functionality of the device can further increase the size and weight. Most medical monitoring applications require a large number of analog input channels. For example, three channels of ECG and up to 16 channels of EEG monitoring create a bulk of unwieldy wires that can significantly limit normal activity and expose user's medical condition. Typical applications include monitoring of athletes and computer assisted rehabilitation. The concept of body area network of intelligent wireless sensors could be a solution of choice for a number of biomedical and monitoring applications.

In this section we present taxonomy of wireless intelligent monitors and survey of existing systems and research projects in the field of intelligent monitors.

3.1. Taxonomy of Wireless Intelligent Monitors

We present two taxonomies of intelligent monitors based on the system mobility (Figure 2) and power efficiency (Figure 3), respectively.

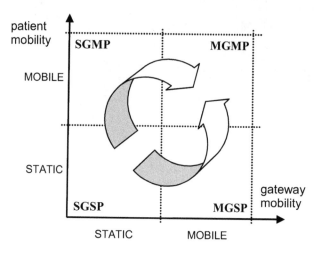

Figure 2. Taxonomy of mobile medical monitors

3.1.1.   Mobile Medical Intelligent Monitors

Ultimate goal of ubiquitous personal health monitoring systems is the user mobility. However, it comes with increased system complexity and power consumption. Moreover, increased complexity and power consumption further increase the size and weight of monitors. As a result, monitors would be less applicable in the real life situations. One possible solution, employed in some systems, would be to use distributed and mobile gateways to reduce power consumption of wireless communication. In the proposed taxonomy we divide systems into four categories and provide some typical examples:

- Static Gateway / Static Patient (**SGSP**), typical of older monitoring systems.
- Static Gateway / Mobile Patient (**SGMP**), wireless in-hospital or home monitoring systems (Bai at al., 1999; Barro et al., 1999; Bauer et al., 2000; Hung and Zhang, 2000; Hung and Zhang, 2002; Sontra, 2004). Another possible future application of SGMP is *health information kiosks* that could be used to collect data from personal monitors on different locations.
- Mobile Gateway / Static Patient (**MGSP**), some telemedical emergency systems (Hoyt et al., 2002; Protocol, 2004).
- Mobile Gateway / Mobile Patient (**MGMP**), distributed wireless monitoring systems (Jovanov at al., 2002; Jovanov at al., 2003).

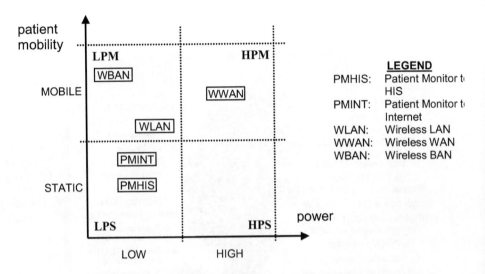

Figure 3. Taxonomy of power efficient mobile medical monitors.

We outlined two current trends in the development of wireless monitors toward the ultimate goal – power efficient system organization with minimum complexity of personal monitors. In the first route, mobility is first achieved on a patient side with the increased complexity of personal monitors, while the second route develops a network of gateways for data collection (Jovanov at al., 2003).

3.1.2.   Power Efficient Medical Intelligent Monitors

Power efficient system organization is emphasized in our second taxonomy, as shown in Figure 3. Typical systems are represented according to the support for patient mobility and power consumption.

We define two broad categories along each of the axes – the patient can be either mobile or static, while the power consumption of the system can be either low or high. More detailed explanation of these taxonomies can be found in a separate paper describing medical monitoring applications for wearable computing (Raskovic et al., 2004).

3.2. Survey of Existing Systems

In this section we present a survey of commercial and research projects in the field of personal intelligent monitors.

Figure 4. Warfighter Physiological Status Monitor (WPSM) [Courtesy of R. Hoyt, US Army Medical Research and Materiel Command/Military Operational Medicine Research Program, Natick, MA]

The U.S. Army Research Institute of Environmental Medicine (USARIEM) and U.S. Army Medical Research and Materiel Command (USAMMRC) are leading the project Warfighter Physiological Status Monitoring (WPSM) (Hoyt et al., 2002; Obusek, 2001). The experimental prototype is presented in Figure 4, and consists of sensors for heart rate, metabolic energy cost of walking, core and skin temperatures, and activity/inactivity. The system is coupled with a module that provides geolocation of a soldier. Data collected by various sensors are transmitted wirelessly to a hub (worn on a soldier's belt) through a low-power PAN. Sensors are expected to be low-cost, capable of collecting data for a few weeks at a time before disposal. Aggregated data can be stored or forwarded to a warfighter's digital fighting system, command center, or in the future, to the Internet. One of the goals of this project is to provide medics with valuable information about wounded soldiers, but the final system is expected to be able to predict the critical aspects of performance under extreme conditions.

Digital Angel Corporation is planning to introduce a medical monitoring system that will be based on their commercially available Digital Angel device. Existing device provides phone and e-mail alerts whenever a person or an asset changes condition or position. By adding biosensors that will report pulse, 3-lead ECG, blood pressure, oxygen level, body temperature, and core temperature they expect to provide physicians with a digital link to the patient that will allow them to quickly transmit bio-data in case of emergency (Digital Angel, 2004).

The *careTrend™ System,* offered by Sensitron, uses combination of Bluetooth and IEEE 802.11b to send patient vital signs data from the point-of-care to the server (Sensitron, 2004) (Figure 5). They also provide HRM – Health Resource Management software that collects, analyzes, trends, presents and stores information. Currently, the company offers blood pressure and pulse monitor, thermometer, weight scale, pulse oximeter, and respiration rate monitor that can be wirelessly connected to the Patient Communication Unit (PCU). The system will also interface with glucose meters and ECG monitors when they become available. Hand-held PCU is used by the caregiver to input pain scores, manage tests and communication, to select patients and to view results. The same unit is used to transfer patient data to the archiving system through the *careTrends* access point.

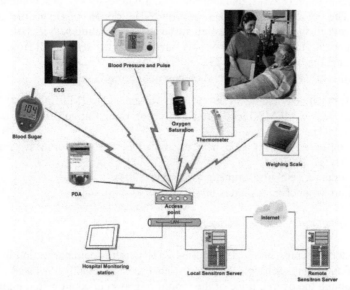

Figure 5. The *careTrend* System [Courtesy of A. Kapoor, Sensitron, San Mateo, CA]

Researchers at Kansas State University are developing a wearable, Bluetooth-enabled portable monitoring health system (Barnes and Warren, 2002). The goal is to provide affordable systems by utilizing plug-and-play sensor units that comply with the common industry standard. Current prototype consists of three units: an Internet-ready base station, a wearable data logger, and a sensor unit. The system will be extended to allow multiple data loggers and multiple sensor units to be connected to each of them.

3.3. Wireless Intelligent Sensors

Although wireless telemetry has been available for a few decades, wireless *intelligent* sensors capable of real-time signal processing have become technologically feasible just recently. Therefore, most wireless intelligent sensors available in the open literature reviewed in this section represent research projects.

Typical example of a commercial system under development is *Symphony™ Diabetes Management System* (Sontra, 2004) from Sontra Medical Corporation. Sontra specializes in continuous non-invasive sensing of glucose and other analytes. This company offers a patch that is based on their core technology providing better skin permeability. Once the sensor/patch is placed on the permeated skin, it will continuously extract interstitial fluid, draw the analytes into the sensor, and measure and calculate the blood glucose concentration. This sensor is a part of Sontra's *Symphony™ Diabetes Management System*. Currently, the system is used for glucose measurements only, but the company plans to add sensors to measure other analytes as well.

Scientists at the d'Arbeloff Laboratory for Information Systems and Technology, MIT, have developed a "ring sensor" that continuously monitors heartbeat rate using photoplethysmograph (PPG) signal and sends data wirelessly to a host computer (Rhee et al., 2001). The device is worn as a finger ring. The double-ring construction provides proper pressure of the sensor on the skin surface, by separating heavier parts of circuitry from the inner ring that holds a sensor only, and shields the sensor unit from the ambient light. In testing, the battery life was 23.3 days of continuous monitoring.

Other currently active research and commercial projects include:

- i-Bean wireless intelligent sensor from Millennial Net (Milennial Net, 2004)
- Implantable MEMS blood pressure sensor from CardioMEMS (CardioMEMS, 2004)
- A non-invasive, fiber optic, heart and breathing rate monitoring system from BTG (BTG, 2004)
- Meridian's TelePulse oximeter (Meridian, 2004)
- e-Care monitoring (e-Care, 2004)

3.3.1. Smart clothes and smart houses

Smart clothes represent a viable alternative to wireless communication in Body Area Network of intelligent sensors. This approach can significantly decrease the price of sensors, although the overall price of the monitoring system could be too high if the goal is non-trivial integration of new sensors into BAN.

There are several groups that currently work on smart clothes and houses. University of Rochester Center for Future Health (University of Rochester, 2004) developed and patented smart socks that detect circulation problems of diabetics and smart bandages that bind to specific DNA sequences of different bacteria and emit detectable light when bacteria are detected. They also work on a concept of Medical Home (Medical Home, 2004), which observes the occupants during their normal activities, looking for the changes in their health condition.

Georgia Institute of Technology, School of Textile and Fiber Engineering (GATECH, 2004) with the *Sensatex* company developed *Smart Shirt* (Smart Shirt, 2004). This research started with the development of a Wearable Motherboard™ for Combat Casualty Care project (Wearable Motherboard, 2004), and recently moved from research to market. Sensatex company plans to market SmartShirt System that will incorporate technology advances in smart clothes, wearable computing, and wireless communication into a product that will allow data collection and monitoring of biometric information.

Intel's Proactive Health Research project (Intel, 2004) incorporates sensor networks, home networks, activity tracking, and ambient displays, to create an environment that will help elderly age gracefully at home. Future phases of this project will concentrate on patients with chronic conditions, and finally on wellness, physical fitness, and mental health.

3.3.2.   Implantable and other inherently wireless sensors

Implantable sensors could solve many problems in the monitoring of chronically ill patients that need constant monitoring, such as diabetic patients (Asada et al., 1998). Obviously, wireless communication is the most natural integration of implanted sensors into personal medical monitoring system. Design of intelligent implanted sensors must take into account a number of different considerations, such as size, power consumption, and power efficient communication. The choice of a communication frequency dictates the size of the antenna and the whole sensor. In addition, reliable transfer of data from the implanted sensors is one of the most important tasks researchers are facing. Power requirements are even more stringent, because it is highly desirable to either ensure very long-life battery operation, or to design an implantable device that could be inductively or optically powered/recharged.

Researchers from the University of Washington, Caltech, and Case Western Reserve University have developed a miniature implantable microcomputer (Diorio and Mavoori, 2003), capable of recording nerve and muscle signals from animals during their normal activity. So far, they implanted their devices in two types of animals – sea slugs and Manduca moths. Similarly to the stationary systems, they use flexible metallic needles, attached to the animal with surgical superglue, to collect signals from nerve bundles, and micromachined silicon probes to record activity of neural assemblies. However, to collect signals from individual neurons, scientist from the University of Washington are working on silicone MEMS probes that will mimic the performance of glass capillaries used on constrained animals. Implantable device consists of variable-gain amplifiers, a system-on-a-chip microcontroller, and a high density memory. In this project, researchers decided to avoid antennas and charge pumps needed for RF-powered devices, due to size constraints of the implantable device. They plan to employ thin-film battery technologies. One-square-centimeter battery manufactured in these technologies weighs 200mg and provides 12 mA peak current.

At the University of Michigan, Center for Wireless Integrated Microsystems, researchers developed BiCMOS wireless stimulator chip (Akin and Najafi, 1994; Ghovanloo and Najafi, 2002) that will be used in conjunction with micromachined passive stimulating microprobes. Those probes were developed at the same university and they are in regular medical use. This will allow wireless and stand-alone operation for an unlimited time period, since the chip uses 4MHz carrier to receive both data and power through inductive coupling. Total power dissipation of the chip is less than 10mW and the area it occupies is about $13mm^2$.

Another example of an RF-powered intelligent sensor is a miniature implantable wireless neural recording device developed at the University of California, Los Angeles (Irazoqui-Pastor et al., 2002). This device records and transmits neural signals. Device size is less than $1cm^2$ and power dissipation is measured at 13.8mW. Tests have shown that the transmitting range is up to 0.5m, and that demodulated signal is highly correlated with the original signals in the range between 5mV and 1.5mV.

Engineering Science and Technology Division at the Oak Ridge National Lab is working on an implantable sensor that will allow non-invasive assessment of blood flow quality for the newly transplanted organs. The implantable unit, about the diameter of a quarter, will provide real-time information by transmitting tissue circulation data to a nearby receiver (Oak Ridge, 2004).

Researchers from the Universities of Glasgow, Edinburgh and Strathclyde are developing a capsule (Astaras et al., 2002) traversing the gastrointestinal tract (part of *IDEAS* - Integrated Diagnostics for Environmental and Analytical Systems project). The sensor gathers data that cannot be collected using traditional medical endoscopy. The device is battery-powered and integrates sensors, processing, and RF bi-directional communication onto a single piece of silicon (current device size is 32×11.5mm). This device uses spread-spectrum based communication subsystem (Aydin et al., 2002).

Given Imaging offers a commercially available system - *Given® Diagnostic System*. Disposable imaging capsule M2A (11×26mm) passes through the gastrointestinal tract and wirelessly streams video images to the receiver worn on a belt. Signal is received through the array of antennas, which are also used to determine the exact location of the capsule (Moore, 2003).

The National Aeronautics and Space Administration (NASA), through their *Sensors 2000!* program, designed another pill-shaped biotelemeter, used originally for space flight applications. Based on those original pills, new generation of pills, which will be used for monitoring the health of fetus, is under development (Mundt et al., 2000). Currently, two different sensor pills are developed – one measures pressure and temperature, while the other one measures pH and temperature. Both pills are still under testing, but plans exist for other pills, that will measure heartbeat rate, ECG, glucose, blood gasses, etc. Each pill is 35mm long and has diameter of 9mm. Transmission frequency is in the biomedical range (174-214 MHz). Low-power operation is achieved by low data rate (a pair of RF bursts at frequency of about 1-5Hz is transmitted) and by using Pulse Interval Modulation (PIM) encoding scheme. Under those conditions, expected transmitter lifetime is several months. Since the device is designed to monitor intra-uterine contractions, which occur over several minutes, this data rate is sufficient. However, real challenge will be transmission of ECG or similar signal that requires higher sampling frequency.

Researchers at the Biomechanics-Laboratory, Oskar-Helene-Heim, Orthopaedic Hospital of the Free University of Berlin developed inductively powered implantable device that measures hip joint forces and temperatures for the hip joint prosthesis (Graichen, 2000). Temperature is measured at ten points, and three semiconductor strain gauges sense the deformation for three-dimensional load measurement. The whole system consists of sensors and two telemetry units. Each of them is based on eight-channel telemetry chip and transmits data at a frequency range of 140-160MHz. External power coil is slipped over the patients leg to allow powering of internal coils, and it's magnetic field (4KHz) is controlled by a personal computer. Force and temperature can be monitored in real time.

Researchers from two Portugal universities developed Subcutaneous Implantable Capsule that uses phonocardiography as a method to monitor changes in heart rate (Torres-Pereira, 2000). Capsule consists of a bimorph piezoelectric sensor, analog signal processing module, a frequency modulated transmitter (30MHz), 3.6V battery, and an antenna. Battery life was measured to be 30 days. Total size of the device is 75mm×17mm and it was tested on sheep.

Medtronic offers Insertable Loop Recorder (Medtronic, 2004), developed in collaboration with Division of Cardiology, University of Western Ontario. It provides up to 14 months of monitoring and data acquisition of critical cardiac events. Up to 40 minutes of history can be stored after an episode. This device weighs 17g, with the approximate volume of 8 mL. Implanted device communicates wirelessly with the activator to initiate storing of previous episode after fainting event and upload ECG data to the computer for analysis.

4. WIRELESS NETWORK OF INTELLIGENT SENSORS (WISE)

We have been developing Wireless Intelligent Sensors (WISE) (Jovanov et al., 2000; Jovanov et al., 2001; Jovanov et al., 2002; Jovanov et al., 2003) for more than four years. WISE are microcontroller-based intelligent sensors of physiological signals for personal medical monitors. Individual sensors communicate with the personal server in BAN as represented in Figure 1. Proposed architecture features hierarchical signal processing and collaboration in telemedical environment (Jovanov et al., 2002). WISE sensors perform local data acquisition and real-time signal processing tasks, such as filtering and feature extraction. BAN is organized as client-server network with single server (Personal Server PS) and multiple clients (WISE sensors). Personal server provides synergy of information from different sensors. For example, accelerometer-based movement sensors (Analog Devices, 2004) could provide the information of user activity in the case of increased heart-bit rate, or changes in ST segment morphology for monitors of silent miocardial ischemia (Wheelock, 1999).

The core of our WISE consists of Texas Instruments' microcontroller MPS430F149 that is responsible for A/D data acquisition and processing. The controller features 16-bit RISC architecture, ultra-low power consumption (less than 1 mA in active mode and ~1 µA in standby mode), 60KB on-chip flash memory, 2KB RAM, AD converter (eight 12-bit channels), integrated UARTs and a 64-pin QFP package. Additional analog channels are used to monitor battery voltage, wireless link quality, and other external analog inputs. Therefore, WISE is capable of reporting the battery status and reporting low-battery warnings to the higher system levels in the system hierarchy.

The Wireless Distributed Data Acquisition System uses Wireless Intelligent Sensors (WISE) as individual Heart Rate Variability (HRV) monitors, and a Mobile Wireless Gateway (MOGUL) (Priddy and Jovanov, 2002). The system is organized in Master-Slave configuration with MOGUL acting as a master. Periodic visits to the training facility with the handheld MOGUL device allow uploading of collected data from individual WISE sensors. The MOGUL moves along the path to come into wireless contact with all WISE devices in the system. During the time that the MOGUL is able to talk to the WISE devices it downloads and catalogues all of the data measurements stored on the WISE devices. The WISE devices that are not in range of the MOGUL's wireless transceiver store all of their measurements into on-board flash memory, waiting to upload them to the MOGUL. It is possible to connect the MOGUL wirelessly to the Internet, and to create true real-time telemedical system.

Mobile gateways are also used to describe and mark events that are important for the group, such as exact start of the exercise, unexpected changes in the plan and schedule, and so on.

We are experimenting with two types of sensors – WISE for ECG signal acquisition (WISE_ECG) and Wireless Heart Rate Monitor (WHRM) for HRV monitoring.

- WISE_ECG uses an off-the-shelf, two-channel bio-amplifier for ECG signal conditioning. The output signals from the bio-amplifier are converted to digital signals using integrated eight channel/12-bit AD converter (Jovanov et al., 2002; Priddy and Jovanov, 2002).

- WHRM uses receiver for Polar heart rate monitor belt to acquire heart rate variability only (sequence of RR intervals with 1 ms resolution). The device provides processing, storage and wireless communication (Jovanov et al., 2003). Wireless heart monitor device has been developed in collaboration with RP Technologies, Huntsville, Alabama (Figure 6).

Connection between WHRM and MOGUL is naturally implemented with some standard PAN wireless technology, such as Bluetooth. However, since the current Bluetooth technology still requires 3-5 times larger power consumption than the standard off-the-shelf Industrial, Scientific, and Medical (ISM) wireless modules, we still use a standard 900 MHz RF modules for wireless communication with a custom power-efficient communication protocol.

A MOGUL device polls all WISE devices that it is configured to poll, using specialized wireless interface. The connection between the MOGUL and the central server could be a standard 802.11b or Bluetooth wireless link. We are currently using Compaq's iPAQ pocket PC as a MOGUL, as shown in Figures 1 and 7.

Most monitoring applications are patient-centered, requiring synchronization only between different sensors on the same patient/subject. In some setups it is necessary to evaluate individual performance of subjects within the group during specific events. That requires synchronization of measurements of individual monitors within the sampling interval of the monitored physiological parameter. This feature is not, to the best of our knowledge, available in the present generation of intelligent monitors. Therefore, we are developing a distributed system of wireless monitors for synchronized monitoring and analysis within the group of subjects under test (Jovanov et al., 2003; Priddy and Jovanov 2002). The proposed system achieves very low power consumption due to the use of mobile gateways (iPAQ PDA) that are used to collect data from individual monitors without interfering with their regular activity.

We have been investigating stress tolerance within a group of individuals during stressful training/activity (Jovanov et al., 2003). Our preliminary results indicate that those individuals who have better stress tolerance also exhibit significantly different patterns of heart rate variability, both before and during stress exposure. These baseline differences in heart rate variability are predictive of actual military and cognitive neuropsychological test performance scores assessed during and after stress exposure (Jovanov et al., 2002; Jovanov et al., 2003).

Figure 6. Wireless Heart Rate Monitor WHRM (Jovanov et al., 2003)

Figure 7. Mobile Gateway (MOGUL) (Jovanov et al., 2002)

We are currently developing applications of WISE sensors for physiological monitoring (Jovanov et al., 2003), portable computer-assisted physical rehabilitation (Milenkovic et al., 2002), and future generation human-computer interfaces (Jovanov et al., 2001) with possible applications to affective computing (Picard, 1997).

5. FUTURE TRENDS

Current technological trend, as outlined in this paper, will clearly allow wider use of wireless intelligent sensors, lower power consumption, and smaller sensor size. The ultimate goal is to have a single-chip intelligent sensor, as proposed by NASA's Jet Propulsion Laboratory (NASA, 2004). They are developing a wireless sensor patch that will incorporate temperature, heart rate, and blood pressure into one miniature Micro Electro-Mechanical Systems (MEMS) device. This device would be comparable in size to an ordinary adhesive bandage, but it would contain sensors integrated with electronic circuitry for processing and transmitting sensor data (Figure 8). No batteries are required because the patch is powered externally using a RF wireless communication link similar to one that is used for implantable brain probes developed at the University of Michigan (Ghovanloo and Najafi, 2002).

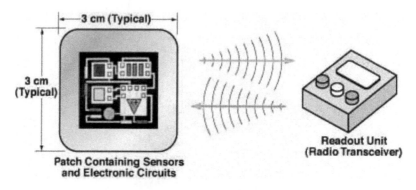

Figure 8. The self-powered wireless sensor patch (NASA, 2004)

6. CONCLUSIONS

Wireless intelligent sensors make possible a new generation of non-invasive, unobtrusive personal medical monitors applicable during normal activity. Decreased power consumption will significantly increase battery life, and even enable externally powered intelligent sensors. Sensor intelligence allows implementation of real-time processing and sophisticated encryption algorithms. On-sensor data processing decreases the amount of energy spent on communication, and allows implementation of power efficient communication protocols.

Another important design issue is protection from eaves-dropping that can disclose user's health condition. The proposed concepts will be able to support strong encryption due to the sensor intelligence. They will feature very low power transmission limited to BAN (less than 2 m range).

In addition to technical issues covered in this paper, faster acceptance of intelligent sensors is also constrained by possible legal issues related to the responsibility for outcome of medical conditions. The manufacturer will want to minimize the chance of being sued if the device does not correctly predict a medical event. It is difficult to determine if there was a malfunction of device, a lack of knowledge of other conditions of the patient, or whether the lawsuit was justified (Martin, 2000). However, we believe that this is the faith of every new technology, and that we will witness increased number of wireless intelligent health monitors in the near future.

7. ACKNOWLEDGMENTS

The WISE project has been partially supported by the University of Alabama in Huntsville, grant 87716. WHRM project has been partially funded by USAMRMC (ERMS#01145005) as a part of "The Warfighter's Stress Response: Telemetric and Noninvasive Assessment" project. We would like to thank Amanda O'Donnell from Naval Aerospace Medical Research Laboratory (NAMRL), Pensacola, Florida, Andy Morgan, Yale University School of Medicine, West Haven, Connecticut, and Frank Andrasik, University of West Florida, Pensacola, FL, for all their help and suggestions. We are grateful to Dr. Reza Adhami from the University of Alabama in Huntsville for discussions regarding signal processing issues in WISE network, and Paul Cox from RP Technologies for help in development and implementation of the WHRM devices and distributed sensor network.

8. REFERENCES

Advanced Medical Electronics Corporation, *Wearable Polysomnograph*, http://www.ame-corp.com/Wearable_Polysomnograph_files/Wearable_Polysomnograph.pdf, August 2004.

Agilent, http://www.agilent.com, August 2004.

Agrawal D.P., Zeng Q., *Introduction to Wireless and Mobile Systems*, Brooks/Cole, 2003.

Akin T. and Najafi K., "A Telemetrically Powered and Controlled Implantable Neural Recording Circuit with CMOS Interface Circuitry," *in Proceedings of IEEE 7th Mediterranean Electrotechnical Conf.*, Antalya, Turkey, April 1994, pp. 545-548

Analog Devices, http://www.analog.com/accelerometers.html, August 2004.

Asada G., et al, "Wireless Integrated Network Sensors: Low Power Systems on a Chip," in *Proceedings of the 1998 European Solid State Circuits Conference*.

Astaras A., Ahmadian M., Aydin N., Cui L., Johannessen E., Tang T-B., Wang L., Arslan T., Beaumont S.P., Flynn B. W., Murray A.F., Reid S.W., Yam P., Cooper J.M., and Cumming D.R.S., "A Miniature Integrated Electronics Sensor Capsule for Real-Time Monitoring of the Gastrointestinal Tract (IDEAS)," *in Proceedings of ICBME 2002: The Bio-Era: New Challenges, New Frontiers*, 4-7 December 2002, Singapore.

Aydin N., Arslan T., and Cumming D.R.S., "Design And Implementation of a Spread Spectrum Based Communication System for an Ingestible Capsule," in *Proceedings of the Second Joint EMBS/BMES Conference*, Houston, TX, USA, October 23-26, 2002, pp. 1773-1774.

Bai J., et al., "A Portable ECG and Blood Pressure Telemonitoring System", *IEEE Engineering in Medicine and Biology*, July/August 1999, pp. 63-70.

Barnes G.E. and Warren S., "A Wearable, Bluetooth-Enabled System for Home Health Care," *Proceedings of the Second Joint EMBS/BMES Conference*, Houston, TX, USA, October 23-26, 2002, pp. 1879-1880.

Barro S., et al., "Intelligent Telemonitoring of Critical Care Patients", *IEEE Engineering in Medicine and Biology*, July/August 1999, pp. 80-88.

Bauer P., Sichitiu M., Istepanian R., and Premaratne K., "The Mobile Patient: Wireless Distributed Sensor Networks for Patient Monitoring and Care," in *Proc. Third International Conference on Information technology in Biomedicine, (ITAB-ITIS2000)*, pp. 17-21, 2000.

BTG: http://www.btgplc.com, August 2004.

CardioMEMS: http://www.cardiomems.com, August 2004.

Cleveland Medical Devices, http://www.clevemed.com, August 2004.

Delin K.A. and Jackson S.P., "Sensor Web for In Situ Exploration of Gaseous Biosignatures," *Proceedings of the 2000 IEEE Aerospace Conf.*, Big Sky, Montana, March, 2000.

Digital Angel Corporation: http://www.digitalangelcorp.com/, August 2004.

Diorio C. and Mavoori J., "Computer Electronics Meet Animal Brains," *IEEE Computer*, vol. 38(1), January 2003, pp. 69-75.

e-Care Project: http://www.ecare-scotland.gov.uk, August 2004.

Ghovanloo M. and Najafi K., "A BiCMOS Wireless Stimulator Chip for Micromachined Stimulating Microprobes," in *Proceedings of the Second Joint EMBS/BMES Conference*, Houston, TX, USA, October 23-26, 2002

Georgia Institute of Technology, School of Textile and Fiber Engineering: http://www.tfe.gatech.edu/, August 2004

Gorlick M., "Electric Suspenders: A Fabric Power Bus and Data Network for Wearable Digital Devices," in *Proc. of the Third Intern. Symp. on Wearable Computers*, San Francisco, CA, October, 1999, pp. 114-121.

Graichen F., Bergmann G., and Rohlmann A., "Implantable Telemetry System for Measurement of Hip Joint Force and Temperature," in J.H. Eiler, D.J. Alcorn, and M.R. Neuman (editors), *Biotelemetry 15: Proceeding of the 15th International Symposium on Biotelemetry*, Juneau, Alaska, USA, International Society on Biotelemetry, Wageningen, The Netherlands, 2000.

Heidenreich P., Ruggerio C., and Massie B., "Effect of a Home Monitoring System on Hospitalization and Resource Use for Patients with Heart Failure," *American Heart Journal*, Vol. 138, No. 4, 1999, pp. 633-640.

Hoyt R.W., Reifman J., Coster T.S., and Buller M.J., "Combat Medical Informatics: Present and Future," *in: Biomedical Informatics: One Discipline.* Isaac S. Kohane, MD, PhD, ed., *Proceedings of the 2002 AMIA Annual Symposium*, ISBN: 1-5603-600-4

Hsu V., Kahn J.M., and Pister K.S.J., "Wireless Communications for Smart Dust", *Electronics Research Laboratory Technical Memorandum Number M98/2*, February, 1998.

Hung K. and Zhang Y.T., "On the Feasibility of the Usage of WAP Devices in Telemedicine," *Information Technology Applications in Biomedicine, ITAB-ITIS 2000*, Arlington, Virginia, November 9-10, 2000.

Hung K. and Zhang Y.T., "Usage of Bluetooth(TM) in Wireless Sensors for Tele-Healthcare," in *Proceedings of the Second Joint EMBS/BMES Conference*, Houston, TX, USA, October 23-26, 2002, pp. 1881-1882.

Irazoqui-Pastor P., Mody I., and Judy J.W., "Transcutaneous RF-Powered Neural Recording Device," in *Proceedings of the Second Joint EMBS/BMES Conference*, Houston, TX, USA, October 23-26, 2002.

Intel, Proactive Health Research: http://www.intel.com/research/prohealth/, August 2004.

Jones V. et al., Monitoring For Healthcare and for Safety in Extreme Environments, in this book.

Jovanov E., Price J., Raskovic D., Kavi K., Martin T., and Adhami R., "Wireless Personal Area Networks in Telemedical Environment," in *Proc. Third International Conference on Information technology in Biomedicine," (ITAB-ITIS2000)*, pp. 22-27, 2000.

Jovanov E., Raskovic D., Price J., Moore A., Chapman J., and Krishnamurthy A., "Patient Monitoring Using Personal Area Networks of Wireless Intelligent Sensors," Biomedical Sciences Instrumentation Vol. 37, in *Proc. 38th Annual Rocky Mountain Bioengineering Symposium, RMBS 2001*, April 2001, Copper Mountain Conference Center, pp. 373-378.

Jovanov E., Raskovic D., and Hormigo R., "Thermistor-based Breathing Sensor for Circadian Rhythm Evaluation," *in Proc. of 38th Annual Rocky Mountain Bioengineering Symposium, RMBS 2001,* April 2001, Copper Mountain Conference Center, pp. 493-497.

E. Jovanov, D. Starcevic, and V. Radivojevic, "Perceptualization of Biomedical Data", chapter in Akay, M., Marsh, A., Eds, "Information Technologies in Medicine, Volume I: Medical Simulation and Education", John Wiley and Sons, 2001.

Jovanov E., O'Donnel A., Morgan A., Priddy B., and Hormigo R., "Prolonged Telemetric Monitoring Of Heart Rate Variability Using Wireless Intelligent Sensors And A Mobile Gateway," *2nd Joint EMBS-BMES*, Houston, Texas, October 2002.

Jovanov E., O'Donnell A., Raskovic D., Cox P., Adhami R., and Andrasik F., "Stress Monitoring Using a Distributed Wireless Intelligent Sensor System," *IEEE Engineering in Medicine and Biology Magazine*, Vol. 22, No.3, May/June 2003, pp. 49-55.

Kilman D.G. and Forslund D.W., Electronic Patient Record for Telematic Applications, in this book.

Martin T., Jovanov E., and Raskovic D., "Issues in Wearable Computing for Medical Monitoring Applications: A Case Study of a Wearable ECG Monitoring Device," *International Symposium on Wearable Computers ISWC 2000*, Atlanta, October 2000.

Medical Home: http://www.futurehealth.rochester.edu/smart_home/index.html, August 2004.

Medtronic: http://www.medtronic.com/reveal/new.html, August 2004.

Meridian Medical Technologies: http://www.meridianmeds.com, August 2004.

Millennial Net: http://www.millennial.net, August 2004.

Milenkovic M., Jovanov E., Chapman J., Raskovic D., and Price J., "An Accelerometer-Based Physical Rehabilitation System," *The 34th Southeastern Symposium on System Theory (SSST2002)*, Huntsville, Alabama, pp. 57-60, 2002.

Moore S.K., "A candid camera for the gut," *IEEE Spectrum*, July 2000, pp. 75; also Given Imaging: www.givenimaging.com, January 2003.

Morse W., "Medical electronics," *IEEE Spectrum*, vol. 34(1), January 1997, pp. 99 –102

Mundt C.W., Ricks R.D., and Hines J.W., "Biotelemeters for Space Flights and Fetal Monitoring," in J.H. Eiler, D.J. Alcorn, and M.R. Neuman (editors), *Biotelemetry 15: Proceeding of the 15th International Symposium on Biotelemetry*, Juneau, Alaska, USA, International Society on Biotelemetry, Wageningen, The Netherlands, 2000.

NASA's Jet Propulsion Laboratory: http://www.nasatech.com/Briefs/Feb00/NPO20651.html, August 2004.

OAK Ridge National Lab: http://www.ornl.gov/Press_Releases/archive/mr20021122-00.html, August 2004.

Obusek J., "Warfighter Physiological Status Monitoring," *The Warrior*, November-December 2001, pp. 6-8.

Picard R.W., *Affective Computing*, MIT Press, Cambridge, 1997.

Priddy B. andJovanov E., "Wireless Distributed Data Acquisition System," *The 34th Southeastern Symposium on System Theory (SSST2002)*, Huntsville, Alabama, pp. 463-466, 2002.

Priddy B. and Jovanov E., Wireless LAN Technologies for Healthcare Applications, in this book.

Protocol (Welch Allyn), http://www.monitoring.welchallyn.com/, August 2004.

Raskovic D., Jovanov E., and Kavi K., "Hierarchical Digital Signal Processing," in *Proc. of 2001 IEEE International Symposium on Intelligent Signal Processing and Communication Systems, ISPACS 2001*, Nashville, TN, Nov. 20-23, 2001.

Raskovic D., Martin T., and Jovanov E., "Medical Monitoring Applications for Wearable Computing," *The Computer Journal*, vol. 47, no. 4, July/August 2004.

Rhee S., Yang B-H., and Asada H.H., "Artifact-Resistant Power-Efficient Design of Finger-Ring Plethysmographic Sensor," *IEEE Transactions on Biomedical Engineering*, vol. 48(7), July 2001, pp. 795-805.

Satava R.M., "Future Technologies for Medical Applications," chapter in Akay, M., Marsh, A., Eds, Information Technologies in Medicine, Volume I: Medical Simulation and Education, New York: John Wiley and Sons, pp. 219-236, 2001.

SARCOS, http://www.sarcos.com, August 2004.

Sensitron: http://www.sensitron.net/technology/theCaretrendsSystem.html, August 2004.

DARPA Sensor Information Technology, http://www.darpa.mil/ipto/programs/sensit/, August 2004.

Smart Shirt, Sensatex: http://www.sensatex.com/smartshirt/, August 2004.

NIST SmartSpaces, http://www.nist.gov/smartspace/smartSpaces/, August 2004.

Sontra Medical Corporation, http://www.sontra.com/, August 2004.

Starcevic D., Obrenovic Z., Jovanov E., and Radivojevic V., "Implementation of Virtual Medical Devices in Internet and Wireless Cellular Networks," *1st IFIP Workshop on Internet Technologies, Applications and Societal Impact*, Wroclaw, Poland, October 2002.

Torres-Pereira L., Torres-Pereira C., and Couto C., "A Novel Subcutaneous Implantable Capsule for Phonocardiography," in J.H. Eiler, D.J. Alcorn, and M.R. Neuman (editors), *Biotelemetry 15: Proceeding of the 15th International Symposium on Biotelemetry*, Juneau, Alaska, USA, International Society on Biotelemetry, Wageningen, The Netherlands, 2000.

University of Rochester Center for Future Health: http://www.centerforfuturehealth.org, August 2004.

Wearable Motherboard[TM], Georgia Institute of Technology: http://www.gtwm.gatech.edu, August 2004.

Wheelock B., "Autonomous Real-Time Detection of Silent Ischemia," M.S. thesis, University of Alabama in Huntsville, Huntsville, 1999.

Widya I., Vierhout P., Jones V.M., Bults R., van Halteren A., Peuscher J., and Konstantas D., Telematic Requirements for Emergency and Disaster Response Derived from Enterprise Models, chapter in: *M-Health: Emerging Mobile Health Systems*, R.H. Istepanian, S. Laxminarayan, and C.S. Pattichis, eds., Chapter 41, Springer Science 2005.

WIRELESS TECHNOLOGIES
FOR HEALTHCARE APPLICATIONS

Brent Priddy[*] and Emil Jovanov

1. INTRODUCTION

Wireless and mobile technologies are widely accepted, influencing many aspects of our daily life (Agrawal and Zeng, 2003; Bing 2000). Wireless systems are designed to provide "anytime, anywhere" service, enabling data entry and data access by medical personnel at the point of care, direct data acquisition from medical devices, patient and device identification, and remote patient management. Wireless technologies will spawn new applications and industries to address healthcare issues in the coming years. Wireless monitoring and communication allows patient mobility and efficient response in emergency situations. It is expected that wireless monitors will be commonly used in hospitals and for home monitoring in the next few years; it is estimated that at least 20% of practicing physicians in the US will use a wireless device other than cell phones and pagers for clinical purposes by the end of 2004 (Health Technology Center, 2003). Basic wireless technologies are now a commodity, widely available in pagers, cordless phones, cell phones, wireless computer networks, ham radio, etc. Increased support for data communications in the third-generation wireless systems outlines a current trend that will facilitate a whole range of new applications and systems.

Wireless communication technologies represent one of the most critical factors for the acceptance of mobile personal monitoring in telemedical systems. We present a survey of existing and future wireless technologies and standards suitable for data transmission in mobile healthcare applications (Agrawal and Zeng, 2003)[0]. There are important issues to be considered when designing a wireless system, including range, bandwidth, power consumption, personal health risks, FCC regulation, interference, security, and applicability to different health applications. Case studies of wireless healthcare applications and in-depth coverage of security, issues are provided in later chapters of this book.

---

[*] Brent Priddy, Electrical and Computer Engineering Dept., The University of Alabama in Huntsville, Huntsville, AL, 35899, USA, Phone: (256) 824 5094, Fax: (256) 824 6632, e-mail: toopriddy@gmail.com

2. WIRELESS TECHNOLOGIES

The main wireless LAN technologies include:

- Optical
- Electromagnetic (EM or Radio)

Use of spectrum of different technologies is represented in Figure 1 and Table 1. Higher frequencies support greater bandwidth of the transmitted signal, at a price of decreased range of transmission. Wireless LANs are typically designed to operate in the industrial, scientific, and medical (ISM) frequency bands because these bands are available for unlicensed operation and it is possible to build low-cost, low-power radios in this frequency range that operate at wireless LAN speeds. To prevent interference, devices operating in the ISM bands must use low-power spread spectrum transmission. The radiated power is limited to 1W in the United States, 100mW in Europe, and 10mW/MHZ in Japan. Different frequency bands are approved for use in the United States, Europe, and Japan and any ISM wireless LAN product must meet the specific requirements for the country in which it is sold. The typical ISM frequency bands are 902 – 928MHz (26MHz available bandwidth), 2.4 – 2.4835GHz (83.5MHz available bandwidth) and 5.725 – 5.85GHz (125MHz available bandwidth).

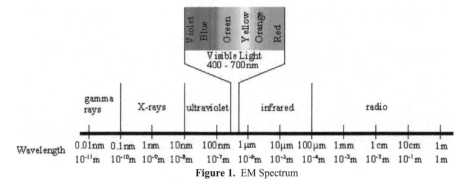

Figure 1. EM Spectrum

Prior to the establishment of the Wireless Medical Telemetry (also known as WMTS), medical telemetry devices generally could be operated on an unlicensed basis on vacant television channels 7-13 (174-216 MHz) and 14-46 (470-668 MHz) or on a licensed but secondary basis to private land mobile radio operations in the 450-470 MHz frequency band. This meant that wireless medical telemetry operations had to accept interference from the primary users of these frequency bands, i.e., the television broadcasters and private land mobile radio licensees.

To ensure that wireless medical telemetry devices can operate free of harmful interference, the FCC decided to establish the WMTS (FCC, 2004)[0]. In a Report and Order released on June 12, 2000, the FCC allocated a total of 14 megahertz of spectrum to WMTS on a primary basis. At the same time, it adopted a number of regulations to ensure that the WMTS frequencies are used effectively and efficiently for their intended medical purpose.

The FCC established a transition period so that existing wireless medical telemetry equipment would not be made obsolete. Beginning October 16, 2002, wireless medical telemetry equipment will not be authorized by the FCC unless it is designed to operate in the WMTS frequency bands.

Table 1. Selected U.S. RF Allocations and Applications

Application		Allocated Frequency (MHz)
Cellular telephone		824-849, 869-894
Industrial, Scientific, and Medical (unlicensed)		902-928, 2450-2500, 5725-5875
Medical Telemetry (WMTS)		174-216, 460-668
Microwave Owens		2400-2483.5
Paging and Radiotelephone Service		152-152.255, 152.495-152.855, 157.755-158.115, 454-455, 459-460, 929-930, 931-932
Personal Communication	broadband	1850-1990
	narrowband	901-902,930-931,940-941
	unlicensed	1910-1930,2390-2400
RFID (Radio Frequency Identification)		902-928, 2400-2500, 5850-5925
Satellite Services	Digital Radio	2320-2345
	Geostat. Fixed	3700-4200
	Geostat. Mobile	1525-1559, 1626.5-1660.5
	GPS	1215-1240, 1350-1400, 1559-1610
	Non-Geostat. Mobile	1610-1626.5, 2483-2500, 5091-5250, 6700-7075
	Non-Geostat. Mobile Nonvoice	137-138,148-150.05, 400.15-401, 455-456, 459-460
Wireless Communications Service		746-764,776-794,2305-2320, 2345-2360
Unlicensed National Information Infrastructure		5000-5350, 5650-5850

Each subsection of this chapter will include a description of the workings of the wireless technology including the range, speed of the transmission, consequences of use, and regulations set by the FCC for the given technology.

3. OPTICAL TECHNOLOGIES

Two main optical technologies are *Infrared (IR)* and *laser* transmission. Both technologies are line-of-sight medium, which significantly limits their range. Infrared (IR) technologies are seen today in remote controls, watches, PDAs, and laptops for convenient and cheap wireless transmission of data. IR transmission uses an IR LED to transmit information and a matching IR photodiode to receive the data. In a typical IR wireless LAN setup, the light that is transmitted from the IR LED must usually be dispersed over a wide range, limiting the ability for many devices to communicate at once. IR communication requires a direct line of sight between the transmitter and receiver (Harter and Bennett, 1993)[0]. IR networks are usually setup in cells or zones of a room or area (Want et al., 1995). These zones are able to be as large as you want with the help of an array of IR transceivers mounted in the ceiling or walls of a room. Most IR devices operate in the 850nm and 920nm wavelength ranges. Speeds of up to 115200 bps and 4 Mbps in the current IrDA 1.0 standard can be achieved with low power consumption (Harter and Bennett, 1993)[0]. IR communication is able to operate at 16 Mbps, using 1W IR laser diodes, but only 4 Mbps has been achieved in most real world applications (Zimmerman, 1995). The typical transmission range of IR communication is 1m but a range of 10m is achievable at a price of consuming hundreds of milliwatts of power.

The Infrared Data Association (IrDA) is an international organization of over 150 companies that creates and promotes interoperable, low-cost infrared data interconnection standards. The IrDA has developed eight standards for short range, line-of-sight infrared data communication. Some of the protocols applicable for mobile health monitoring devices would be IrLAN, IrCOMM, IrOBEX, and IrDA Lite. IrLAN is a standard for a protocol used in infrared access to LANs. IrCOMM is a standard for serial and parallel port emulation for legacy applications, such as printing and serial devices such as modems. The IrOBEX standard specifies a method of object exchange, much like the HTTP protocol. Finally, IrDA Lite is a slimmed down version of the IrDA standards for smaller devices, but still implements compatibility with the full IrDA implementations (Bing, 2000)[0].

IR communications suffers negligible degradation from RF transmissions. Photodiodes have enough sensitivity to detect a 1 $\mu W/cm^2$ IR signal. An IR transmitter capable of sending data to a photodiode at 25m would have to transmit at a power of 1W (easily achievable with an array of transmitters). One drawback of IR communication is that there should be a direct line of sight in-between the transmitter and receiver. IR communication can work when bounced off of reflective surfaces, but at a cost of throughput.

IR communication is unregulated by the FCC although current health and safety guidelines in many countries setout "safe" and "average" levels of IR light that are equivalent to sunlight (Harter and Bennett, 1993)[0].

4. EM TECHNOLOGIES

There are two categories of Electromagnetic field technologies:

- Far-field EM
- Near-field EM

The separation of EM technologies into two categories depends on what frequencies you are considering. The definition of when Electromagnetic Radiation is considered as "near" or "far" varies depending on the wavelength of the signal. We will consider, for health applications, the near-field EM technologies are only applicable for Body Area Networks (BAN), a physiological network of sensors, or for communication with implanted sensors.

4.1. Far-Field EM Technologies

RF communication has been around for a long time and continues to be a leading form of transmission of voice and data. RF technology is seen in our pagers, cell phones, wireless LAN devices, handheld radios, satellite radios, etc.

4.1.1. Pager and Cell Phone Networks

The pager network has a bit rate of 512, 1200, and 2400 bps in the 450 MHz and 900 MHz frequency range (FCC, 2004). The newer FLEX protocol has a bit rate of 1600, 3200, and 6400 bps, allowing bi-directional communication, and extends the pager's battery life by minimizing the pager's power consumption (Motorola, 2004)[0]. Piconets use the 418MHz frequency range to transmit information for short distances at 40 kbps (Bennett et al., 1997). Alphanumeric pagers with two-way messaging and notification features are now commodities. The SkyTel paging system, which offers guaranteed delivery across the nation, is a prime example of how two-way paging is widely in use today. SkyTel uses Motorola's ReFLEX 25 and ReFLEX 50 two-way response system. ReFLEX provides much more effective use of the restricted paging bandwidth, resulting in higher throughput. Mobile health monitoring devices, through the use of the pager network, would have the ability to have an always-on connection to alert medical emergency facilities.

Cell phones come in a variety of flavors today: FDMA, AMPS, CDMA, TDMA and GSM. Frequency Division Multiple Access (FDMA) is the simplest technique. FDMA divides the RF spectrum into fixed frequency channels, one for each call; the original U.S. analog cell phone network uses a version of FDMA that typically provides 45 call channels per cell. AMPS, CDMA, and TDMA operate in the 800MHz band and are capable of operating at 49.6 kbps over 10 miles (Agrawal and Zeng, 2003). Time Division Multiple Access (TDMA) is a variation on FDMA that is used in digital networks. Each phone is assigned a time slot within a particular RF channel. This allows any channel to carry multiple simultaneous calls. In the United States, the meaning of TDMA is more specific: it refers to an implementation of TDMA that provides three time slots per channel.

Code Division Multiple Access (CDMA) broadcasts portions of the digital signal using a large number of frequencies at the same time. This means that a low-power signal can be spread across a wide range of frequencies, reducing the chance that interference will result in a dropped call. A large number of low-power signals can use this wide bandwidth at the same time. CDMA can carry 10 or 20 times as many calls per cellular base station as comparable TDMA implementations. Cordless phones operating at 900MHz can operate up to 100m and the technology has the ability to transmit at speeds of 57.6 kbps. Other 900MHz research technologies are capable of operating at 200 kbps (Ayoz, 2001). Other digital standard, such as personal communication services (PCS), use smaller cells than typical digital cellular networks, thus requiring lower-powered RF transceivers in the phone, but at a consequence, it requires more cells to cover a given area. There isn't a single PCS standard: some carriers use CDMA, some use TDMA, and others use a combination of those two technologies.

In more than 85 countries around the world, including Europe and the United States, cell phones use another standard for spectrum sharing, modulation, and data encoding that is called GSM, or the Global System for Mobile Communications. GSM uses TDMA technology, but in a way that offers eight or ten time slots per channel. GSM allows call data to be compressed, using less RF bandwidth. This means that GSM service providers reduce costs by supporting more phones per base station. GSM can deliver 9.6 and 28.8 kbps and will even deliver 115.2 kbps in the near future over a range of 4 miles.

4.1.2. 802.11 Wireless LAN

The 802.11 wireless LAN standard took a relatively long time to complete, 7 years. This standard was a joint collaboration from many wireless vendors. It defines an interface between a wireless client and an access point, as well as an interface between wireless clients. The 802.11 standard defines a physical (PHY) and a medium access control (MAC) layer, which are used to perform various functions, including: fragmentation, error recovery, mobility management, and power conservation. Most all of the other 802.X standards (Ethernet and token ring for example) define the same types of layers, but do not have the complexity or the functionality defined in the 802.11 standard. The PHY layer controls the characteristics of the wireless medium, the type of spread spectrum used, and the method for transmitting, the type of coding and modulation used. It is primarily concerned with the link management, and adapting to changing conditions of the wireless link. The MAC sublayer defines a method for basic access for multiple mobile wireless nodes, as well as defining packet formats for fragmentation and encryption.   The MAC sublayer's main concerns are synchronization, power management, association and reassociation.

The IEEE 802.11 standard uses three types of spread spectrum communication to allow to cope with different settings, such as outside, rooms and buildings. The three options are: direct-sequence spread spectrum (DSSS), frequency-hopping spread spectrum (FHSS), and diffuse infrared (DFIR). There are several data rates that can be employed in different circumstances from 1 Mbps to 54 Mbps, but the 1 Mbps and 2 Mbps data rates are guaranteed for the range of the product.

Wireless LANs, 802.11(a, b, and g), are being used today to transmit data at speeds up to 54 Mbps over distances of 300m (Dornan, 2002). The 802.11 series of specifications for wireless devices covers both the 2.4GHz and 5.2GHz ISM bands. There are only a handful of chipsets that support the 802.11 (a and b) standards and have many mature drivers written for them. Most 802.11 chipsets use 300mA when transmitting and 250mA when receiving data, which is not desirable for low power systems. The 5 GHz devices will require more transmit power to achieve the same range as 2.4 GHz adapters. Wi-Fi™ is the abbreviation for Wireless Fidelity, which describes the IEEE 802.11b standard. The non-profit organization WECA (Wireless Ethernet Compatibility Alliance) was formed in 1999. The goal of this organization is to certify interoperability of Wi-Fi™ products and hence promote Wi-Fi™ as the global wireless LAN standard.

4.1.3. Bluetooth

The Bluetooth interest group (SIG) was formed in 1998 by major computer (Intel, IBM, Toshiba) and cellular telephone companies (Nokia, Ericsson) to provide wireless connectivity among mobile PCs, cellular phones, and other electronic devices (Bluetooth, 2004). The aim was to make connectivity simple, seamless and intuitive.

The Bluetooth group has developed a standard that supports both voice and data traffic and provides a universal bridge to existing data networks. Bluetooth can also be used to form ad hoc networks of devices that are particularly important for health monitoring systems. Bluetooth, like IEEE 802.11, is another specification for a RF device using the 2.402 - 2.480GHz frequency range. Bluetooth is a specification for the packet protocols and link management protocols involved with communicating from device to device. It is designed to provide a low-cost and robust voice and data communications channel. The Bluetooth specification is running in the unlicensed band of FCC regulated frequencies and is required to have a less than 1W transmission power. This requirement keeps the range of Bluetooth devices' communication to 10m. The Bluetooth specification defines two power levels: a low power level that covers a small personal area within a room, and a high power level that can cover a medium range, such as the area within home.

Bluetooth supports two types of connections:

- Synchronous connection oriented (SCO);
- Asynchronous connectionless (ACL).

SCO packets are transmitted over reserved slots at 1 Mbps and a hopping rate of 1600 hop/s. This gives maximum packet length of approximately 78 bytes. A packet normally covers a single slot, but can be extended to cover up to five slots. An asynchronous connection supports 721 kbps in the forward direction and 57.6 kbps in the reverse direction. Alternatively, it can support a symmetric connection of 432.6 kbps.

The main features of Bluetooth, relevant to healthcare applications, include:

- Fast frequency hopping to minimize interference. Bluetooth uses a higher spread spectrum hopping frequency than the 802.11 series of RF devices therefore requiring smaller packet sizes. The smaller packet restricts the throughput of Bluetooth communication.
- Fast acknowledgments to decrease coding overhead for links.
- Flexible packet types that support a wide application range.
- 8 devices per piconet, 10 piconets per coverage area.

IEEE Standard 802.15.1 "Wireless MAC and PHY Specifications for Wireless Personal Area Networks (WPANs)" was primarily adapted from portions of the Bluetooth wireless specification. IEEE licensed wireless technology from the Bluetooth SIG, Inc., to adapt a portion of the Bluetooth specification as a foundation for the IEEE Standard 802.15.1-2002. The approved IEEE 802.15.1 standard is fully compatible with the Bluetooth v1.1 specification.

4.1.4. ZigBee

The latest emerging standard for personal area network applications is ZigBee (IEEE Standard 802.15.4) (ZigBee, 2004). It is a very low power, low data rate standard (250 kbps and 20 kbps), with 255 devices/network, and dual PHY (2.4GHz and 868/915 MHz).

4.1.5. Ultrawide Band

Ultra-wideband (UWB) is a very promising wireless radio communication technology, particularly after recent FCC authorization in February 2002. Unlike conventional radio systems, UWB radio systems transmit a series of extremely narrow (0.2-1.5 ns) and low power (~50μW) pulses (Fullerton, 1989 and 1990; Time Domain, 2004; Pulse Link, 2004)[0]. Therefore, it is also known as an "impulse carrier radio" or "zero carrier radio" technology. The formal definition of a UWB system by the Defense Advanced Research Projects Agency (DARPA) is the system that has relative bandwidth greater than 0.25 or bandwidth greater than 1 GHz. Formally, UWB is defined to be:

$$BW_{rel} = \frac{2(f_H - f_L)}{(f_H + f_L)} > 0.25$$

where f_H and f_L are upper and lower frequency ranges respectively (Taylor, 1995)[0]. A typical TM-UWB system has a center frequency of 2 GHz, with the bandwidth of 2 GHz, which results in a relative bandwidth of about 2, which has several order of magnitude wider bandwidth than a symmetric communication system based on sinusoids. An FM broadcast would have a relative bandwidth in the order of 0.002. Due to its large bandwidth, energy can be distributed such that spectral power density remains extremely small which makes it possible to share the spectrum with other users without exceeding the limit determined by the Part 15 of the FCC regulations (-41dBm/MHz).

There are two basic methods of generating UWB signals:

- TM-UWB (Time Modulated UWB). TM-UWB transmitters emit a series of short monocycle wavelets with precisely pulse-to-pulse intervals between 25 and 1000ns. A UWB system is implemented using a series of pulses that are placed in pre-determined temporal positions, known as a code. A code plays a critical role in shaping the spectrum of the TM-UWB system. If these pulses were transmitted at a regular interval, the resultant spectrum would contain a comb line structure separated by the pulse repetition rate. Therefore, a code must be designed to meet a set of criteria such as smooth spectrum, a high auto-correlation property, and a large channelization capacity. The receiver demodulates the information by combining a sequence of pulses after they are cross-correlated with a template signal and integrated over an integration interval. Secure communication with greatly reduced probability of detection is a result of a wide, smooth spectrum with low level of spectral density. Basic modulation techniques of UWB are:
    Pulse position modulation, positioning a pulse 'earlier' or 'later' from the reference position to encode either a '0' or '1'.
    Pulse polarity modulation, using positive or negative pulses.
- DSC-UWB (Direct Sequence Phase Coded UWB). A pseudo-noise sequence allows channelization, spectrum spreading, and modulation.

In addition to communication, UWB could be used in monitoring applications for precise position location. A picosecond precision timer enables precise time modulation, pseudo-noise coding, and distance determination. The inherent penetration property of UWB technology could be used for non-contact sensors in cardiology, obstetrics, and pulmology (McEwan, 1996 and 1998; Staderini 2002).

UWB provides a natural solution for the range of problems in portable medical monitoring applications (Kim, 2001). The main advantages of UWB communication for medical monitoring applications include:

- Good immunity to noise and interference.
- Innate security mechanisms, data encoded by the very nature of TM UWB communication.
- Efficient channelization, which will be increasingly important with the present trend of spectrum pollution. This allows multiple sensors to be operational at the same time without interfering with each other.
- High available bandwidth.
- Possible sensor localization. That allows patient localization in the case of medical emergencies, but also a position of rescue personnel.
- Low power consumption. Unlike existing wireless communication system, UWB transmitter is not active all the time during transmission. It is active only when the pulse is transmitted, while high bandwidth allows short transmission time in the case of physiological monitors.

Obstacles for wider use of UWB technology in medical monitoring include:

- It is still in infancy without the necessary level of integration for sufficiently small monitoring system integration.
- It requires a relatively large antenna for implantable sensors.
- Too much asymmetric receiver/transmitter complexity. A UWB receiver is significantly more complex than the transmitter.

Recent FCC approval of UWB technology (FCC, 2004) will enable commercial wireless applications. Time Domain is developing chipsets to facilitate UWB system development. The PulsON 200 chipset has been designed to support link rates up to 40 Mbps (TimeDomain, 2004)[0]. There is a direct tradeoff between data rates and range. An application that requires lower data rates will have increased range. The PulsON 300 has been designed to deliver in excess of 100 Mbps, with scalability up to 500 Mbps.

4.1.6. Magnetic Induction

When most people think of Magnetic Induction technologies, they think of battery chargers for their watches, Personal Data Assistants (PDA), hearing-aids, etc. Aura Communications is selling a system using Magnetic Induction for communication purposes (AURA, 2001). Magnetic induction technologies work on the same principals as RF technologies, except for the fact that the transmitted signal does not include an electric field, only a modulated magnetic field. Magnetic induction does not yet have all of the hype that RF technologies have, with respect to RF radiation and how it causes cancer. The requirements for safe operation of communications equipment using Magnetic Induction are well within the regulations for magnetic fields in most countries.

The attenuation of a magnetic field is greater than an electromagnetic field. The signal strength of a magnetic field dissipates as $1/distance^3$ instead of $1/distance^2$ as with RF technologies. Consequently, this technology has a range of up to 3 meters. The system architecture of a Magnetic Induction communications device is almost exactly the same as an RF device. The operating frequency and power are lower than that seen in a typical RF device. Magnetic Induction works by causing a quazi-static magnetic field around a transmitting coil. The receiver's will have a modulated current induced on its coil that is incident to the transmitter's magnetic flux density. This system uses roughly 10mm coils for antennas. These coils are middle grounded to minimize the E-field generated by the coil. This system only works when the receiver and transmitter coils are oriented in the same direction, because of the orthogonal properties of magnetic fields. Some techniques have been developed by Aura Communications to use three coils mounted in all three axis to measure the receive signal strength and pickup the signal (AURA, 2004).

The system devised by Aura Communications is a system-on-a-chip requiring only 3 coils for antennas. The three coils take up a 10mm by 24mm by 10mm space. The maximum transmission speed of Aura Communications' device, LibertyLink, is 204.8 kbps. The device only draws 4mA when transmitting data and 7mA when transmitting voice. Typical RF solutions would require up to 50mA for the same transmission speed and range. This low current draw enables the use of a small battery thus a smaller package size than RF technologies.

The FCC does have restrictions about Magnetic Field limits for devices that are – 1.5dBµA at 10m. The frequency bands that are unlicensed are 12.29-12.293MHz, 12.51975-12.52025MHz, 12.57675-12.57725MHz, and 13.36-13.41Mhz. Commercial devices are required to undergo EMI testing.

4.1.7. Satellite Systems

Satellite systems are launched into orbit for several purposes, including communications. There are several satellites that are in Geostationary Earth Orbit (GEO), Low Earth Orbit (LEO), and Medium Earth Orbit (MEO).

GEO satellites operate at an altitude of 35,768km. Some are used for large footprint mobile phone coverage, about 34% of the earth's surface. At this altitude, there is latency for voice and data communication of 275ms. The uplink frequencies to the satellite are 1635 to 1660.5MHz and the downlink frequencies are 1530 to 1559MHz.

LEO satellites operate at an altitude of 500km to 1,500km and are separated into two types: Big LEO and Little LEO. Little LEO satellites are primarily used for two-way messaging and positioning information. These satellites have an uplink frequency range of 148 to 150.05MHz and a downlink frequency range of 137 to 138MHz allowing them to have a bit rate as the order of 1 kbps. Big LEO satellites have an uplink frequency range of 1610 to 1626.5MHz and a downlink frequency range of 2483.5 to 2500MHz. These satellites offer voice, data, fax, and paging services to many handheld devices, depending on the throughput desired. The Iridium Big LEO satellite has a bit rate of 2.4 kbps and serves 4000 users, whereas the Teledesic Big LEO satellite has a download bit rate of 64 Mbps and an upload bit rate of 2Mbps, servicing 2500 users. The Latency of LEO satellites is 5 to 10ms.

MEO satellites operate at an altitude of 5,000km to 12,000km and have similar uplink and downlink frequency ranges as LEO systems. These systems aid in providing global coverage of satellite voice and data communication systems. The latency between the earth and these satellites is about 70 to 80ms (Agrawal and Zeng, 2003)[0].

Latency is a big issue with satellite systems because humans are able to discern a lag in speech when there is a greater than 200ms latency involved with two-way conversations. Satellite receivers must be sophisticated enough to be able to transmit on the high frequency bands that the satellites operate on, thus making them depend on higher amounts of power.

4.2. Near-Field EM Technologies

This technology is used to exchange information between devices on or very close to the human body using capacitively coupled picoamp currents (Zimmerman 1995; Gupta 2000; Post, 1997). Near-field EM uses a less than 1MHz carrier frequency to transmit information across the skin. Many times this technology is used synonymously with Personal Area Networks (PAN) or BodyNetworks.

This communication technology has the potential to work at 417 kbps but has only been tested at 2400 bps (Zimmerman, 1995). More sophisticated modulation techniques could be used to achieve a data rate closer to the 417 kbps level. The area of the conductor, which should be in contact with the skin, of a PAN device is 25 to 80mm long (the size of a watch or credit card). The power needed to transmit information increases as you increase the carrier frequency from 0.1MHz to 1MHz. In a system using a 330kHz carrier frequency at 30V with a 10pF electrode capacitance would consume 1.5mW. The energy stored in the capacitor is actually recycled by using a resonant inductance-capacitance (LC) tank circuit. Communication is limited to the body that the PAN devices are attached to, unless one person comes in physical contact with another via a handshake or some other means. This technology enables the exchange of business cards or other personal information during a handshake. Multiple near-field EM medical devices could communicate using the human body as the communication medium.

The FCC does not regulate near-field PAN devices since the field strength lays 89dB below the field strength regulations.

5. DISCUSSION

Each of the technologies described in this chapter has their own niche that they work well in. For example, TV remote controllers only use IR communication technology. However, infrared devices must have a line of sight for communication. IR communications is also hindered when used in sunlight. Most wireless LANs rely heavily on RF technologies, though IR networking is starting to be used by PDAs, cell phones and more recently, larger computer networks.

RF communication is the most widely used form of wireless communication and is not restricted to line of site communications like IR. Its only downfall is the amount of power required to transmit information. Devices using RF technologies are subject to interference in the form of signal degradation and multipath propagation. Multipath propagation is the result of the electromagnetic waves bouncing off of the surfaces of the device's surroundings, which make an echo of the signal.

Newer technologies like UWB and Electric Field Sensing are going to push themselves into the areas that IR networking and wired channels, like cell phone headsets, are now operating. UWB is a step forward in reducing transmission power to make RF available for low power battery powered devices.

Magnetic Induction communication is poised to take over very short-range communications. There are already products for cell phones and PDAs that are using this technology. The only downfall for this type of communication is the short range of communications in comparison to RF and IR technologies.

BodyNets or networks using Electric Field Sensing techniques are poised to take over body communications. This technology is suited for communication between PDAs, watches, cell phones, and cell phone headsets. Handshaking could enable the parties involved to exchange business cards, their day schedule, and any small amount of information that they want to exchange. This technology is not suitable for out of body communications like the other technologies mentioned in this paper.

There is a general health concern with RF emissions, particularly in the case of devices worn on the body. However, the penetration depth of skin of RF waves at 2.4GHz is 1.5cm and for 900MHz is 2.5cm. There is a large difference between the transmission powers of your conventional microwave oven and transmitting devices like cell phones, Bluetooth devices, and pagers. Yet, as we are constantly around such devices we do need to be aware of their long lasting effects on the body.

A Summary of relevant features of different wireless technologies is given in Table 2.

Table 2. Summary of wireless technology features

Technology	Range	Speed	Regulations
Infrared (IR)	10m	< 4 Mbps	Food and drug administration safe limits for IR light.
Pager Network (RF)	4.8km	< 9600 bps	FCC
Cell Phone Network	4.8km	< 115.2 kbps	FCC
802.11 Wireless LAN	300m	< 54 Mbps	FCC
Bluetooth	10m	721 kbps	FCC
Ultra-wide Band RF	3m	< 100 Mbps	FCC
Near-field EM	<1m	< 2400 bps	FCC (but not applicable in most cases)
Magnetic Induction	<3m	204.8 kbps	FCC (but not applicable in most cases)

6. REFERENCES

D.P. Agrawal, Q. Zeng, Introduction to Wireless and Mobile Systems, Brooks/Cole, 2003.
AURA Communications, http://www.auracomm.com/ (August 2004)
S. Ayoz, H. Tuncer, Low Power 200kb/s 868MHz/915MHz Radio Module,
     http://www.uk.research.att.com/868MHz/ (October 2001).
F. Bennett, D. Clarke, J.B. Evans, A. Hopper, A. Jones, D. Leask, Piconet: Embedded mobile networking, *IEEE Personal Communications Magazine*, 4(5), October 1997, pp. 8-15.
B. Bing, High-Speed Wireless ATM and LANs, Artech House, Inc., 2000.
Bluetooth, http://www.bluetooth.com/ (August 2004).
A. Dornan, LANs with No Wires, but Strings Still Attached, *Network Magazine*, 2/2002, pp. 44-47.
Federal Communications Commission, http://www.fcc.gov/ (August 2004).
L.W. Fullerton, "Time Domain Radio Transmission System", US Patent 4,813,057, Mar 14, 1989; US Patent 4,979,186, Dec 18, 1990.

P.   Gupta,   Personal   Area   Networks: Say   it   and   you   are   connected, http://www.wirelessdevnet.com/channels/bluetooth/features/pans.html (2000).

A. Harter and F. Bennett, *Low Bandwidth Infra-Red Networks and Protocols for Mobile Communications Devices*, Number 93-5, Olivetti Research Laboratory, October 1993, available online: ftp://ftp.orl.co.uk/pub/docs/ORL/tr.93.5.ps.Z. (August 2004)

Health Technology Center, *The Future of Wireless in Healthcare*, Technology Forecast Report, http://www.healthtechcenter.org (September 2003).

J. Kim, E. Jovanov, "Biomedical Applications of Ultra Wide Band Personal Area Networks," ISPACS'2001 the 9th International Symposium on Intelligent Signal Processing and Communications Systems, Nashville, Tennessee, November 2001.

T.E. McEwan, "Body monitoring and imaging apparatus and method," US Patent 5,573,012, November 12, 1996; US Patent 5,766,208, June 16, 1998.

Motorolla FLEX TechnologyOverview, http://www.mot.com/wireless-semi/ (August 2004), also on-line article: D.R. Gonzales, FLEX Protocol Delivers De Facto Messaging Standard for Handhelds, http://www.techonline.com/community/ed_resource/feature_article/8012 (August 2004).

E.R. Post, M. Reynolds, M. Gray, J. Paradiso, N. Gershenfeld, Intrabody buses for data and power, First International Symposium on Wearable Computers, 1997, pp. 52 –55.

Pulse Link, http://www.pulselink.net/ (August 2004).

E. M. Staderini, UWB Radars in Medicine, *IEEE AESS Systems Magazine*, Jan 2002, pp. 13-18.

Taylor, J. E. (ed), Introduction to Ultra-Wideband Radar System, CRC, 1995.

Time Domain, http://www.timedomain.com (August 2004).

R. Want, B.N. Schilit, N.I. Adams, R. Gold, K. Petersen, D. Goldberg, J.R. Ellis, M. Weiser, *The ParcTab Ubiquitous Computing Experiment*, Palo Alto Xerox Research Center, 1995, http://sandbox.parc.xerox.com/parctab/ (August 2004).

ZigBee, http://www.zigbee.org/ (August 2004)

T.G. Zimmerman, *Personal Area Networks (PAN): Near-Field Intra-Body Communication*, Master's Thesis, MIT Media Laboratory, September 1995.

WIRELESS COMMUNICATION TECHNOLOGIES FOR MOBILE HEALTHCARE APPLICATIONS: EXPERIENCES AND EVALUATION OF SECURITY RELATED ISSUES

Manolis Tsiknakis*, Apostolos Traganitis, Manolis Spanakis, and Stelios C. Orphanoudakis

1. INTRODUCTION

The healthcare environment is currently changing with increased emphasis on prevention and early detection of disease, primary care, intermittent healthcare services provided by medical centers of excellence, home care, and continuity of care. This requires the provision of personalized healthcare services based on best practices and evidence-based medicine.

In such a dynamic environment, Information and Communication Technologies (ICT) are taking on a leading role and are currently having a significant impact on the practice of healthcare at all levels. The traditional healthcare delivery structures flatten. Instead of the three or four level hierarchy (primary, secondary, tertiary, and university hospital level care), only two are likely to exist in the near future, i.e. centers of excellence specializing in technology-intensive procedures and primary care (front line) facilities. At the same time, part of the responsibility of care is shifting into the hands of the citizen, with emphasis on wellness or health maintenance.

Other important trends in healthcare include citizen mobility, the movement towards shared or integrated care in which the single doctor-patient relationship is giving way to one in which an individual's healthcare is the responsibility of a team of professionals across organizational boundaries within the healthcare system. This is being accompanied by a very significant growth in home care, which is becoming increasingly feasible even

* Manolis Tsiknakis, Institute of Computer Science (ICS), Foundation for Research and Technology – Hellas (FORTH), PO Box 1385, GR 711 10,Heraklion, Crete, Greece, E-mail: tsiknaki@ics.forth.gr

for seriously ill patients, through sophisticated e-health services facilitated by intelligent sensors, monitoring devices, hand-held or wearable technologies, the Internet and wireless broadband communications. Within such an environment, the need for a single I-EHR for every citizen becomes the cornerstone for supporting continuity of care and the evolving, novel e-health and m-health services of today.

Particular attention is displayed worldwide towards electronic records management and mobile-health solutions as a way to manage clinical decision making at the point of care. Advances in communications technologies (both wired and wireless) have greatly enhanced information exchange between medical practitioners and facilities. New technologies, including Bluetooth, IEEE 802.11x and Ultra-Wideband (UWB), offer improved methods for collecting and exchanging EHR information wirelessly.

Whether it is via wireless LAN, wide area network or cell phone, wireless connections are an integral part of communication among healthcare providers. But, despite the real advances in standards, performance and prices that wireless technologies have made the last couple of years, it is not always possible to put all the pieces together to form a secure environment.

In this Chapter we primarily discuss the security features and shortcomings of the current and emerging wireless technologies with emphasis on their application in the Health sector. We also present a description of the security mechanisms of the most popular wireless technologies and analyze the performance of their main security components – encryption, authentication and data integrity. We present measures that can be taken in order to provide strong security for wireless communications users. Our analysis is based on experiences with a number of communications infrastructures based on wireless network connections employed for the delivery of e-health and m-health services that were designed and implemented for home care, hospital care and pre-hospital health emergency care environments within the context of HYGEIAnet.

2. MOBILE HEALTH SERVICES IN HYGEIANET

HYGEIAnet, represents a conscious and systematic effort toward the design, development and deployment of advanced e-health and m-health services at various levels of the healthcare hierarchy, including primary care, pre-hospital health emergency management, and hospital care in the island of Crete, Greece. Specifically, as shown in Figure 1, e-health and m-health services for timely and effective patient management are routinely employed in pre-hospital emergency services, supporting synchronous and asynchronous collaboration of medical professionals in primary Healthcare facilities and specialists at the Regional Hospital. In addition they are routinely used in the domains of Cardiology and Radiology as well as for supporting the remote management (tele-management of patients at home) of asthmatic patients. Finally, e-health and m-health services providing access to an integrated, lifelong electronic health record (I-EHR) is being used to support continuity of care across organizational boundaries.

HYGEIAnet is fundamentally designed to bring timely health information to, and to facilitate communication among, those wishing to make informed health decisions for themselves, their families, their patients, and their communities. It has been established as a scaleable RHIN and is based on an open architecture and tools for the integration of specialized autonomous applications, and a Health Information Infrastructure (HII) which supports these applications and provides the relevant services.

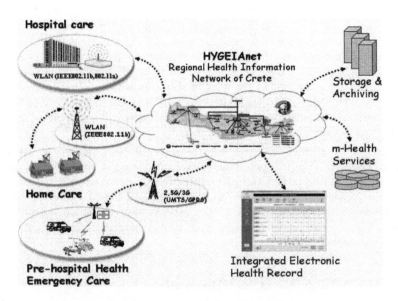

Figure 1. m-services and technologies in HYGEIAnet

From the technological point of view, a diverse set of state of the art technologies have been developed and deployed as part of the required scalable HII of HYGEIAnet (Tsiknakis et al., 2002). These include fixed, as well as wireless and mobile communications, distributed computing and middleware (i.e. ActiveX/COM, CORBA), web technologies and services (i.e. XML, SOAP, UDDI), peer-to-peer computing, H.323, X.500, X.509, and also a number of Healthcare specific standards for the implementation of the required systems and services.

The deployment and use of HYGEIAnet services has demonstrated significant economic, clinical and access to care benefits. Specifically, referrals of patients to the specialists were significantly reduced, since expert opinion was available through services provided over the RHIN, thus reducing unnecessary traveling and hospitalization. Loss of workdays of family members and relatives has been reduced and duplication of unnecessary examinations has also been avoided. Access to care has greatly improved in all categories where e-health services have been employed. The findings to date are extremely encouraging and suggest that use of e-health services benefit not only the patients, but also health professionals by fostering their collaboration and serving as a tool for continuing education. In one of the HYGEIAnet services, i.e. telecardiology, 10% of the cardiac patients of a rural area served by a primary Health centre were involved in a teleconsultation session during a period of 6 months, thus making medical expertise instantly available to remote and isolated populations. Furthermore, 65% of pre-hospital health emergency episodes have been managed by paramedics. Given that the first 60 minutes (the golden hour) are the most critical regarding the long-term patient outcome, the ability to remotely monitor the patient and guide the paramedical staff in their management of the patient is obviously facilitating access to care and, most importantly, access to care by a specialist.

The evaluation of HYGEIAnet services also revealed significant benefits with respect to quality of care. Specifically, regarding the e-health services for the remote management of chronic diseases (Traganitis et al., 2001), significant benefits have been reported, based on extended clinical trials conducted for the purpose of evaluating the clinical effectiveness of the service. The evaluation of clinical results reveals the effectiveness of the platform, as reported in detail in ref. (Guillen et al., 2002), and suggest a relatively high rate of health improvement, and can therefore be considered as strong evidence for the effectiveness and efficiency of e-health in supporting the delivery of tele-services to citizens with chronic diseases.

With regard to e-health services in Cardiology, the evaluation study also revealed significant benefits. Detailed preliminary evaluation results have been reported in (Chronaki et al., 2001). The service had a rather strong diagnostic impact: only in 9 out of 21 cases the patient was immediately transferred to the hospital. Since the health care centers are in remote locations and all the patients seen would normally have been referred and transported to the Regional hospital for evaluation, this represents a clear saving of time, cost and resources.

3. NEED FOR A TRUSTED COMPUTER BASE

A formal risk analysis of the e-Health and m-Health services in HYGEIAnet revealed that user confidence in relation to the trust and security infrastructure of the RHIN is an important factor influencing user adoption. Available evidence form analyzing the adoption rates of HYGEIAnet users' indicates that user adoption is accelerated when users are convinced that a trusted computing base is in place.

In computer security terminology, the *trusted computing base* is the set of all hardware, software and procedural components that enforce the security policy. Within a general security model (Blobel et al., 1999), security issues can be described as concepts of either or both communications and application security with their related security services, allowing one to focus on the relevant area of interest.

Regarding the general security model distinguishing between communication security and application security, strong authentication is needed for security services within both concepts. Other security services which are common to both domains are availability, accountability, non-repudiation and confidentiality. Services specific to communication security are identification, authentication, integrity and certification, while security services specific to application security are authorization, access control, data quality and audit.

The trusted computing base of HYGEIAnet includes computer security mechanisms to enforce user authentication and access control, communications security mechanisms to restrict access to information in transit across a network, statistical security mechanisms to ensure that records used in research and audit do not possess sufficient residual information for patients to be identified and availability mechanisms such as backup procedures to ensure that records are not accidentally deleted.

This chapter does not intend to deal with the issues related to application security, or the clinical and ethical aspects of information security. We will focus on the communications security and especially on communications security involving wireless and mobile communications. We will be evaluating wireless and mobile communication technologies and standards with regard to the basic security services of identification, confidentiality, authentication, integrity, availability and accountability they offer, which are the dominant security consideration in Healthcare.

4. OVERVIEW OF WIRELESS TECHNOLOGY

With the term wireless technologies, we describe the technologies that permit processor-based devices (personal computers, laptops, personal digital assistants, medical devices and more) to communicate without any physical connection; that is without requiring any network cabling or networking devices.

Wireless technologies have been around for a long time but only recently, with the worldwide acceptance of the cellular telephony and the proliferation of computers and wireless networking, their adoption in many working environments has been accelerated. Today the continual growth of wireless technology is driven by the ease of installation, flexibility, and most of all the mobility demanded by the users looking for significant gains in efficiency, accuracy, and lower business costs.

4.1. Wireless Networks

Wireless network is the information transport mechanism between devices that are not connected with wires or other cables. There are many different network types able to be deployed today. They can be divided into three major categories. Namely, (a) wireless networks connecting two devices, (b) wireless networks connecting more than two devices and (c) wired infrastructures (wireless LANs connected through an Access Pont (AP) to a fixed network).

Another way of characterizing wireless networks is based on their coverage area, as shown graphically in Figure 2. They can be categorized as Wireless Wide Area Networks (WWAN), Wireless Local Area Networks (WLAN) and Wireless Personal Area Networks (WPAN). The WWANs represent wireless networks using wide coverage area technologies. The WLANs represent wireless networks, deployed usually within the limits of a neighborhood or a building, and such systems are IEEE802.11x, Hyperlan, Hi-Fi, HomeRF and much more. The WPANs represent wireless networks that cover an area of a few meters, and such systems are deployed by Bluetooth and Infrared technologies.

Another category of wireless technologies that is recently emerging, the Wireless Body Area Networks (WBAN) needs to be mentioned. Wireless Body Area Networks are deployed around a human body and are used in order to allow the communication of cloth embedded sensors and devices that a man is carrying. They are increasingly useful in m-health services supporting illness prevention, wellness monitoring and personal health management services.

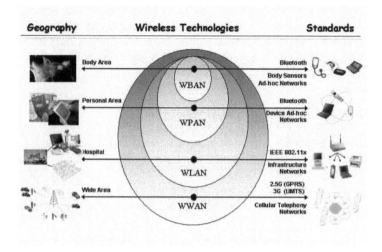

Figure 2. Taxonomy of Wireless Technologies

All these technologies receive and transmit data through electromagnetic waves at the ISM (Industrial, Scientific, and Medical) frequency bands. The ISM bands are 11 frequency bands from 6.78 MHz to 245 GHz allocated by FCC for unrestricted use for industrial, scientific and medical purposes with low power equipment.

Our discussion will now be focused on the security related issues of two specific technologies, i.e. IEEE's 802.11x WLAN standard and Bluetooth WPAN protocol operating at the ISM frequency band of 2,4 GHz, which are the two wireless technologies most widely used in state of the art m-health applications.

5. WIRELESS COMMUNICATION STANDARDS

Over recent years, the market for wireless communications has experienced tremendous growth. Many standards in the area of wireless networks have been proposed. The principal advantages of standards are to encourage mass production and to allow the interoperability between products from multiple vendors.

In this section we discuss the most common standards proposed in the area of WLANs and WPANS. IEEE has proposed the standard 802.11x, where x denotes its "flavors" developed by the various working groups on 802.11. It describes the protocol specifications of wireless local area networks. Similarly the Bluetooth SIG has proposed Bluetooth, a wireless Personal and Body Area Network protocol, which is today the dominant standard in the area of ad-hoc wireless networking. With the term ad hoc we refer to those networks that do not have a specific hierarchy and infrastructure, but are designed to dynamically connect devices, such as laptops, pda's, cell phones, forming groups of inter-connected devices. These networks are called ad-hoc due to their shifting network topology, maintaining random network configurations, relying on a system of mobile nodes, connected by wireless links to enable communication among them.

The main reason we are focusing on these two standards is the fact that they are the most common, they are supported by major telecommunications industries and they have already found their way in a number of applications in the medical health sector. They are serving respectively the area of Wireless Local Area Networks and Wireless Personal Area Networks. Even though these two standards have different architectural implementations, and different security features they have some commonalities. They both support security features like, authentication, confidentiality and integrity, and they are both implementing security only over the wireless links. We discuss the security issues, and all the known vulnerabilities and threats that they present.

5.1. IEEE 802.11x

The IEEE 802.11 is a physical and data link layer protocol specifying a Wireless Local Area Network (WLAN). It is a subset of the IEEE 802 family of network protocols, such as the 802.2 (Token Ring) and 802.3 (Ethernet). It specifies wireless networking using spread spectrum, frequency hopping or infrared technologies (Prasad et al., 2002). The 802.11x standard describes the Medium Access Control (MAC) and Physical layer (PHY) functionality for wireless connectivity of fixed, portable, and mobile stations. The mobile stations speed can be up to some tens of miles per hour.

The IEEE's 802.11 standard was designed for supporting transmission speeds between 1Mbps and 2Mbps, 802.11b can provide wireless links up to 11Mbps, while the latest 802.11a can go up to 54Mbps. All the above protocols operate at the ISM (Industrial, Scientific, and Medical) unleashed frequency band. The first two operate at the 2,4GHz frequency band while the last one operates on the 5GHz. A variation of the 802.11b standard specifies transmission speeds of 22Mbps or even 44Mbps, whereas the most recent standard 802.11g specifies transmission speeds of 54Mbps

Some of the protocol's features are the following: (a) Support of asynchronous and time-bounded delivery services; (b) Continuity of service within extended areas via a distribution system such as Ethernet; (c) Transmission rates from 1Mbps to 55Mbps; (d) Support of mass market applications; (e) Multicast and broadcast services; (e) Network management services; (f) Authentication services; (g) Registration services; (h) Operating environments: inside buildings, offices, banks, shops, malls, hospitals, houses, hotels, outdoors, across campuses and more; (i) Power management; (j) Security of the whole coverage area; and (k) Wireless IP addressing;

5.2. Bluetooth

Bluetooth is a new standard for a wireless personal area network (WPAN) designed to replace the cables connecting devices to computer based systems like PCs, laptops, PDAs, or to other networks. It allows the communication with other devices with a Bluetooth interface, forming groups of ad-hoc device networks.

The Bluetooth specification was announced in February 1998 by a consortium consisted by major commercial companies. The main purpose of the proposed protocol was to bridge various standards, provide functional integration, and produce a universal interface, among devices, services, programs, and network applications. Bluetooth is based on a cheap, small, easily programmed, low-power radio transceiver that can be embedded in almost any kind of devices.

The operational frequencies band of Bluetooth is in the ISM band at the 2.45 GHz. The Bluetooth specification designates spread-spectrum communication using frequency hopping (SS-FH) sequence with Gaussian Frequency Shift Keying (GFSK) modulation.

The Bluetooth architecture uses a combination of circuit and packet switching technologies. It can support either one asynchronous data channel and up to three simultaneous synchronous speech channels, or one channel that transfers asynchronous data and synchronous speech simultaneously. The Link Manager is an essential part of the Bluetooth architecture. It uses Link Manager Protocol (LMP) to configure, authenticate and handle the connections between Bluetooth devices.

6. SECURITY PROBLEMS AND THREATS IN WIRELESS NETWORKS

6.1. Overview of Wireless Security

All computer and communication systems face security threats that can compromise the services provided by the systems, and/or the data stored on or transmitted between systems. The most common threats are denial-of-service attack, interception, bad manipulation, masquerading, and repudiation. A detailed taxonomy of computer attacks with applications to wireless networks can be found in (Bishop et al, 1996; Lough et al., 2001).

Security in general but especially in the Healthcare domain has traditionally focused on ensuring the correct disclosure (or confidentiality) of data, the complete integrity of the data, and the availability of the data when needed (that is, the service is not denied). Wireless systems involve a third spatial dimension not seen in traditional networks. Attacks will not be limited to three dimensions only through wires, but will encompass a fourth dimension: the spectrum. In an enterprise network which includes a wireless network, authentication of users and confidentiality of communications are more difficult to achieve than they are with a wired network.

In modern network topologies, physical boundaries between public and private networks no longer exist. Wireless networks and Internet-based VPNs (virtual private networks) are two examples of this, and the security implications are apparent. First, whether a user has the necessary permissions for access to a system can no longer be assumed based on physical location, as with a wired LAN in a secure facility. Second, data is broadcasted using radio frequencies, which can travel beyond the control of an organization. The information traversing the wireless network is susceptible to eavesdropping. These problems represent a violation of two fundamental network security policies: authentication and confidentiality.

6.2. IEEE802.11b security

The 802.11 specifications include services for link-level authentication and an optional privacy (i.e., confidentiality) service termed Wired Equivalent Privacy (WEP), designed to protect the confidentiality of link layer traffic. The security provisions of the standard present no claims about secure integrity assurance, they simply specify a frame check sequence error control mechanism. Research reports and press accounts have reported serious flaws in the authentication service and the WEP algorithm, issues that we discuss at the end of this part.

The three basic security features of IEEE's 802.11 wireless Local Area Networks standard are: authentication, confidentiality and integrity. These features are provided by the WEP protocol in order to protect wireless link level data during transmission, and provide a secure operating environment for the wireless links of the infrastructure's clients and access points.

6.2.1. Authentication

One of the goals of WEP is to ensure the identity among communicating stations, and provide access control to those stations authorized to use the specific network resources and deny access to all others. The IEEE802.11b specification describes two ways of validating wireless stations attempting to access the wireless resource and through there to gain access to the rest of the communication network (wireless or wired).

The first way is based on cryptography and is using a shared key authentication with RC4 in order to pass through the authentication phase successfully. It is based on a shared, with the client, secret that the authenticator, i.e. the access point, verifies with a challenge-response scheme. First the client makes an authentication request in order to start the process and the authenticator creates a random challenge and sends it to the wireless client. Then the client encrypts the received challenge using RC4 stream cipher algorithm with the cryptographic key that owns and sends back the result to the access point. The access point decrypts the result, and permits the access to the client only if the result is the same as the random request that he first created and sends back to the client the results.

The second way is a non-cryptographic approach using two different identity based methods in order to identify and authenticate a wireless client joining the network. These two methods are (a) the Open System Architecture, which accepts a client to join the network if it responds an empty Service Set Identifier (SSID) and (b) the Close System Architecture that accepts clients that transmit the correct SSID, a 32-byte string, of the network. This type of authentication is based only on an identification scheme, it does not include any security mechanism and mutual authentication, and practically does not offer any robust security scheme against malicious clients trying to access the network resources.

6.2.2. Confidentiality

The 802.11b supports confidentiality through the use of cryptographic techniques only for the wireless interface, with RC4 as the symmetric key stream cipher algorithm, for the creation of the ciphered data sequence. The transmitted data is the plain text data added modulo2 (XOR) with the pseudo random key-stream from RC4. This WEP technique protects data from disclosure attacks during the wireless transmission. This technique is applied at the lower level of the network stack of 802.11b standard in order to protect all the transmitted data packets (TCP/IP, UDP/IP, IPX, HTTP and more).

The WEP supports key sizes from 40 to 104 bits, with the latter becoming a 128bin RC4 key after adding a 24bit Initialization Vector. This is done following the general principle that larger key size means increased security of a cryptographic algorithm. The scope is to avoid the brute-force attacks, and this appears to be possible when key sizes are longer than 80bits. Unfortunately researches have shown that WEP is a not so strong cryptographic technique with many vulnerabilities.

6.2.3. Integrity

The integrity related security service of the IEEE802.11b standard foresees data integrity for the messages exchanged between wireless hosts and access points, by rejecting messages that have been changed by an unauthorized host. This security service prevents attacks that are launched by someone located in the middle of two communicating hosts and is known as the "man in the middle attack". In order to accomplish this protection scheme it uses a simple encrypted Cyclic Redundancy Code (CRC). A CRC code is calculated for each packet's payload and it is attached at the end of its packet forming the actual plaintext data encrypted the RC4 in order to provide the cipher text finally transmitted. On the other end, after decrypting the received data, the CRC of the packet payload is recomputed. Then the receiver compares the transmitted CRC with the one produced on its side. If the two CRC's do not match, which indicates an integrity violation, the packet is discarded. This technique is also vulnerable to attacks irrespective of the key size.

6.2.4. Threats related to the IEEE's 802.11x Security Standard

Many problems have been identified with IEEE's 802.11b security mechanisms. Some of them are due to the fact that WEP as a mechanism has problems because it uses a cryptographic algorithm (RC4) that has many weaknesses and is vulnerable to many passive and active attacks. We will try to give a short description for some of the problems with the existing 802.11b security scheme as identified during the course of our work and/or reported in the scientific literature.

Security options, even with their weaknesses, are not usually enabled by the users in commercial installations, making the wireless network infrastructures vulnerable to attacks. The malicious users usually seek unprotected wireless network installations in order to gain access to the network through them. For example most of the time the authentication scheme requires only the SSID credential in order to obtain access to network resources.

The Initialization Vectors, which are part of the WEP key are short and static. This produces key streams with repeated parts over time, making possible the decryption of the ciphertext. Combined with the fact that WEP's description is known, and the weakness of RC4, the resistance to adversaries' attacks is very limited. Also all the cryptographic keys used in the RC4 algorithm are shared among users and the fundamental requirement of security concerning secrecy about cryptographic keys is not satisfied, thus making it easier to compromise security.

WEP does not have any key management mechanisms. It does not describe how the various keys are shared among users and if and how the keys are to be updated or redefined, when invalidated or discovered by an attacker. The protocol just assumes that all clients (nodes in the network) own the necessary cryptographic keys.

The WEP system has a poor integrity mechanism. It uses CRC32 linear block code in order to provide cryptographic integrity. It is possible even with the encrypted CRC32 linear algorithm, to change the content of a message with out the need to change the attached CRC. This means that a malicious user can change the payload of a packet, destroying the contained data, and send it over to the real recipient, who is unable to ensure if this is a valid message or not.

The security mechanism authenticates only the devices and not the users of the devices. So if an authenticated device is been used by an unauthorized user everything will appear normal and the user will have access to the rest of the network infrastructure.

The authentication mechanism of the protocol is a one-way authentication and not a mutual authentication, since only the access point authenticates the client. The wireless client does not have any way of knowing if the access point is the one that it is claiming to be. This lack of mutual authentication makes both ends subject to man in the middle attacks, very common in wireless communications.

6.3. Bluetooth security

The Bluetooth wireless technology in order to secure its hosts from attacks, in addition to the built-in encryption and authentication mechanisms, it uses frequency-hopping schemes with 1600 hops/sec with an automatic output power adaptation to reduce the range exactly to the minimum necessary. The security features of this technology are based on the following security mechanisms: authentication, which prevents spoofing and unwanted access to critical data and functions, and encryption, which prevents eavesdropping and maintains link privacy. In order for all the devices within the ad-hoc network to satisfy their varying demands on data security, privacy and flexibility, security is implemented, in the Bluetooth Generic Access Profile, with the following three modes:

(a) *Security Mode1*: non-secure, no security procedures are initiated, and the security mechanisms of authentication and encryption are bypassed.

(b) *Security Mode2*: service level enforced security. The security procedures are initiated after the wireless channel activation at the Logical Link Control and Adaptation Protocol level which resides at the data link layer and provides connection oriented and connectionless data services to all the upper layers of the protocol. So, this security mode enables us to decide if a certain device has authorization to use a certain service.

(c) *Security Mode3*: link level enforced security, where the Bluetooth device that wants to pass security procedures activates of all them prior to channel activation. In this mode we are able to support both authentication, mutual or not, and encryption, based on a secret key (128bits long) that is shared by a pair of devices, with the pairing procedure starting when two devices try to establish a connection for the first time.

Different security levels exist for the Bluetooth devices themselves, i.e. "trusted devices" and "un-trusted devices". The trusted devices have a fixed relationship with the rest of the network and an unrestricted access to all services, while the un-trusted devices maintain temporal relationships with the network and have restricted access privileges.

For all network services three security levels are defined providing independent security services to devices: services that require authorization and authentication, services that require authentication only and services that are open to all devices.

6.3.1. Authentication

The Bluetooth authentication scheme uses a challenge-response scheme between two devices that are named as the claimant and the verifier. The later has the responsibility of verifying the identity of the other one who is trying to prove its identity. The protocol is based on the fact that both devices share the same Bluetooth link-keys, which are secret to everybody else. The authentication scheme is as follows: first the claimant transmits a

48-bit address to the verifier, which creates and transmits a 128-bit random challenge back to the claimant. Then the verifier and the claimant compute an authentication response using the E1 algorithm, with inputs the received address, the link keys and the random challenge. The claimant returns the computed result (named SRES) to the verifier. Finally, the verifier checks if the claimants SRES matches the one that he has computed. If the comparison proves that the two SRES are identical the verifier continues with the establishment of the connection with the other side, otherwise he stops all communication with this host. If for some reason the authentication fails, there is a period of time that the claimant must wait before making another attempt to be authenticated. This time interval is increased exponentially to avoid the success of repeated attempts to be authenticated using trial and error techniques in order to find the link-keys. The waiting time decreases exponentially to a minimum when no failed authentication attempts are made during a time period. However we must say that this does not provide security from offline attacks that are trying to search exhaustively for the secret keys.

6.3.2.   Confidentiality

Bluetooth in addition to authentication service provides a confidentiality service in order to deal with the possibility of having eavesdropping attacks on the air interface of the wireless network, using encryption on the payloads of the transmitted packets. The procedure is based on stream cipher E0, which is re-synchronized for every payload. The E0 stream cipher consists of the payload key generator that generates the key used to encrypt/decrypt the payload using exclusive-OR. The encrypt/decrypt function takes as input the secret keys, the master's identity, a random number, and a slot number. The last one changes for every packet and so even though all the other remain static the ciphering engine is reinitialized for each packet ciphering.

The encryption key used is produced by an internal key generator based on the link-key, a random number, and an ACO value (a 96bit authenticated cipher parameter). The important thing is that during the whole process of ciphering or deciphering these secret credentials are not transmitted outside the Bluetooth device.

The Bluetooth standard specifies three different modes for confidentiality depending on whether a device has a semi-permanent link key or a master key. If the unit key or a combination key is used, broadcast traffic is not encrypted and individually addressed traffic can either be encrypted or not. If a master key is used, there are three possible modes. In encryption mode 1, nothing is encrypted. In encryption mode 2, broadcast traffic is not encrypted, but the individually addressed traffic is encrypted with the master key. And in encryption mode 3, all traffic is encrypted with the master key.

6.3.3.   Threats related to the Bluetooth Security Standard

In this sub-section we discuss some known problems related to Bluetooth's security scheme described in the previous section and present some known threats and attacks.

The E0 stream cipher with 128-bit key length can be broken in proportion to 2^{64} steps in some circumstances, using a divide-and-conquer type of attack. Hence, a better, more robust encryption algorithm is necessary to ensure security.

The use of Personal Identification Numbers (PIN) codes, in the generation of the link keys and encryption keys, in the security scheme can also cause some problems. Firstly, the keys are only four digits long so they can be easily guessed. Secondly in a large Bluetooth network with many users, problems will appear with the generation and distribution of all the necessary PINs, because there is no key management protocol. This is certain to cause scalability and flexibility problems that may lead to security problems. Finally, the PINs are suffering from all the typical security problems of common passwords, users usually make them simple in order to remember them, write them down in paper, do not change them over time, share them with other people and so on.

Authentication and encryption are based on the assumption that the link key is the participants' shared secret. All other information used in the procedures is public. The problem with this key scheme is that, within a Bluetooth piconet, the unit key of a device can be shared and reused. As a result a malicious host can obtain it, during the authentication or encryption phase, allowing this host to calculate the encryption key of the piconet using a faked device address. This will then enable the host to listen to the network traffic, or to authenticate himself to other devices posing as someone else.

Additionally, the Bluetooth protocol supports only device authentication and not user authentication. Application-level security is not employed at all. In the authentication procedure a device has the opportunity to retry a failed authentication procedure repeatedly. The protocol has not defined an upper limit for the repeated authentication attempts of a device, so offline attacks are a possibility. The challenge response's pseudorandom generator may produce a static periodic number which will reduce the effectiveness of the Bluetooth authentication scheme. The master key of a Bluetooth network is shared since it is broadcasted to all the hosts. Battery draining with denial of service attacks is another problem, against which Bluetooth has no protection.

There are a number of attacks reported in the literature describing ways of penetrating, through Bluetooth's security mechanisms, into the network. These threats involve loss of confidentiality, where an unauthorized device gains access to certain network resources and manages to fake itself as an authorized and trusted host, or loss of integrity, where malicious users manage to gain access to secure data and change them, delete them, redirect them from the original receiver or archive them, or even just sniff them just to analyze the traffic patterns, the user behavior and so on.

7. SECURITY RELEVANT CONCEPTS OF SAFETY

7.1. Health Considerations

Significant research has also been done on harmful effects of electromagnetic radiation on people, focusing mainly on the effects that radar and other systems with huge transmission power levels have. Initially, after the appearance of wireless cell phones, there was no major concern about the effect of the electromagnetic radiation to human health. With the development of personal communication systems, where users have to carry a mobile terminal with them most of the time, safety issues gained great importance, and then it was obvious that it was not only necessary to secure a system from malicious users but to secure a system from harming its users.

Although relevant R&D work to date has been focused mainly in the frequency bands used by the mobile phones, 900Hz-2000Hz, some work was also be done for higher frequencies, which are used for wireless broadband communications. Available research evidence indicates that the issue is not only the absorption of the RF power inside the head, but also on the long term influence of the human brain, due to the antenna's radiation pattern and power output.

In WLANs, antennas do not radiate near the human body, thus allowing the setting of less restrictive power limitations than mobile telephones. The standards for safety levels have already been set in Europe and Unites States and manufactures must build products taking in to account these limitations. Scientists and researchers try by evaluating SAR (amount of power dissipated per mass unit) levels inside the head or the human body, to determine the maximum permitted power levels. We must say that the actual problem with a WLAN system is not the effect of radiation since the electromagnetic signal's frequency is very low to affect human cells, but rather the effect of the thermal energy created when the radio waves interact with the human cells.

7.2. Interference on other wireless devices

All the wireless network equipment, Access Points, NIC cards, etc must meet the relevant standards introduced and enforced by the European Commission to indicate compliance with the standards. All the medical devices used in healthcare for the diagnosis, prevention, monitoring, or treatment of illness or handicap, operate in the same frequency band with wireless networks today, since the M in the ISM band designation stands for Medical. Extensive tests have shown that no significant levels of interference were detected from cordless devices, local area networks or cellular base stations. In USA and in Europe expert's groups are formed in order to come up with guidelines and rules of how wireless local area network and wireless personal area networks should be used within hospitals, or other medical facilities.

Wireless LAN devices radiate at about 5% the power of a mobile phone. Therefore, no serious problems are expected to be encountered with a WLAN transceiver that is located at least one meter away of any other medical device. We must say that, since wireless networks are becoming widely adopted, medical device manufacturers are building devices shielded from this kind of radiation interference in order to ensure their proper functioning under all circumstances. The Bluetooth protocol is designed to minimize the interference with other devices, through its low power consumption, and to be very tolerant to surrounding noise.

In any case, Health organizations need to formulate local policies, taking into account the types of medical devices in use and the communications equipment to be used as well as the local environment and the manufacturer's technical guidelines. Before committing to the use of a wireless communication infrastructure, tests related to the field strength of typical base transceivers must be done, so as to ensure that all wireless network devices operate within their safe power limits.

8. DISCUSSION AND CONCLUSIONS

Wireless communication technologies represent one of the many critical factors that influence the widespread deployment of a range of novel e-health and m-health services.

In this chapter we have presented the security features of the two most common standards in the area of wireless networks, their threats and vulnerabilities related to confidentiality, integrity and availability of personal health information data, in the light of experience in developing and evaluating a variety of e-health and m-health services in the context of HYGEIAnet, the Regional Health Information Network of Crete.

As discussed, despite the real advances in standards, in performance and prices that wireless technologies have made during the last couple of years, it is not always possible to put all the pieces together to form a secure environment. Security in wireless networks is a major issue today since the medium is reachable by anybody, while at the same time one needs to make sure that nobody will be able to obtain, understand, destroy or even fake the data that we like to send and receive through the network infrastructure in the context of advanced m-Health services. Users need to be sure that the systems and services they use are safe. As a result the questions one needs to answer today are: "can we use these modern technologies for providing advanced m-Health services and if we can, can we guarantee security and safety?"

The first question is easy to answer since, as discussed, current trends in health telematics, and especially in the fields of e-health and m-health necessitate the use of wireless technologies in order to support medical collaboration and patient management at the point of need. Evaluation of a variety of e-Health and m-Health services in the context of HYGEIAnet, the regional health information network of Crete, reveals that the large scale deployment of novel e-health and m-health services of clinical relevance has significant economic, clinical and access to care benefits. User adoption, although a gradual process and a process affected by a number of diverse factors, is significantly dependent on the end-users' trust regarding the underlying computer and communication insfrastructure.

The answer to the second question, related to security, is more difficult. Available wireless solutions do provide the necessary security mechanisms, but they do not provide ways of orchestrating these mechanisms easily in creating a secure network infrastructure, as a result there are a number of attacks reported in the literature describing ways of penetrating the available security mechanisms. These threats involve loss of confidentiality or loss of integrity.

So in order to develop mobile health services while at the same time guaranteeing the communication security aspects required, software and communication engineers must have a deep understanding of the security features of the wireless standards involved, their vulnerabilities and the various threads that these create. This understanding, when available, enables developers and network operators to work closely together to install the security mechanisms and, more important, in collaboration with end-users to create policies, procedures and guidelines enabling the creation of a trusted computer base and novel, clinically effective mobile health care applications and services.

9. REFERENCES

Bishop M. and Bailey D., September 1996, A critical analysis of vulnerability taxonomies. The University of California, Davis, Technical Report CSE-96-11.

Blobel, B, P. Pharow, F. Roger-France, 1999, Security analysis and design based on a general conceptual security model and UML. In: Performance Computing and Networking, P. Sloot, M. Bubak, A. Hoekstra, B. Hertzberger, ed., Lecture Notes in Computer Sciences 1593, Springer, Berlin, Heidelberg, New York, pp. 919-930.

Chronaki C., et al, 2001, Preliminary results from the deployment of integrated teleconsultation services in rural Crete. Proceedings of the IEEE Computers in Cardiology 2001 Conference, Amsterdam, pp. 671-674.

Guillén S., et al, 2002, User satisfaction with home telecare based on broadband communication, Journal of Telemedicine and Telecare, Volume 8, Number 2, pp. 81-90.

Lough D.L., 2001, A taxonomy of computer attacks with applications to wireless network, Virginia Polytechnic Institute and State University, Doctoral Dissertation.

Prasad N., Prasad A., 2002, WLAN systems and wireless IP for next generation communications. Artech House, Boston, London.

Traganitis A., et al, 2001, Home monitoring and personal health management services in a regional health telematics network. In Proceedings of EMBC 2001, 23rd Annual International Conference of the IEEE Engineering in Medicine and Biology Society, Istanbul, Vol. 4, pp. 3575-3578.

Tsiknakis M., Katehakis D. G. and Orphanoudakis S. C., 2002, An open, component-based information infrastructure for integrated health information networks, International Journal of Medical Informatics, Vol. 68:1-3, pp. 3-26.

SECURE MOBILE HEALTH SYSTEMS: PRINCIPLES AND SOLUTIONS

Milan Marković[*], Zoran Savić and Branko Kovačević

1. INTRODUCTION

The low-cost nature of the Internet coupled with the ease of making transactions has led to an explosive growth in e-business but trust in this medium is still a major concern. E-security is the foundation that enables trust in e-business (Opliger, 1998, Ford and Baum, 2001). In this sense, main cryptographic aspects of modern TCP/IP computer networks are: digital signature technology based on asymmetrical cryptographic algorithms, data confidentiality by applying symmetrical cryptographic systems, and PKI system – Public Key Infrastructure.

The Internet is also changing the way the healthcare industry does business. It offers astounding opportunities to share information between healthcare professionals and to reduce the costly paper trail. However, organizations must create secure architecture to protect the privacy of patient records since main security requirements in mobile healthcare include privacy and integrity of information related to patients. Such information includes information related to person, medical service given, and e.g. social status, and should be kept out of reach of unauthorized persons. Healthcare security benefits are: protect patient confidentiality from network-based violations, securely provide information to remote physicians, partners, and branch offices, and comply with government regulations on network security.

This chapter is devoted to the emerging topic in domain of healthcare systems – Mobile Healthcare security based on Public Key Infrastructure (PKI) systems. First, we consider possible vulnerabilities of the mobile healthcare systems and possible techniques to eliminate them. We signify that only a general and multi-layered security infrastructure could cope with possible attacks to e-healthcare and mobile healthcare systems. We evaluate security mechanisms on application, transport and network layers of ISO/OSI

---

[*] Milan Marković, Delta banka a.d., Milentija Popovića 7b, 11070 Belgrade, Mathematical Institute SANU, Kneza Mihaila 35, 11001 Belgrade.

reference model and give examples of the today most popular security protocols applied in each of the mentioned layers. These mechanisms include confidentiality protection based on symmetrical cryptographic algorithms and digital signature technology based on asymmetrical algorithms for authentication, integrity protection and non-repudiation. User strong authentication procedures based on digital certificates and mobile PKI systems are especially emphasized in this chapter. We evaluate and signify differences between software-only, hardware-only and combined software and hardware security systems. Also, ubiquitous smart cards and hardware security modules are considered. At the end, we give the brief description of the main components of the PKI systems, emphasizing Certification Authority and its role in establishing a cryptographically unique identity of the valid system users based on X.509 digital certificates. Also, security considerations of the communication and data handling need deal with the exchange of information with mobile healthcare equipment are given.

2. POTENTIAL VULNERABILITIES IN HEALTHCARE SYSTEMS

The Internet has revolutionized the ways in which companies do business, since the Internet Protocol (IP) is undeniably efficient, inexpensive and flexible. However, the existing methods used to route IP packets leave them vulnerable to a range of security risks such as spoofing, sniffing and session hijacking and provide no form of non-repudiation for contractual or monetary transactions. Besides securing the internal environment, organizations need to secure communications between remote offices, business partners, customers and travelling and telecommuting employees. Transmitting messages over the Internet or Intranet to these different entities poses an obvious risk, given the lack of protection provided by the existing Internet backbone. Control and management of security and access between these different entities in a company's business environment is of paramount importance. Without security, both public and private networks are susceptible to unauthorized monitoring and access. Internal attacks might be a result of minimal or nonexistent intranet security. Risks from outside the private network originate from connections to the Internet and extranets. Password-based user access controls alone do not protect data transmitted across a network.

In a sequel, we will give a list of common types of network attacks that are applied to fixed and mobile TCP/IP computer networks. A good overview of these vulnerabilities could be found from Microsoft resources (www.microsoft.com). The same attacks could be applied in the electronic and mobile healthcare systems, i.e. in systems where healthcare data travels through some Internet/Intranet/Extranet networks based on TCP/IP protocols. These systems include: transmission systems between medical institutions, preparing and storing electronic medical prescriptions via Internet, electronic medical records, mobile healthcare systems (rural doctors, emergency help, etc.), and transmission systems between patients and doctors.

Without security measures and controls in place, your data might be subjected to an attack. Some attacks are passive in that information is only monitored. Other attacks are active and information is altered with intent to corrupt or destroy the data or the network itself. Your networks and data are vulnerable to any of the following types of attacks if you do not have a security plan in place.

2.1 Eavesdropping

In general, the majority of network communications occur in a plaintext (unencrypted) format, which allows an attacker who has gained access to data paths in a network to monitor and interpret (read) the traffic. When an attacker is eavesdropping on communications, it is referred to as sniffing or snooping. The ability of an eavesdropper to monitor the network is generally the biggest security problem that administrators face in an enterprise. Without strong encryption services that are based on cryptography, data can be read by others as it traverses the network.

2.2 Data modification

After an attacker has read data, the next logical step is often to modify it. An attacker can modify the data in the packet without the knowledge of the sender or receiver.

2.3 Identity spoofing (IP address spoofing)

Most networks and operating systems use the IP address to identify a computer as being valid on a network. In some cases, it is possible for an IP address to be falsely used. This is known as identity spoofing. An attacker might use special programs to construct IP packets that appear to originate from valid addresses inside an organization intranet. After gaining access to the network with a valid IP address, the attacker can modify, reroute, or delete data. The attacker can also conduct other types of attacks, as described in the following sections.

2.4 Password-based attacks

A commonality among most operating systems and network security plans is password-based access control. Access to both a computer and network resources are determined by a user name and password. Earlier versions of operating system components did not always protect identity information as it was passed through the network for validation. This might allow an eavesdropper to determine a valid user name and password and use it to gain access to the network by posing as a valid user. When an attacker finds and accesses a valid user account, the attacker has the same rights as the actual user. For example, if the user has administrator rights, the attacker can create additional accounts for access at a later time.

After gaining access to a network with a valid account, an attacker can do any of the following:
- obtain lists of valid users and computer names and network information,
- modify server and network configurations, including access controls and routing tables,
- modify, reroute, or delete data.

2.5 Denial-of-service attack

Unlike a password-based attack, the denial-of-service attack prevents normal use of a computer or network by valid users. After gaining access to a network, an attacker can do any of the following:

- distract information systems staff so that they do not immediately detect the intrusion. This gives an attacker the opportunity to make additional attacks,
- send invalid data to applications or network services, causing applications or services to close or operate abnormally,
- send a flood of traffic until a computer or an entire network is shut down,
- block traffic, which results in a loss of access to network resources by authorized users.

2.6 Man-in-the-middle attack

As the name indicates, a man-in-the-middle attack occurs when someone between two users who are communicating is actively monitoring, capturing, and controlling the communication without the knowledge of the users. For example, an attacker can negotiate encryption keys with both users. Each user then sends encrypted data to the attacker, who can decrypt the data. When computers are communicating at low levels of the network layer, the computers might not be able to determine with which computers they are exchanging data.

2.7 Compromised-key attack

A key is a secret code or number required to encrypt, decrypt, or validate secured information. Although determining a key is a difficult and resource-intensive process for an attacker, it is possible. After an attacker determines a key, that key is referred to as a compromised key. An attacker uses the compromised key to gain access to a secured communication without the sender or receiver being aware of the attack. With the compromised key, the attacker can decrypt or modify data. The attacker can also attempt to use the compromised key to compute additional keys, which might allow access to other secured communications.

2.8 Sniffer attack

A sniffer is an application or device that can read, monitor, and capture network data exchanges and packets. If the packets are not encrypted, a sniffer provides a full view of the data that is inside of the packet. Even encapsulated (tunneled) packets can be opened and read if they are not encrypted. Using a sniffer, an attacker can do the following:

- analyze a network and access information, eventually causing the network to stop responding or become corrupted,
- read private communications.

2.9 Application-layer attack

An application-layer attack targets application servers by causing a fault in a server's operating system or applications. This results in the attacker gaining the ability to bypass normal access controls. The attacker takes advantage of this situation, gaining control of an application, system, or network, and can do any of the following:
- read, add, delete, or modify data or an operating system,
- introduce a virus that uses computers and software applications to copy viruses throughout the network,
- introduce a sniffer program to analyze the network and gain information that can eventually be used to cause the network to stop responding or become corrupted,
- abnormally close data applications or operating systems,
- disable other security controls to enable future attacks.

3. MULTILAYERED SECURITY INFRASTRUCTURE IN HEALTHCARE SYSTEMS

Like in all others electronic business systems, key security features that should be included in modern medical computer networks are: user and data authentication, data integrity, non-repudiation, and confidentiality. This means that in secure healthcare systems, the following features must be realized:
- ❑ strong user authentication both for doctors and other medical employees, as well as for patients,
- ❑ integrity of medical data transferred either via wired or wireless IP networks should be ensured and that
- ❑ the non-repudiation function should be implemented.

These features are implemented by using digital signature technology based on asymmetrical cryptographic algorithms. Besides, the confidentiality and privacy protection of transferred data must be preserved during whole transmission and they are done by using symmetrical cryptographic algorithms. Also, strong user authentication techniques based on smart cards should be implemented.

In this Section, we will give the overview of modern security mechanisms with particular emphasis on their use in medical electronic business systems and mobile healthcare systems. The considered security mechanisms are based on Public Key Infrastructure (PKI), digital certificates, digital signature technology, confidentiality protection, privacy protection, strong user authentication procedures and smart card technology. An overview of these techniques is given in (MarkoviA 2002).

In order to preserve the potential malicious attacks to the particular network, the multilayered security architecture has to be implemented. Modern computer networks security systems consist of security mechanisms on three different ISO/OSI reference model layers:
- ❑ Application level security (end-to-end security) based on strong user authentication, digital signature, confidentiality protection, digital certificates and hardware tokens (e.g. smart cards),

▪❑ Transport level security based on establishment of a cryptographic tunnel (symmetric cryptography) between network nodes and strong node authentication procedure,

▪❑ Network IP level security providing bulk security mechanisms on network level between network nodes – protection from the external network attacks.

These layers are projected in a way that a vulnerability of the one layer could not compromise the other layers and then the whole system is not vulnerable.

3.1 Application level security mechanisms

Application level security mechanisms are based on asymmetrical and symmetrical cryptographic systems, which realize the following functions:

▪❑ Authenticity of the relying parties (asymmetrical systems),

▪❑ Integrity protection of transmitted data (asymmetrical systems),

▪❑ Non-repudiation (asymmetrical systems),

▪❑ Confidentiality protection on application level (symmetrical systems).

The most popular protocols in domain of application level security are: S/MIME, PGP, Kerberos, proxy servers on application level, SET, crypto APIs for client-server applications, etc. Most of these protocols are based on PKI X.509 digital certificates, digital signature technology based on asymmetrical algorithms (e.g. RSA) and confidentiality protection based on symmetrical algorithms (e.g. DES, 3DES, IDEA, AES, etc.) (Schneier, 1996). Most of the modern application level security protocols, such as: S/MIME and crypto APIs in client-server applications are based on digital signature see Fig. 1, and digital envelope technology, see Fig. 2.

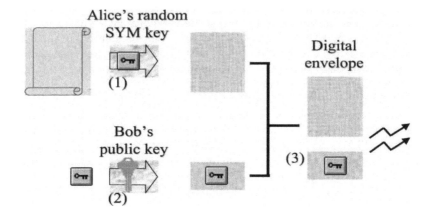

Figure 1. Digital signature and signature verification processes

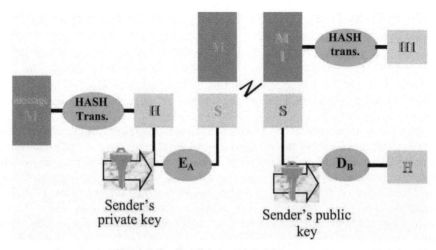

Figure 2. Sending digital enveloped message

In modern e-commerce and e-business systems, RSA algorithm is mainly used according to PKCS#1 standard. PKCS#1 standard (RSA laboratories) describes a method for encrypting data using the RSA public-key cryptosystem. Its intended use is in the construction of digital signatures and digital envelopes, according to the syntax described in PKCS#7 standard. There is a lot of work on optimization of RSA algorithm implementation in hardware security module (HSM) based on signal processor (MarkoviA et al., 2002, UnkaševiA et al., 2001, DjordjeviA et al. 2002, MarkoviA et al., 2003).For digital signatures, see Fig. 1, the content to be signed is first reduced to a message digest with a message-digest algorithm (such as MD5), and then an octet string containing the message digest is encrypted with the RSA private key operation of the signer of the content. The content and the encrypted message digest are represented together according to the syntax in PKCS#7 to yield a digital signature.

For digital envelopes, see Fig. 2, the content to be enveloped is first encrypted by a symmetric encryption key with a symmetric encryption algorithm (such as DES, 3DES, IDEA, AES, ...), and then the symmetric encryption key is encrypted with the RSA public key of the intended recipient of the content. The encrypted content and the encrypted symmetric encryption key are represented together according to the syntax in PKCS#7 to yield a digital envelope.

Security systems on application level consist also of the user authentication procedure which could be one, two or three-component authentication procedure.

3.2 Transport level security mechanisms

Security mechanisms on transport level generally include confidentiality protection of transmitted data based on symmetrical cryptographic algorithms. These systems are mostly based on establishing the cryptographic tunnel between two network nodes on transport level. The establishment of the tunnel is preceded by strong authentication procedures. In this sense, the systems are based both on symmetrical algorithms for realization of cryptographic tunnel and a bilateral challenge-response authentication procedure based on asymmetrical algorithms and PKI digital certificates for authentication of the nodes and for establishing the symmetrical session key for this tunnel session. The transport level security system is mostly used for communication protection between client with Internet browser programs (Internet Explorer, Netscape Navigator, etc.) and WEB server, and the most popular protocols are: SOCKS (used earlier), SSL/TLS and WTLS. Between them, the most popular is SSL (Secure Sockets Layer) protocol, which is used for protection between client browser program and WEB server. Furthermore, SSL is the most popular and the far widest used security protocol today.

SSL protocol consists of two phases: authentication phase with bilateral exchanging of PKI digital certificates of the WEB server and the client (optional) and establishing the symmetrical session key and secure communication based on symmetrical algorithm and established session key (cryptographic tunnel). SSL protocol is placed just below the application layer of the ISO/OSI reference model and just on top of the TCP/IP layer (transport layer). This means that SSL is not necessarily used only under the HTTP protocol but could be used also under some other application level protocols, such as: POP3, SMTP, etc.

WTLS (Wireless Transport Layer Security) protocol is a kind of wireless version of SSL protocol and serves for transport level protection between microbrowsers on WAP (Wireless Application Protocol) enabled GSM mobile phones and WAP servers, based on the same principles and functionality as the SSL protocol. This way, WTLS protocol is intended to use for secure communication in wireless networks (GSM), and is implemented in most of microbrowsers and WAP servers. WTLS protocol uses special digital certificates for wireless communication (WAPCert). A simplified block diagram of the WAP protocol is shown on Fig. 3.

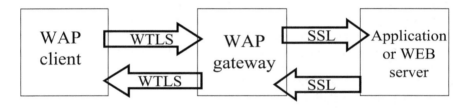

Figure 3. Wireless Transport Layer Security (WTLS) protocol

3.3 Network level security mechanisms

Network level security mechanisms include security mechanisms implemented in communication devices, firewalls, operating system security mechanisms, etc. These methods represent the basis for realization of Virtual Private Networks (VPN). Security protection is achieved by encrypting the complete IP traffic (link encryption) between two network nodes. The most popular network layer security protocols are: IPSec (AH, ESP), packet filtering and network tunnelling protocols, and the widest used is IPSec. Like transport level security protocols, IPSec consists also of network node authentication based on asymmetrical cryptographic algorithms and link encryption based on symmetrical algorithms. IPSec represents a group of protocols consisting of AH – Authentication Header, ESP – Encapsulated Security Payload and IKE – Internet Key Exchange protocols in transport and tunnel mode. AH is used for authentication IP packets, ESP is used for encryption the payload of the IP packets and IKE is used for authentication and IPSec session key establishment. The most secure IPSec protocol is ESP in tunnel mode, since attacker does not know internal addresses (source and destination) – only addresses of IPSec gateways could be seen externally.

Firewalls could be computers, routers, workstations and their main characteristics is to define which information and services of internal network could be accessed from the external world and who from internal network is allowed to use information and services from the external network. Firewalls are mostly installed at breakpoints between insecure external networks and secure internal network. Depending of the needs, firewalls consist of the one or more functional components from the following set: packet filter, application level gateway, and circuit level gateway. There are four important examples of firewalls: Packet Filtering Firewall, Dual-Homed Firewall (with two network interface), Screened Host Firewall, Screened Subnet Firewall (with DeMilitarized Zone – DMZ between internal and external networks).

4. PUBLIC KEY INFRASTRUCTURE (PKI) SYSTEMS

Public-key cryptography uses a combination of public and private keys, digital signature, digital certificates, and trusted third party Certification Authorities (CA), to meet the major requirements of e-business security. Before applying the security mechanisms you need the answers for the following questions: Who is your CA? Where do you store your private key? How do you know that the private key of the person or server you want to talk to is secure? Where do you find certificates?

A public-key infrastructure (PKI) provides the answers to the above questions. In the sense of X.509 standard, the PKI system is defined as the set of hardware, software, people and procedures needed to create, manage, store, distribute and revoke certificates based on public-key cryptography.

PKI system provides a reliable organizational, logical and technical security environment for realization of the four main security functions of the e-business systems: authenticity, data integrity protection, non-repudiation and data confidentiality protection. PKI systems are based on digital certificates as unique cryptographic based electronic IDs of relying parties in some computer networks. A content of the X.509 digital certificate is displayed on Fig. 4.

PKI system consists of the following components: Certification Authority (CA) – responsible for issuing and revoking certificates, Registration Authorities (RAs) – responsible for acquiring certificate requests and checking the identity of the certificate holders, Systems for certificate distribution – responsible for delivering the certificates to their holders, Certificate holders (subjects) – people, machines or software agents that have been issued with certificates, CP, CPS, user agreements and other basic CA documents, systems for publication of issued certificates and Certificate Revocation Lists (CRLs), and PKI application (secure WEB transactions, secure E-mail, secure FTP, VPN, secure Internet payment, secure document management system – secure digital archives, access control system, working time control system)

Format version
Serial number
Digital signature algorithm identifier
Name of the Certification Authority
Validity time
Owner of digital certificate - user
Owner's asymmetrical Public Key
Extensions
CERTIFICATE'S DIGITAL SIGNATURE FROM CERTIFICATION AUTHORITY

Figure 4. Content of the X.509 digital certificate

The method defined in X.509 for revoking certificates involves the use of a certificate revocation list (CRL). This list identifies revoked certificates and is signed and timestamped by the CA. Normally, each certificate is identified by a unique serial number that is assigned when the CA issues it. The CA publishes the CRL, at regular intervals, into the same public repository (e.g. LDAP) as the certificate themselves.

There are several types of CA: corporate CAs, closed user group (CUG) CA, CA of vertical industries, and public CAs. Regarding the implementation approach, CAs could be divided to: outsourced CA – when some organization use certification services from the earlier established CA, and insourced CA – when some organization establishes its own CA services. In all cases, all CA organization mostly used the CA software-hardware technology from the established CA technology vendors, such as: Baltimore, Entrust, Utimaco, SmartTrust, RSA Data Security, NetSeT, etc.

In the following, a brief description is given of the generic model of the Certification Authority software-hardware system which is realized as a web multitier architecture. The described system is similar to the most modern and most secure PKI systems today. Also, some possible variants of system realization depending on the set of the requests that should be fulfilled are discussed. This generic CA represents a solution which could be fully customized to be adapted to the customer requirements.

4.1 Main features of the generic CA system

The generic CA is a WEB-based Certification Authority system which could support both closed PKI systems with strictly defined users of usually only one or two different user profiles, as well as public PKI systems with more user profiles and more different ways of user registration. The generic CA system represents the public CA system fully customizable to the particular requests of different users. Main features of the generic CA system are the following:

- ❑ The system fulfils all worldwide PKI standards and could be customized according to both adding new features and customizing the applied cryptographic algorithms.
- ❑ The generic CA is WEB multitier CA application which is based on smart cards for users.

- ▪❑ Generic CA system supports different database servers, such as: MS SQL Oracle and IBM DB2.
- ▪❑ The generic CA supports a working system with one asymmetrical keypair, with two keypairs and combined system.
- ▪❑ The generic CA supports a hierarchical PKI structure and has the off-line Root CA and more on-line Intermediate CAs. As a rule, each user profile should have its Intermediate CA server.
- ▪❑ The generic CA supports different ways of the user registration, such as: through registration authorities (RA) and RAO operators, as well as directly (for specific user profiles) via WEB CA server.
- ▪❑ The generic CA has implemented a procedure of distributed responsibilities (secret sharing, necessity of presence of number of specific users) in sense of creating the Root CA asymmetrical private key for generating the new Intermediate CA certificate.
- ▪❑ The generic CA has a support for life cycle certificate management (renewal, suspension, revocation).
- ▪❑ The generic CA has possibilities for electronic personalization of the smart cards and this could be done by client themselves, RAO or CAO.
- ▪❑ The generic CA system has a support for printing PIN code (lettershop) for accessing the cards which should be sent to the user separately from the smart card.
- ▪❑ The generic CA system provides the printing of different reports depending of the user needs.

4.2 System architecture of the generic CA

A system architecture of the described generic CA system is given on Fig. 5. What missing on the Fig. 5 are application servers from different business processes which use the generic CA system. For example, WEB server in DMZ zone could be a business WEB server which will eventually realize strong authentication procedure of the users with smart cards, issued by the described generic CA system.

As it could be seen from the Fig. 5, the generic CA system consists of OnLine and OffLine parts. OffLine part represents RootCA which is used only in rare cases when the Root CA asymmetrical private key should be created for a purpose of generating a new Intermediate CA certificate in hierarchical structure shown on Fig. 6, which is the most popular in the modern PKI systems. Root CA is located in totally separated room from the rest part of the CA where there exist a vault in which the Root CA asymmetrical private key's individual parts are securely stored. These parts are used according to the defined procedure of "distributed responsibilities" (or "secret sharing") in cases of generating new Intermediate CA certificates (this procedure is called "CA ceremony"). Eventually, the Root CA could be also in the same room (if necessary) as the OnLine CA but, as mandatory request, outside the LAN network and with mandatory vault for storing the Root CA private key parts.

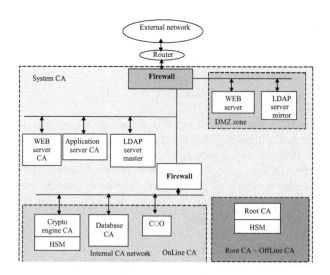

Figure 5: A simplified network configuration of the generic CA system

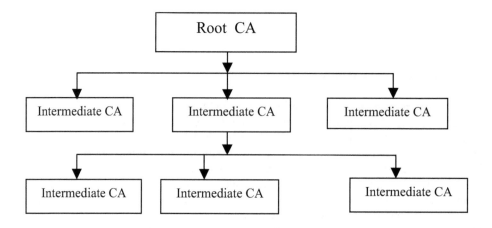

Figure 6: Modern hierarchical structure of the Certification Authorities

In the CA ceremony procedure, it must be present a corresponding minimal number of special CA employees who have access to the corresponding individual parts of the private key, stored on smart cards in special separated boxes of the vault. Namely, a corresponding pre-defined number of smart cards must be present in order to create the Root CA private key in HSM device of the Root CA server, fully in accordance with General and Internal CA practises. After that, a new Intermediate CA asymmetrical keypair is generated in Root CA's HSM and the Intermediate CA certificate is created by a digital signature applying the Root CA private key. The encrypted private key and certificate of the Intermediate CA will be programmed into the new smart card (Intermediate CA smart card) which will be installed into the HSM device of the new

Crypto Engine server, intended for use as a OnLine CA for this Intermediate CA system. After that, Root CA private key will be deleted from the Root CA HSM device and the smart cards with Root CA private key parts will be returned to the vault.

As it could be concluded, it is possible that more Intermediate CA simultaneously work in OnLine working mode, i.e. that more Intermediate CA Crypto Engine servers are activated in the OnLine working mode for digital certificate generation (e.g. Intermediate CA for different kind of medical institutions). OnLine and OffLine parts of the generic CA system should support the using of the HSM modules, see Fig. 5. In DMZ zone, besides WEB server, there is a LDAP mirror server which serves for publishing the CRL and ARL lists, as well as for eventual publication of issued digital certificates. This server is a copy of the master LDAP server which is located in the internal zone.

It should be mentioned that described architecture of the generic CA system could be one example of possible realization of modern CA system and that actual implementations are more or less different depending on the way of key generation for users, the way of distribution for keys and certificates, as well as on ways of CRL publishing. However, although there are differences, basic principles and concepts of the modern certification authorities are the same as in the described example.

4.3 Possible user registration methods

The generic CA system supports different methods for user registration, and the system is fully flexible to support different requests regarding ways of user registration according to the adopted documents (Certificate Policy and CPS) which defines the appropriate user profiles (both for individuals and legal persons). The system supports the issuing of digital certificates on different media (smart cards, mini CD, etc.) and enable functioning in the system with one or two asymmetrical keypairs.

Some examples of possible realization of the user registration methods for issuing digital certificates and their distribution to the users on smart cards will explained in the following.

4.3.1 User registration methods in the system with two asymmetrical keypairs

This method could be described in the following way:
1. A user should come to RA (for example in some medical institution) and apply for issuing digital certificate in the presence of RAO.
2. RAO is connecting to WEB CA server and fill a special WEB page form for issuing the first asymmetrical keypair for this user. The filled form is digitally signed by the the RAO smart card and is sent to the WEB CA server through secure http communication channel (encrypted on application level by digital envelope technology + SSL on the transport level).
3. WEB CA server verifies digital signature of the RAO and, if it is O.K., proceeds this request to the appropriate intermediate CA server (with appropriate Intermediate CA) where the first asymmetrical keypair is automatically generated for this user and is stored into the CA database (encrypted asymmetrical private key and user certificate). The encrypted private key is stored also into special key archive server database for eventual later official purposes.

4. CAO is responsible for: electronic personalization of the previously printed smart cards with generated private key and user certificate, generating the PIN code for the user and all these are stored in CA database for given user (PUK code is also stored in encrypted form and could only be known by the help of CAO application). It is possible to generate a special activation code for the user for some purposes and this is also then stored in the CA database.

5. CAO prints the PIN code (and activation code if exists) into PIN mailer. User address, or RA address where the request is acquired, is printed on the internal side of the envelope. The electronically personalized user smart card with one asymmetrical keypair is put into the second envelope. Namely, the options are that smart card and PIN envelope should go directly to the user or the user must come again to RA to pick up his card.

6. All activities of CAO are stored in a LOG file which is digitally signed by the CAO. Also, CAO enables printing of the appropriate reports which could be requested by the security administrator.

7. If the user receive smart card and PIN code on the home address, he should access the WEB CA server and make additional personalization of his card with the second asymmetrical keypair which is mandatory generated on the card. Namely, by using the software module for the second keypair enrollment in the environment of Internet browser program, the user starts generating keypair on the card, extract public key from the card in the form of the PKCS#10 certificate service request (which is self-signed) and submit it to the WEB CA server, by using already obtained PIN code for the strong authentication purpose.

8. WEB CA server proceeds this request to the appropriate Intermediate CA server which automatically generates digital certificate and store it into CA database, connecting it to the previously generated keypair (the first keypair). This certificate is simultaneously sent back to the user via WEB CA server, who, based on his client application, put it into smart card. This way, the process of additional personalization is finalized.

An alternative to the previous procedure is that user smart card is sent to the RA (instead directly to the user) and after that, RAO calls the user and makes the additional personalization by using the RAO software application in the presence of the user. This option is more secure since there is a secure link between the RAO and WEB CA and the request is sent through the secure crypto tunnel but the previous option is more flexible. The generic CA, depending on business needs, should support both options.

4.3.2 User registration methods in the one keypair system

These methods are realized similar to the previously described procedure of additional personalization of the second asymmetrical keypair. Namely, only a difference compared to the previous case is that there are no keypair generation at CA and consequently there are no separate archive servers for key storage. There are several options:

First option:

1. User comes to RAO and applies for digital certificate and smart card. RAO creates a request on a previously defined way and proceeds it to the WEB CA via secure link on application and transport level. This procedure could be done also by the user himself via WEB CA.

2. In CA, the request is only archived in database and correpsonding data are generated (PIN code and activation code optionally).
3. CAO realizes the electronic personalization of previously printed smart card with generated data (partial personalization). Such personalized smart card and activation data are separately sent to the user.
4. When the user receive the card and activation data, he proceeds with the enrollemnt for the on-card keypair according to the previously described procedure for additional personalization.

Second option:

1. CAO makes in advance electronic personalization of smart cards with generic PIN codes and sends them to the RAO.
2. User comes to RAO and applies for digital certificate and smart card and received them with already personalized generic parameters.
3. User then makes enrollment for the on-card keypair on the previously described way for the additional personalization. In this case, it might be needed to enable a user capability to change the username also besides the password.

4.4 Certificate Life-Cycle Management

In the generic CA system, the certificate life-cycle management is implemented which comprises of the following procedures:

- ▪❑ Certificate renewal,
- ▪❑ Certificate suspension and reactivation,
- ▪❑ Certificate revocation.

These functions are implemented in accordance with Certificate Policy and Certificate Practise Statement of this CA system. In this case, the user will be enabled to make certificate renewal by himself, while the suspension and revocation will be done exclusively by the RAO and CA employees, according to the written procedure in Certificate Policy and CPS.

5. SMART CARDS AND HARDWARE SECURITY MODULES

Software only security solutions are not safe and are very vulnerable to some attacks (e.g. Trojan horse). There are several reasons why SW only security systems are not suitable: certificate and private key are stored on conventional media which is not secure, consumers are tied to their PC and are thus not mobile, and consumers are to manage certificates, which is not simple.

Hardware security modules (HSM) represent very important security issue of the modern computer networks. Main purposes of the HSM are twofold: increasing the overall system security and accelerating cryptographic functions (asymmetric and symmetric algorithms, key generation, etc.). HSMs are intended mainly for use in server applications and, optionally for client sides too in case of specialized information systems (government, military, police) (MarkoviAet al., 2001). For large individual usage, smart cards are more suitable as hardware security modules. However, for large usages, the best approach is in the combination of SW and smart card solutions for best performance. Namely, smart card increases security and SW increases the total processing speed. In this sense, the most suitable large-scale solution consists of: SW for bulk symmetric data encryption/decryption and smart card for digital envelop retrieval and digital signature generation.

In modern and most secure PKI systems, two asymmetrical keypairs are used: one for digital envelope retrieval and the other for digital signature generation. In short, smart cards are credit-card sized plastic card with an embedded computer chip. There are several types of smart cards. Regarding the processing power, smart cards could be divided into the following categories: memory cards – containing a memory chip only with non-programmable logic, microprocessor's chip card with internal memory, microprocessor's chip card with internal memory including additional PKI capabilities (with additional RSA, 3DES, and RNG (Random Number Generator) coprocessors – called PKI smart cards). Regarding the physical contacts of the chip, smart cards could be: contact cards – chip with electrical interface, contactless cards – chip with electromagnetic interface, combo cards – with two chips: one with contact and one with contactless interface, and dual interface chip cards - with chip that have two interfaces: electrical and electromagnetic. Regarding the chip operating system, smart cards could be: proprietary operating system smart cards (with a single or multiple application capabilities (MULTOS)), and JAVA smart cards. Also, smart cards could be divided regarding the power of the implemented microprocessors (8-bit, 16-bit or 32-bit) or regarding the amount of the available memory (EEPROM) (16 KB – 128 KB).

Modern PKI strong secured information systems are mostly based on PKI smart cards for end users. Today's PKI smart cards are mostly based on 8-bit microprocessors (based on the well-known Intel 80C51 microcontroller) with smaller amounts of 16-bit and 32-bit microprocessors. However, it is clear that 8-bit smart card microprocessors will be forgotten very soon and that the market will move toward more power microprocessors. Also, there is a clear move toward JAVA smart cards instead of previously used proprietary OS smart cards. JAVA smart cards enable both multiapplication and easier customization of the existing applications.

Smart cards provide many advantages: financial institutions use smart cards to provide secure access to information and can also be used to conduct secure transactions over the Internet, GSM SIM cards are actually smart cards containing security and subscription information and could be used for user personalization on the network, thus enabling billing information, etc., and smart cards in PKI environment provide user strong authentication based on X.509 digital certificates, as well as confidentiality and non-repudiation. For the purpose of non-repudiation function, the asymmetric signature key pair and actual signature must be generated on the card.

Smart cards used in PKI systems provide a secure and portable way to store the private cryptographic keys and corresponding X.509 digital certificates. The smart card enhances the PKI security by enforcing an extra authentication layer at the end-user level. This extra authentication layer, coupled with the fact that cryptographic keys generated on the card never leave the card, adds an important additional security layer which increases the security of the overall solution. Actually, PKI smart cards with two X.509 digital certificates and two private asymmetric keys stored (for digital envelope retrieval/identification and for digital signature), where signature keypair is generated on the card, represents the most up-to-date security solution for large scale users which provides all four main security functions in modern information systems: authentication (X.509 digital certificate), data integrity (digital signature), non-repudiation (digital signature by asymmetric key generated and stored on the card), and confidentiality (based on asymmetric private key for digital envelope retrieval).

6. HEALTHCARE SECURITY MECHANISMS

This Section deals with the basics of security mechanisms in healthcare systems. Key players in healthcare systems are: medical organizations (hospitals, clinics, pharmaceutical organizations), insurance organizations, healthcare professionals (doctors, physicians, nurses, pharmacist, etc.), and patients – end users. Most modern healthcare systems represent information systems based on TCP/IP computer networks and their work fast move toward the electronic business in healthcare industry – electronic healthcare (e-healthcare). In this environment, security mechanisms for e-business must be implemented with necessary adaptation to the healthcare environments. There are a lot of technical and security issues for these systems that include, between the others: electronic patient record or electronic health record (EHR) must be fully private, central database of patient electronic records must be enabled for use from all players (medical organizations, professionals, insurance, patients), privacy protection of the patient records, secure communications between all players in the system, electronic order entry, enabling mobile healthcare, HIPAA compliance, etc. Security mechanisms that are necessary to be implemented in these e-healthcare systems are: strong user authentication procedure, digital signature technology, confidentiality protection of data in the system on the application, transport and network layers, privacy protection of the patient personal data, strong protection of the central healthcare database based on multiple firewall architecture, and PKI systems, which issue X.509 digital certificates for all users of the system (healthcare professionals and patients) - digital identities (IDs) for the users.

6.1. Strong user authentication

There are several types of user authentication procedures that could be based on the following components: Username/Password – PIN code – something that user know, hardware token – something that user has, and biometric characteristic (e.g. fingerprint) – something that user is.

Regarding the above components there are several types of authentication procedures which combine some of them, such as:

- ❑ Username/password based authentication – weak authentication,
- ❑ Username + dynamic password (one-time password) obtained by appropriate hardware token – stronger than previous one but not in the class of strong user authentication procedures,
- ❑ Username + dynamic password obtained by appropriate hardware token + challenge-response procedure – strong user authentication procedure,
- ❑ Username/password or PIN code + PKI smart card + bilateral challenge response procedure based on PKI X.509 digital certificate and asymmetrical cryptographic techniques – strong user authentication procedure (stronger than the previous one),
- ❑ Username/password or PIN code + PKI smart card + biometric characteristic checking + bilateral challenge response procedure based on PKI X.509 digital

certificate and asymmetrical cryptographic techniques – the strongest user authentication procedure.

In other words, the class of strong user authentication procedures consists of the two or more component procedures and use of the bilateral challenge-response procedure.

Modern healthcare information systems must be based on the strong authentication procedure.

6.2. Digital signature technology

It should be pointed again that the state-of-the-art solution for all the three security functions, authenticity, data integrity and non-repudiation, could be today achieved only by use of the PKI smart cards with signature generation on the card and when the signature private key is generated on the card and never leaves the card. In the modern healthcare systems, healthcare professionals, as well as the patients, should use the smart cards as the hardware tokens.

6.3. Confidentiality protection

Since data that is transmitted through the e-healthcare system contain very sensitive, often personal patient's data, confidentiality must be fully preserved. This is done by using digital envelope technology based on symmetrical cryptographic techniques and PKCS#7 file format. This technology is based on digital certificate, symmetrical algorithms for encryption of data and asymmetrical algorithms for protection of symmetric key. This technology is mainly used for application level protection and it mainly represents the protection from internal attacks to the system.

However, the transport and network level protection should be also used in the system in order to prevent external attacks. Namely, besides the application level protection based on digital signature and envelope technologies, that are based on end users smart cards, the transport level (SSL) and network level (IPSec, VPNs) security mechanisms should be used. In other words, it is recommended that security mechanisms on more than one level should be used. In this sense, the application level protection should be mandatory in combination with one or two additional level protection, transport or network based, depending of system characteristics, type of application and connection, required system throughput, other technical requirements, etc.

6.4. Privacy protection of the personal patient data

It is already emphasized that the main issue of the e-healthcare system is to protect privacy of the patient personal medical data – now processed and stored in the electronic form. This means that data should be protected on the whole path in the system, i.e. from the medical professional workstation to the central database. In other words, unauthorized access to the data should be protected in the entire electronic healthcare system.

6.5. Protection of the central database

As we already mentioned, the central healthcare database should be maximally protected from internal and external attacks. Normally, the new designed healthcare application should be multi-tier (three or more tiers) applications that could be WEB based or client-server based applications. Modern trends move toward web based applications with pretty thin clients (Opliger, 2000). Client part of application should prepare data (e.g. offline) which includes applying of the appropriate security mechanisms (digital signature and digital envelope based on smart cards) and send this data (in online mode) through web browser interface to the web site (e.g. healthcare WEB portal) of the central (or other) location. Before sending data to the web portal, the enduser must authenticate him on the web portal by using strong authentication PKI procedure based on smart card and digital certificates. Modern trends move toward establishing web portal for different medical organizations (or for central point) with single-sign-on capabilities of end-user authentication. This means that administration of the valid users should be centralized and that users cannot do any action if they are not strongly authenticated before through adequate single-sign-on function.

A generic model of the central healthcare site is proposed on Fig. 7. In this model, we could see four different parts of the central medical site:

- external part for accessing to the system which is on the one side of Firewall (connected to one particular interface of the firewall),
- DMZ – DeMilitarized Zone with some general purpose servers, such as: mail, ftp, http, as well as with the WEB healthcare portal,
- Internal part with different applications servers – middle tiers of different multitier medical applications, and
- internal part where the most sensitive parts of the system (e.g. central database) are located.

In this model, multiple firewall architecture is applied. Between the external part and one or more DMZs, the commercial firewall (mostly based on packet filtering techniques) could be applied.

However, for protection of the most sensitive part of the system – the central database, some firewall of the application proxy level gateway type should be applied (SaviAet al., 2001, SaviAand MarkoviА 2003). The best protection will be achieved if this second firewall will be proprietary made from the system designer – not commercial one.

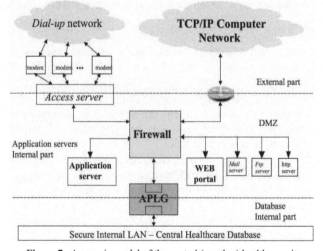

Figure 7. A generic model of the central (or other) healthcare site

6.6. Healthcare PKI systems

To enable application of the all previously mentioned security mechanisms, the appropriate PKI system must be established before. The healthcare PKI system has the following characteristics: it is based on X.509 digital certificates as digital IDs for valid users of the system, central point of the PKI system is Certification Authority (CA), and CA issues digital certificates on smart cards (patient and healthcare professionals' smart cards). In integrated healthcare medical systems, the CA could be truly centralized, centralized with hierarchical CA structure or decentralized. In the truly centralized system, there is only one CA (most often at some state healthcare authority) who issues all digital certificates for all kind of end-users (patients, healthcare professionals, insurance employees, etc.). In this system, individual medical organizations are not independent in defining its own PKI policy but must conform to the global healthcare PKI policy. In the centralized system with hierarchical CA structure, there is a root CA at the healthcare authority and several levels of intermediate CAs. These intermediate CAs will be for different kind of end users and for individual medical organizations (some clinic that has a large information system). The advantage of this architecture is that medical organizations are independent in creation of their own PKI subsystems and that each of the users group is under one certificate management system. However, since all of the intermediate CAs is under the one centralized root CA, compatibility of communications between parties belonging to different intermediate CAs is completely achieved. Decentralized CA structure could be used in the case that all medical organizations have their own PKI subsystem, independent of some centralized authority. This system provides independency but there is an issue regarding the communications between parties that does not belong to the same CA. In this case, this could be only achieved by applying the cross-certification procedure between the CAs. The trend is that modern healthcare information systems are based on hierarchical PKI security infrastructure and digital certificates stored on cards.

7. HIPAA REQUIREMENTS FOR TECHNICAL SECURITY, SERVICES AND MECHANISMS

One of the significant efforts to define technical requirements for Security of Healthcare Systems is represented by HIPAA (Health Insurance Portability and Accountability) Act (HIPAA, 1996, www.checkpoint.com). By HIPAA act the government encourages the use of electronic transactions and address privacy concerns. HIPAA include provision to protect patient privacy and to standardize the security of electronically stored and transmitted data. The security rules are currently in proposal status. They may be categorized in six areas.

Access Control - Each entity is required to maintain mechanism for access control that restricts access to resources and allows access only by privileged entities with a business need for it. Possible types of access control include mandatory access control, discretionary access control, time of day, subject-object separation etc. each entity should define a context-based, role based and/or used based access policy and to adopt procedures to implement need-to-know principle.

Audit Controls - System activity logging is required in order to recreate pertinent system events and action taken by system users and administrators. An audit procedure is required to examine logged information in order to identify questionable access activities, investigate breaches, respond to potential weaknesses and assess the security program.

Authorization Control - Each entity must implement a mechanism to authorize the privileged use of protected health information available via system and applications. The mechanism must limit these privileges to the maximum practical extent in accordance with professional needs. It is important to maintain individual accountability for actions taken by several different groups (shared, generic, trusted).

Data Authentication - It is necessary to provide adequate measures that protect health information in the possession of each entity from unauthorized alternation or destruction. Data authentication methods include, but are not limited to, the use of checksums, double keying, message authentication codes and digital signatures. It is recommended that technical integrity controls should be employed for critical automated functions such as physicians' orders and prescriptions.

Entity Authentication - Entities (person, system or process) must be authenticated prior accessing protected health information. Authentication process may occur through a trusted process such as the provision of a secret password, a personal identification number (PIN) or a token. The stronger authentication protocols may be needed to access to critical health information such as bi-directional challenge–response or certificate based authentication. Automatic log offs, or inactivity time-outs can help enforce authentication by precluding other individuals from accessing unattended sessions. The separate area is consideration of alternative authentication methods based on biometric characteristics of the individuals. The advanced authentication systems include secure architectures such as RADIUS, TACACS, PKI or PKI derived methods.

Communications/Network Controls - If entities use external communication systems, such as PST network, GSM network or open network, such as Internet, they are required to use adequate safeguards to protect health information that traverse them. These services use some form of encryption that is obligatory when using open networks. Additionally it is necessary to protect information systems from intruders attempting to exploit external communication points such as Internet host systems and telephone switches. Encryption is often processor intensive and tends to be system dependant and impedes performances. Hardware based encryption is generally costly but fast because it does not require CPU cycles.

8. MOBILE HEALTHCARE SECURITY SYSTEMS

Mobile healthcare systems represent subsystems of the global healthcare information systems and could not be defined as a single technology area, but as a combination of three components: a handheld or mobile computing device, wireless connecting technology, and a centralized information system. The main difference of the mobile healthcare system compared to the traditional "wired" healthcare system is in mobility of their users. In this sense, at the heart of the mobile healthcare systems are smart cards as hardware tokens personalized for given user that include: healthcare insurance card (HIC) for citizens, and professional card for healthcare professionals (doctors, physicians, nurses, insurance employees, state and authority employees in healthcare sector, etc.). However, instead of using these smart cards at medical offices, ambulances, hospitals, clinics, etc., in the mobile healthcare systems, these cards are used in mobile environment by using some Mobile Healthcare Computing Devices (MHCD).

8.1. MHCD

For the purpose of data capturing, Internet connection and to read and store data to the central database in mobile environment, the healthcare professionals should use MHCD that could be in different shapes and forms, such as: laptops, palmtops, personal digital assistants (PDAs), mobile pnones, mobile communicators and wearable computers (Zabrek, 2001). Healthcare professionals use MHCD in the same healthcare information system as the "wired" users do, but in the mobile conditions, such as: doctor periodical visits of the patients in the hospital (MHCD are used for data capturing, checking the stored patient data, making the prescriptions, etc.), home practicing, rural doctors in low populated areas, and emergency healthcare. In all of these cases, there is also electronic order entry for insurance companies. As we mentioned before, all of the users in the mobile healthcare systems should have smart cards with digital certificates, issues by the appropriate CA in this healthcare systems. In this sense, MHCD should have appropriate smart card reader for reading and storing data from the cards. In spite of this, some possible MHCD could be: laptop computer with PCMCIA-based smart card reader for use at doctor visiting inside the hospital, for rural doctors and home visiting, PDA with card reader and for emergency healthcare teams, mobile phone with SIM cards and SIM toolkit enabled for emergency purposes, smart card enabled hardware tokens with dynamic password and bilateral challenge-response user authentication for patients authentication from public kiosks or Internet cafés, wearable computers, like Xybernaut and ViaII based products, for all of the mobile healthcare cases. It is clear that driving force for enabling mobile health, and in general e-healthcare, is the smart card and digital certificate for the purpose of the strong authentication of all end users in the system. The smart cards for medical purposes are used widely in the world, e.g. in USA, Australia, UK, Taiwan, Slovenia, etc. (examples are given in Bowen, 2002, Chepesiuk, 2001, www.gieseckedevrient.com).

There are several types of the healthcare smart cards, but the following three types are the most important: Patient Data Card (PDC), Health Insurance Card (HIC), and Health Professional Card (HPC).

8.1.1. Patient Data Card (PDC)

A PDC is a mobile data carrier held by the patient. It stores current, accurate patient health information. Data typically stored on PDC include patient ID – X.509 digital certificate, insurance information, emergency record, disease history and electronic prescriptions. For patients, data mobility means that they have complete, up-to-date and accurate medical information in a handy format. They gain peace of mind through knowing that critical data is readily accessible in an emergency. Even in normal situation, complete data may help to ensure the most appropriate treatment. Health card information prevents duplication of medical tests and examinations. For healthcare professionals, the benefits of the new data mobility are the confidence that comes from knowing they are working with accurate, up-to-date patient data. With this data, it is easier to find the best treatment and avoid the risk of prescribing potentially dangerous combinations of drugs. Typical data contents of PDC: patient's personal data (name, address, date of birth, phone number), X.509 digital certificate issued by the appropriate CA, patient's insurance data (insurance company, insurance number, validity), small personal medical database for the patient (up-datable information about diseases, treatments), emergency data for the patient (blood group, allergies, diabetes, pace maker), prescription data (medication, dosage, date of prescription), electronic purse for eventual small amount payments in the healthcare system, and patient PIN and cryptographic keys for mutual authentication between the card and smart card reader.

It should be pointed out that personal sensitive data of the patient should be secured on the PDC by use the encryption technique (privacy protection). In this sense, it is not necessary that PDC should be a PKI smart card. However, if something should be digitally signed by the patient, it is mandatory that PDC is the PKI smart card and that digital signature is done on the card with a private asymmetric key generated and never leaves the card.

8.1.2. Health Professional Card (HPC)

A HPC is an individually programmed access authorization card held by the health professionals. It gives him/her the right to read/or writes specific data fields on a PDC and it also carry a digital certificate and appropriate cryptographic keys for secure communication. Patients are guaranteed privacy and data security by sophisticated access rules, held on the HPC, that prevent unauthorized access to their medical data and/or tampering with stored or communicated information. For health professionals, HPCs open the doors to secure and fast electronic information exchange between health professionals and with other partners in the e-healthcare system. Typical HPC data contents: health professional's identification data (name, address, phone number, …), X.509 digital certificate, individual access rights profile of the health professional to read and/or write data fields of patient cards, PIN for access to the card and cryptographic keys for mutual authentication between the card and smart card reader, and asymmetric keys for performing digital signature and digital envelope retrieval functions. Based on the above, it is clear the HPC must be a PKI smart card.

8.1.3. Health Insurance Card (HIC)

A HIC is and ID card with a purely administrative function. It is held by the insured person and stores his/her personal data, the insurance company ID and data confirming insurance coverage. In some systems, PDC and HIC could be integrated in one PDC card. Patients benefit from streamlined admission and administration procedures in doctors' offices and hospitals. HICs increase patient satisfaction and cut down paperwork. Medical insurance claims are processed quickly and accurately. Health professionals profit from fast, easy, accurate and cost-effective data management. This means less paperwork, reduced transaction costs and more efficient billing. Insurance companies benefit from electronic data processing, from fast, simplified, error free processing of insurance claims and from better-cost transparency. Typical HIC data contents: patient's personal data (name, address, date of birth, phone number, …), patient's insurance data (insurance company, insurance number, validity,...), insurance company ID, and insurance coverage.

Therefore, it is clear that PDC and HIC could be one PDC card and that PDC and HPC is minimal set of the ID cards in the secured e-healthcare systems.

8.2. Wireless connecting technology

As for the connection of the MHCD to the central database, some of the wireless technology is used. For short-range communications (e.g. inside the hospital or clinic) wireless LAN technology based on IEEE 802.11 standard is used (Padget et al., 1995, Andren, 1998, BoškoviA and MarkoviA 1999, 2000). For long-range communications, GSM cellular networks are very successfully used.

Wireless LAN technology has many advantages to the wired LANs in many conditions and, in e-healthcare environment, it is often used during the doctors' regular and emergent visits inside the hospital or other medical organizations. This technology is mostly used with laptops as MHCD and serves for secure and mobile connection with central database. This technology does not limit the use of application, transport and network layer security mechanisms based on smart cards and software on the laptops. Instead, this adds a new security level in the wireless domain combining spread spectrum transmission technology (frequency hopping or direct sequence) and data information security. This is called WEP (Wired Equivalent Privacy) and is based on RC4 stream cipher symmetric algorithms (www.verniernetworks.com, 2002). However, the main point of security, as we emphasized before, should be on application layer security layer. Wireless LAN technology also could be integrated Microsoft 200x network domain controller security and secure Windows 200x logon features. GSM technology is used for long-range communications and could be exploited from home visiting, rural doctors, emergency team, etc. This could be exploited by using laptops with GSM connectivity or GSM mobile phone itself. This means that, besides the short-range wireless LAN technology, laptops could be used with GSM mobile phones for establishing long-range dial-up connections to the access server of the central system.

On the other hand, WAP-enabled GSM phones could be used for establishing emergent connections by using the phone itself. However, a microbrowser implemented in GSM mobile phone is not yet capable for serious communications based on WAP protocol. There are limitations in sense of transmitting and receiving data from the central database site and this kind of communications could be used only for short messages and emergency sending some information. For the purpose of the security, in this case, there is a WTLS protocol that is similar to the SSL, as explained in subchapter 3.2. Besides, there are SIM cards that enable SIM toolkits and provide some additional features and functions by using SMS services. These features include a lot of functions for additional GSM services that could include some central database retrieval of short information through this communication channel. In these functions, some of security functions could be implemented like dynamic password, challenge-response authentication and digital signature technology. The main advantage of this approach is because the mobile phone acts as a simultaneous smart card reader and MHCD. As for the network security of the GSM communications, there are networks with implemented frequency hopping and data security option (A5 symmetrical algorithm).

Both systems upgrades performances with introducing GPRS high speed connections and will be certainly upgraded more seriously when 3G wireless communication services (EDGE, UMTS) will be established. With introducing these features (e.g. GPRS), mobile phones could use more higher and performant protocols for communications (e.g. HTTP) instead of the WAP.

8.3. Central information system

Central mobile healthcare information systems are the same central system as in the classical e-healthcare system. In fact, this is one global healthcare central database information system that provide desktop and mobile users and that have different e-healthcare channels, such as: HTTP communication channel through WEB portal, GPRS communication, WAP communication channel for WAP-enabled mobile phones, SMS channel, GSM voice channel, etc.

9. CONCLUSIONS

In this Chapter, the modern computer security systems are analyzed and their application in electronic healthcare systems is emphasized. It is concluded that only multilayered security architecture could cope with potential internal and external attacks to the modern computer networks. The most frequently used security mechanisms on the application, transport and network layers are analyzed. It is concluded that more than one layer should be covered by the appropriate security mechanisms in order to achieve high quality cryptography protection of the system. There is also an analysis of the electronic healthcare systems with special emphasis to the mobile healthcare systems. The analysis is done regarding security challenges and appropriate security mechanisms for coping with potential vulnerabilities. It is concluded that, between many specific conditions, in the e-healthcare systems security mechanisms should be distributed on the client side, communication side and central database side, and that, in each of the parts, appropriate security measures should be applied. Central points of both e-healthcare and mobile healthcare systems are smart cards for end users (citizens, healthcare professionals, etc.) that could be used for applying digital signature and digital envelope technology and the central PKI system. This applies especially for doctor, and other healthcare professionals, smart cards. For the patient – end users, main point is that data should be protected from unauthorized use (privacy protection). Differences that divide e-healthcare systems from the mobile healthcare system are in the way of using some form of mobile healthcare devices (MHCD) and wireless connecting technology (wireless LAN or GSM). Security measures and techniques applied in mobile healthcare systems are actually the same as in the e-healthcare system because this is the same global healthcare system with different communication channels.

Based on the presented material, it could be concluded for high secure healthcare system that the best solution is based on PKI smart cards, digital signature and digital envelope technology, as well as X.509 digital certificates issued by the appropriate healthcare Certification Authority. These security schemes are most complex to implement but, at the same time, they represent the preferred mechanisms for securing modern e-healthcare and mobile healthcare systems.

10. REFERENCES

Andren, C., 1998, IEEE 802.11 wireless LAN: can we use it for multimedia?, *IEEE Multimedia,* April-June pp. 84.89.

BoškoviА B., MarkoviА M., 1999, Low-Bit Rate Speech Coding in Mobile Satellite Systems, in *Advances in Intelligent Systems and Computer Science*, Nikos Mastorakis, Ed., World Scientific Engineering Society, pp. 57.62.

BoškoviА B., MarkoviА M., 2000, On spread spectrum modulation techniques applied in IEEE 802.11 wireless LAN standard, *Proc. of IEEE/AFCEA Intern. Conf. EUROCOMM 2000*, Information Systems for Enhanced Public Safety and Security, pp. 238.241.

Bowen, C, 2002, Welfare agencies seek benefits from chip cards, *Card Technology*, January, pp. 40.44.

Chepesiuk, R., Davis, D., 2001, Chip card remedies for health data ills, *Card Technology*, April, pp. 58.64.

DjorđeviА G., UnkaševiА T., MarkoviА M., 2002, Optimization of modular reduction procedure in RSA algorithm implementation on assembler of TMS320C54x signal processors, *DSP 2002*, July, Santorini, Greece.

Ford, W., Baum, M.S., 2001, *Secure Electronic Commerce: Building the Infrastructure for Digital Signatures and Encryption,* Second Edition, Prentice Hall PTR, Upper Saddle River, NJ 07458.

Healthcare Insurance Portability and Accountability Act, 1996, HIPAA Requirements for Technical Security, Services and Mechanisms.

IEEE 802.11 standards web site, http://stdsbbs.ieee.org/ groups/802/11/index.html.

LaMaire, R.O., Krishna, A., Bhagwat, P., Panian, J., 1996, Wireless LANs and Mobile Networking: Standards and Future Directions, *IEEE Communications Magazine*, August, pp. 86.94.

MarkoviА M., 2002, Cryptographic Techniques and Security Protocols in Modern TCP/IP Computer Networks, Short-Tutorial, in *Proc. of ICEST 2002*, Oct., 1-4.

MarkoviА M., DjorđeviА G., UnkaševiА T., 2002, Influence of key length in possible optimization of RSA algorithm implementation on signal processor, in *Proc. of ICEST 2002*, Oct., 1-4, pp. 23.26.

MarkoviА M., ĐorđeviА G., UnkaševiА T., 2003, On Optimizing RSA Algorithm Implementation on Signal Processor Regarding Asymmetric Private Key Length, in *Proceedings of WISP 2003*, Budapest, Sept. 2003, pp. 73-77.

MarkoviА M., SaviА Z., ObrenoviА Ž., NikoliА A. , 2001, A PC Cryptographic Coprocessor Based on TI Signal Processor and Smart Card System, *Communications and Multimedia Security Issues of the New Century*, R. Steinmetz, J. Dittman, M. Steinebach, Eds., Kluwer Ac. Publishers, pp. 383.393.

MarkoviА M., UnkaševiА T., DjorđeviА G., 2002, RSA algorithm optimization on assembler of TI TMS320C54x signal processors, in *Proc. of EUSIPCO 2002*, Toulouse, France, Sept. 3-6.

Oppliger, R., 1998, *Internet and Intranet Security*, Artech House, ISBN 0-89006-829.

Oppliger, R., 2000, *Security Technologies for the World Wide Web*, Artech House, Boston, London.

Padget, J.E., Guenther, C.G., and Hattori, T., 1995, Overview of wireless personal communications, *IEEE Communications Magazine*, January, pp. 28.41.

RSA Laboratories, *PKCS standards*.

SaviА Z., NikoliА A., MarkoviА M., 2001, Cryptographic proxy gateways in securing TCP/IP computer networks, *Information Security Solution Europe, ISSE 2001*, London, UK.

SaviА Z., MarkoviА M., 2003, Development of Secure Web Financial Services in Serbia, in *Proceedings of ISSE 2003*, October 7-10, 2003.

Schneier, B., 1996, *Applied Cryptography, Second Edition, Protocols, Algorithms and Source Code in C*, Joh Wiley & Sons, Inc., New York, Chichester, Brisbane, Toronto, Singapore.

UnkaševiА T., MarkoviА M., DjorđeviА G., 2001, Optimization of RSA algorithm implementation on TI TMS320C54x signal processors based on a modified Karatsuba-Offman's algorithm, in *Proc. of ECMCS'2001*, 11-13 September, Budapest.

UnkaševiА T., MarkoviА M., DjorđeviА G., 2001, Optimization of RSA algorithm implementation on TI TMS320C54x signal processors," in *Proc. of TELSIKS'2001*, September 19-21, Niš, pp. 603.606.

Zabrek, E., 2001, Mobile healthcare enterprize solutions with the pocket PC, white paper, www.microsoft.com.

www.microsoft.com, Security issues with IP,

www.checkpoint.com, Healthcare security,

www.giseckedevrient.com, Taiwan launches intelligent health insurance card, press release, 2002.

www.verniernetworks.com, 2002, Wireless security: protecting your 802.11 netwok, white paper, September.

COMPUTATIONAL AND WIRELESS MODELING FOR COLLABORATIVE VIRTUAL MEDICAL TEAMS

George Samaras[*], Dimosthenis Georgiadis, and Andreas Pitsillides

1. INTRODUCTION

Mobile wireless computing brings about a new paradigm of distributed computing in which communications may be achieved through wireless networks and users can continue computing even as they relocate from one support environment to another. The impact of wireless computing on system design goes beyond the networking level and directly affects data management and data computational paradigms. Infrastructure research on communication and networking is essential for realizing wireless and mobile systems. Equally important, however, is the design and implementation of data management applications for these systems (e.g., wireless virtual teams) a task directly affected by the characteristics of the wireless medium, the limitations of mobile computing devices, and the resulting mobility of resources and computation.

These limitations require a more flexible computational model than the one supporting wireline communications. This is because most of the assumptions that influence the definition and evolution of the traditional client/server model for distributed computing (Gray and Reuter, 1993) are no longer valid. These assumptions include: (a) fast, reliable and cheap communications, (b) robust and resource rich devices, and finally (c) stationary and fixed locations of the participating devices.

Wireless networks, however, are more expensive, offer less bandwidth, and are less reliable than wireline networks. Consequently, network connectivity is weak and often intermittent. Moreover, mobile elements must be light and small to be easily carried around and thus in general have less resources than static elements, including memory, screen size and disk capacity. These results in an asymmetry in mobile computing systems: fixed hosts have sufficient resources, while mobile elements are resource poor. Furthermore, since mobile elements must rely for their operation on the finite energy

---
[*]George Samaras, Department of Computer Science, University of Cyprus, Kallipoleos 75, P.O. Box 20537, CY1678 Nicosia, CYPRUS, Email: cssamara@ucy.ac.cy

provided by batteries, energy consumption is a major concern. Mobile elements are also easier to be accidentally damaged, stolen or lost. Finally, having mobile hosts roaming around, changing their location and therefore their point of attachment to the fixed network places new requirements on system design. The center of activity, the system load and the notion of locality changes dynamically. The search cost to locate mobile elements is added to the cost of each communication involving them. Moreover, as mobile elements enter regions in which communication is provided by different means, connectivity becomes highly variable in performance and reliability.

These limitations of mobility and wireless communications result in a form of distributed computing with drastically different connectivity assumptions than in traditional distributed systems. Thus, it is necessary to employ new computational paradigms to achieve a usable system. These new software models must cope with the special characteristics and limitations of the mobile/wireless environment. They should also provide efficient access to both existing and new applications which is a key requirement for the wide acceptance of mobile computing. In summary, the appropriate mobile computing models must efficiently deal with:

- *disconnections*, to allow the mobile unit to operate even when disconnected
- *weak connectivity*, to alleviate the effect of low bandwidth
- *both light-weight and heavy-weight mobile clients*, i.e., clients with different resource capacities
- *dynamic adjustment of the mobile host functionality*, i.e., the type and degree of functionality assign to mobile hosts
- *the variability of the wireless environment*
- *multiple types and application models*, including: current, and future TCP/IP applications, and emerging mobile application paradigms.

To this end, a number of models have been introduced recently including the *client/agent/server* (Badrinath et al., 1993; Fox et al., 1996; Oracle, 1997, Zenel and Duchamp, 1997; Tennenhouse et al., 1996), the *client/intercept* (Samaras and Pitsillides, 1997; Housel et al., 1996), the *peer-to-peer*(Reiher et al., 1996; Athan and Duchamp, 1993), and the *mobile agent* (Chess et al., 1995; White, 1996; Mobile Agents, Aglets Workbench) models.

This chapter identifies the shortcomings of the wireline client/server model and its inability to support mobile computing, presents various software models for mobile wireless computing and discusses and compares the degree in which each one of them supports mobility and wireless connectivity. The requirements pertaining to wireless/mobile software models are clearly stated and the positive and negative aspects of the various models are presented in relation to these requirements. Defining a taxonomy of models and providing a methodology for building software for mobile computing is critical, since it provides the framework for the development of future applications and a yardstick to measure their effectiveness in addressing the challenges of mobile/wireless computing.

Virtual teams are then presented in general terms and medical virtual teams in more specific term. We aim to ask the question *"which is the best computational model for creating collaborative virtual medical teams?"* A multilevel analysis is provided within the context of collaboration, virtual teams and healthcare requirements. Finally we proposed as the computational model most suitable for the "collaborative healthcare environment" the Client/Agent/Server and the Client/Intercept ones

The remainder of this chapter is organized as follows. Section 2 introduces the underlying wireless architecture, discusses the challenges presented by the mobile/wireless environment and the inability of the classical client/server model to deal with them. Section 3 presents the various mobile software models and how each one of them copes with the limitations of wireless connectivity and mobility. It also provides a taxonomy and a comparison of the identified models. Section 4 presents virtual teams and wireless medical teams and their requirements. Section 5 describes the computational models for such teams. Section 6 concludes the chapter.

2. REQUIREMENTS OF SOFTWARE MODELS FOR MOBILE WIRELESS COMPUTING

There is a need to make a distinction between mobility and wireless computing. Mobility can exist without wireless communication by having, for example, a process surfing the fixed network to optimize its computations. Wireless communications simply exacerbate the complexity of mobile computing. Wireless computing does not necessarily imply mobility either, since wireless devices may be stationary as well.

2.1. Network Infrastructure

The networking infrastructure for providing ubiquitous wireless communication coverage, known as Personal Communication Network (PCN), is expected to be partially based on the existing digital cellular architecture (see Figure 1 (Imielinksi and Badrinath, 1994). In this architecture, the whole geographic area is divided into cells with a mobile service base station (MSS) in the center of the cell. The MSS is serving the mobile host through wireless communications and is attached to the fixed network. The mobile host can communicate with other units, mobile or static, only through the base station of the cell in which it resides. Cell sizes vary widely (Forman and Zahorjan, 1994). For instance, a satellite beam can cover an area more than 400 miles in diameter, a cellular transceiver has a range of few miles, and wireless LANs just cover communication in a building.

As a mobile host (MH) moves, it may cross the boundary of a cell, and enter an area covered by a different base station. This process is called handoff. During handoff an MH may be disconnected from the network for a finite, but arbitrary period of time. Thus, to maintain the computations at the mobile host uninterrupted, handoffs must be executed in a timely fashion. While this is straightforward for voice communications, due to the higher loss of information that can be tolerated, it becomes very complicated for data transfer where the information loss must be extremely low.

The characteristics of mobile hosts vary. The PCN architecture must be able to support both smaller units, that are called light-weight clients, and larger more powerful units, called heavy-weight clients. A mobile host is battery depended. Receiving and sending messages as well as CPU processing and disk and memory access drain the battery that has to be recharged. Such battery power restrictions cause mobile hosts to be frequently, voluntarily or unwillingly, disconnected (powered off). To preserve energy, especially important for heavy weight mobile units, the mobile host may also enter a special mode of operation called *doze mode*. While in this mode, the clock speed is reduced and no user computation is performed. A specific event or signal can later awake the "sleeping" client.

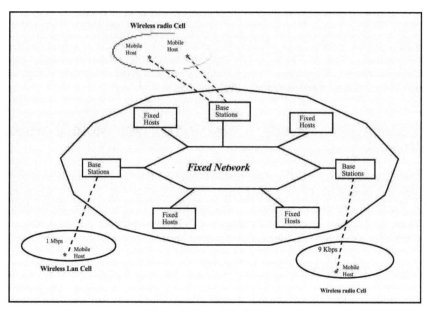

Figure 1. Network Architecture

It is speculated that ubiquitous communications will be provided by PCNs in a hybrid fashion: heavily populated areas will be covered by cheap base stations of small radius (picocells); less populated areas will be covered by bigger radius base stations and farm land, remote areas and highways will be covered by satellites that will provide the bridge between these different islands of population density. This architecture increases the challenges for the definition of a PCN since it now has to take into consideration the different power and transmission requirements impose by these different islands of communications.

2.2. Characteristics of Mobile Wireless Computing

The mobile computing environment is constrained in many ways. The restrictions induced by mobile wireless computing can be categorized (Imielinksi and Badrinath, 1994; Forman and Zahorjan, 1994; Satyanarayanan, 1996; Alonso and Korth, 1993) in those induced by (a) the wireless medium, (b) mobility and (c) the portability of wireless elements. These restrictions have a great impact on the design and structure of mobile computing applications and motivate the development of new computing models. Specific models put greater emphasis on some of the characteristics while ignoring others, depending on the application for which they are built and the specific requirements of the particular environment.

Wireless Medium. Wireless communications face many obstacles because the surrounding environment interacts with the signal. Thus, while the growth in wired network bandwidth has been tremendous (in current technology Ethernet provides 100 Mbps, FDDI 100 Mbps and ATM 622 Mbps), products for wireless communication achieve only 19 Kbps for packet radio communications, and 9-14 Kbps for cellular telephony. Similarly, the typical bandwidth of wireless LANs ranges from 250 Kbps to 2 Mbps and it is expected to increase to 10 Mbps (Mathias, 1994). Disruptions may be hidden from the wireless application by error recovery procedures in the link or higher layer communications protocols. However, this often results in excessive retransmissions (and extra cost), not to mention user frustration that may result in requesting resubmissions. In practice, the radio transmission error rate is so high that the effective bandwidth is limited to less than 10 Kbps (Miller, 1994). Furthermore, a wireless link is shared among all the mobile terminals within the range of the wireless base station. Consequently, the deliverable bandwidth per user is even lower. Thus, bandwidth is a scarce resource. Finally, an additional reason that makes bandwidth consumption a major concern of mobile computing designs is that data transmission over the air is currently monetarily expensive. The above limitations of the wireless medium result in a form of distributed computing with drastically different connectivity assumptions than in traditional distributed systems, where disconnections are rare. In addition, the PCN infrastructure provides *variant connectivity* since wireless technologies vary on the degree of bandwidth and reliability they provide.

Mobility. The location of mobile elements and therefore their point of attachment to the fixed network change as they move. The consequences of mobility are numerous. First, mobility introduces various forms of heterogeneity. For instance, outdoors, a mobile client may have to rely on low bandwidth networks, while inside a building it may be offered reliable high-bandwidth connectivity or even operate connected via wireline connections. Moreover, there may be areas with no adequate coverage resulting in disconnections of various durations. There may be also variability in the provision of specific services, such as in the type of available printers or weather reports. Furthermore, the resources available to a mobile element vary, for example, a docked computer has more memory or is equipped with a larger screen. Mobility also results in distributed systems whose configurations are no longer static. Topology, the center of activity, the system load, and the notion of locality change dynamically. Finally, the location of moving objects changes with time. Thus, the search cost to locate mobile elements is added to the cost of each communication involving them.

Portability of Mobile Elements. Mobile elements must be light and small to be easily carried around. Such considerations, in conjunction with a given cost and level of technology, will keep mobile elements having less resources than static elements, including memory, screen size and disk capacity. Furthermore, mobile elements rely for their operation on the finite energy provided by batteries. Even with advances in battery technology, the concern for power consumption will not cease to exist. This concern spans various levels in hardware and software design. For example, software systems must take into consideration in their design doze mode operation. Finally, mobile elements are easier to be accidentally damaged, stolen, or lost. Thus, mobile elements are less secure and reliable than static elements.

2.3. Issues and Challenges for Deriving Mobile Computing Models

The characteristics of the mobile and wireless environment place a number of constraints that define the desired capabilities of each software model.

Functionality of Mobile Units. An important consideration in designing software architectures is the type of functionality assigned to mobile hosts. Mobile units are still characterized as unreliable and prone to hard failures, i.e., theft, loss or accidental damage. Mobile elements are also resource-poor relative to static hosts. These reasons justify treating the mobile unit as a dumb terminal running just a user-interface. The InfoPad (Narayanaswamy et al., 1996) and ParcTab (Schilit et al., 1993) projects employ such a dump terminal approach and off-load all functionality from the mobile unit to the fixed network. On the other hand, slow and unreliable networks argue for placing more functionality at the mobile hosts so that they are less dependent on remote servers. Although, there is no consensus yet on the specific role the mobile host should play in distributed computation, the above contradictory considerations motivate the development of models that provide for a flexible adjustment of the functionality assigned to mobile clients.

In general, how is functionally adjusted between mobile and static hosts is a function of the capabilities of the mobile hosts and connectivity. Light-weight clients (i.e., vulnerable and poor in resources) argue for more dependency on the fixed network while heavy-weight clients (i.e., secure and rich in resources) for less dependency.

Disconnections. Disconnections can be categorized in various ways. First, disconnections may be voluntary, e.g., when deliberately avoiding network access to reduce cost, power consumption, or bandwidth use, or forced e.g., when entering a region where there is no network coverage. Then, disconnections may be predictable or sudden. For example, voluntary disconnections are predictable. Other predictable disconnections include those that can be detected by changes in the signal strength, by predicting the battery lifetime, or by utilizing knowledge of the bandwidth distribution. Finally, disconnections can be categorized based on their duration. Very short disconnections, such as those resulting from handoffs, can be masked by the hardware or low-level software. Other disconnections may either be handled at various levels, e.g., by the file system or an application, or may be made visible to the user. *Since disconnections are very common, supporting disconnected operation, that is allowing the mobile unit (or application) to operate even when disconnected, is a central design goal in deriving mobile computing models.*

Weak Connectivity. Weak connectivity is the cumulative effect of the unreliable and relatively low bandwidth that wireless media offer, mobility itself and the number of mobile units sharing the media within a particular cell. For example, since the number of the mobile units can dynamically change due to mobility, the effective data rate is similarly affected often resulting in weak connectivity. Since weak connectivity is a fact of life in wireless environments, measures must be taken by the various computing models so that it does not seriously affect performance.

Mobility. The physical mobility of portable devices has several consequences. For example, the distribution of servers on a fixed network might need to change dynamically to facilitate more efficient data access for the mobile clients.

Adaptivity. The mobile environment is a dynamically changing one. Connectivity conditions vary from total disconnections to full connectivity. In addition, the resources available to mobile computers are not static either, for instance a "docked" mobile computer may have access to a larger display or memory. Furthermore, the location of mobile elements changes and so does the network configuration and the center of computational activity. Thus, a mobile system is presented with resources of varying number and quality. Consequently, a desired property of software systems for mobile computing is their ability to adapt to the constantly changing environmental conditions (Forman and Zahorjan, 1994; Katz, 1994; Satyanarayanan, 1996; Joseph et al., 1997).

But how can adaptivity be captured and realized? A possible answer is by varying the partition of duties between the client and the server. For instance, during disconnection, a mobile client works autonomously, while during periods of strong connectivity, the client depends heavily on the fixed network sparing its scarce local resources. Another way to realize adaptivity is by varying the quality of data available at the clients depending on connectivity.

Upward Compatibility. The various computing models should be flexible enough to accommodate not only a single new type of application, but multiple types and application models, including: current and emerging TCP/IP applications, terminal emulation, applications using dialup services and new mobile application paradigms. A key point for the wide acceptance of mobile computing is efficient access to applications (Pitoura and Samaras, 1998). This includes access to legacy and existing applications with minimum effort and cost, and cost effective and reliable support for new applications.

2.4. The Inadequacy of the Traditional Client/Server Model

With client/server (c/s) computing in its simplest form, an application component executing in one computing system, called the client, requests a service from an application component executing in another computing system, called the server. In a wireless setting, the mobile host acts as the client requesting services from servers located at the fixed network (Figure 2). The traditional client/server model is based on assumptions that are no longer justifiable in mobile computing such as: static, relatively resource-rich and reliable clients as well as reliable and low-latency communications. Furthermore, there is a rigid division of responsibilities among the client and the server and the inability to fulfill them implies the end of execution.

In particular, in the wireline c/s paradigm disconnections and weak connectivity are rare events and such as receive no special treatment. In the cases of disconnections, commonly detected as communication time-outs, the execution of the application is usually discontinued. Similarly, weak connectivity seriously affects the application by resulting into unacceptable performance because the environment did not anticipate it.

Adaptivity is viewed as an answer to the variable characteristics of the wireless media and the capabilities of the mobile client. Facilitating adaptivity in the wireline c/s model implies that the client and server site of the application must be designed accordingly. That is, the application must be able to monitor the environment and dynamically adjust its code at both the server and the client side. However, in general, c/s applications never face the need for adaptivity and traditionally are uninterested to changes of the environment. One of the few environments in which the need for wireline adaptivity appeared is within the context of WWW where due to its popularity, bandwidth consumption increased dramatically and weak connectivity became a serious issue. Web proxies and a number of optimizations (Fox and Brewer, 1996) were used to alleviate the problem. Similar approaches are proposed for the mobile/wireless environment.

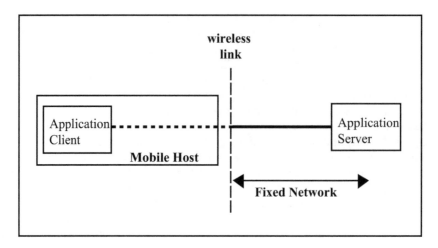

Figure 2. Traditional Client/Server Model

While mobility is central to the design of mobile computational models, in the c/s paradigm the location of both clients and servers is fixed simplifying the requirement of locating and communicating with each other. Furthermore, in wireless settings, there should not be a clear division of functionality between the mobile client and the server at the fixed network. To this end, M. Satyanarayanan (Satyanarayanan, 1996) talks about an extended client/server model where the roles of clients and servers are temporarily blurred. The most evident such case is during disconnections, when a disconnected client has to emulate the functionality of a server to continue operating.

Thus, the traditional client/server model must be extended to provide special components responsible for implementing appropriate support for wireless mobile computing and imposing minimum changes in clients and servers.

3. SOFTWARE MODELS FOR MOBILE COMPUTING

Devising appropriate models for structuring applications that involve wireless elements is an issue central to developing software for mobile computing. In this section, we elaborate on such models, their strengths and weaknesses. First, we consider extensions to the popular client/server model and then the employment of mobile agents.

3.1. Extended Client/Server Models

Extensions of the client/server model are merely based in the introduction of agents placed between the mobile client and the fixed server. The agents alleviate both the constraints of the wireless link, by performing various communication optimizations, and of any resource constraints, by undertaking part of the functionality of resource-poor mobile clients. The degree in which this is achieved depends on the placement and the functionality of the agents.

3.1.1. The Client/Agent/Server Model

A popular extension to the traditional client-server model is a three-tier or client/agent/server (c/a/s) model (Badrinath et al., 1993; Fox et al., 1996; Oracle, 1997; Zenel and Duchamp, 1997; Tennenhouse et al., 1996), that uses messaging and queuing infrastructure for communications from the mobile client to the agent and from the agent to the server (Figure 3). In its most generic form, an *agent* or *proxy* is just a surrogate of the client on the fixed network.

This three-tier architecture somewhat alleviates the impact of the limited bandwidth and the poor reliability of the wireless link by continuously maintaining the client's presence on the fixed network via the agent. Agents split the interaction between mobile clients and fixed servers in two parts, one between the client and the agent, and one between the agent and the server. Different protocols can be used for each part of the interaction and each part of the interaction may be executed independently of the other.

The c/a/s model is most appropriate for light-weight clients of limited computational power and resources. It moves responsibilities from the client to the agent. For example, the agent can manage a complex client request with only the final result transmitted to the client. Located on the fixed network, the agent has access to high bandwidth links and large computational resources that it can use to benefit its client. Similarly, performing caching at the agent can minimize long round trips on the network and thus reduce application response time. To further enhance the client's benefit, the agent can be used to off-load significant computational burden from the server. For example, the agent instead of the server can perform data compression.

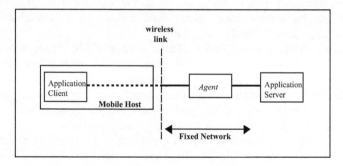

Figure 3. Client-Server based Models: Client/Agent/Server

In general, how the agent's resources are used, that is the functionality of the agent, greatly depends on the specific application the client needs to support. Minimum functionality at the agent includes support for messaging and queuing for communication between the mobile client and the agent so as to be able to support weak connectivity and disconnections. The agent can also assume a more active and intelligent role (Oracle, 1997; Tennenhouse et al., 1996), in which case, when application-specific predefined events occur, it notifies the client appropriately. Furthermore, to reduce the load on the mobile unit and thus preserve valuable resources (i.e., battery life), the agent might be responsible for starting/stopping specific functions at the mobile unit or for executing client specific services.

To deal with disconnections, a mobile client can submit its requests to the agent and wait to retrieve the results when connection is re-establish. In the meantime, any requests to the disconnected client can be queued at the agent to be transferred upon reconnection. The agent can be used in a similar way to preserve battery life. The client submits its requests to the agent, enters the doze mode, and waits the agent to wake it up when the response becomes available.

The agent can effectively handle weak connectivity in a variety of ways. The agent can employ a number of optimization techniques to minimize the size of data to be transmitted to the client depending on the type of data and on the specific application. The agent can manipulate the data prior to their transmissions to the client (Fox et al., 1996; Zenel and Duchamp, 1997; Tennenhouse et al., 1996), by changing their transmission order so that the most important information is transferred first, by performing data specific lossy compression that tailors content to the specific constraints of the client, or by batching together multiple replies.

Taxonomy of the C/A/S Architecture and Other Features. Agents are used in a variety of forms and roles in this particular architecture. At one extreme, an agent acts as the complete surrogate of a mobile host on the fixed network. In this case, any communication to and from the mobile host goes through the mobile host's agent. The surrogate role of an agent can be generalized by having an agent acting as a surrogate of multiple mobile hosts (Fox et al., 1996). At the other extreme, the agent is attached to a specific service or application, e.g., web browsing (Housel et al., 1996) or database access (Oracle, 1997). Any client's request and server's reply associated with this application is communicated through the service-specific agent. In this scenario, a mobile host must be associated with as many agents as the services it needs access to. This second case can be symmetrically generalized by having a service-specific agent servicing multiple clients (Housel et al., 1996).

While the client/agent/server model offers a number of advantages, it fails to sustain the current computation at the mobile client during periods of disconnection. Furthermore, although the server notices no changes, the model still requires changes to the client code for the development of the client/agent interaction rendering the execution and maintenance of legacy applications problematic. Adaptivity is not well supported by this model since it requires the participation of the light-weight client site application which must be specifically design to deal with it. Finally, the agent can directly optimize only data transmission over the wireless link from the fixed network to the mobile client and not in the opposite direction.

3.1.2. The Client/Intercept/Server Model

To address the shortcomings of the client/agent/server (c/a/s) model, references (Samaras and Pitsillides, 1997) and (Housel et al., 1996) propose the deployment of an agent that will run at the end-user mobile device along with the agent of the c/a/s model that runs within the wireline network (Figure 4). This client-side agent intercepts client's request and together with the server-side agent performs optimizations to reduce data transmission over the wireless link, improve data availability and sustain the mobile computation uninterrupted. From the point of view of the client, the client-side agent appears as the local server proxy that is co-resident with the client. Since the pair of agents is virtually inserted in the data path between the client and the server, the model is also called (Samaras and Pitsillides, 1997; Housel et al., 1996) client/intercept/server instead of client/agent/agent/server model. This model is more appropriate for heavy-weight clients with enough computational power and secondary storage to support the client-side agent

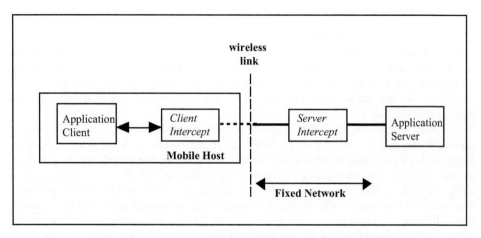

Figure 4. Client-Server based Models: Client/Intercept/Server Model.

The model provides a clear distinction and separation of responsibilities between the client- and the server- side agents. The communication protocol between the two agents can facilitate highly effective data reduction and protocol optimization. The existence of the agent pair also facilitates adaptivity since the two agents can dynamically divide the workload among them based on the various environmental conditions.

The model offers flexibility in handling disconnections. For instance, a local cache may be maintained at the client-side agent to allow it to work autonomously. The cache can be used to satisfy any client's requirements for data during disconnections. Cache misses may be queued by the client-side agent to be served upon reconnection. Similarly, requests to the client can be queued at the server-side agent and transferred to the client upon reconnection. Weak connectivity can also be handled in a variety of ways. For example, relocating computation from the client-side agent to the server-side agent or vice versa can minimize the effect of weak connectivity. Background prefetching to the client-side agent can reduce communication during weak connectivity. During periods of strong connectivity, on the other hand, the client-side agent can heavily depend on the server-side agent sparing its scarce local resources, for example, to require a high degree of data fidelity or no/low compression of the transmitted data.

The intercept model provides upward compatibility since it is transparent to both the client and the server. Therefore, the agent pair can be employed with any client application. Legacy and existing applications can be executed as before since the agent pair shields them from the limitations of mobility and the wireless media. The main weakness of the model is that every application requires development work at both the server and the client site. However, there is no need to develop a pair of agents for every instance of an application. Instead, since the functionality and optimizations performed by the agent pair is generic enough, it is only required to develop a different pair of agents per application type e.g., file, database, or web application. For example, prefetching documents at the cache of the client-side agent follows the same principles independently of the specific type of web-application.

3.2. Mobile Agent Models

Besides the functional components of a mobile application, the organization of data is also important. Data may be organized as collections of objects. In this case, objects become the unit of information exchange between mobile and static hosts. Objects encapsulate not only pure data but also information regarding their manipulation, such as operations for accessing them. Incorporating active computations with objects and making them mobile leads to mobile agents.

Mobile agents are processes dispatched from a source computer to accomplish a specified task (Chess et al., 1995; J.E.White, 1996). Each *mobile agent* is a computation along with its own data and execution state. In this sense, the mobile agent paradigm extends the RPC communication mechanism according to which a message sent by a client is just a procedure call. After its submission, the mobile agent proceeds autonomously and independently of the sending client. When the agent reaches a server, it is delivered to an agent execution environment. Then, if the agent possesses necessary authentication credentials, its executable parts are started. To accomplish its task, the mobile agent can transport itself to another server, spawn new agents, or interact with other agents. Upon completion, the mobile agent delivers the results to the sending client or to another server.

The mobile agent model supports disconnected operation. During a brief connection service, a mobile client submits an agent to the fixed network and then disconnects. The agent proceeds independently to accomplish the delegated task. When the task is completed, the agent waits till reconnection to submit the result to the mobile client. Conversely, a mobile agent may be loaded from the fixed network onto a laptop before disconnection. The agent acts as a surrogate for the application allowing interaction with the user even during disconnections. Weak connectivity is also supported by the model since the overall communication traffic through the wireless link is reduced from a possibly large number of messages to the submission of a single agent and then of its result. In addition, by letting mobile hosts submit agents, the burden of computation is shifted from the resource-poor mobile hosts to the fixed network. Mobility is inherent in the model. Mobile agents migrate not only to find the required resources but also to follow mobile clients. Finally, mobile agents provide the flexibility to load functionality to and from a mobile host depending on bandwidth and other available resources. And thus, materializing high degree of adaptivity.

Mobile Agents and the Client/Server Variations. The mobile agent computational paradigm is not orthogonal to the client/server model and its extensions. For example, the agent of the c/a/s model may be seen as a form of a static agent, i.e., an agent lacking the ability to migrate to other servers. Alternatively, one can implement the agent of the c/a/s model as a mobile agent and let it move on the fixed network following its associated client.

Mobile agents can be used in conjunction with agents located at the fixed network. Let us call the agents at the fixed network proxies for clarity. In this scenario, a client submits a general mobile agent to a server-side proxy. The server-side proxy refines and extends the mobile agent before launching it to servers on the network. When the mobile agent finishes its task, it first communicates its results to the proxy. The proxy filters out any unnecessary information and transmits to the mobile client only the relevant data. Such an approach entails enhancing proxies with capabilities to process mobile agents. Building on this approach, a proxy may be *programmable*, that is extended with the ability to execute mobile agents submitted to it by clients or servers. Such an approach is in accordance with current research on *active networks* (Tennenhouse et al., 1996). In Section 4, we will use mobile agents to implement all models.

3.3. Comparison of the Models

The agents, placed between the mobile client and the fixed server, aim to alleviate the constraints of mobile wireless computing. Table 1 summarized some of their types. Instead of a single pair of agents, multiple pairs of agents handling mobility at different levels may be inserted on the path between the mobile clients and the fixed server. Such agents can cooperate in various ways. Agents at lower layers may convey information to agents at higher layers and vice-versa. For instance, a transport–layer agent that queues RPC replies can cooperate with an application-layer agent to delete unwanted messages form the queue of pending requests or re-order the queue based on user-defined priorities. Such an agent hierarchy may prove very flexible in building software for mobile wireless computing.

Table 1. Agent Types

Agent Type	Property
Surrogate of the mobile host	The complete functionality of the mobile host is off-loaded to the agent
Service (Application) specific	Implements the mobile-specific part of a service/application
	For example, a web-agent or a database agent
Mobile	Can be relocated to a number of sites
Programmable	Understands and can execute mobile agents

The major characteristics/capabilities of each model pertaining to the requirements defined in section 3.1 are summarized in Table 2. The combined model, where mobile agents are transmitted between a server-side and an agent-side agent, provides additional functionality.

Table 2. Summary of the Proposed Models

	Client/Server	Client/Agent /Server	Client/Intercept /Server	Mobile-Agents
Disconnections	NO	Partly Handled at the server (e.g., by queuing requests)	YES Handled at both the client (e.g., by caching) and the server (e.g., by queuing requests).	Partly By the asynchronous submission of mobile agents to the fixed network
Weak Connectivity	NO	Partly Only agent to client transmission - - by performing various optimizations at the server-side agent	YES Both agent to client and agent to client – by performing various optimizations	Partly By replacing message transmissions with submission of agents
Adaptivity	NO	Medium At the server-side	High At both the server and the client	High
Type of Clients	Heavy-weight	Light-weight	Heavy-weight	Light-weight
Functionality Division	Strict	Partly flexible	Flexible Dynamically adjusted between the server and the client agents	Flexible By loading mobile agents on and off the mobile host
Mobility	Not aware	Not aware	Aware - not aware	Inherent
Upward Compatibility	No	Server-side only	Server and client side	No

4. COLLABORATION CONCEPTS FOR MEDICAL VIRTUAL TEAMS

Collaboration and virtual teams refer to the notion that a team of professionals decide to collaborate over the internet and thus create a virtual team that eliminates the need of physical presence. The collaboration requirements are based on identified scenarios of collaboration analysed using UML. These scenarios can identify the communication and collaborative requirements a computational model must support in an integrated fashion. Creating virtual collaborative teams for the healthcare environment can further introduce new and more specialized requirements. The collaborative functionality of artifacts may be also enhanced with embedded hybrid intelligence as well.

4.1. Understanding Virtual Teams

The usual image of a working group in a company contains meetings, encounters in the hallway, getting together for lunch and dropping into one another's offices. People that work together sometimes doesn't meet one another especially when they are in different places or spread out around the world or even housed in different parts of the same city. Many teams today never meet face-to-face, but work together only online. Face-to-face interactions among people from the same organization are the typical old model of teamwork (Lipnack and Stamps, 2000).

In short one might described a virtual team as a group of people who work interdependently with a shared purpose across space, time, and organization boundaries using technology. The main benefit is that in a virtual team you can do what you can't do alone. The challenge of our time is to learn to work in virtual teams and networks while retaining the benefits of earlier forms. In time, virtual teams will become the natural way to work. Since many virtual teams do meet periodically, or a few times, or at least once, they also find themselves in the conventional face-to-face setting. It is generally agreed that a good virtual team is, in a way, a good team.

Four words capture the essence of virtual teams: people, purpose, links, and time. People populate and lead small groups and teams of every kind at every level, from the executive suite to the subcommittees of the local school's parent association. Purpose holds groups together, which for teams mean a focus on tasks that makes work progressing from goals to results. Links are the channels, interactions, and relationships. The greatest difference between in-the-same-place teams and virtual ones lies in the nature and variety of their links. Time is a dimension common to schedules, milestones, calendars, processes, and life cycles (Lipnack and Stamps, 2000).

Sometimes virtual teams fail. One major reason why many virtual teams fail is because they overlook the implications of the obvious differences in their working environments. People do not make accommodation for how different it really is when they and their colleagues no longer work face-to-face. Teams fail when they do not adjust to this new reality by closing the virtual gap. So why put up with the trouble of working across boundaries? Because, when the virtual team is successful, the business performance is improved dramatically. In brief, the cost can reduced by cutting travel costs and time, the cycle time can be shorten by moving from serial to parallel processes and establishing better communications, the innovation can be increased by permitting more diverse participation, stimulating product and process creativity, and encouraging new business development synergies and finally we can succeed leverage of learning by capturing knowledge in the natural course of doing the work, gaining wider access to expertise, and sharing best practices (Lipnack and Stamps, 2000).

4.2. Definition of Virtual Teams

Even though interactivity is often presented as a key characteristic of a computer-mediated communication system, the emphasis is often on the computer-human interaction rather than on human-to-human computer-mediated interaction (Panteli and Dibben, 2001). The latter is particularly important since virtual teams are effective not only because of technological advancements but also and most importantly because individuals are able to interact and thus constructively engage in knowledge sharing and creation in the increasingly emergent virtual work environments.

During the last few years there has been an increasing volume of literature on virtual organizations and virtual teams. This body of research generally agrees that virtual teams consist of a collection of geographically dispersed individuals who work on a joint project or common tasks and communicate electronically (Jaervepaa and Leidner, 1998). For example, Lipnack and Stamps (Lipnack and Stamps, 2000) define a virtual team as "a group of people who **interact** through inter-dependent tasks guided by a common purpose" that "works across space, time and organizational boundaries with links strengthened by **webs of communication technologies**". Indeed, virtual teams have been presented in the popular literature as a communication intensive and a computer-mediated linked type of group. Electronic data interchange, computer-supported cooperative work, group support systems, as well as email, videoconferencing and teleconferencing facilities, to name but a few, enable people based in different locations to communicate and coordinate their actions with great speed and effectiveness.

We are interested here in understanding and identifying the fined requirements that defined the computational model for supporting such teams, and more specifically medical virtual teams, in the wireless environment.

4.3. Formalizing Virtual Teams

To this definition then were added 5 new criteria: (1) members work apart more than they do at the same location, (2) they are based on telecommunications technologies as computer networks, email, groupware, and phones, (3) members solve problems and make decisions jointly, (4) members are mutually accountable for results and (5) members in such organizations can be brought together for a finite time to solve specific tasks or problems (Tiwana, 2000).

The members of a virtual team can emanate from the same or different organism. These members can also be in the same labor space or in different states. Trying to categorize all the virtual team types, we led to eight likely scenarios (Figure 5).

The above figure presents a clear classification of the various types and forms of virtual teams based on the dimensions of time, place and organization. Eight cases are identified, in the four, the members of virtual team, belong in the same organism, in other four cases the members are in the same labor space and in other four, the members work simultaneously (Georgiadis, 2002).

4.4. Medical Virtual Teams

Medical Virtual Teams (MVTs) are Virtual Teams build around medical objectives and the most of the team members are medical specialists (e.g. doctors, nurses, physiotherapists, and psychologists). Medical Virtual Teams can be categorized into three major categories based on the distance between the team members. It is interesting to note that the distance among team members defined the urgency of the collaboration.

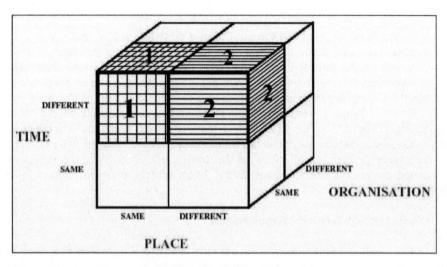

Figure 5. Virtual Team Types

4.4.1. Short Distance Medical Virtual Teams *(Hospital Medical Virtual Teams)*

The Short Distance Medical Virtual Teams (SDs) are Medical Virtual Teams (MVTs) that are limited into the boundaries of a building, like a hospital. On figure 5 we can see the SDs as the number 1 virtual team type. In this case all the team members are collaborating mostly wirelessly (via a wireless LAN). In case of an emergency, the existence of the MVT ceased. The team members after they alerted for the emergency, they meet for the best settlement of the problem. This type of MVT requires a less synchronous model and because of that, the team members are unknown before any action taken (shift based teams).

4.4.2. Long Distance Medical Virtual Teams (*Home Care Medical Virtual Teams*)

The Long Distance Medical Virtual Teams (LDs) are MVTs that their team members works mostly separated and the lifetime of the time is a bit extended. On figure 5 we can see the LDs as the number 2 virtual team type. In this case all the team members are collaborating wirelessly (via a wireless WAN). Due to the limitations of the wireless link this MVT model requires an asynchronous model which in essence guarantees the continuous running of the team. For that reason the team members are well known before any homecare visit (subject based teams). If a team member is unavailable, the system will record the request and when the team member will be available, the request will be served. In case of extreme emergency a backup member is usually assign. A good example of a LD is a home care medical virtual team such as DITIS. DITIS, supports the activities of PASYKAF (Cypriot Association of Cancer Patients and Friends), who run a national home-based healthcare service for cancer patients living all over Cyprus (Pitsillides et al., 1999).

4.4.3. Variable Distance Medical Virtual Teams *(Emergency Medical Virtual Teams)*

Variable Distance Medical Virtual Teams (VDs) are MVTs that can be change between SDs and LDs depending on the situation that will occur. On figure 5 the VDs can vary between the number 1 & 2 virtual team types. In this case all the team members are collaborating wirelessly apart from few exceptions. An example of a VD is a MVT that handles Emergency medical occurrences, such as the MVTs of the Kardio-T project. In the Kardio-T project, persons having a heart condition, are monitored and for better monitoring of the MVT. This MVT model requires a high synchronous model. As the first model, the team members are unknown before any action taken (shift based teams).

The main problem in all types of the MVTs is that all the properties that we mentioned above, can change dynamically. Every system must be tolerant in every change that may occur.

4.5. Discussion and Technical Requirements

In the previous section, we mentioned in brief the preferences that categorize the MVTs. In more detail, these preferences that make these categorizations are the collaboration model, the identity of the team members, the type of the network that the MVT will be active on, the distance between the team members and finally the devices that the member will use.

Before the creation of the MVT, it must be clear the needs of collaboration between the team members. This analysis, will determine if the MVT will use a synchronous or asynchronous collaboration model. For instance if we conduct an analysis that will show that the system requires rapid requests and responses then we will use definitely a synchronous collaboration model. Otherwise we prefer the asynchronous collaboration model, if the need for rapid messaging is limited, like the case of DITIS.

Our next concern should be the type of the network and the type of Medical Devices we will use. These two preferences will determine the application model that we will use. As we saw above, all the types of the MVTs are using wireless network. Therefore, as we mentioned in a previous section, it would be better to use an application model that supports an agent to the fixed network that will maintain the user's instance all the time. There are two models for this functionality; the Client/Agent/Server and the Client/Intercept/Server Model. Considering the capabilities of the mobile devices that it will be used, we choose between the two application models. Nowadays, most of the Medical Mobile Devices are small in resources and size. This reason makes the Client/Intercept/Model inappropriate.

Another characteristic is the boundaries that the MVTs will act. This is very important characteristic, because as we saw above the MVTs types are categorized mainly by the distance between the team members. The distance, between others, determines many things like the cost of supporting the MVT (which is vital to the MVT's existence) and the need for synchronous or asynchronous model.

Finally the identity of the team members affects the type of the MVT. When we say identity of the team member we mean the advantage of knowing the person in the other side. It's better to know the other users, because the impersonal relationships are proved to be ineffective.

Summarizing the above, we can compare the three models of the MVTs with the following table (Table 3).

The continuous present of the collaboration even during disconnections members is a must in all the types of virtual teams.

Table 3. Medical Mobile Medical Teams Properties.

	SD	LD	VD
Collaboration Model	Synchronous	Asynchronous	Synchronous
Medical Mobile Devices	Small-Mid Devices	Small Devices	Med-High Devices
Team Member Identity	Shift Based	Subject Based	Shift Based
Boundaries	Within a Building	None	Vary
Network	Wireless	Wireless	Wireless
Suggested Application Model	Client/Agent/Server	Client/Agent/Server	Client/Agent/Server or Client/Intercept/Server

5. THE COMPUTATIONAL MODEL FOR WIRELESS VIRTUAL MEDICAL TEAMS

In such unpredictable environment, like the wireless one, it is natural that want systems that will be persistent and fault tolerant, even when disconnections occurs. The question that rises is: *"which is the best combination of technologies for creating collaborative virtual medical teams?"* To better answer the above question its best to split the problem into different layers. The first one is the technological layer, in which we examine the technologies that we'll use for the wireless networking and the types of the devices that the team members will use. The second layer is the application architecture layer. In this layer we examine the proper application model that we'll use, depending on the decisions that we'll make in the first layer. Finally, there is a third layer in which we examine the way that the collaborative virtual team will be implemented. This layer is related with the other two layers. Thus, these three layers are dependent to each other, so the procedures of the analysis and deployment start together for all three of the layers (Georgiadis, 2002).

The technological layer must be examined after the analysis of the third layer, because we have to understand first the nature of the virtual teams. We can easily exclude the technologies that they depend on being in direct line of sight or technologies that work within few meters (e.g. IrDA, Bluetooth), because when we talk about virtual teams, we assume that a respective distance between the team members, exists. Two of the things that we have to determine are the immediateness of the communication and the real distance between the team members. The distance between the team members reflect on the type of the wireless network that we will choose. If the virtual team moves within the boundaries of a hospital a WLAN may be suitable. If the team members work all over the country, we would use GSM type of networks. The immediateness of the communication determines the type of the connection that we'll use. For example if we want to send information to the team members all the time we'll choose a connection that the users will be always online (e.g. GPRS) (Georgiadis, 2002).

In the application architecture layer, appropriate models are the Client/Agent/Server and the Client/Intercept/Server, depending of the resources that the mobile devices will have. If the team members possess mobile devices with many resources, then we can have a Client-side Agent for handling disconnections that may occur. The certain thing is that we'll have a Server-side Agent. This agent will not only exist for sustaining the user's existence in the network, but also for sustaining the user's existence in the virtual medical team (Georgiadis, 2002).

In the final third layer, we have to guarantee that every team member will be equal the other team members in the virtual team anytime. Thus, we can use an agent for each user, representing his in every action. The agent will be always online and connected with the other members of his virtual team. Users can communicate with each other through their agents. The agent receives the message from the senders' agent and forwards it to his user for advising. If the user isn't connected, the agent waits for the user to connect or responds by himself depending on his responsibilities (Georgiadis, 2002).

A system that applies successfully the technologies that mentioned above is DITIS (Collaborative Virtual Medical team for home healthcare of cancer patients) (Pitsillides et al., 1999). In DITIS the Client/Agent/Server application model is used. Every user is represented by a mobile agent. Thus the virtual medical team exists anytime.

6. CONCLUSIONS

To deal with the characteristics of mobile computing, especially with wireless connectivity and small devices, various extensions of the client/server model have been proposed. Such extensions advocate the use of proxies or middleware components. Proxies of the mobile host residing at the fixed network, called *server-side* proxies, perform various optimizations to alleviate the effects of wireless connectivity such as message compression and re-ordering. Server-side proxies may also perform computations and *collaborative tasks* in lieu of their mobile client. Proxies at the mobile client undertake the part of the client protocol that relates to mobile computing thus providing transparent adaptation to mobility. They also support client caching and communication optimizations for the messages sent from the client to the fixed server. Finally, mobile agents have been used with client/server models and their extensions. Such agents are initiated at the mobile host, launched at the fixed network to perform a specified task, and return, if necessary, to the mobile host with the results.

We presented these software models and studied them within the context of collaboration, virtual teams and healthcare requirements. Finally we proposed as the computational model most suitable for the "collaborative healthcare environment" the Client/Agent/Server and the Client/Intercept ones.

7. REFERENCES

Aglets Workbench, by IBM Japan Research Group. Web site: <http:/ aglets.trl.ibm.co.jp>.

Alonso R., Korth H.F., Database System Issues in Nomadic Computing, *Proceedings of the 1993 SIGMOD Conference*, Washington, D.C., May, 1993.

Amrit T., "From Intuition to Institution: Supporting Collaborative Diagnoses in Telemedicine Teams", Proceedings of the 33rd Hawaii International Conference on System Sciences – 2000

Andreas Pitsillides, George Samaras, Marios Dikaiakos, Eleni Christodoulou, *DITIS: Collaborative Virtual Medical team for home healthcare of cancer patients*, Conference on the Information Society and Telematics Applications, Catania, Italy, 16-18 April 1999.

Athan A., Duchamp D., Agent-Mediated Message Passing for Constrained Environments, Proceedings *USENIX Symposium on Mobile and Location-Independent Computing*, pp. 103--1070, Cambridge, Massachusetts, August, 1993.

Badrinath B. R., Bakre A., Imielinski T., Marantz R., Handling Mobile Clients: A Case for Indirect Interaction, *Proceedings of the 4th Workshop on Workstation Operating Systems*, Aigen, Austria, October, 1993.

Barbara D., Imielinski T., Sleepers and Workaholics: Caching Strategies in Mobile Environments, *Proceedings of the ACM SIGMOD Intl. Conference on Management of Data (SIGMOD 94)*, pp. 1-12, 1994.

Barelos D., Pitoura E., Samaras G., "*Mobile Agents Procedures: Metacomputing in Java*", University of Ioannina, Computer Science, Technical Report 27-98, Sept. 1998.

Bartlett J. F., Experienece with Wireless World Wide Web Clients, *Proceedings of the IEEE COMPCON*, San Francisco, CA, March, 1995.

Berners-Lee T., Caililiau R., Luotonen A., Nielsen H.F., Secret A., The World-Wide Web, *Journal Communications of the ACM*, Vol. 37, No. 8, pp.76-82, August, 1994.

Chadwick D., *A quick guide to wireless communication standards,* Matt Publishing Limited (2002).

Chess D., Grosof B., Harrison C., Levine D., Parris C., Tsudik G., Itinerant Agents for Mobile Computing, *Journal IEEE Personal Communications*, Vol. 2, No. 5, October, 1995.

Christoyiannis D.C., Papamichalopoulos A., Pouliou E., Kyprianou Th., Nanas S., Tsanakas I., Rasidakis A., Roussos Ch., "*Designing and Internet-based Collaborative Environment for Cystic Fibrosis Treatment*", Summary Proceedings of Euromednet '98, Nicosia, Cyprus, March 1998.

Demers A., et al., The Bayou Architecture: Support for Data Sharing among Mobile Users. *Proc. Workshop on Mobile Computing Systems and Applications,* IEEE, December, 1994.

Forman G.H., Zahorjan J., The Challenges of Mobile Computing, *IEEE Computer*, Vol. 27, No. 6, pp. 38-47, April, 1994.

Fox A., Brewer E.A., Reducing WWW Latency and Bandwidth Requirements by Real-Time Distillation, *Proceedings of the 5th International World Wide Web Conference*, Paris, France, May 1996.

Fox A., Gribble S.D., Brewer E.A., Amir E., Adapting to Network and Client Variability via On-Demand Dynamic Distillation, *Proceedings of the ASPLOS-VII*, Cambridge, MA, October, 1996.

Georgiadis T. D., "Dynamic Creation of Collaborating System and Database Management, Using Mobile Agents.", Master thesis supervised by George Samaras, Computer Science Department, University of Cyprus, June 2002

Georgiadis T.D., "*Mobile Agents in the Service of Telemedicine.*", Undergraduate thesis supervised by George Samaras, Computer Science Department, University of Cyprus, June 2000.

Gray, J., and Reuter A., *Transaction Processing: Concepts and Techniques*. Morgan Kaufman Publishers, Inc., 1993.

Greem R., Article: Java Access to SQL Databases. Canadian Mind Products, 1997.

Gruber R., Kaashoek F., Liskov B., Shrira L., Disconnected Operations in the Thor Object-Oriented Database System. *Proc. Mobile Computing Systems and Applications,* IEEE, Los Alamitos, CA, USA, p 51-56, 1995.

Harrison C.G., Chessm D.M., Kershenbaum A., Mobile Agents: are they a good idea? Research Report, IBM Research Division, March 1995.

Housel B.C., Samaras G., Lindquist D.B., WebExpress: A Client/Intercept Based System for Optimizing Web Browsing in a Wireless Environment, *Journal ACM/Baltzer Mobile Networking and Applications (MONET), Special Issue on Mobile Networking on the Internet*. To appear. Also, University of Cyprus, CS-TR 96-18, December 1996.

Imielinksi T., Badrinath B.R., Wireless Mobile Computing: Challenges in Data Management, *Journal Communications of the ACM*, Vol. 37, No. 10, October, 1994.

Ioannidis J., Maguire G.Q. Jr, The Design and Implementation of a Mobile Internetworking Architecture, *Proceedings of the 1993 Winter Usenix*, pp. 491-502, San Diego, California, January, 1993.

Jepson B., Database Connectivity: The Lure of Java, Java Report. Whiley Computer Publishing, 1997.

Jepson B., Java Database Programming. Wiley Computer Publ., 1997.

Joseph A.D., Tauber J.A., Kaashoek M.F., Mobile Computing with the Rover Toolkit, *Journal IEEE Transactions on Computers*, February, 1997.

Katz R.H., Adaptation and Mobility in Wireless Information Systems, *Journal IEEE Personal Communications*, Vol. 1, pp. 6-17, 1994.

Lipnack J., Stamps J., " *VIRTUAL TEAMS: People Working Across Boundaries with Technology"*, Second Edition, Published by John Wiley & Sons, New York 2000.

Mathias C.J., New LAN Gear Snaps Unseen Desktop Chains, Data Communications, volume 23, No. 5, pp.75-80, March, 1994.

Miller K., Cellular Essentials for Wireless Data Transmission, *Data Communications*, Vol. 23, No. 5, pp. 61-67, March, 1994.

Mobile Agents, www.agent.org/ and www.cs.umbc.edu/agents.

Narayanaswamy S., Seshan S. et. al, Application and Network Support for InfoPad, *Journal IEEE Personal Communications Magazine*, March, 1996.

Noble B. D., Price M., Satyanarayanan M., A Programming Interface for Application-Aware Adaptation in Mobile Computing, *Journal Computing Systems*, Winter, Vol. 8, No. 4, 1996.

Oracle, Oracle Mobile Agents Technical Product Summary, www.oracle.com/products/ networking/mobile/agents/html/, June 1997.

Panayiotou C., Samaras G., Pitoura E. and Evripidou P., "Parallel Web Computing Using Java Threads and Java Mobile Agents", submitted for publication. Also technical report TR-99-3, University of Cyprus, February 1999.

Panteli N., and Dibben M.R., (2001), Reflections on Mobile communication systems, Futures, 33/5, March, 379-391

Papastavrou S., Samaras G., Pitoura E., *"Mobile Agents for WWW Distributed Database Access"*, Proc. 15th International Data Engineering Conference, Sydney, Australia, March 1999.

Peonides N., Ove Arup and Partners, *Wireless Communications*, AIA NY Information Technology Seminar, 12th September 2000.

Pitoura E., Samaras G., *"Data Management for Mobile Computing"*, Kluwer Academic Publishers, ISBN 0-7923-8053-3, 1998.

Pitsillides A., Pattichis C., Pitsillides B., Kioupi S., "Tele-Homenursing: A cooperative model for patient care in the home", Comprehensive Cancer Care: Focus on cancer pain, Limassol, Cyprus, 28-31 May 1997, pp 48. (summary proceedings).

Pitsillides A., Samaras G., Dikaiakos M., Olympios K., Christodoulou E., *"DITIS, Collaborative Virtual Medical team for home healthcare of cancer patients"*, Re-engineering Cyprus for the digital age: Tele-Medicine, Nicosia, Cyprus, 17-19 Dec. 1999.

Radmanabhan V.N., Mogul J.C., Improving HTTP Latency, *Journal of Computer Networks and ISDN Systems*, Vol. 28, No. 1, December 1995.

Reiher P., Popek J., Gunter M., Salomone J., Ratner D., Peer-to-Peer Reconciliation Based Replication for Mobile Computers, *Proceedings of the European Conference on Object Oriented Programming 2nd Workshop on Mobility and Replication*, June 1996.

Samaras G., Pitsillides A., Client/Intercept: a Computational Model for Wireless Environments, *Proceedings of the 4th International Conference on Telecommunications (ICT'97)*, Melbourne, Australia, April 1997.

Sarker S., Lau F., Sahay S., *"Building an Inductive Theory of Collaboration in Virtual Teams: An Adapted Grouped Theory Approach"*, Proceedings of the 33rd Hawaii International Conference on System Sciences – 2000

Sarker S., Lau F., Sahay S., *"Building an Inductive Theory of Collaboration in Virtual Teams: An Adapted Grouped Theory Approach"*, Proceedings of the 33rd Hawaii International Conference on System Sciences – 2000.

Satyanarayanan M., Fundamental Challenges in Mobile Computing, *Proceedings of the 15th ACM Symposium on Principles of Distributed Computing*, Philadelphia, PA, May, 1996.

Satyanarayanan M., Kistler M., Mummert L., Ebling M.R., Kumar P., Lu Q., Experience with Disconnected Operations in a mobile Computing Environment. *Proc. 1993 Usenix Symposium on Mobile and Location Independent Computing*, CA, November, pp 11-28, Cambridge, MA, August 1993.

Satyanarayanan M., Mobile Information Access, *Journal IEEE Personal Communications*, Vol. 3, No. 1, February 1996.

Schilit B.N., Adams N., Gold R., Tso M., Want R., The ParcTab Mobile Computing System, *Proceedings of the 4th IEEE Workshop on Workstation Operating Ssytems (WWOS-IV)*, pp. 34-39, October, 1993.

Series of articles, *"Individual to collaborative cognition: a paradigm shift?"*, Artificial Intelligence in Medicine, (12), Elsevier, 1998.

Shepherd M., Zitner D., Watters C., "Medical Portals: Web-Based Access to Medical Information", Proceedings of the 33rd Hawaii International Conference on System Sciences – 2000

Spyrou C., Samaras G., "Mobile Agents to Support Views for Wireless Clients", Technical Report TR-99-5, University of Cyprus, February 1999.

Tennenhouse D.L., Smith J.M., Sincoskie W.D., Minden G.J., A Survey of Active Network Research, *Journal IEEE Communication Magazine*, Vol. 35, No. 1, pp. 80--86, January, 1996.

Tiwana A., "From Intuition to Institution: Supporting Collaborative Diagnoses in Telemedicine Teams", Proceedings of the 33rd Hawaii International Conference on System Sciences – 2000

Tyson J., *How Wireless Networking Works*, (Howstuffworks, Inc 2002); http://www.howstuffworks.com

Weissman Lauzac S., Chrysanthis P.K.. ``Programming Views for Mobile Database Clients." Proceedings of the 9th DEXA Conference and Workshop on Database and Expert Systems Applications: Mobility in Databases and Distributed Systems, Vienna, Austria, Aug. 1998.

White J.E., Mobile Agents, General Magic White Paper, www.genmagic.com/agents, 1996.

Wolfson O., Sistal P., Dao S., Narayanan K., Raj R., View Maintenance in Mobile Computing, Sigmod Records, 1995.

Zenel B. and Duchamp D., General Purpose Proxies: Solved and Unsolved Problems, Proceedings of the Hot-OS VI, 1997.

Zenel B., Duchamp D., Intelligent Communication Filtering for Limited Bandwidth Environments, Proceedings of the 5th IEEE Workshop on Hot Topics in Operating Systems (HOT-OS V), Rosario WA, May 1995.

II. SMART MOBILE APPLICATIONS FOR HEALTH PROFESSIONALS

Andreas Lymberis

Section Editor

SECTION OVERVIEW

Andreas Lymberis[*]

The ongoing evolution of wireless technology and personal mobile devices capabilities is changing gradually the way healthcare practitioners perform their jobs and the way care is delivered. The growth and acceptance of mobile information technology at the point of care, coupled with the promise and convenience of data on demand, creates opportunities for enhanced prevention and patient care and safety.

The international Information and Communication Technologies (ICT) community is devoting, during the last years, a great effort to drill down from the vision and the promises of wireless and mobile technologies and provide practical application solutions. Research and development activities focuses on wireless networking in medical settings, including advanced technologies for data gathering, omni-directional transfer of vital information and broadband applications. A significant effort is also put on the development, of powerful and user-friendly handheld computers and of applications that integrate handheld devices such as Personal Digital Assistants (PDA), into existing systems for access to a myriad of remotely based health information systems and interactive data/order entry, encompassing legacy-based information systems. For example, the U.S. Army Telemedicine & Advanced Technology Research Center (TATRC), USAMRMC, is investigating the utilisation of wireless medical digital assistants in department of defence settings (MDAs, handheld computing devices used in point-of-care medical applications). The TATRC Wireless Medical Enterprise Working Group (WMEWG) was formed to facilitate management of this research program and to provide intra- and interoffice co-ordination of these R&D efforts. Another example is the large promotion of the education and use of PDAs in healthcare and medical schools in Japan. The development of Smart Mobile & Wireless networks, systems and applications in health has being strongly supported also in Europe through private and public funds and in particular through the research and technology development programmes of the European Commission.

---

[*]Andreas Lymberis, European Commission, Directorate General Information Society, Av. De Beaulieu 31, 1160 Brussels.

Currently a growing number of health professional and particularly physicians are making handheld computers part of their daily life. Common applications of handheld devices range from clinical to administrative tasks such as, point of care assistance (e.g. drug information, clinical guidelines and decision aids), patient information (e.g. clinical results), administrative functions (e.g. electronic prescription and scheduling) and medical education. Those who are using them in clinical practice feel that they could make safer decisions and improve productivity and interactions with patients.

Those we adopted the "wait and see" approach were either not familiarised with modern technology or had concerns about physical constraints or about matching with personal preferences. Concerns exist also with the security of patients' information, the interoperability and the reliability of devices and mobile applications as a whole. It is a fact that many mobile health applications are still not fully validated and proven benefits today and most is unable to integrate well with existing information systems. Most vendors of mobile applications are in the earliest stages of rolling out their products. Training of health professionals and support from physician group practices, hospitals and other health care organizations, including managed care and public health organizations, could improve acceptance of handheld devices and mobile applications and be advantageous for all users and stakeholders.

Within the Information Society Technologies (IST) programme of the 5th Framework Programme of the European Commission (1999-2002), projects in the area of "applications relating to health" aimed at promoting the "ambient intelligence vision" through the R&D of mobile & wireless systems and applications, enabling doctors and paramedics to access remotely available best medical practices and patients' medical files and provide healthcare services through embedded decision support. The work is being mainly focused on the integration of intelligent interfaces e.g. voice recognition and natural language understanding, medical S/W as well as mobile multimedia workstations exploiting the potential offered by new generations of mobile communications i.e. GPRS and UMTS. To improve the performance of the individual projects and increase the overall impact for the sector, the projects have been clustered in order to maximise the possibilities for interaction with other projects and stakeholders. Clustering activities include support of common interest topics e.g. identification of industrial/technical problems and solutions, fostering standardisation and interoperability, benchmarking and best practices, assessment and technical validation, user awareness and identification of future R&D.

This section includes some of the success stories of European research and development in mobile health systems and applications for health professionals. These projects resulted in enhanced secure and user friendly medical mobile devices and applications for, accessing and processing patient data (MOBI-DEV, DOCMEM), supporting day-by-day activities of medical staff in hospital through collaboration and wireless access to patient records, (WARD in HAND), providing computerised medical decision support (SMARTIE), and delivering mobile technologies for homecare and care at the point of need (MOBIHEALTH, MOEBIUS, DITIS). In addition, advanced developments is human computer interaction through natural medical language understanding are presented. The project MEMO has been granted by the EC to support clustering activities in the area of mobile systems and applications for health professionals in Europe, such as, dissemination and promotion of mobile medical devices, technology watch, interoperability and requirements for business models.

The papers of this section confirm that great opportunities for mobile and hand-held technology are there, but enormous challenges, in respect with personal issues and the devices themselves, are there, too. The results of the projects will hopefully have an impact on patient and health professional mobility, as well as on healthcare developments in the European Union as expressed in a report delivered by the High Level Group in December 2003.

THE MEMO PROJECT –
AN ACCOMPANYING MEASURE FOR MEDICAL
MOBILE DEVICES

Clive Tristram[*]

1. INTRODUCTION

The MEMO project is a project funded by the European commission, as an accompanying measure to existing medical Mobile devices projects. In this chapter we describe the MEMO project, give short overviews of each of the associated projects and draw conclusions from them and from the research being undertaken by the project as to the existing state of the market for Medical Mobile devices and potential avenues for future development.

Central to the MEMO project is the definition of a Medical Mobile Device (MMD). The definition used is:

1. A handheld computer (frequently referred to as a personal data system), which is used to extend the reach of existing and future IT infrastructures in the medical field or to provide stand-alone support directly to healthcare professionals in undertaking their professional tasks.
2. The devices will normally have communications capabilities using any of the existing or emergent communication technologies.
3. The device may be linked to other measuring devices such glucose meters or Blood Pressure monitors.
4. The device may or may not include native capability to support voice and data in the healthcare applications
5. The definition is restricted to:
   a. applications for healthcare professionals
   b. the use of standard interfaces to applications and communications in the healthcare domain
   c. a focus on well-established healthcare platforms and emphasis on cost benefit of solutions

---
[*] Clive Tristram, ETS Tristram Clive, Les Rives, 86460 Availles Limouzine, France

      d. it excludes devices which are not Internet enabled, those concerned with domains external to healthcare and healthcare administration.

2. INTRODUCTION

MEMO has two principal objectives both of which are supported by its designated activities. These activities are divided into work packages which will achieve the following results:
- To support the rapid uptake of mobile devices in healthcare
- To actively encourage European projects to build on the existing work with medical mobile devices

1. A web portal which supports both objectives in that it contains information under the following topics
   a. Current news of developments in the technology, marketing and use of MMDs
   b. Interviews with users and vendors of MMD solutions
   c. Calendar of events associated with MMDs
   d. Information about the activities of the associated projects
   e. Educational information to support better understanding of the capabilities of the technology incorporated into MMDs and for planning their purchase and deployment
   f. Reviews of publications dedicated to MMDs
   g. Discussion forums to allow the interchange of news between those interested in or involved in MMDs
   h. An observatory of the activities sponsored by funding agencies in Europe.
   i. Links to sites of interest to those involved in MMDs
2. A monthly electronic magazine is to be published which summarises the previous month's news and provides an editorial understanding half the implications of this news together with an analysis of the current research undertaken by the project.
3. The publication of a roadmap of technical developments to support those creating MMD use and those intending to make use of them in healthcare.
4. The publication of a report which will indicate which structures and activities are necessary for governments and funding organisations to support the effective implementation of MMD use in healthcare and to realise their benefits.
5. Publish a report on interoperability issues between the partners' products, together with recommendations for resolving them.

6. The publication of a validated evaluation method for ascertaining the impact of MMD use on healthcare practice.

It is intended that all of these results will be presented Conferences and in articles for magazines so that's the widest possible number of organisations and people can gain advantage from the work of the MEMO consortium.

In addition the MEMO consortium will make considerable efforts to expand the number of partners involved within it so that the value of the interchange between the partners can be increased and the power of the European organisations involved in MMD is enhanced.

The interaction between the project components are shown below:

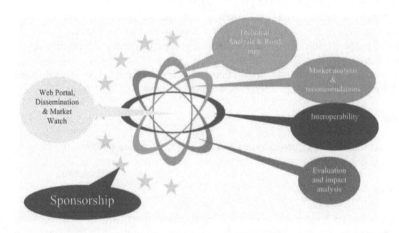

3. CURRENT STATUS

3.1 The web portal

At the time of writing, the web portal has been implemented (http://www.med-mobile.org/) using open-source software[*] and some specially written packages. It contains nearly all the output of the project, including

1. 100s of news articles, references to events and interviews which are syndicated to other web portals;
2. Over 150 web links to sites having information about medical mobile devices. These include sites in French, German, English and Spanish.
3. Many papers and presentations are stored and can be downloaded directly.
4. A report in HTML form on the technical aspects of MMDs
5. A set of tutorials designed to assist organisations and individuals understand and implement medical mobile devices. These cover the following topics

---

[*] Open source software used includes PHPNuke, MySQL, PHP, and specially written software made available to the open source community.

     a. Speech recognition architecture
     b. Natural Language Understanding
     c. Mobile devices and web services
6. An observatory in which European projects using handheld devices are summarized together with links if relevant to the base information. The countries currently covered are:
     a. Italy
     b. France
     c. United Kingdom
     d. Germany
     e. Spain
7. A set of interviews with significant players in the world of MMDs
8. A questionnaire to collect market information from end users and suppliers

Further additions are planned to achieve the project objectives. These include automatic collection of material for the observatory and discussion groups to enable information to be exchanged on a pan-European basis.

The only failure so far has been with the planned electronic magazine which has been developed but its transmission ran into problems over SPAM regulations

3.2 Dissemination

The work of dissemination has been pursued actively by the various partners resulting in presentations at a number of international conferences.

3.3 Market Analysis

The analysis of the market and its constraints has been derived from structured interviews, web research, questionnaires, a conference on handheld devices and literature. The conclusion of this work is that the market has significant potential but that this is currently, largely unrealised outside that of using MMDs for agendas and relatively simple tick type record keeping. It may be that the integration with voice on SMARTPHONEs will more successful.

3.4 Interoperability

The work being undertaken to examine the interoperability issues between the partners' systems is ongoing at the time of writing this report using the ISO 7 layer model as the framework for analysis. Problems have been discovered in many aspects of the applications, from the user interface, to the messaging standards selected and the method of interfacing to existing hospital and other systems. It will result in recommendation for several standards.

3.5 Evaluation method

The project has produced an evaluation method for mobile devices. This has been published and can be found in the downloads section of the web portal. The method is based on the evaluation of the device and applications in the healthcare domain defined by the supplier. The evaluation is holistic and includes analysis of technical, clinical, administrative and economic impact. The results will be expressed descriptively as quantitative analysis is not possible at this early stage of market development. In the latter part of 2003 and early 2004 it will be validated on the associated projects.   Results will be published on the portal (http://www.med-mobile.org)

4   CONCLUSIONS OF THE MEMO PROJECT

4.1 State of the market

The market is at a very early-stage of its evolution. The evidence for this may be found in the large number of funded projects against the small number of actual implementations.

The majority of actual implementations are to be found in the United States and to large extent these are driven by medical schools, the need for accurate prescribing and better use of time. The main implementations in Europe are for routine work lists (or agendas) and simple data collection not involving large amounts of data entry.

There are very few "mass market" examples in Europe of complete MMDs integrated into other healthcare systems.

This underpins the need for an effective business model which will support investment decisions by the potential purchasers and to assist policy makers in establishing guidelines by funding bodies to ensure that the potential of MMDs are realised.

4.2 Technical conclusions

There remain to be solved many technical constraints; these include screen size, resolution, power and battery life.

The user interface is a major impediment to widespread use of these handheld devices and much hope has been invested in the use of speech recognition. The assembled research of the project shows that, at present, this is best limited to single word or simple phrase recognition used to select from long lists (where the list is bigger than the screen). This comes as a result of the poor quality of the incorporated microphones, the lack of sufficient power to enable continuous speech recognition and the lack of user experience in using speech as a data input medium. Where continuous speech recognition is possible it is included in an architecture in which speech passes from the MMD to a server where speech recognition is performed and if necessary, natural language understanding. This later may be used to ensure that the medical representation is both correct and appropriate to the context. Even with these limitations however there remain considerable opportunities for improved user interfaces using speech recognition.   The single word/simple phrase method is being investigated by the MEMO project as an implementation for SMARTIE.

Security remains a major issue for authentication of the user as well as encryption of transmitted data. The associated projects have investigated the use of smartcards holding encryption codes and contacts have been made with other suppliers to identify the feasibility of adding biometric based smartcards to MMDs, but these are currently very expensive.

Initial investigations into remote communications architectures and technologies indicate that for many cases GPRS will be an adequate substitute for UMTS based devices for the foreseeable future. Local communications are shown to work well with both WiFi and Bluetooth. The difference in usage being dependent upon distance and members of devices connected. This balance between these two technologies is currently in great flux with a swing towards WiFi for MMDs in hospitals and major mass market players introducing WiFi hotspots.

4.3 Justification for the use of MMDs

Initial summary conclusions support the need for this project. They have reinforced the original understanding of the benefits that MMDs can bring to the healthcare market. These are:
1. Bringing the support inherent in the clinical information systems to the patient;
2. Enabling patients to collect data about their condition and transfer this directly for the use of clinicians;
3. Reduction in medical errors by collecting data at the point at which it is generated, i.e. during the patient consultation;
4. Supporting the way in which clinicians operate. Their work is mobile and it is essential that the data is available and collected wherever they are;
5. Assisting in ensuring compliance to clinical guidelines.
6. Making available to clinicians evidence based data, obtained from local and remote sources at the time and place at which clinical decisions are made.

4.4 Work remaining

The project team will validate the tentative conclusions summarised above and supplement them when completing all the items described in the introduction. In particular a very large effort will be expended in a ensuring that as many other European projects and European organisations as possible can gain from the cumulative experience of the MEMO consortium.

5. ASSOCIATED PROJECTS

MEMO is an accompanying measure to various existing European funded projects. In this section each of the projects are described in summary form. The projects range from the use stand-alone PDAs to complex integration of PDAs with medical centres, internet and GPRS. In the middle lies the integration of handheld computers with hospital systems for the better treatment of patients within the hospital environment.

Each project is described in greater detail elsewhere in this book so the descriptions focus on the key aspects and the results which further the understanding of the use of MMDs.

5.1 SMARTIE

The rationale behind the SMARTIE project is to bring to a clinician in an easily usable manner tools and calculations published in the extensive medical literature. It does this to help reduce medical error, make available medical knowledge in a manner which is easy and safe to use, this aids the introduction of evidence based medicine and supports the "just-in-time" learning concept so necessary in such an extensive field of knowledge as medicine. It is does this by developing collections of multi-platform medical software applications from peer reviewed articles in the medical literature. These tools can provide clinicians the critical information needed to manage and inform the patient and citizen.

The process of selection and development starts with the identification of the domain and selection of potential tools by clinical experts. This selection is followed extensive research of the medical literature to translate the textual description into a storyboard which is used by clinical experts to validate the work done prior to entry into the technical phase.

The storyboard is converted into an XML file and this becomes the basis for automatic creation of software applications to run on PALM, the Internet, Compaq IPAQ, and PC platforms.

The technical products are then tested in a clinical environment and then posted on the SMARTIE web site for download under an open source license.

Trials in North Wales, Barcelona and Düsseldorf have confirmed the superiority of PDA based calculators and tools for ensuring conformance to guidelines when compared to paper but to gain widespread acceptability there remain issues of cost and user interface.

The project was completed in July 2003. The results can be seen at http://www.ist-smartie.org.

During the course of the project considerable expertise has been gained in translating the knowledge contained in the technical literature of medicine into technical applications on computers. This has proved difficult and challenging but creates the foundation for a new sub industry - translating clinically proven approaches into clinically approved computer applications.

5.2 WARD IN HAND

The objective of WardInHand is to support the day-by-day activities of doctors and nurses within a hospital ward by providing a tool for workgroup collaboration and wireless access to the patient's clinical records. The project enables access to be made to the information system from the patient's bedside through a PDA client. Ward-In-Hand complements Hospital legacy systems by providing access to existing tools, updating data in real time and making this data available to doctors and nurses. By achieving its objectives it adds mobility and ubiquitous computing to currently available Hospital Information Systems.

The project was completed, as planned, in March 2002, and the three pilot installations, in Genoa (Italy), Barcelona (Spain) and Offenbach (Germany) have been validated and rated by the respective users after about three months of trials. The specialisations were endocrinology, pneumonology and cardiology.

Feedback from the users has been favourable in terms of general system characteristics, functionality, and usability.

In one case (Genoa) a very significant saving/improvement was achieved.

Trials at the pilot hospitals have demonstrated that implementation of WardInHand, provides the maximum benefits when internal operational systems are redefined to take advantage of the new technology. If this is done productivity can be increased by up to 40 percent.

5.3 MOBIDEV

The two key issues confronting the MobiDev team were how to improve the mobility of users in their use of existing systems and to improve their ease of use by professionals.

These two objectives were achieved by building an integrated platform for Healthcare professionals which comprises PDA clients and a traditional server architecture. The heart of the project is the integration with the existing Hospital Information Systems. Security is provided with digital signatures and encrypted transmissions.

The ease of the use issue was attacked on two fronts firstly through the development of a special user interface on the Palm handheld computers and secondly through the use of natural language understanding on the server of text dictated into the handheld computers.

Mobile communications are provided by using Bluetooth within the hospital and UMTS externally. Synchronisation is achieved over TCP/IP and extensive use is made of Intranet and Extranet technologies.

The project was completed in July of 2003. The PC and Palm applications have been completed and the majority of the E-signature applications delivered. Natural language input is operational for prescribing and admitting patients in Spanish, English and Italian. The integration with existing hospital systems and the validation phase has been completed and the final application is being piloted in four countries.

5.4 MOEBIUS

This project was conceived with the intention of providing support to patients regardless of location and clinically validating the results. It provides patients with diagnostic tools and equipment from which data can be transmitted to a data triage algorithm. The data is then passed to a medical contact centre which acts as a communications hub and nurse centre. The data is then the made available to the patient's doctor.

The main results come from the obese risk reduction study which started in September 2001. This showed good cooperation between the involved Healthcare providing teams and good compliance by the patients. The patients transmitted bio-data daily using the handheld devices with which they were provided. There was a high acceptance of system by the patients. Three hours per patient was required for teaching and initial support.

The clinical objectives for the risk reduction trial under the headings of mortality, Blood Pressure reduction, angina symptom reduction, reduced lipids and decreased possibility of developing diabetes were only partially tested during the timescale of the project, but extensions will produce these results in the coming months.

The results achieved particularly the cooperation between the Healthcare providing teams and the patients have encouraged the project team to progress and expand the trials.

Technically the equipment worked very well throughout. The only change was that GPRS was used in place of UMTS which was not available. The functionality of GPRS was more than adequate for the trials

5.5 DITIS

The objective was to create an environment in which a patient-centric virtual healthcare team is established and maintained for domiciliary and hospital care of cancer patients. The original plans were to accomplish this by developing a flexible communications and data architecture accessible by the healthcare team regardless of location.

An internet based architecture was created in which all data was stored on an SQL server accessed by JDBC. The user interface was dependent on the user device and this could be HTML to a PC or PDA device or WAP to a mobile phone.

The focus of the work was the care of cancer patients in their homes and all interactions with the patient were recorded from what-ever user device was available to the central database and thus made available to all other team members.

The project has received additional funding from several sources to expand its work outside Cyprus. This it is doing under a phased developed plan in conjunction with its commercial sponsors.

MEDICAL NATURAL LANGUAGE PROCESSING ENHANCING DRUG ORDERING AND CODING

Mariana Casella Dos Santos[*], Frank Montyne, and Christoffel Dhaen

1. INTRODUCTION

The foundation of Natural Language Processing (NLP) did not surface until the late seventies, when it was discovered that semantic patterns in natural language could be expressed in mathematical formulas. NLP is the technology that allows computers to analyze and generate natural languages, such as English and Dutch. Applications of NLP include search and retrieval, information extraction, data warehousing amongst others.

One of the tasks that an NLP system can perform is the parsing of a sentence to determine its syntax. Determining the meaning (semantics) of a sentence is a much harder task to perform. When an NLP system can achieve this, we talk about Natural Language Understanding (NLU).

"Understanding" language means, amongst other things, recognizing which entities in a given domain of reality a word or phrase stands for, and knowing how to link these entities together in a meaningful way. An approach for a computer to understand natural language, is to create an environment that is a formal (i.e. unique, commonly accepted) representation of a specific domain, which is called an ontology. An ontology consists on its basic level of a group of concepts, representing entities in a specific domain of reality,organized in a network of relationships. To put it another way, an ontology represents knowledge about the world, in the case discussed in this paper the medical world, in a form a computer can understand.

---

[*] Mariana Casella Dos Santos, Language and Computing; Maaltecenter Blok A, Derbystraat 79, Sint-Denijs-Westrem, Belgium

2. ONTOLOGY MANAGEMENT SYSTEMS

2.1. Introduction

To create, manage and integrate an ontology in an NLP application, one needs a terminology and ontology management system (OMS).

Since these are currently not available for managing large formal ontologies at minimum processing times, researchers and industries have to build and maintain their own environment. Management systems for smaller ontologies have been developed, e.g. ODE (Blázquez et al., 1998), WebOnto (Dominguez et al., 1998), Ontolingua (Farquhar et al., 1996), HoZo (Kozaki et al., 2000), JOE (Mahalingam et al., 1997), Protégé (Noy et al., 2000), OntoSaurus (Preece et al., 2001), but none of these are capable to deal with enormous and complex ontologies.

2.2. LinKFactory®

To resolve the aforementioned problems (initially for the medical environment) Language and Computing (L&C, 1998) created the OMS LinKFactory® (Ceusters et al., 2001) implementing a knowledge representation and compatible reasoning mechanism in a database structure.

2.2.1. General Description

LinKFactory® is especially designed for large scale, complex, multilingual, formal language-independent ontologies. "Language-independent" has to be understood in terms of independency from any specific language - such as English, French and Dutch - but not from language as a medium of communication.

The fact that the ontologies are language-independent has some major consequences on the type of applications that can run on top of them. It will, for example, be much easier to search for relevant information on the web (or a thesaurus): the search can be done in one language in free text. This free text search will be linked to language-independent concepts (based on the semantics) that will be the basis for the information retrieval. Since terms in several languages are attached to the concepts using a linguistic ontology (Ceusters et al., 1997), also relevant info in other languages can be retrieved, while semantically irrelevant information will not appear in the list of results.

2.2.2 System Architecture of LinKFactory®

LinKFactory® stores the data in a relational database (currently Oracle is used). Access to the database is abstracted away by a set of functions that are "natural" when dealing with ontologies, e.g. get_children, find_path, join_concepts, get_terms_for_concept.

The relational database is accessed by the OMS at server-side through a standardized application interface, which makes the OMS database independent as long as this API is maintained (Oracle, Sybase, SQL Server have been tested).

The OMS at server-side, which was written in platform independent Java code (Windows, Solaris, Unix and Linux tested), is also accessible through a standardized API and this allows developers to build applications on top of the semantic database without requiring intimate knowledge of its internal structure.

We finally settled on RMI (Remote Method Invocation) as our technology of choice because of its simplicity and proven robustness. This means that our server-side component is a Java Application that extends java.rmi.Remote. The application requires an RMI registry (a sort of Domain Name Server for RMI servers) to be running in order for it to be able to register itself and for clients to be able to connect to the RMI server.

The LinKFactory® system consists of 2 major components (Fig. 1), the LinKFactory® Server, and the LinKFactory® Workbench (client-side component). The LinKFactory® Workbench allows the user to browse and model the LinKBase® data.

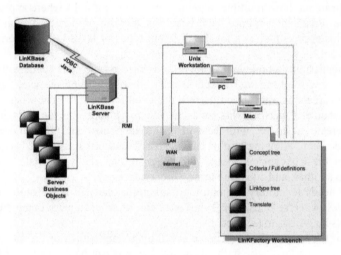

Figure 1. LinKFactory® components

The workbench is a dynamic framework for the LinKFactory® Beans. Each bean has its own specific functionality and limited view to the underlying formal ontology, but combining a set of beans in the workbench can provide the user with a powerful tool to view an manage the data stored in the semantic database. The workbench provides the user with an optimal flexibility to create a customized tool to view and manage the data in the ontology.

Different views on the semantic network are implemented as Java beans. Examples are: Concept Tree, Concept Criteria and Full Definitions, LinkType Tree, Criteria List, Term List, Search Pane, Properties Panel and Reverse Relations. The LinKFactory® framework is implemented in 100% pure Java code. The modular design is done using Java beans organized and linked in a freely configurable workspace. The beans themselves can also be used outside this workspace, so software-developers can integrate them as (static) components into their programs.

When modelling the ontology, several quality assurance mechanisms are built in: versioning, user tracking, user hierarchies, formal sanctioning with possibility to overrule, sibling-detection, link type hierarchy, etc.

2.2.3 Dynamic Layouts in LinKFactory® Workbench

Each user can create multiple views on the semantic network using the beans available. These views are called 'layouts'. Each layout can consist of several frames on which the beans can be laid out.

Creating a new layout or adding new frames to an existing layout are all simple actions the user can select from the menu.

Each frame can be divided into several bean-spaces where beans can be placed. The user can select any of the beans available, and simply drag and drop them into the desired area of the workspace. Once the user has placed the desired beans in the layout, he can create links between the several beans, again, simply by means of drag and drop.

Each bean can have multiple incoming and outgoing links where appropriate. Beans can also be linked inter-frame. Each bean has specific properties, which can be set at runtime. This approach allows for the different types of tasks to be performed using the optimal layout.

The functionality of a layout is not only defined by the used beans within the layout, but also by how the beans are linked to each other, providing the user with an optimal flexibility to create a customized interface designed for the task(s) he wishes to perform.

For instance, a person who needs to add more terms to the knowledgebase will probably create a layout with several Translate-beans, and link those beans to a ConceptTree bean, allowing him to view the different terms associated with the selected concepts in the ConceptTree bean. This way, he can easily translate terms from one language to another, and add more terms where necessary.

If one needs to modify the ontology structure, one will make more use of several FullDefTrees, ReverseTrees, LinkTypeTrees, etc, in order to get a complete overview of the concepts he wishes to (re)model.

Each of the layouts defined can be saved by a simple click of the mouse, which will store the layout as Java code in the database. Layouts can be defined on different levels: Organization, Group and User. If the user wants to reuse any of saved layouts, he only needs to select the desired layout from the menu.

2.2.4 Specifications of the Available Beans

2.2.4a. General. The different beans provide information on and a view of different parts of the ontologies built. All of these beans can be linked to each other. A bar on top of each bean shows the other beans the bean has been linked to and also the direction of the link. Other items in the bean bar are the bean label, the button to display/edit the bean properties and the possibility to refresh the bean contents. Optional items (dependent on the kind of bean) are a shortcut to the link type filter property and a dragable item possibility.

2.2.4b. Most important beans. The ConceptTree bean (Fig. 2) provides the user with a view to the hierarchical relations in the semantic network of concepts. As concepts can have multiple parents (network structure) and the representation is a tree-view, the network structure is split up into the matching tree representations. Modifications to the structure can be made by means of drag and drop. The functionalities of the ConceptTree bean include search by knowledge name, search by terms, modify hierarchy and history of searches. The bean properties provide a way to specify the number of siblings to display, the font, the child depth, the number of children to display, the preferred language, the parent depth and the leaf-node child depth.

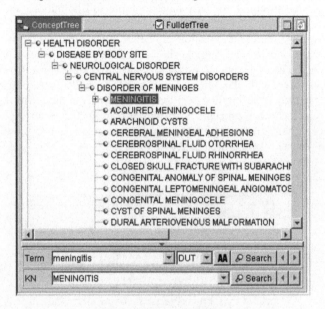

Figure 2. ConceptTree bean

A second important bean is the Full Definition bean (Fig. 3). This bean shows the user the hierarchical and non-hierarchical relations a concept has with other concepts. These relations are sub classed in the relations explicitly specified for this concept (beneath the node labeled CRITERIA), and the implicit relations (beneath the node labeled INHERITED CRITERIA). It also shows the full definitions for a concept, i.e. the sufficient criteria to uniquely identify this concept. Explicit relations and full definitions can be added, removed or modified by drag and drop.

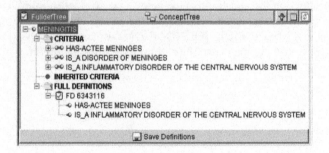

Figure 3. Full Definition bean

We introduced the notion of concept-definition and concept-criteria, which allows us to group a number of concept-criteria (essentially relationships) to form a full definition. In this way a concept could not only have multiple full definitions and loose concept-criteria, but also the definitions could overlap.

Concept-criteria represent a relationship between two concepts by use of a link type.

The ReverseConcept bean (Fig. 4) shows the relations other concepts have with the selected concept. The node labeled Reverse ConceptCriteria shows the explicit relations other concepts have with the selected concept. The node labeled Inherited Reverse ConceptCriteria shows the implicit relations other concepts have with this concept, i.e. the explicit relations other concepts have with a concept that is an explicit child of the selected concept, hence the concepts have an implicit relation with the featured concept. The inherited reverse relations are not shown by default.

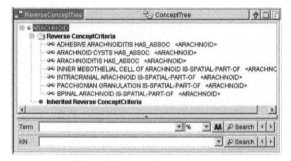

Figure 4. Reverse Concept bean

The LinkType Tree bean (Fig. 5) provides the user with a view to the hierarchical relations in the semantic network of link types. As link types can have multiple parents (network structure) and the representation is a tree-view, the network structure is split up into the matching tree representation. Modifications to the structure can be made by means of drag and drop. Link types were deemed to have a hierarchy just like concepts so we added the LINKTYPE_TREE to represent this; this simple construct suffices because there is only a hierarchical parent-child relationship between link types. This hierarchy will have an effect on the constraints (see below) because when a link type is used in a concept-criterion it automatically implies all the parent-link types are used. E.g. when there is a link HAS-BONAFIDE-BOUNDARY and it has HAS-BOUNDARY as parent then that parent is also implied when the child is used in a relationship.

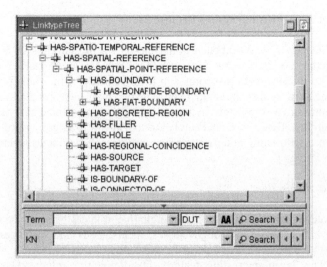

Figure 5. LinkType Tree bean

The Translate bean (Fig. 6) shows the list of terms related to the selected concept, the selected link type or the selected criterion in a certain language. Terms can be added, modified or removed. Several Translate beans can be viewed simultaneously giving terms in different languages, all linked to the language-independent concept.

Other available beans include Concept Properties bean, LinkType Properties bean, Criteria bean, Bookmark bean and others. All of these beans can be selected by the user and linked to each other, as such creating a powerful environment for browsing and editing large ontologies.

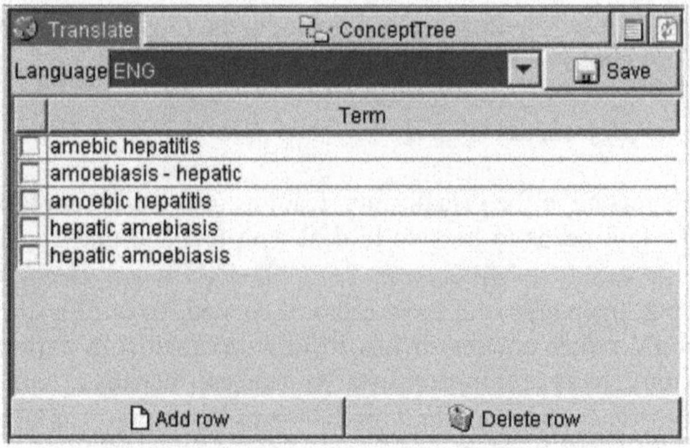

Figure 6. Translate bean

2.3. Comparing LinKFactory® to Protégé 2000

The LinKFactory® Workbench is not the only tool that aims to facilitate working with knowledge bases; there are a number of other tools on the market, one of which is Protégé 2000. Because both programs aim to solve some of the same problems it might be interesting to make a small comparison. The two tools have certain things in common: they are both written completely in Java, they both support some sort of mechanism for easy extensibility (Beans in LinKFactory® and Tabs in Protégé) and of course they are both tools for constructing and maintaining ontologies. Nevertheless there are substantial differences.

2.3.1. GUI customisation

The Protégé tool offers the user one single (hierarchical) view of the knowledge base and allows him to browse through classes, meta-classes and instances one at a time presenting the user with automatically generated forms based on a specific class definition. This is an interesting feature considering the arduous task of knowledge acquisition but LinKFactory® goes a step further. Because of the highly customizable interface, users are able to construct user interfaces that offer multiple views on the same knowledge base. Several ConceptTree beans can be combined on one form to allow the user to keep track of different areas of the knowledge base without requiring a lot of clicking and browsing. The idea is that there are different kinds of users that have different needs; translators will want to deal differently with the knowledge base than modellers and LinKFactory® accommodates that.

2.3.2. Multi-user

LinKFactory® was conceived from the ground up to be a multi-user tool that can deal with a highly concurrent environment thanks to its three-tier architecture. Multiple users can work on the same data without fear of corruption or interference. Protégé was developed for single-user environments and is now moving into the multi-user world but still has many limitations and does not allow concurrent editing.

2.3.3. Large knowledge bases

This issue is related to the visualisation features of the different tools: when data sets start to grow the tool needs to be able to deal with this and not overwhelm the user. In LinKFactory®, a variety of filters can be applied on a per Bean basis, limiting for example the depth, the height and the breadth of concept-trees. It is also possible to filter out relations you'd rather not see or just to limit yourself to a particular relation-type. This visualisation power combined with a database backend that easily scales for knowledge bases upwards of several million entities makes LinKFactory® ideal for handling large data sets. Protégé in contrast suffers from both a lack of good filtering (which makes the interface extremely hard to navigate in large datasets) and from the architectural issue that all the data is stored in a custom file-format on the client machine. As a result, client-side memory and resource requirements explode as the size of your data set grows.

2.3.4. Conclusion

While both tools have their strengths we feel that LinKFactory® is better equipped to deal with more demanding environments where collaborative authoring/multi-user authoring, scalability and advanced visualisation are a must.

2.4. LinKBase®: a large formal ontology built with LinKFactory®

Since the initial focus of L&C was the medical world, we started to construct a formal representation of that world by means of LinKFactory®. LinKBase® is a large multilingual medico-linguistic knowledge base covering most parts of healthcare. It contains **2.2 million medical and general language concepts, 653 different relationtypes that correspond to linguistic functional roles, 5.8 million links and 2.8 million** terms. This makes this ontology currently the biggest in its field, showing the ontology size LinKFactory® can manage.

Terms can be stored in different languages and are linked to language-independent concepts, criteria and link types with an intersection table, allowing us to define both homonyms (1 term that has several different meanings or linked concepts/criteria/link types) and synonyms (multiple terms associated with 1 concept/criteria/link type).

LinKBase® has an IS_A hierarchy without loops. This is called a 'directed acyclic graph'. It means that no concept can be a child (of a child of a child…) of itself. In this hierarchy, it is not presumed that the children of one parent are mutually exclusive. In some cases, they are, what can be made explicit by using the DISJOINT link. In most cases, they are not.

A number of medical coding/classification systems are crossmapped via the ontology, e.g. ICD-9-CM, ICD-10, SNOMED-RT/CT, OPCS4, ICPC, UMLS and MedDRA. Closely related with the mentioned intersection table (TERM_CCL) is the SOURCE and SOURCE_OBJECT construction. SOURCE is a table that stores a number of medical coding/classification systems that classify medical concepts according to their own hierarchy and are used throughout the medical world. By using SOURCE_OBJECT we can link TERM_CCL records with certain sources and assign a code to that combination. We now have the possibility to translate from existing formal medical hierarchies to our own conceptual structure and back, which is a very powerful feature.

Reasoning in LinKBase® is based on extended description logics, while a more axiomatic approach is being implemented.

2.5. Future Directions

LinKFactory is fully compatible with OWL DL. LinKFactory now uses a subset of the expressiveness of OWL DL to perform real-time reasoning to aid the modeler. Future development is targeted towards an even more intelligent way of managing the ontology and using, if possible, the full expressiveness of OWL DL to perform real-time reasoning during the modeling-process.

Another future goal is to integrate unsupervised learning capacities into the LinkFactory®. Using maximum-entropy models on large amounts of free texts, we are currently able to infer head-modifier relationships automatically from huge text corpora.

The goal is now to find out how the head-modifier relationships can be "named" by using information from the LinkBase®. This possibly will lead to an optimal collaboration amongst statistical and symbolic methods.

3. STATE OF THE ART IN MEDICAL ONTOLOGIES

3.1. Introduction

In this part, the most important medical ontologies/terminologies will be described and compared, highlighting strengths, weaknesses and evolution. We realize we don't deal technically with the same kind of systems, but since they all serve more or less the same purpose, we can compare them.

3.2. LinKBase®

This ontology has been discussed under 2.4.

3.3. UMLS®

The Unified Medical Language System® (UMLS®, 2003) is owned by the U.S. National Library of Medicine. It includes the Metathesaurus® and the Semantic Network. Strictly speaking it is not an ontology. The Metathesaurus® consists of a large number of coding/classification systems that have been "linked" together primarily using lexical techniques.   Because this is mainly term-based matching rather than concept-based matching, links are often inaccurate and/or circular. The Semantic Network has 135 semantic types and 54 relationships. The Metathesaurus® concepts are mapped onto the Semantic Network to form an "ontology-like" system. It contains 875,255 concepts and 1,815,280 terms. UMLS® doesn't contain general language concepts (the word "to be" for example). It is primarily English, though it includes classification systems in several other languages. There are a lot of loops in the hierarchy, because of the merge of different systems.

UMLS® has been used as a research tool, but has not been very successful commercially. It is possible that with more work, it could become more complete and accurate, and ontology-like. But this would require investment from the U.S. government and/or the company Apelon (2002).

3.4. GALEN

GALEN (2002) stands for Generalized Architecture for Languages, Encyclopedias and Nomenclatures. It was the outcome of an academic medical ontology project that originated at the University of Manchester in England. It is a medical terminology server for supporting the development of clinical coding schemes. The content of GALEN has been released along with a GRAIL specification as an open source product. GRAIL stands for GALEN Representation And Integration Language for concept modelling to build the GALEN Common Reference Model. GRAIL is a description logic-based system that provides for a fair degree of logical reasoning. But it is not very scalable to large ontologies.

It is a noble idea, but the open source model might be a less good match for a medical ontology at this point in time. Even disregarding eventual implementation issues, there just isn't a critical mass of developers, users, and applications that you would need to make this ontology approach successful.

3.5. SNOMED-CT

This classification system arose out of merger of the U.S. College of American Pathologists' SNOMED-RT with the U.K. NHS's Clinical Terms (SNOMED, 2002). SNOMED-RT stands for Systematized Nomenclature of Medicine, Reference Terminology. SNOMED-CT contains approximately 300,000 medical concepts in a description logic framework. It includes crossmappings to several classification systems, including ICD9-CM and ICD-O. Like UMLS, it has been widely used for research purposes. In the U.K., it is likely that the National Health Service in some form will mandate SNOMED coding of clinical encounters.

SNOMED will continue to be updated, mainly in the form of more complete content and more accurate modelling. Long-term success will depend on either discovering a business model or else obtaining long-term governmental support.

3.6. Other Terminology Systems

There are a number of medical terminology systems in the world, created for a variety of different purposes. These include ICD, ICPC, MedDRA, CPT, Medcin, etc. All of these have some loose hierarchical structure, but none of them can properly be called ontologies.

3.7. Comparison of the Different Systems

Currently, LinKBase® is the largest ontology in the healthcare domain (table 1). The main difference with the other medical ontologies or terminologies is that it is language-independent but that it recognizes language as a medium of communication. It is also the only system with a real semantic representation.

Table 1. Comparison of different medical ontology/terminology systems

Medical ontologies	#[a] Concepts	# Link types	# Links	# Terms	Links to other CC[b]
LinKBase®	2.2 million	623	5.8 million	3.8 million	Yes
UMLS®	875,255[c]	54	[d]	1,815,280	Yes
GALEN[c]					No
SNOMED-CT	300,000	[d]	[d]	[e]	Yes

[a] Number of
[b] Coding/classification systems
[c] 135 semantic types
[d] No numbers found
[e] No recent numbers found

UMLS® contains fewer concepts and terms than LinKBase®. The concepts in the latter are more structured in the semantic network through more link types and links. UMLS® doesn't contain general language concepts, e.g. the word "to be". LinKBase®

does. UMLS incorporates knowledge from different classification systems with different modelling leading to some logical inconsistencies.

The GALEN medical ontology is much smaller than LinKBase®, and contains mainly high-level concepts.

The main conceptual difference from LinKBase® is that SNOMED is intended for use as a reference terminology for information sharing, reporting, etc. This creates a requirement for high stability, for example. The content of LinKBase®, on the other hand, is intended to stay in the background, and serves to support other applications, i.e. NLU, e.g. semantic indexing and information retrieval and extraction. LinKBase® therefore can be continually updated and improved, without serious adverse effects to its users. SNOMED also has a more limited content than LinKBase®, having both fewer medical and non-medical concepts.

3.8. Conclusion

Concerning this state of the art in medical ontologies/terminologies, it can be stated that there is a variety of systems available, but only a few can be called ontologies. Of these, the LinKBase® is the largest, language independent semantic network and it can be easily used as a base for NLU applications.

4. NLU APPLICATIONS FOR MEDICAL MOBILE DEVICES

The aforementioned technology can be applied in NLU tools for Medical Mobile Devices (MMDs). These tools are middleware, i.e. they have to be integrated in an end user application of a third party.

A first example is FreePharma®. This is software plug-in that analyzes drug prescription information expressed in free natural language (written or spoken) and structures it automatically for subsequent integration in host applications. FreePharma® can derive the dose, route, frequency etc. from natural language descriptions and common language descriptions.

Software is available with drug-drug interactions, drug-allergy interactions etc, but all of these databases/applications need structured input. Prescriptions however are made in human language. Hence, most applications use data entry forms combined with pick lists and point and click interfaces to obtain structured data input from the very beginning. This is time-consuming and quite often very impractical and resisted by many doctors.

Figure 7 shows a medication ordering workflow integrated in the Hospital Information System (HIS). A doctor at the patient's bedside can use e.g. a Personal Digital Assistant (PDA) to dictate an order. This is sent to a language technology server for converting the WAV file to digital text, which on its turn is processed to a structured output in an XML message that can be communicated to the pharmacy and the ward, and integrated in e.g. the patient's medication DB of an electronic patient record system (Fig. 8).

Each year medication errors cause hundreds of thousands of patient deaths annually (140,000 deaths per year in the US alone) and cost billions of dollars. The goal of using FreePharma® is to speed up this workflow and make it more accurate, especially in reducing the prescription mistakes, and make drug information available for analysis, internal communication and research purposes. Application areas include Data Entry, including Physician Order Entry (POE) and entry from office notes or discharge summaries, and retrospective analysis for error reduction and cost control.

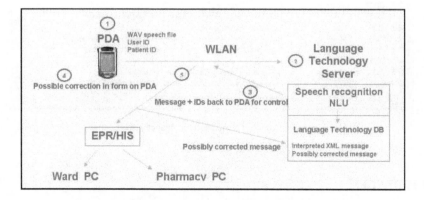

Figure 7. Medication ordering workflow in a hospital environment

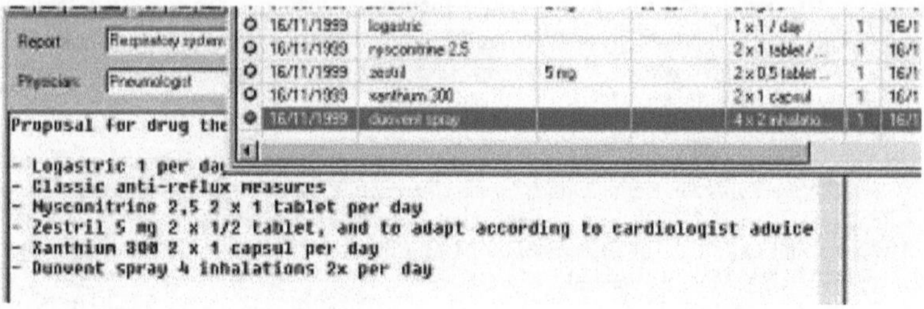

Figure 8. Left: unstructured medication information as in discharge letter; right: output of FreePharma® in medication table in EPR

Another NLU application for MMDs is the semi-automatic coding tool FastCode® (Cools et al., 2002). Medical code tables and international classification systems can contain several thousand codes. Searching for one can be unpleasant and time-consuming. Fastcode® offers a very fast and accurate solution using semantic technology. LinKBase® contains the medical terms linked to most of the existent classification systems (cf. 2.4) on the basis of their conceptual meaning. Fastcode® analyses the meaning of the input words and performs a search based on the related concepts as stored in the ontology. This approach solves problems related to synonyms, homonyms, compound words, orthography and multiple spelling possibilities. Thus the tool transforms narrative expressions (diagnosis, clinical findings, procedures) in natural language into classification system labels and codes, without the user having to be knowledgeable of the labels of the coding/classification system. The output of this coding consists of a set of relevant codes from which the user can choose (Fig. 9).

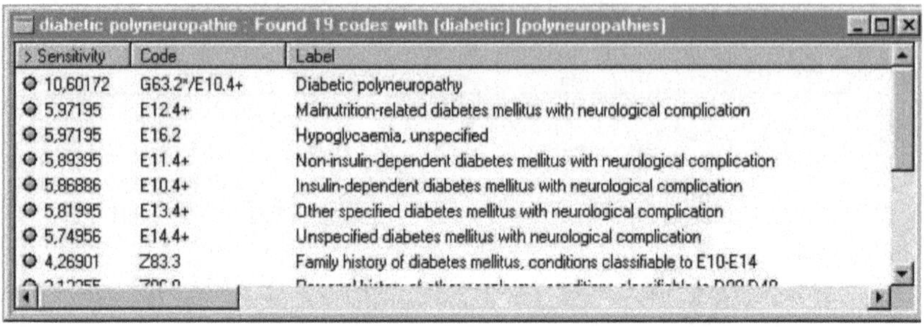

Figure 9. Output of Fastcode® for "diabetic neuropathy"

5. REFERENCES

Apelon, 2002; http://www.apelon.com/about/customers.htm.

Blázquez, M., Fernández, M., García-Pinar, J.M., and Gómez, A., 1998, Building ontologies at the knowledge level using the ontology design environment, in: *Proceedings of the 11th Banff Knowledge Acquisition Workshop*, Banff, Alberta, Canada.

Ceusters, W., Buekens, F., De Moor, G., Waagmeester, A., 1997, The distinction between linguistic and conceptual semantics in medical terminology and its implication for NLP-based knowledge acquisition, in: *Proceedings of IMIA WG6, Conference on Natural Language and Medical Concept Representation*, C. Chute, ed., IMIA WG6, Jacksonville, pp. 71-80.

Ceusters, W., Martens, P., Dhaen, C., Terzic, B., 2001, LinKFactory®: an advanced formal ontology management system, in: *Proceedings of K-CAP - First International Conference on Knowledge Capture - 2001 Workshop: Interactive Tools for Knowledge Capture: Interactive capture of ontological knowledge*, Victoria B.C., Canada (October 20, 2001); http://www.isi.edu/~blythe/kcap-interaction.

Cocchiarella, N.B., 1991, Formal Ontology, in: *Handbook of Metaphysics and Ontology*, H. Burkhardt and B. Smith, ed., Philosophia Verlag, Munich, pp. 640-647.

Cools, H., Lorré, J., Laga, M., Mommarts, J., 2002, Natural Language Processing in Medical Coding; http://www.landcglobal.com.

Dominguez, J., Tadzebao, and WebOnto, 1998, Discussing, browsing, and editing ontologies on the Web, in: *Proceedings of the 11th Banff Knowledge Acquisition Workshop*, Banff, Alberta, Canada.

Farquhar, A., Fikes, R., and Rice, J., 1996, The Ontolingua Server: a tool for collaborative ontology construction, in: *Proceedings of the 10th Banff Knowledge Acquisition Workshop*, Banff, Canada.

GALEN, 2002; http://www.opengalen.org/.

Kozaki, K., Kitamura, Y., Ikeda, M., and Mizoguchi, R., 2000, Development of an environment for building ontologies which is based on a fundamental consideration of relationship and role, in: *Proceedings of The Sixth Pacific Knowledge Acquisition Workshop*, Sydney, Australia, pp. 205-221.

L&C, 1998; http://www.landcglobal.com.

Mahalingam, K., Huhns, M., 1997, An ontology tool for query formulation in an agent-based context, in: *Proceedings of the 2nd IFCIS International Conference on Cooperative Information Systems*, Columbia, SC, USA.

Noy, N.F., Fergerson, R.W., and Musen, M.A., 2000, The knowledge model of Protege-2000: combining interoperability and flexibility, in: *Proceedings of EKAW 2000 International Conference on Knowledge Engineering and Knowledge Management*, Juan-les-Pins, France.

Preece, A., Flett, A., Sleeman, D., Curry, D., Meany, N., and Perry, P., 2001, Better knowledge management through knowledge engineering, *IEEE Intelligent Systems* **16**:1.

SNOMED, 2002; http://www.snomed.org/snomedct_txt.html.

UMLS®, 2003, UMLS® Knowledge Sources, National Library of Medicine, U.S.A (January, 2003), pp. 3-7; http://www.nlm.nih.gov/research/umls.

MOBI-DEV: MOBILE DEVICES FOR HEALTHCARE APPLICATIONS

Roberto Altieri, Francesca Incardona, Harris Kirkilis, and Roberto Ricci *

1. INTRODUCTION

Healthcare institutions everywhere, today, are attempting to respond to demands to reduce cost and improve the effectiveness of care delivery. Often spurred by a combination of economic and regulatory pressures, hospitals and healthcare providers search for new solutions that can address core business needs and manage the huge volumes of security-sensitive data that are involved.

While most large healthcare institutions have already automated back office systems that handle functions such as patient billing and inventory management, the front end of care delivery has remained dominated by paper printouts, islands of digital automation, and the need to manually re-enter information into clinical systems. Not surprisingly, most providers seek solutions to close the information gaps surrounding care delivery. Such solutions should enable to eliminate paper-based medical records and improve access to patient records, clinical results, pharmaceutical databases and information about new treatment advances.

The emergence of desktop systems helped partly address this need. However, PC-based systems also introduced several key drawbacks. Most important for healthcare providers, they do not allow healthcare practitioners to efficiently document care as they move from patient to patient. When examined or treated, patients are physically in their bed, either in the hospital or at home; on the other hand, the desktop PC for entering or retrieving clinical information is usually somewhere else. As a result, healthcare professionals are requested to either move patients for examination or

Roberto Ricci, Informa srl, via dei Magazzini Generali, 31 – 00154, Rome, Italy. E-mail: r.ricci@informacro.info

treatment where the desktop PC is located – supposing that it is feasible - or to memorise all clinical information related to their work and enter it into the information system afterwards. In both cases the shortcomings are so big, that very often healthcare professionals decline to adopt this type of workflow.

In this respect, mobile solutions can provide an invaluable additional worth, by literally accompanying healthcare professionals in the performance of their jobs. Thanks to the Internet, many doctors and nurses have already grown familiar with information technology, and are attracted by the promise of anytime, anywhere access. Furthermore, the installation of wireless LAN infrastructure can be far less expensive and disruptive compared to conventional wired networks. With the emergence of wireless LAN (Local Area Network) standards such as 802.11a and 802.11b, the opportunity is here today for a new generation of open, plug-and-play solutions that can work on virtually any wireless-enabled handheld device yet, not interfere with existing medical electronic devices already in place[*].

Handheld, wireless applications can enable doctors and nurses to gain ready access to complete patient treatment histories, and gain access to the right information at the right time to prescribe the right course of treatment. Three main aspects should be considered in evaluating the potentially paramount impact of mobile solutions in the healthcare sector:

1. Mobility: mobile devices make some activities possible, that would be unfeasible otherwise, such as accessing a huge amount of medical information in the palm of one's hand, connecting to relevant DB's for data input/output at anytime from everywhere, checking medical references, working with electronic medical record systems.

2. Time saving: the workflow reengineering implied by the adoption of handheld computers enables to save a considerable amount of time - a particularly valuable resource for busy people like clinical practitioners - in activities such as appointment scheduling, billing and other administrative tasks, test-result reporting, interaction with pharmacies.

3. Reduction of error rates: according to a recent survey, damages for the patient and adverse events to drugs are significantly due to errors in the prescribing/transcribing/administering process. (Arici et al., 2002) In ordinary clinical practise based on manual procedures, adverse events to drugs due to errors in the processing of prescribing information have an incidence close to 30% of total occurrences, to be compared with the 5.3% due to the sole prescription errors (Bates et al., 1995; Bates, 1996). These figures could be reduced sensibly – completely, in the case of transcription errors - by using mobile devices (Bates et al. 1998, Nightingale et al., 2000).

Among the applications that could specifically benefit from mobile deployment are:

*   Mobile access to latest drug reference databases, gaining access to the most up-to-date prescribed treatments.
*   Electronic access to patient records, reducing demand for paper records.
*   "e-Prescribing," enabling prescriptions to be written and verified against potential drug interactions.

---

[*] Medical devices operate in the sub-1GHz frequency range, far below the 2.5 GHz and 5 Ghz devices supported by 802.11x-based technologies.

- Patient/drug verification, allowing practitioners to ensure the appropriate medicine is being administered to the correct patient.
- Prescription formulary reference, providing the ability to electronically identify the most economic pharmaceuticals for a patient.
- Electronic billing for in-home healthcare workers.

Basing its strategy on the soundness of the scenario just outlined, Informa, an Italian Contract Research Organisation with consolidated experience in developing software applications for clinical research, started in 2000 the Palmhospital project, with a contribution from GlaxoSmithKline and in partnership with Nergal, an Italian high-tech company specialised in the design and development of PDA solutions. Palmhospital was initially meant to provide the sole "Ospedali Riuniti di Bergamo" (a major hospital institution in North Italy) with a palm-based system for prescription and dispensing, but the system was so favourably welcomed, that it has been later adopted by other 14 hospitals all over Italy.

Encouraged by the success of Palmhospital, Informa promoted in 2001 a European project, Mobi-Dev, that aimed at realising a pioneering mobile platform specifically suited to healthcare providers needs. The Mobi-Dev platform supplements an otherwise standard PDA with innovative technologies such as wireless connectivity support, natural language understanding and smart card support for user authentication and data encryption, providing users with the ability to remotely access relevant data sources either interactively or in synchronised mode.

In the next section we provide a brief overview of state-of-the-art handheld device technology. The rest of the paper describes specifically the Mobi-Dev system, detailing the philosophy that inspired the project, presents the validation methodology of project's results.

2. MOBILE - HANDHELD SYSTEMS: STATE OF THE ART

The most important platforms currently available on the market are:
- Palm OS
- Pocket PC/Windows CE
- Symbian
- Embedded Linux

PalmOS-based devices, namely Palm's own Palm Pilot systems, have historically dominated market share for years. According to NPD Intelect, in August 2000 handhelds based on Palm OS carried over 85% market share, while PocketPC devices were a distant second with approximately 10% market share. Since then, however, PocketPC-based devices have been gaining larger and larger market shares. With Compaq's new iPaq product line, Compaq has sprung from negligible market share into third place behind Palm and Handspring. Also, Microsoft's recent initiatives propelled a larger user base towards PocketPC.

The fundamental difference between Palm and Microsoft is that Palm licenses its OS and carries a line of handhelds, while Microsoft's PocketPC provides an operating environment for hardware manufactured by other companies such as Compaq, NEC, HP, and Casio. Palm devices inherit a lot from the single-purpose schedule-management devices that preceded them. The concept was to extend these platforms with a more general-purpose operating system, add better integration with desktop PCs and allow the user to add applications to the device. This incremental improvement has proven very successful, since it largely retained the ease of use of the dedicated device while adding some degree of expandability and capability as a general-purpose computing platform.

PDAs based on PocketPC/Windows CE came from a very different direction. The model here was to bring as much of the functionality of a desktop PC as possible to a handheld device. It's very much a programmable computer that spends most of its time performing the functions of a dedicated device. The OS was rewritten from the ground up for these devices, but the application programming interfaces (API) that power it are largely identical to those present on any desktop Microsoft Windows-powered PC.

As for performances, for many corporate markets, simplicity, price, and functionality override the need for more extensive multimedia features. However, in healthcare and other markets, the requirements are more complex, particularly since the device must service many functions. Extended battery life and operating power are also key characteristics of a good healthcare handheld, but in addition, flexibility and multimedia options are similarly important. Since handhelds are used in healthcare to address a variety of needs, from e-prescribing to lab results to e-detailing, different features are required to fulfill each of the different needs. Physicians must carry handhelds around all day on their rotation, thus the handhelds must be compact and weigh as little as possible. Multimedia is becoming increasingly important, e.g. for allowing physicians to fully utilize the convenience of speech beyond dictation.

Among the many rival devices to PDAs, smartphones, wireless phones with integrated PDAs, are playing a dominant role in device manufacturers' efforts. Although Symbian OS emerged as the worldwide leader with its integrated wireless technology interface, security, and support for voice and multimedia applications, PalmOS and PocketPC have made their moves to partner and expanded their capabilities. As smartphones develop and more services are made available, these devices may become very useful to healthcare, but currently palm-sized handhelds are better options due to the greater quantity and quality of available applications, as well as their superior integration capability with wireless LANs.

Embedded Linux devices are in early stage, but they promise to be a valid alternative to commercial platforms, primarily because of the well known advantages of open source solutions – flexibility, reliability, vast basin of developers, free software availability, short time-to-market for custom applications.

Restricting our analysis to Palm and Pocket PC devices, the following general considerations can be made:

- PalmOS-based handhelds tend to be smaller and lighter than their PocketPC-based counterparts. While they have less power and memory, Palm and other compatible third-party applications generally require lower system resources, so battery life on these handhelds is longer on average.
- PocketPC-based devices generally have larger, colour displays with better lighting and resolution than PalmOS-based handhelds. PocketPC handhelds have greater multimedia support—these are typically equipped with Microsoft Media Player, and hardware such as microphone, speakers, and a stereo ear-phone jack. In addition, these devices boost greater memory, faster processor speeds, and enhanced storage capacity, often at the cost of a higher power consumption and much shorter battery life.

3. MOBI-DEV'S APPROACH TO M-HEALTH

Mobi-Dev ("Mobile Devices for Healthcare Applications", IST-2000-26402, http://www.mobi-dev.arakne.it) is a European Commission co-funded project, started in January 2001 and lasting 30 months, which has been involving healthcare institutions, universities and IT companies from Italy, Spain, Greece, Belgium and the United Kingdom.

Mobi-Dev's main purpose is to provide an enabling, next generation mobile communication tool for healthcare professionals, offering new opportunities to improve the quality of healthcare services. In order to achieve this ultimate goal, three priority objectives were identified:

1. Mobility: provide users with the ability of anywhere, anytime direct access to clinical information systems, both in and out of the usual workplace.
2. Ease of use: enable the input of data by voice, meaning by that not simple voice transcription but natural language understanding with automatic extraction of structured information to be eventually recorded onto the Hospital Information System (HIS).
3. Security and confidentiality: ensure cryptographic protection for data transmission and management and authentication through smart cards.

Mobi-Dev solution is naturally addressed to hospital medical and nursing staff, who need to stay connected with the HIS in their work either in the hospital or in home-hospitalisation, and more generally to all practitioners, who could benefit by connecting with diagnostic laboratories, pharmacies and with their own clinical databases, while providing home visiting service. Fig. 1 presents typical usage conditions, showing the connectivity domains covered by the wireless technologies used in Mobi-Dev. Effective correspondence to users' needs was ensured by involving representative healthcare institutions in the specification and test phases of the project. Also, considering the delicacy of the matter involved, a strict policy of attention to security and confidentiality aspects, as reflected in the EU and local regulations, was adopted as a major guideline.

Mobi-Dev custom-built User Interface (UI) was designed so as to allow users to perform easily and efficiently a number of tasks related to their daily work:

- edit medical/nursing history;
- handle pharmaceutical prescriptions by voice;
- benefit from a "red flag" control system, checking against the input of wrong or out-of-range values - e.g. for vital signs or lab results;
- receive and give orders related to medical/nursing care;
- receive "early warnings".

Beside accessing Mobi-Dev specific functions, the handheld device can also be used to perform general purpose activities not directly concerned with Mobi-Dev core application, such as storing and retrieving medical information files and pharmaceutical formularies, installing useful medical programs, exchanging e-mails and browsing the Internet.

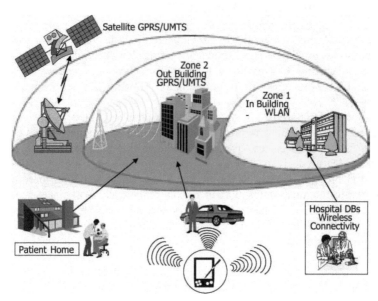

Figure 1. Mobi-Dev's connectivity capabilities: healthcare practitioners can use the Mobi-Dev palm device to connect to the hospital DB, by exploiting WLAN connectivity if in the hospital or GPRS/UMTS mobile network in all other situations.

4. OVERVIEW OF THE MOBI-DEV SYSTEM

The analysis of the user requirements, collected at the beginning of the project by an internal panel of qualified end users (hospital institutions and university polyclinics), drew to the specification of two different technical solutions, respectively called Web Mobi-Dev and Synch Mobi-Dev. The two solutions closely match different user needs, beside reflecting specific constraints and local restrictions. In particular, the decision to implement a synchronised version of the system, with no language understanding support, was initially taken to meet specific internal desiderata. This was motivated both by market limitations - the unavailability of speech to text translators for the Greek language - and Greek pilot site's peculiar needs - namely the possibility to use the mobile system on a hospital boat in discontinuous wireless network coverage conditions. The project consortium determined that these requirements reflected a quite common operational condition, characterized by an unreliable or poorly performing network infrastructure. This consideration led to regard the implementation of a version of the system based on synchronisation as a favourable opportunity, enabling to open commercial perspectives in an otherwise inaccessible segment of the market.

In summary, the two Mobi-Dev specifications were arranged to meet the following sets of broad requirements:

1.   Web Mobi-Dev
   –   Natural Language Understanding (NLU)
   –   Wireless web access to the Hospital Information System (HIS)
   –   Digital signature based on smart cards

       –   Portability
2.   Synch Mobi-Dev
       –   Remote wireless synchronisation with the HIS
       –   Digital signature based on smart cards

Both implementations are compliant with Mobi-Dev security guidelines, which can be summarised as follows:

- DB Access control lists based on users' profiles matching professional roles
- Use of mirror DB's, to avoid interference with already operational information managing systems
- XML data protection via VPN or SSL protocols, to ensure confidentiality
- User authentication via smart card
- Traceability of data paternity on the DB, to ensure non-repudiability
- Secure DBMS

4.1. Web Mobi-Dev

Web Mobi-Dev is based on a client server architecture (see Fig. 2), that enables wirelessly connected handheld devices to securely access, through a middleware software and an ad-hoc web interface, data stored into a generic relational DB. Specific applications installed on the palm device enable to authenticate users by comparing user credentials with encrypted data stored on a smart card, perform data input by voice, initiate a wireless SSL session for data transmission and safely record data onto a remote DB. In implementing the architecture, the approach has been followed to realise a flexible and portable system, based on open standards whenever possible, namely:

- JDBC DB interface, enabling the use of open source DBMS.
- Dynamic XML, allowing to develop a flexible system made-up of highly independent components. Considered as a great leap forward for Web-based content, XML (eXtensible Markup Language) is a widely used protocol for defining, validating, and sharing document formats on the Web.
- Java, ensuring portability to different hardware platforms.
- Open source software, such as Tomcat, the servlet container that is used in the official Reference Implementation for the Java Servlet and JavaServer Pages technologies.

On client side, an iPAQ 3870 PocketPC 2002 palm device is used, even though all required software modules could easily be ported to different PDA's and operating system. The handheld device is supplemented with some hardware plug-ins (depending on the initial equipment: a Bluetooth or WLAN transceiver, a GPRS/UMTS module, a microphone, a smart card reader) and some specifically developed software applications (a voice recording module that transforms voice input into WAV files, a WAV dispatcher module, a module that accesses the smart card content for user identification and data signing).

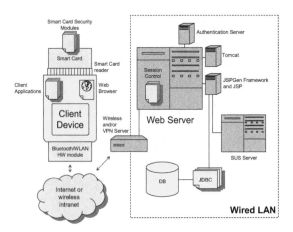

Figure 2. Web Mobi-Dev general architecture

On server side, a Java Servlet Pages (JSP) server is responsible of interfacing structured data sources - typically, the HIS - and returning to the client dynamically formatted HTML pages. The JSP pages are created by using a tool that allows developing the logic and the "look & feel" of the system separately. This allows developers to produce a different look for each client by coding the application logic only once.

Natural language understanding is ensured by a separate specific Speech Understanding System (SUS) server, which processes the WAV files and returns XML files with the voice input transposed in structured text form. A middleware servlet running on the JSP server is responsible of interfacing the clients with the SUS, by rerouting WAV files to the SUS and sending back HTML pages produced by processing returned XML data.

A typical case of use of Web Mobi-Dev is the following:

A user – hospital doctor, nurse, general practitioner – needs to access the Mobi-Dev information system while she is by a patient's bed, either in the hospital or at patient's home. Depending on the wireless network coverage of the specific location at hand, she initiates a GPRS/UMTS session or simply switches on her WLAN connected PDA. She launches the Mobi-Dev log-in application, plugs a smart card with her personal identification token onto the PDA's reader and enters her credentials as required. The credentials are compared with data securely stored on the smart card and, in case of positive upshot, the Mobi-Dev application automatically launches and redirects the PDA standard browser to the user's personal page on the Mobi-Dev web server. Starting from such page, the user interactively accesses the information she's interested in, with possible limitations that reflect her user profile (see Fig. 3 for an example of a typical page). Whatever new data she inputs is automatically marked with her personal identification "stamp", thus allowing to trace back the paternity of each piece of data if needed. She's given the ability to input new a drug prescription by voice dictation by just pressing a key on the PDA. Her words are interpreted and returned as structured text on the PDA display, where she can make changes by hand before eventually recording the data onto the information system.

THERAPIES - EXAMIN.

NAME	SURNAME	BED N.
paperino	paolino	1

Drug		StartDate
Aceclidina vl		2002-09-05
Premarin mg1.25		2002-09-05

Examin.	StartDate	SuspDate
Bilancio acanche	2002-09-05	NP
Bilancio acanche	2002-09-05	1996-05-02

Figure 3. A typical Web Mobi-Dev page as seen on a practitioner's PDA. The page reports patient's details and examinations, as well as the list of drugs the patient is being currently given. Underlined items link to related forms or pages with further information. Icons in the bottom allow to perform specific actions such as accessing the medical folder, assigning a new therapy, discharging the patient etc.

4.2. Synch Mobi-Dev

Synch Mobi-Dev differs from Web Mobi-Dev in the following respects:

- It provides no speech recognition functionalities. This is mainly due to the contingency that there is no solution available on the market yet for capturing and managing voice data in Greek. Nothing impedes to implement this functionality on the system at a further step, if required.
- It uses remote synchronisation providing WWAN independency, meaning that data are selectively moving from/to the target database, to/from the distributed personal and/or PDA database systems.

The synchronisation mechanism requires the temporary storage of data on the PDA, which in turn ensures the ability to reliably access data, even when not connected to a network, rapid response times and a better use of computing resources, due to the workload being distributed onto different platforms. No single point of failure can bring down the system as a whole and connection costs are minimized.

The target DB selectively synchronises data with an "UltraLite DB" running on the handheld device – a Compaq's iPAQ H3670 operated by Windows CE 3.0. Synchronisation, implemented using Sybase's MobiLink Synchronization Technology, is made possible through a wireless connection between the two systems.

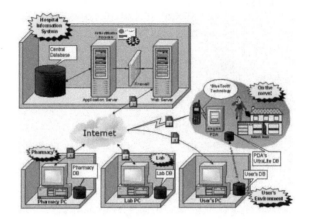

Figure 4. High level architecture of the Synch Mobi-Dev system

As an alternative, the UltraLite DB can be synchronised to a user's personal DB, running on his/her desktop PC and containing a replica of the subset of HIS data concerning his/her patients only.

Pharmacies and analysis laboratories can also be involved in the more complex scenario illustrated in Fig. 4, where a pharmacy and a laboratory are supposed to maintain DB's with replicated subsets of the HIS data - regarding patients' prescriptions and lab tests respectively. Whenever the hospital practitioner prescribes a drug or orders a lab test for a patient, a prescription or lab test code number is automatically generated, stored onto the PDA and queued to be recorded on the HIS during the next synchronisation process. After synchronising the PDA with the HIS, an e-mail notifies the pharmacy or the laboratory that there is a new assignment to be carried out and provides the corresponding code number. Details about the prescription or lab test can be retrieved by synchronising the pharmacy or laboratory PC with the HIS. This operation also enables to mark pending assignments as executed and to upload lab test results. After a further synchronisation process, an alarm mechanism implemented on the PDA informs the practitioner that the lab test he/she ordered has been carried out, allowing him/her to see the results.

5. VALIDATION METHODOLOGY

The most relevant assessment criteria that have been taken into account in planning the validation of project results are:

- *Relevance*, concerning the perceived need for the new service - the basic rationale for making the Mobi-Dev new service available in order to more effectively meet the needs and demands of the healthcare community.
- *Effectiveness*, emphasizing evaluation of system outputs and immediate results of the new service and considering its success in achieving pre-determined objectives.

- *Process evaluation*, referred to the tracking of service activities. Here attention is focused on the degree at which service implementation complies with the pre-determined objectives and plan.
- *Efficiency evaluation*, concerning the balance between the results obtained from the service and the resources expended for its operation.
- *Impact evaluation*, concerning the long term implications of the new service – changes observed over time in characteristics that the service is designed to influence.

The validation plan covers all aspects of the Mobi-Dev system: hardware, software and network components; controlled functions and associated documentation; all typologies of system users. This approach is in accordance with the regulatory framework, presiding over the use of computerised systems within GxP (GCP, GMP, GLP, GDP) environments. Healthcare authorities regard non-validated systems as unacceptable and require that validation is applied not only to system design and internal testing, but also to SOPs (Standard Operating Procedures) that govern system training, use, management, security etc. The validation process is designed as an on-going procedure, that covers the whole system's life cycle, complying with GxP at every stage.

One of the most important issues addressed by validation is system security. The relevance of security and confidentiality for ICT solutions addressed to the healthcare sector is founded on two general yet simple principles, both originating in law:

- Right to be treated fairly and with dignity - Human Rights Act
- Right to confidentiality - Data Protection Act

According to these principles, patients have the right to expect that information about their health status is handled discreetly, not divulged unnecessarily and only used either in the course of care delivering or in the planning and management of healthcare services.

As far as Mobi-Dev is concerned, this approach has a number of consequences:

- Accuracy in managing information is a major issue. For example, the handheld device must provide users with the ability to maintain an audit trail on what has been requested/ordered for a particular patient. It is also important to be able to determine in real time whether the requester has the authority to make a particular request. This is particularly important as long as the operational conditions involve the requesting of drugs, especially controlled medications.
- It must be possible to ensure that the handheld device is capable of screening out some types of information, depending on the access rights of who is using it – typically, if she is a doctor or a nurse. In this respect, it must be noted that, at least for Web Mobi-Dev, no security issue would result from the handheld device being left unattended, lost or stolen. Indeed, the device itself is not being holding any data, as the application uses a standard browser for accessing the DB and any active session is timed out after a period of inactivity.

All of these issues and queries have been addressed in the validation and the first results – validation is still in course at the moment of writing this paper – show a substantial compliance with the guidelines identified in a antecedent phase of the project.

In order to yield information on either non-quantifiable characteristics of the system or intangible aspects of the Mobi-Dev service, such as those related to performance or costs, a standard approach has been followed, based on questionnaires.

All users involved in the validation in the four pilot sites have been requested to respond to a series of pre-established questionnaires. An additional group of health professionals, who do not participate directly in the pilot phase, will also be contacted in order to get their opinion on the Mobi-Dev service. The results of the questionnaires, after being collected and processed, are planned to be discussed by ad hoc groups of users for a better and in-depth interpretation.

In order to achieve a methodologically sound evaluation, the following steps have been planned:

- Identification of representative samples of target users in each pilot site.
- Stratification of these samples into defined categories, based on their specific characteristics, namely general practitioners, specialists and nurses.
- Adoption of suitable statistical methodologies in case of numerically significant samples.

In such a scenario an effort has been made to randomise the users into control and a study groups, although, for the level of assessment required in this case, such an distinction cannot always be made.

6. OPEN PROBLEMS AND REMAINING BARRIERS

The main obstacles to the penetration of mobile solutions based on handheld devices and wireless technology, such as Mobi-Dev, in the healthcare sector can be roughly classified into objective and subjective impediments.

Objective obstacles have to do with deficiencies and limitations of current technology resources, with particular regard to interoperability, compliance to widely accepted standards and security. Interoperability and standards issues are especially relevant, as they often impede or make it a tough job to integrate mobile solutions into the framework of existing legacy systems currently used in healthcare institutions for information storing and management purposes. To help overcome this difficulty, the adoption of open standards and cross-platform technology is advised wherever possible. Access control and security are further crucial points to be considered for the assessment of the eventual acceptance of mobile solutions by the healthcare community and should not be underestimated. The confidential nature of the data to be accessed obviously requires that access rights be carefully matched against user authentication parameters. Wired network solutions usually approach the problem by adopting security models based on digital certificates and electronic signatures. It is far from being ensured that technological solutions based on standard private key infrastructures can be ported on the mobile devices currently available on the market. On the contrary, the first results obtained on this respect in Mobi-Dev show a substantial immaturity of PKI technology for mobile platforms – problematic support for smart card readers, unreliable functioning of available cryptoAPI libraries for digital certificate management, pocket browsers not enabling client authentication in SSL sessions etc. These problems should be overcome as long as new PDA models and OS versions are made available.

Subjective obstacles have to do with the reluctance of healthcare professionals – often busy people with techno-phobic attitudes and little time left for technical training – to abandon traditional mostly paper-based workflows for adopting new practices based on highly technologic appliances such as new generation PDAs. This is especially true in case of applications that require for their operation excessively awkward or lengthy procedures. The experience accumulated so far evidences the importance of parameters such as user-friendliness, lightness and functionality in application design, even at the cost of realizing less attractive user interfaces.

A further major obstacle to the adoption of advanced mobile solutions in the healthcare sector is related to the need to comply with usually very restrictive regulations, conceived so as to ensure decision traceability in each step of care delivery and guarantee the patients against privacy violations. This is a quite delicate issue that can often only be dealt with by promoting actions aimed at awakening in the legislators the necessary appreciation for the benefits deriving to the healthcare system from the adoption of information technology mature solutions.

7. CONCLUSIONS

The technology sector of healthcare is entering into a new evolutionary phase. The medical community has an ethic obligation to the public to provide the safest, most effective healthcare delivery system possible. Today, this is hardly achievable without the use of computer technology at the point-of-care.

Optimal medical practice will depend more and more on computers. Outcomes analysis that will shape medical treatments in the future will plausibly be derived directly from the data entered into mobile devices. If a global database could eventually be obtained using point-of-care devices, existing methods for tracking clinical outcomes would become obsolete. Optimal management of disease prevention and management could be achieved. In addition, these devices, with appropriate software, will allow for foolproof coding. Using mobile solutions, physicians and hospitals can capture additional earned revenue that can amount to millions of lost healthcare euro per year.

We will all move forward and healthcare will continue along its technological path. Older techno-phobic physicians will be retiring, and the young doctors who are learning computer programming in elementary school will be taking over. The process will be incremental but inexorable and mobile devices will remain crucial in the inevitable transformation to a paperless, wireless healthcare world. We firmly believe that Mobi-Dev has the momentum to contribute to shorten the time needed to achieve this ultimate goal.

8. REFERENCES

Arici C., Ripamonti D., Suter F., 2002, Why hospital doctors don't use computers, (18 December 2002), http://bmj.com/cgi/eletters/325/7372/1090#27934

Bates D., 1996, Medication errors. How common are they and what can be done to prevent them?, *Drug saf.,* 15(5):303.

Bates D., Boyle D.L., Vader Vliet M., Schneider J., Leape L., 1995, Relationship between medication errors and adverse drug events, *J. Gen. Intern. Med.,* 10(4):199.

Bates D., Leape L., Cullen D., Larird N., Petersen L., Teich J., Burdick E., Hickey M., Kleefield S., Shea B., Vander Vliet M., Seger D., 1998, Effect of computerized Physician Order Entry and a Team Intervention on Prevention of serious Medication Erros, *JAMA,* **280**:1311.

Nightingale P. G., Adu D., Richards T., Peters M., 2000, Implementation of rules based computerised bedside prescribing and administration: intervention study. *BMJ,* **320**:750.

WARDINHAND

Salvatore Virtuoso[*]

1. INTRODUCTION

WardInHand is an acronym for "Mobile workflow support and information distribution in hospitals via voice-operated, wireless-networked handheld PCs", a EU-funded IST project (IST – 1999 -10479) started in January 2000 and completed in spring 2002. Participants were three IT companies (TXT e-solutions, Italy; BMT, UK; and RT, Greece); the Department of Informatics and Information Sciences - University of Genoa (Italy) and three Healthcare organizations as end-users (the Department of Endocrinology and Metabolism – University of Genoa, Italy; Corporaciò Sanitària Clinic – Barcelona, Spain; and Staedtische Kliniken Offenbach, Germany).

WardInHand aims to support the day-by-day activities of doctors and nurses within a hospital ward by providing a tool for workgroup collaboration and wireless access to the patient's clinical records. The project is based on accessing the information system from the patient's bedside through a PDA client. Ward-In-Hand is not intended to replace or compete with Hospital legacy systems, it is rather meant to complement them. It exchanges information with existing tools, updates data in real time and makes them available to doctors and nurses, adding mobility and ubiquitous computing as new dimensions to currently available Hospital Information Systems.

2. THE OBJECTIVES AND THE APPROACH

The problem addressed by WardInHand is to significantly reduce the inconveniences of paper handling and repeated transcriptions during the process of delivering healthcare services to patients in a hospital ward. Processes associated to services delivered in a hospital involve a large amount of information collected from the patient, exchanged by the several actors concerned, filed for legal purposes, etc. Examples are prescription of analysis and collection of the results; prescription, execution and monitoring of

---

[*] Salvatore Virtuoso, TXT e-Solutions SpA, Milan, Italy

treatments; coordination among doctors and nurses, performing and monitoring them; exchange of information between the department and the other units of the hospital (lab services, logistics, etc.). Most often, processing of this information is based on paper, with subsequent transcriptions made by different actors in different places at different times. In other industry sectors, such as the manufacturing industry, the effort to minimize losses in efficiency and quality intrinsic to this kind of practices has led since long time to adoption of real time data collection and production tracking systems, that gather data at the time and the place where it is generated, and organize the flow of information with limited human intervention.

In a traditional environment, clinical information is often recorded on paper documents (the clinical record) and notes are taken on notebook by nurses while doctors perform their visit tours and then transcribed to other paper files for planning treatments, exams and drug distribution. The result is often inefficient, as this activity drains valuable time from the hospital personnel, and can also introduce errors and inaccuracies, either due to the transcription process or to the fact that information is not updated in real time.

Even when systems for electronic records management exist, manual input and transcription is not eliminated. Moreover, such systems are often located in fixed positions within the wards, forcing people to move there for data entry and data retrieval, leading to errors, omissions, duplication of information and adding huge overheads to doctors' and nurses' work.

In WardInHand, the entire process is centred around the patients' clinical records and involves activities such as retrieving information about the patients, including results of analysis, monitoring its progresses, prescribing treatments, executing and monitoring them, etc. It also comprises activities such as: exchanging information among doctors, among nurses and between doctors and nurses; co-ordinating and synchronising activities; ensuring that the necessary drugs, consumables are available when needed; etc.

The need for transcription is then virtually eliminated by providing the Hospital personnel with direct access to information at any time and in any place they are by using PDAs as well as their desktop systems.

In summary, the major objectives of WardInHand are:

- To increase the quality of the services provided by the hospitals to patients,
  - o   By reducing the possibility of error in data transcription
  - o   By enforcing quality and safety standards
  - o   By providing better and more timely information to Healthcare professionals
  - o   By improving the control on access to clinical information and on task execution responsibility
- To reduce costs by increasing the productivity
  - o   Less time spent in low added value activities
  - o   More efficient team communication and synchronization
- To increase the satisfaction of the Hospital's personnel

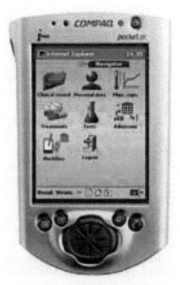

Figure 1. WardInHand main page

The architecture

In a scenario where WardInHand is installed, doctors and nurses are equipped with personal handheld PCs linked via wireless networks to a server. The server maintains in a repository all clinical records and all relevant information about patients in the ward, including clinical records, results of tests and clinical analyses, etc. The server is connected through the Hospital's intranet to other hospital systems, such as systems running in the analysis labs, treatment rooms or the logistic department.

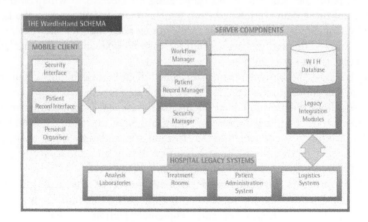

Figure 2. General system architecture

Figure 2 above illustrates the client-server architecture of WardInHand. The functional modules included in the client (left) just show that users access the corresponding functional modules on the server (right), as the design has been intentionally made to support "thin clients", for security reasons and to ensure application longevity.

Handheld PCs and desktop systems – as long as they are connected to the intranet - completely replace traditional paper-based management of clinical records, allowing, e.g., doctors to access and update information on the patient status, to prescribe drugs, treatments and clinical analysis, etc. Data is transmitted to the server in real time, allowing other doctors or nurses in the ward to always get the most up-to-date information.

The server runs a workflow application, which, based on the information collected triggers actions updates a "personal organizer" of each doctor and nurses, automatically assigning and scheduling tasks for them. Nurses will use handheld PCs to confirm that treatments have been performed, so that clinical records of patients are updated in real-time. Doctors and nurses will also be allowed to exchange informal messages and information via an internal E-mail service available via handheld PCs, regardless of their location in the hospital unit.

Offline workflow design tools to edit review and implement Hospital and Ward based workflow processes.

The WardInHand system is composed of four major components:

- The Patient Record Manager (PRM)
- The Personal Organizer (PO)
- The Security Manager
- The Legacy System Integration layer (LSIL)

The **PRM** is responsible for the overall management information about patients, including their personal details, vital signs, drug treatments, orders and results, discharge and clinical history. It keeps a local database containing information about patients currently in the ward, where data are loaded at the time of patient admission and removed as the patient is discharged. Part or all (depending on the HIS to which WardInHand is connected to) of these data are fetched from the Hospital System through the LSIL, using a unique key (for example the social security number) to identify the patient while the remaining is entered manually. This is often the case for special (for that particular ward) information that is not kept centrally.

During his stay, the patient record is enriched of new information about treatments, vital signs etc. and stored in the database until the patient is released.

PRM uses this database to present the user with relevant information primarily using a wireless LAN and the attached mobile devices, although the same data can be accessed from regular desktop PCs connected to the Hospital's intranet. A typical usage pattern for a doctor is to enter reports via a PC in his office during an admission visit, and take a handheld with him during the daily visits to patients in the ward.

The PRM provides summary and detailed views of the data. In the current implementation, there are specific screens for the following functions/data views:

- Login, ward and language selection
- List of patients in the selected ward

- Personal data of the selected patient
- Physiological signs (summary and detail)
- Drugs and treatments (view and prescription)
- Tests (prescription and results)
- Admission report (anamnesis etc.)
- Annotations (text and voice)
- Discharge report

The Human Computer Interface (HCI) of the PRM heavily uses graphics (icons, symbols, etc.) as well as menus and lists to facilitate navigation through the screens. Data is entered either by the standard (the one that comes with the handheld) virtual keyboard or the special one developed for WardInHand that uses special predictive algorithms and custom dictionaries to minimize pen strokes. Navigation and data entry can also be done "by voice", albeit not to the extent originally envisioned, using the handheld as a "microphone", and letting a speech recognition engine installed at server side converting the audio stream to a text string.

The Personal Organizer **(PO)** is the other main module accessible to the user. It can be considered as a personal agenda that keeps the list of the "things to do" for each individual user.

WardInHand allows the system administrators to define workflows, that is a formal description of the practices in use at that particular ward, by using an offline configuration tool called Workflow Designer. Workflows define relationships between actors (e.g. doctors, nurses, etc.), and variables (i.e. a body temperature, etc.), establishing sequences of tasks to be performed. Examples are "Patient admission", "Blood examination", etc.

A "blood examination", for example, can be a workflow, that can only be started by a doctor, where a nurse will have to book the test at the lab, another will have to prepare the labels, then another will take the blood sample, and so on. It is important to note that only when a task is accomplished, the following one can be started.

In this example, when the doctor starts the workflow "blood examination", a new task "book a test at the lab" is immediately listed in a nurse personal organizer, and only when this task has been performed (the nurse checks the activity out), a new task "prepare labels" is scheduled. The mechanism goes on until the all the steps of the workflow are complete.

At user level, only the activities for that particular user are listed; the system administrator has a special tool, called "Workflow Administrator" that provides a global view of all outstanding workflows, that allows to perform maintenance activities.

The **Security Manager** performs the authentication of the users when they access the system and guarantee that data is transmitted securely between clients and server. This is accomplished using digital certificates, which are used for encoding and decrypting transmitted data. Therefore unauthorised devices will be incapable of interacting with the server, and any data that is intercepted during its transmission will be encoded and therefore meaningless.

The Security Manager also allows user accounts to be created within WardInHand, and have authority assigned to them. The authority of the user (e.g. doctor, nurse) is then used by the other WardInHand modules to determine what operations an individual user can perform.

The Legacy System Integration Layer allows WardInHand to link with Hospital Systems for data required within the application. This reduces the need for duplication of data between Hospital Information Systems and WardInHand. The integration module is very flexible and can link to almost any current Hospital Patient Information System. The integration module can import and export in four main formats:

- Open Database Connectivity (ODBC);
- Text delimited;
- XML;
- HL7 (patient summary information).

ODBC allows users to send data using the standard database connectivity format, XML is an emerging standard that is very flexible in terms of structure for data to be imported and exported between systems. HL7 is the emerging standard for data transfer in Europe and already the de facto standard in the UK.

The results

The project was completed, as planned, in March 2002, and the three pilot installations, in Genoa (Italy), Barcelona (Spain) and Offenbach (Germany) have been validated and rated by the respective users after about three months of trials. The ward's specializations were, respectively, endocrinology, pneumology and cardiology.

At the end of the final validation phase, lasted three months, an evaluation of the results obtained was made, based on objective measurement of the time required to perform some typical operation (such as patient admission, drug and exam prescriptions, etc.) before and after the installation of WardInHand. A questionnaire to assess other aspects not easy to measure objectively, such as ward personnel satisfaction, was administered as well.

The feedbacks from the users have expressed favourable rates (all above the neutral point) in terms of general system characteristics, functionality, and usability, and this result is even more significant when considering that users could not fully experience all of the system features (as WardInHand was in a pre-commercial phase anyway).

In the figures below, the evaluation of the three pilot sites is summarized (the acronyms identify the three pilot sites: SM – San Martino (Genoa, Italy); OF – Offenbach (Germany); CSC – Corporaciò Sanitaria Clinic (Barcelona, Spain).

The overall results measured were encouraging, and in one case (Genoa) a very significant saving/improvement was achieved. A deep analysis of these results have shown that the reasons of these apparent discrepancies lie with a different starting point from a HIS support point of view, that shows higher improvements when the initial support is lower.

The overall system features of the WIH system were evaluated by means of a questionnaire administered at the three testing sites at the end of the trial field test. This questionnaire included 17 questions and a score scale to answer them (From 1 to 5; 1-Strongly Disagree, 2-Somewhat Disagree, 3-Neutral, 4-Somewhat Agree, 5-Strongly Agree).

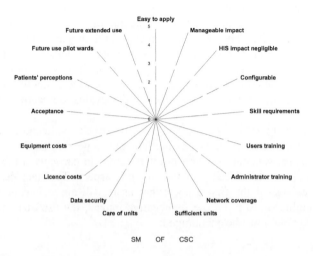

Figure 3. End Users evaluation (Overall System Features)

The questions explored the impact of installing the WIH system from different perspectives: physical installation, existing IS, system maintenance, options of design, patients' perception and future use. Among the key findings, it is worth to note that from a user perspective the impact (in a broad sense: integration, organizational aspects, user training, etc.) of introducing a new IS component is a critical task that goes beyond the pure technological aspect.

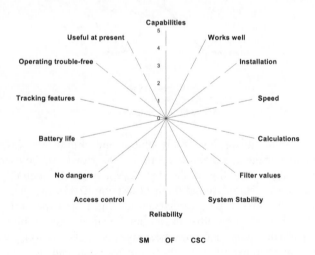

Figure 4. End user evaluation (System Functionality)

The functionality of the system was also evaluated by means of a questionnaire that included 14 questions and was administered at the end of the field trial test. In general, the functionalities satisfied the user requirements, and in particular users positively appreciated other important system functionalities: good responsiveness (4.00), correct handling of the boundary values (4.00), stability and reliability were convincing (4.33 in both cases) and no potential dangers to clinical conditions were identified (4.67). Some bugs still present in the system at the time of the evaluation were also brought to attention.

The three sites, with an overall score of 3.74, positively evaluated system usability. However, not all the different characteristics got the same results. The weight of the handheld, for instance, was considered a problem (2.67) in particular in the case of SM and OF. Both sites reported a score of two for this aspect. This fact deserves special consideration since one of the first aspects that users will appreciate in a PDA based solution is its weight. So, to minimise its negative impact, the use of lighter models of PDA (if/when available) must be encouraged.

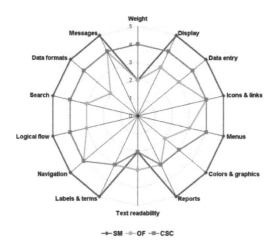

Figure 5. System usability

The size, in particular the display size, resolution and form factor, was not a problem for the users with an average acceptability score of 4.00. However, the users were not so much satisfied with the readability of the text (2.33). Also, the legibility of symbols, graphics and colours was scored around the neutral point (3.33). In other words, though technically feasible, the final display format did not fulfil the expectations of users on text readability that should not be difficult to improve since other usability factors were well accepted. In particular icons and links meaning was considered very good (4.33) and the opinion on the appropriateness of toolbars, menus, commands and options was also positive (4.00).

Also, it should be noted that the software was easy to manage by the users: it was easy to navigate (4.33), it had a logical and consistent flow (4.00) and it was easy to access the information the users where looking for (4.00). Data formats were judged as consistent (3.67) though the score differed in the case of OF (2.00) with respect to the other sites (CSC, 4.00; SM, 5.00). The situation was similar concerning the readability of labels and terms (overall score 3.67) but in this case two sites (OF and CSC) evaluated it around the neutral point. This suggests the need to review this part and provide a solution more in line with the one adopted for the messages of the application that got very positive scores (4.33).

Summary

Although the technologies used in the project (handheld devices, wireless LAN, Speech recognition etc.) are by themselves innovative, the most significant advance is related to the fact that WardInHand uses these technologies to support the operational ward processes. While ICT investments in the healthcare sector are still largely concentrated in the administrative functions, the focus here is on the key issues of clinical data management and collaborative tools with the goal of improving quality, increase productivity and reduce the overall costs.

Having ubiquitous and online access to clinical information, healthcare professional can now make more efficient and accurate decisions. Data entry and transcription errors are virtually eliminated, with immediate impact on quality (in the broadest sense: level of service, better satisfaction of professionals and patients). Unnecessary work and duplications are eliminated, with immediate fallout on efficient usage of resources. As the automation of the work flows improves the communication among actors and helps action coordination, delays are reduced, and this can lead to shortening of the length of stays.

The project successfully demonstrated that adoption of mobile technologies to bring information close to where it is most needed (i.e., at patient bedside) brings significant benefits both from an efficiency and quality point of view.

Reaching the objectives of reduced costs, improved productivity and higher satisfaction of the Healthcare professionals, however, is more complex than just automate the distribution of information to doctors and nurses, and involves the integration of organizations and system within and outside the ward itself, as well as some degree of reengineering of the care delivery process.

3. REFERENCES

Ancona M., Dodero G., Gianuzzi V., Locati S., Romagnoli A., "An Integrated Environment for Scientific Data Entry and Management on mobile systems", Proceedings of 1st Software Engineering Workshop, JSCC, Punta Arenas, Nov.2001.

Dodero G., Gianuzzi V., Coscia E., Virtuoso S., "Wireless networking with a PDA: the Ward-In-Hand project", Proc. Workshop on "CORBA and XML: towards a bio informatics integrated network environment", Genova, May 2001, pp.115-118.

Masui T., "POBox: an efficient text input method for handheld and ubiquitous computing", Proc.HUC99 (Int. Symp. on Handheld and Ubiquitous Computing), Lecture Notes in Computer Science, 1707, Springer-Verlag, pp.289-300, 1999.

Pascoe J., "Adding generic Contextual Capabilities to Wearable Computers", Proc. 2nd Int.Symp. On Wearable Computers, Oct.19-20, pp.92-131, 1998.

Stephanidis C., "From User Interfaces for All to an Information Society for All: recent achievements and future challenges". Proc. 6th ERCIM workshop "User interfaces for all", Florence, Oct. 2000.

Workflow Management Coalition standards, http://www.wfmc.org

DOCMEM – MOBILE ACCESS TO ELECTRONIC HEALTH RECORDS

Najia Eddabbeh[*] and Benoit Drion

1. INTRODUCTION

DOCMEM is a European R&D project conducted between July 2001 and June 2003 with the co-funding of the European Union in the frame of the IST Programme (Project IST-2000-25318).

It targets ambulatory and in-hospital medical doctors and offered them an advanced interactive environment for getting a ubiquitous, permanent and intelligent access to patients' medical files. In spite of common requirements, hospital and ambulatory contexts differ, for instance by the healthcare information system they use. A hospital practitioner interfaces with the always-operating information system of his hospital department, whereas an ambulatory practitioner may need to copy on a mobile device the data from information system lying on his PC.

The present chapter presents first the research and development dimension of the project, the applications adapted to the two healthcare contexts of in-hospital and ambulatory practices, and finally their validation through pilot projects.

2. RESEARCH AND DEVELOPMENT ACTIVITIES

Considering that most inadequate decisions made by practitioners are due to information lack, the availability of patients' medical information at the time and point of care has emerged as a key need to improve quality of healthcare. Moreover, quality of care is also often associated to increasing needs for collaboration among healthcare professionals to deal with a patient.

---

[*] Najia Eddabbeh, BFC, 196 rue Houdan, F-92330 Sceaux (France), Tel : +33 1 41 87 96 20, email : najia.eddabbeh@business-flow.com

To support these rising collaborative practices in healthcare, the DOCMEM project is positioned as an online service (a Portal) providing practitioners with a mobile, permanent and intelligent access to Electronic Health Records (EHR):

- DOCMEM enables practitioners to store on a secured server the medical information of their patients and to get access to it at the points of care through wireless information devices.
- With DOCMEM, the practitioner is also able to communicate patient's information with other independent healthcare professionals or with a hospital.
- DOCMEM also offers an intuitive and flexible means for practitioners to deal with patients' medical information thanks to facilities like speech recognition and virtual representation of a patient's body.

The DOCMEM Portal implementation consists of two sets of services emerging from the three main underlying research lines:

- A **Secure Online Patients' Files Service** that allows the storage and exchange of multimedia medical data in standard way. This is the core of the system that includes, in particular, a secure environment where structured patients' files are stored. Medical data is stored into a repository constituting a part of the central database which is **compliant with the CEN 13606 standard**. The central DB includes three parts:
  - The patient medical records under a standardised structure (EHR), consist of general and medical patient's information organized historically: patient identification and registration details, practitioner's summary (with a minimal set of structured and multimedia data), description of current episode of care and actions taken , information on patient's medical history.
  - The terminology and patterns terms
  - The Practitioners' repository giving information about practitioners using DOCMEM

  The EHR solution developed in the DOCMEM project is compliant with the CEN 13606 standard and has been implemented for Relational Data Bases. It will be offered as an Open source Implementation to any user interested in developing EHR based on the CEN.

- A **Secure Mobile Access Service** enables healthcare professionals to access the service via wireless information devices. This subsystem allows access to the DOCMEM services for the future generation of mobile communication and PDAs. Mobile multimedia terminals requires specific development for securing access with smart card on the mobiles. The work performed in this area within the project aims at ensuring that the DOCMEM services is accessible through E-GPRS and is scalable to future UMTS networks. Wireless connections are used within hospitals (Remote radio Bluetooth technologies). A rendering component automatically adapts the User Interface to the capabilities and limitations of specific user mobile devices with the goal to offer the maximum DOCMEM functionalities allowed by the device.

- An **Intelligent User Interface** permits to the users to access to the services. A Web interface generates pages, forms, XSL and other templates. This interface can use a secure process of Public Key Interface (PKI) for authenticating practitioners. Specialized Intelligent User Interaction facilities were developed to enhance this basic Web interface:
    - Practitioners can control access to patients' files through **voice recognition** and pattern matching techniques, both with patients' identification and with the CEN13606 recommended terminology. A Voice Assistant is used in order to match terms used by the medical doctors (voice) with the EHR and medical terminology used by the DOCMEM services.
    - Patient's medical information can be accessed through a **virtual representation of a human body**. The use of the W3C 3D-XML recommendations were considered for these developments. For this facility, the equipment is a multimedia mobile GPRS terminal.

3. MOBILE HEALTH APPLICATIONS

The DOCMEM Service supports several kinds of healthcare professionals in different contexts of use: City practice, In-hospital practice, and City-hospital collaboration.

- Independent city practitioners rely on DOCMEM to :
    - Archive their patients' files on the DOCMEM Portal through a secured Web connection
    - Access patients' files and capture new patients' data when visiting patients outside the practice through wireless information devices. The Wireless devices rely on secured mobile telecommunication.
    - Exchange confidential patients' data with other healthcare professionals in a secured way
- Hospital practitioners use PDAs and radio communication at the point of care, in bedrooms, operations and service wards. Practitioners can thus get access through a DOCMEM Portal to patients' files stored in the hospital medical software. PDAs offer an advanced user interface for controlling access to medical data through voice and graphical interaction.
- Ambulatory practitioners of a geographic area rely on the DOCMEM Portal to get access to patients' medical data available in the local hospital medical software. Such collaboration is needed by independent city practitioners when visiting patients or by ambulatory staff, for instance in ambulances for emergency care.

The DOCMEM Key benefits in the three different contexts of use are the Electronic Health Record (EHR) availability at the point and time of care, the medical data usability on wireless information devices with advanced speech and graphic control features, the standardised storage and exchange of medical data among practitioners and the secured access to EHR.

The DOCMEM platform consists of :

- ❑ A Portal Manager that provides the user interface for multiple delivery channels
- ❑ A Services Manager that controls secure processes for getting access to the various services and data
- ❑ An Exchange module that handles asynchronous communication and data exchange
- ❑ A Data/XML Manager that manages the access to the central database
- ❑ A Central database that gathers patient medical records, terminology and user-related data

The architecture and the tools used to develop the DOCMEM platform are further described hereafter.

Portal Manager:

It contains two main sub-modules

- *Secure Mobile Access*: this subsystem allows access to the DOCMEM system for the future generation of mobile communication and PDAs. The work performed in this area within the project aims at ensuring that the DOCMEM services will be accessible through E-GPRS and UMTS as soon as these technologies start being deployed by European mobile communication operators. Wireless connections are used within hospitals (Remote radio Bluetooth technologies). The secure mobile access provides connection to the web interface to assure processing of data.

DOCMEM Architecture

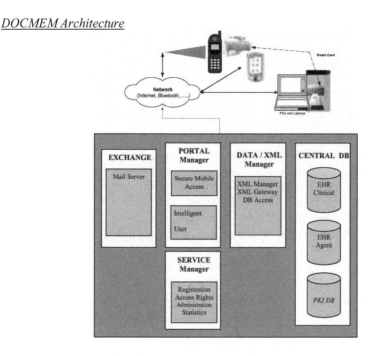

Figure 1. DOCMEM architecture

- *Intelligent User Interface:* the purpose of this subsystem is to provide users' access to the services. It can be acceded by PCs, laptops and other mobile devices, through Internet or Virtual Private Network (VPN). Specialized Intelligent User Interaction facilities were developed to enhance the basic Web interface:
  - A *Patients' Files Assistant* enabling practitioners to control access to patients' files; thanks to voice recognition and to pattern matching both with patients' identification and with the CEN13606 recommended terminology.
  - A *Virtual Patient* facility enabling access and presentation of a patient's medical information through a virtual representation of a person's body. The use of the W3C 3D-XML recommendations will be considered for these developments. (For this facility, the target equipment is a multimedia mobile E-GPRS or UMTS terminal).

The intelligent user Interface communicates to the DB access through the Data Management subsystem. A direct access by VPN can be made to the XML handling sub-module. A connection is made to the Exchange sub-system for sending and receiving messages.

Services Manager:

This subsystem represents the core of DOCMEM, It includes the web service management and a secured process of PKI (Public Keys Interface) with its own Data Base (DB) containing the public keys of the practitioners using DOCMEM. Services processed are:

- Administration interface
- Practitioners' Registration and management
- Patients file access management and Replication of common information for other DOCMEM servers
- Access rights management and secure physical protection using the PKI process
- SNMP monitoring and reporting and Statistics processing

The online services are in permanent connection with the web interface in order to produce the web services. A connection is also made to the DB access within the Data Management subsystem.

Exchange Module: this subsystem is able to send and receive e-mail, fax and SMS; it is compliant with the intelligent user interface. It can also transmit data from the DB through the XML handling sub-module. The Exchange subsystem will carefully takes into account legal constraints.

Data/XML Manager: the role of this Subsystem is to manage the access to the central DB. An XML sub-module transforms the data in order to be XML compliant with the central DB and the networking.

A direct connection to the XML sub-module is made for practitioners using VPN. Communication facilities are developed in XML/EDI to implement CEN 13606 compliant messages adapted to the two target populations:

- Ambulatory practitioners need import/export facilities in their IS to build copies/archives of their files in the DOCMEM server.
- Hospital IS uses similar XML/EDI messages but only for creating transient files on the DOCMEM server that acts as a Gateway between the HIS and mobile healthcare professionals
- Both hospital professionals and ambulatory practitioners need occasionally to exchange medical data with other professionals; this module offers them a secure communication channel.

In addition to the above features, this service also enables access to External Web Services relating to healthcare information provision, in particular those concerning the standardization/revision of Good Practice Protocols.

Central DB: the DOCMEM Data Base is composed of three parts:

- The patient medical records under a standardised structure (EHR) based on the CEN 13606 standard, they consist of general and medical patient's information organized historically:
    - Patient identification details (name, address, age, sex) and Patient registration details (own general practitioner, identification number)
    - Practitioner's summary (text)
    - A minimal set of structured and multimedia data
    - Definition of the conditions requiring the current episode of care
    - Indication of actions taken (or to be taken) to address the clinical condition
    - Previous history of intervention
- The terminology and patterns terms
- The Practitioners' repository giving information about practitioners using DOCMEM.

4. CLINICAL / MEDICAL VALIDATION - IMPLEMENTATION

In the DOCMEM project, pilot operations was run in 3 countries (France, Italy and Greece) in order to validate the solution. The Services implemented will show the practical usage of the platform developed. A significant share of resources will be devoted to the communication activity required to raise the interest of practitioners in order to ensure progressive subscription of pilot users in the three pilot regions.

The DOCMEM outcome should therefore be tailored to the requirements and needs of 3 pilot contexts:

- The French pilot "Mutuelle d'Assurance du Corps de Santé Français", targets independent ambulatory practitioners in **city practice**. The targeted practitioners will use DOCMEM platform in order to Archive their patients' medical files, to get access to their patients' medical files while visiting their clientele and to transfer information to colleagues or to National Healthcare establishments. At the beginning the users of DOCMEM system will be General Practitioners and Cardiologists. This enables the French pilot to work on both the issues concerning the General Practitioner, who is the main contributor to patients' care and treatment, and the specialist who is only called upon for very specific health complaints.

The French pilot organisation aims at involving first 10 ambulatory practitioners for the verification phase and to progressively commit up to 70 by the end of the project. The communication activity will be organised as a "marketing oriented" campaign targeting ambulatory practitioners that use one of the two EHR software supported by DOCMEM in France.

- The Greek pilot "Athens Medical Centre", mainly focuses on the **in-hospital** use of the system for obtaining clinical data at the time of need, even from the ambulance and the patient's bed. The Greek pilot organisation belongs to a private group gathering 8 hospitals. The main focus of DOCMEM for this group is to improve the quality of service by providing practitioners with patients' medical files whenever and wherever needed and by ensuring a seamless exchange of patients' information among practitioners of the group, especially when a patient is transferred between medical services.

In the first validation phase, the pilot will focus on one large hospital, involving 1 medical department for the verification phase and progressively committing up to 5 by the end of the project. The communication activity will be organised as a change management programme within the pilot hospital in order to raise awareness and commitment of employed practitioners.

- The Italian pilot "Azienda USL di Parma", targets a rural area characterised by the extension of the territory. In this area, the access to the laboratory test results as well as to the patient medical records is difficult and quite often late, for the needs of General Practitioners. The pilot focuses on **city-hospital collaboration** and aims at improving the transfer of the patients' test results from the hospital laboratory to the city practitioners. This will facilitate the work of General Practitioners in shortening the time for delivering the tests results from the Laboratory. Another objective is to provide General Practitioners with the medical files of their patients whenever needed.

The Italian pilot organisation aims at involving first the Laboratory department of one hospital together with 5 ambulatory practitioners for the verification phase and to progressively commit up to 30 ambulatory practitioners by the end of the project. The communication activity will be organised as a mix of a marketing campaign/change management programme vis-à-vis regional ambulatory practitioners working closely with the hospital.

As of January 2003, the pilot services implementation was not yet finalised and the validation has not yet started but the main validation objectives have already been defined. For each pilot, the online service will be supplemented with proximity services of network animation and support. In this context, DOCMEM will be assessed with respect to usability and usefulness. DOCMEM validation will be completed during summer 2003.

5. REFERENCES

DOCMEM - D2.2 Platform specification
DOCMEM - D3.1 Platform Development
DOCMEM - D4.1 Specification of the DOCMEM Services

SMARTIE: SMART MEDICAL APPLICATIONS REPOSITORY OF TOOLS FOR INFORMED EXPERT DECISION

Juan J. Sancho[*], Susan Clamp, Christian Ohmann, José.E.Lauzàn, Petros Papachristou, Jean P.Thierry, Angela Dunbar, Julio Bonis, Sergio Rodríguez, Chris Kirke, Innes Reid, Hans-Peter Eich, and Clive Tristram

1. INTRODUCTION

1.1 Project Aims

The aim of the SMARTIE project is to create a large number of computerised decision support tools (DSTs) to help clinicians with decision making in routine clinical practice.

1.2 Project Overview

SMARTIE is part of the Information Society Technologies research program funded by the European Commission. The project has a total duration of 30 months, ending June 2003. Project partners include experts in all aspects of computerised decision support development, from research and clinical, through software and systems development, to quality management and dissemination. Decision support is being developed in several medical specialties: Gastroenterology, Endocrinology, Emergency Medicine and Intensive Care Medicine, Paediatrics, Nutrition and Quality of Life with plans to expand to other specialties in the future.

---

[*]Juan J. Sancho, MD. PhD, Servei de Cirurgia General i Digestiva Passeig Marítim, 25-29. 08003 Barcelona, Spain

1.3 Type of Decision Support to be Computerized

MedNotes™ contain existing clinical knowledge represented in the form of a decision support tool. The term decision support tool may be defined as anything that can be used to aid clinical decision making. This broad definition encompasses a large range from simple clinical calculators, through management algorithms, to complex expert systems.

2. METHODS

2.1 Decision Support Tools

MedNotes™ cover an ample spectrum of critical information needed by the practicing clinicians to take or refine decisions, from simple calculators to full expert systems. Some examples and categories are represented in Table 1.

Table 1. Types of applications developed

Type of Application	Example
Clinical calculators	Body Mass Index, Creatinine clearance, TNM staging
Checklists	Signs of deep vein thrombosis
Scoring systems	Child-Pugh, Ranson-Imrie, APACHE II, SAPS II, SOFA, Alvarado.
Decision trees	Tests to perform in front of a hyperamylasemia to diagnose an acute pancreatitis. Guidelines for the treatment of Gastro Esophageall Reflux Disease.
Expert systems	Acute Abdominal Pain

2.2 Knowledge Acquisition

Two main processes are used to accurately acquire the knowledge behind the MedNotes™: a) Identification and selection of Decision Support Tools (DSTs) and b) Creation of the knowledge base content

For the identification and prioritisation of the DSTs, a panel of international experts has been recruited. The profile of each expert is a senior working clinician, a member of national/ international associations, an author of SCI indexed publications, and a specialist with an interest in informatics and best clinical practice. Currently, the project SMARTIE has enrolled 57 experts from UK, Spain, Germany, France, Italy, Ireland, Netherlands, Belgium, Portugal, Norway, Switzerland, Check Republic and Austria.

2.3 Selection of Decision Support Tools for Computerization

The role of these experts is to identify existing decision support tools in clinical use, assist in the selection of DSTs for development, contribute to gather and formalise the knowledge from publications and approve the methodology for validation. In order to identify suitable DSTs for development into MedNotes™, a database of existing decision support tools has been created containing all identified decision support tools potentially suitable for computerisation. The DSTs were identified from literature searches, searches of the World Wide Web and recommendation by clinical experts. There are over 1900 DSTs in the database. This database is available from the WEB under restricted access, so experts around the world can contribute to the project.

The clinical experts objectively assess each decision support tool for suitability for computerisation according to defined criteria. The results of this process are compiled and used to select DSTs.

In addition to this process, several specific clinical problems have been identified to act as focuses for the evaluation phase of the project. A concerted effort is being made to create comprehensive packages of decision support tools that may help in the management of these clinical problems. An example is the package for acute abdominal pain, which will include severity scores, management algorithms, guidelines for imaging, preoperative checklists and DSTs to guide the assessment and management of postoperative complications.

The capture and formalization of the knowledge is facilitated by the creation of storyboards for each application where the expert actually draws the screens of the application and adds the calculations, logic tree and additional help for the users.

2.4 Software Engineering

MedNotes™, the SMARTIE applications are generated from the storyboards in 3 main languages (English, Spanish, French). Selected applications are also being translated to German, Greek and Czech. Then they are compiled for 4 platforms: stand alone MS Windows, PalmOS and Pocket PC, Web version (either to download or to run them directly inside an internet web browser).

The representation of the medical knowledge is precise, non-ambiguous, human-readable, computable and computing platform independent. In order to accomplish this, a double layer of XML documents for each application is created with clear separation of the knowledge description and the user interface elements.

3.   VARIFICATION & VALIDATION STRATEGY

3.1 Interface Verification

The interface components, using RAD (Rapid Application Development) tools, were tested by clinical experts to verify the feasibility for the task of each interface component. The bed-test applications were filled (as an example) with the knowledge contents from the prototypes that the consortium already created to assess the potential of the SMARTIE idea. All MedNotes™ in all languages, on all platforms share the same user-friendly, wizard-like interface.

3.2 Contents Verification

Once the Expert Panels have generated the algorithm and help contents for each MedNote™, selected users groups analyse the format of the input/output proposed, the consistency and completeness of the explanations and the pertinence of the sources selected. This is a quick review and acts as a user-audit for the output for the clinical quality of the applications.

3.3 Verification of Pre-release (Beta) Versions

The testing and re-coding of the α-versions and β-versions of the assembled MedNotes™ requires two thirds of the total time (and effort) for the development of each individual application. The testing is conducted in an industry standard fashion with pre-build check-lists for each phase of the testing that allow for consistent reporting and management of the versions and modifications. The checklists are generated by the co-ordinating partner. The general set-up follows a modified Yourdon system adapted for the production line strategy adopted by SMARTIE. The most complex MedNotes™ follow a strictly adapted methodology for in depth testing. Similarly, the technical porting to alternative platforms undergoes parallel testing lines. Finally, each localised version undergos an abbreviated contents and final version testing, mirroring the commercial strategy to test localised versions of general applications.

3.4 End-User Validation

The real validation phase (as opposed to the previous "test and verification" process) begins as soon as the stand alone version of the MedNote™ is created. The SMARTIE website serves as a central place to obtain the MedNotes™ for all platforms, languages, and clinical specialties. A wider user group was created and is constantly expanding to include early adopter. The Users & Validation Partner manages an iterative process of validation on the real use and impact of each MedNote™ under real conditions. The feedback is then passed to the developers and co-ordinating partners who, in turn, take the necessary actions to improve the released applications and those in the production line.

3.4.1   Clinical Trail Evaluation

Specific testing for patient outcomes and clinical practice impact produced by the use of MedNotes™ has been designed following the methodology of a clinical trial (as if a MedNote™ were a new diagnostic tool) comparing clinicians using and clinicians not-using MedNote™ applications.

MedNotes™ are currently implemented in 8 hospitals throughout Europe. Three hospitals in Spain: Hospital Clínic de Barcelona, Hospital del Mar, Hospital Parc Tauli de Sabadell. Three hospitals in the UK: Gwynedd Hospital, Glan Clwyd Hospital, Wrexham Maelor Hospital, and two hospitals in Germany: Department of General- and Trauma Surgery (Heinrich-Heine University Duesseldorf), Department of Visceral and Vascular Surgery (University of Cologne).

These rigorous clinical trials are currently underway and focused to evaluate economic, access, and quality of care benefits. We estimate that MedNotes™ will support access to care through providing professionals decision support tools which may otherwise not be used in clinical practice and encouraging compliance and greater use of clinical guidelines. This is illustrated through our study addressing the clinical problem of upper gastro-intestinal bleeding (UGB). In this trial, on the modified Rockall score and internal clinical guidelines associated for each risk group are all compiled in Palm® -based UGB package of MedNotes™. The uptake of the guidelines with MedNotes™ is compared to the uptake with paper forms and the compliance to the guideline is studied. We are also studying the impact of MedNotes™ on acute abdominal pain routines, and evaluating clinical improvement of enteral-/parenteral- nutrition in patients on intensive care units bycomputer-supported calculation of nutrition on hand-held devices.

4.   DISCUSSION

4.1 Innovation

MedNotes™ provide healthcare professionals with decision support tools which are just as versatile and mobile as they are. Available in 3 languages on 4 different platforms (Pocket PC, Palm, Web, PC), these high quality mini-wizards generate the results of complex calculations and algorithms already published and clinically certified in scientific literature.

The applications are conceived with the "Just-in-Time Learning" concept in mind, to assist in immediate clinical decision taking by the user. Few or no software medical applications fully embrace today this innovative concept. SMARTIE develops a "collection" of modular applications (as opposed to a monolithic single piece of software) with different degrees of complexity in order to demonstrate that the approach can be used as a single solution for very different kinds of decision support applications.

The role of physicians is -in fact- to *change the prognosis* of a disease. Nevertheless, few medical doctors actually know the prognosis of a disease affecting a particular patient. Few and cumbersome applications provide this kind of information for the different therapeutic options available for a specific disease. A significant number of MedNotes™ are devoted to produce an innovative change in this area fulfilling this hole in medical knowledge. An essential technical "innovation" in today's software development and distribution is the ability to be ubiquitous. MedNotes™ are developed in more than one platform, available to be downloaded (or run) from a WEB site. The adherence to Open Source methodology for development, licensing and distribution is a dramatic innovative step in this direction.

A substantial role of today's health care professionals is to give personalised information to patients about their particular class, form and stage of the disease. Scientific measurement of the real impact of any decision support tool on the clinical practice is unfortunately lacking. Validation is one of the key issues of SMARTIE, involving the users in the process –from choosing the scope of the applications, to shaping the interface, and to testing each MedNote™.

4.2 DST's in current use

The knowledge in clinical practice is represented by the knowledge that some experts have about the best practice management. This knowledge has not been ever properly (universally, consistently, timely and updated) disseminated to the working clinician. The know-how about the best management practice for many diseases remains the privilege of few and the real user of this knowledge either does not apply it at all or uses outdated knowledge. The Computer Assisted Decision Support Systems (CADSS) –successors of the Expert Systems of the 80's, tried to partially solve this absurd situation without any uniform success.

Very few CADSSs have been used in routine clinical practice, although scoring systems for various measures of prognosis or severity of a condition are used. Even a dedicated palmtop calculator has been built and distributed to calculate a single severity score for ICU patients! (APACHE II score). Most of these however are done by hand. The main problem has been the integration of computer packages into routine clinical use. Some problems occur with attitudes of clinicians towards these but the other significant problems have been the user interface and accessibility of systems for busy clinicians. There is no issue with the take up of new technology in medicine if it is perceived as useful (i.e.: the use of EKGs, MRI, CT etc).

There is currently an apparent paradox in the medical decision support field. On the one hand visionary leaders recognise the overwhelming need to support medical decision making via electronic patient records, automated alerts and reminders, and clinical guidelines that are clearly documented, constantly maintained, and delivered just-in-time and on-point. They view such technologies as strategic weapons in the fight to reduce waste, poor quality, and preventable expense in a health care industry that has grown worldwide.

On the clinical side, evidence-based medicine and the randomised clinical trial are producing new practice guidelines at a quickening rate. The strict view of AIM (Artificial Intelligence in Medicine) is that after a quarter of a century of trying it has failed to achieve its vision of software supplanting clinical expertise. There has been no paradigm shift in the practice of medicine as a result of expert systems, and the dominant modes of performance support are still non-automated. The field is entering a new age of prolific evidence and knowledge generation. The SMARTIE project addresses key issues that have previously prevented computerised decision support from being incorporated into everyday clinical practice. There is substantial clinician involvement throughout the project to ensure that the decision support provided is clinically useful and acceptable and that it integrates easily into working patterns. We hypothesise that the large number of computerised DSTs, as well as the comprehensive coverage of certain key clinical problems, will ensure that clinicians become familiar and comfortable with using decision support in their routine practice. A strong emphasis has been put on quality assurance and reliability of the computerised DSTs, further inspiring clinician confidence and increasing use.

5. SUMMARY

SMARTIE aims to develop and demonstrate tools for capturing, representing and transferring best practice knowledge in several clinical specialities. Its objective is to solve the lack of homogeneity and availability across EU countries in the application of best practice guidelines by means of a single interface for decision support agents along with re-engineering pre-existing knowledge based systems. Ultra friendly user interface, multi-platform support and integration with existing Clinical Information Systems are the main technological features. Strong user driven involvement selecting, testing and validating applications and knowledge extracted from Evidence Based sources and validated by EU experts are its key characteristics. The conceptual framework is the Learning-Just-in-Time paradigm. MedNotes™ are a huge collection of medical decision support tools running from the Web, PC, and PDA (Personal Digital Assistants) devices, tested and tried in real world situations and licensed under Open Source.

6. REFERENCES

Ohmann, H.P. Eich, E. Keim and the COPERNICUS Abdominal Pain Study Group. Multilingual decision-support for the diagnosis of acute abdominal pain. An European Concerted Action (COPERNICUS 555). E. Keravnou, C. Garbay, R. Baud, J. Wyatt (Eds.) Proceedings of the 6t^h Conference on Artificial Intelligence in Medicine Europe, AIME'97.
Grenoble, France, March 23-26, 1997, 393-397.
Q. Yang, C. Ohmann, H.P. Eich. Information technology and diagnosis in acute abdominal pain: results of a multicenter observational prospective study in Eastern Europe (Euro-AAP). In: N. Victor et al. (ed): Medical Informatics, Biostatistics and Epidemiology for Efficient Health Care and Medical Research. Urban & Vogel, München, 1999, S. 88-89
Sancho JJ, Planas I, Doménech D, Martín-Baranera M, Palau J, Sanz F. IMASIS. A Multicenter Hospital Information System – Experience in Barcelona. En: Iakovidis I, Maglavera S, Trakatellis, User Acceptance of Health Telematics Applications. Looking for convincing cases. Amsterdam. IOS Press. 1988; 35-42.
Silverman, B.G., "Computer Reminders and Alerts," IEEE Computer, v.30. n.1, Jan.1997
Silverman, B.G., Murray, A., "Collaborative Human Judgment Theory: Issues for Improving Knowledge Based Systems, in Full-Sized Knowledge Based Systems Workshop Proceedings, Washington DC: GWU Tech Report, May 1990.
Sancho JJ, Sanz F. Learning Just-in-Time in Medical Informatics. En: Iakovidis I, Maglavera S, Trakatellis, Information Technologies for Medical Informatics Education. Amsterdam. IOS Press. 1999.
Ohmann C, Eich HP, Sancho JJ, Díaz C, Faba G, Oliveri N, Clamp S, Cavanillas JM, Coello E. European and Latin-American countries associated in a networked database of outstanding guidelines in unusual

clinical cases. En: Kokol P et al, eds. Medical Informatics Europe'99. Amsterdam. IOS Press. 1999; 59-63.

TELEMEDICINE AS A NEW POSSIBILITY TO IMPROVE HEALTH CARE DELIVERY

Hans Rudolf Fischer, Serge Reichlin, Jean-Pierre Gutzwiller, Anthony Dyson, and Christoph Beglinger[*]

Part I: BASICS OF TELEMEDICINE

1. INTRODUCTION

Telemedicine is defined as the delivery of health-care and sharing of medical knowledge over a distance using telecommunication systems. The term telemedicine is usually associated with modern telecommunication systems: transmission of electronic medical records and images, remote monitoring of a patient's vital parameters, tele-conferencing and interactive tele-teaching.

2. MOBILE COMMUNICATION TECHNOLOGIES

The rapid technological progress in the last decade, with the deployment of high-speed, high-bandwidth telecommunication systems around the world and the development of devices capable of capturing and transmitting images or other data in digital form, has led to a surge of interest in telemedicine.

With the World Wide Web booming, the general public has gained easy access to a variety of information, some of which was initially intended for professionals only. This has brought with it security concerns. These issues can be avoided by maintaining the information in a private network, isolated from the public Internet, i.e. an intranet. Security is achieved at the cost of accessibility. The solution to this dilemma is the

[*] Christoph Beglinger, Clinic of Gastroenterology, University Hospital Basel, Petersgraben 4, 49031 Basel, Switzerland, E-mail: beglinger@ tmr .ch or cbeglinger@uhbs.ch

extranet, i.e. the use of Internet communication paradigm to allow secure access to private information to closed user groups over the public Internet. Typical extranet application scenarios involve employees of an organization (e.g. employees from a hospital) that need to access sensitive information or services, when physically far from organization premises. In telemedicine this might comprise communication between physicians or between a patient and his physician.

Also, personal mobility is playing an ever-increasing role in modern lifestyle. This applies to both professional and leisure activities. Enter the mobile extranet. In this concept, two principles are merged: the need for personal and/or terminal mobility and the need for access to information and services. Telemedicine is a field that stands to benefit greatly from the application of the mobile extranet.

3. PATIENT-FOCUSED HEALTH-CARE

Increasingly, patients are managed at home, in part because the cost of in-patient care is major concern in various countries. In many instances in the past, however, providing medical care at home just resulted in a shift of the costs rather than cost savings - complicated diseases require sophisticated treatment strategies with frequent home visits. Many people are seeing tele-homecare as a potential solution to the dilemma. Using low-cost equipment and regular telecommunication networks, the level of care of a patient at home can be improved through increasing the frequency of contacts at a much lower cost per contact as compared to physical visit.

Recently, attempts have been made to develop a new strategy for he telemedical monitoring of patients in the form of medical contact centres, offering their services to patients and doctors around the clock. These teleconsultation centres (MCC) act as interfaces between the patient, the treating physician (GP) and the medical competence centre (CoC e.g. teaching hospital) (figure 1). Communication between the patient, treating doctor and medical competence centre is established by different means using a mobile extranet as well as a conventional or mobile telephones. Clinical as well as technical feedback to patients and doctor-to-doctor dialogue is provided. Both patient and data triage are facilitated.

In this context we used the so called "Integrated mobile healthcare solution an management disease"-mode, which integrate diagnosis, therapy and nursing as showed in figure 2.

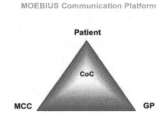

Figure 1. Organization of the communication

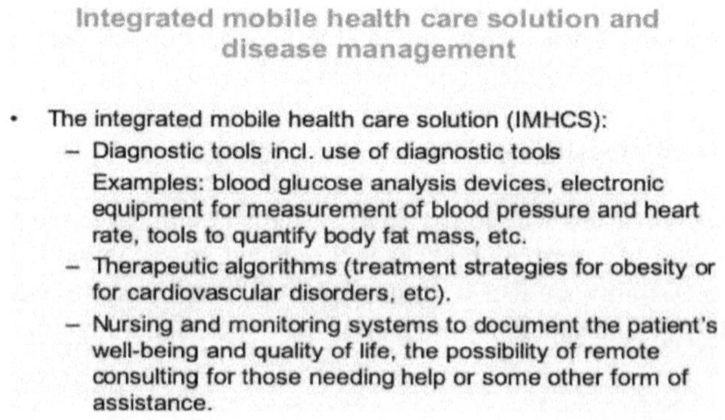

Integrated mobile health care solution and
disease management

- The integrated mobile health care solution (IMHCS):
  - Diagnostic tools incl. use of diagnostic tools
    Examples: blood glucose analysis devices, electronic
    equipment for measurement of blood pressure and heart
    rate, tools to quantify body fat mass, etc.
  - Therapeutic algorithms (treatment strategies for obesity or
    for cardiovascular disorders, etc).
  - Nursing and monitoring systems to document the patient's
    well-being and quality of life, the possibility of remote
    consulting for those needing help or some other form of
    assistance.

Figure 2. Integrated mobile health care solution

4. THE MOEBIUS PROJECT

With this background in mind, the information society technology project MOEBIUS (Mobile Extranet-Based Integrated User services) was planned. MOEBIUS aims to test the above-mentioned integrated mobile extranet concept: it entails a modern IT infrastructure along with an integrated communications tool for healthcare applications. The culmination of the MOEBIUS project is the realization of a clinical feasibility study using mobile technology in different aspects of patient-focused health-care.

Here we report on a specific clinical trial that was designed to gain experience in networked health-care delivery. The trial involved a group a young obese patients that participated in weight loss program.

The mobile extranet used in the trial was based on General Packed Radio Service (GPRS), as it was the most attractive mobile data transfer technology available. Due again to the modular infrastructure of the system, new solutions such as cellular IP and Universal Mobile Telecommunication Systems (UMTS) can be adopted as soon as they become available.

5. THE TECHNICAL SOLUTION OF MOEBIUS, INTRODUCTION AND DEFINITIONS

Clinical trials are one of the most important tools in the development process of a new drug. Legal offices and ethical reasons formed standard rules for development, well known as Good Clinical Practice (GCP). The last years introduced a new paradigm in information technology supporting clinical trials. While in former times case report forms (questionnaires) on paper have been used, today many trials use so called remote data entry systems (RDE systems), that is questionnaires that can be answered using a standard PC or a laptop. Data is then submitted to the trial database via standard telephone/ISDN phone lines or via an internet connection or data is collected and transmitted to the trial database in a bulk once in a while.

5.1 Remote data entry (RDE) systems

There are many reasons why to use RDE systems in clinical trials. Just to mention a few, increased speed of information exchange, reduced errors, improved process control and enhanced accessibility are listed. In this paper we won't stress too much on the RDE system itself. But for understanding the rest of the paper, some important points have to be mentioned.

In general, RDE systems consist of a simple structure: on the backend a database serves as the main data storage. This database could be complex when serving different trials at the same time. When designing and implementing this part, the important rules given by GCP and legal offices (e.g. FDA, BfArm, IKS) must be followed for getting results of the trial officially accepted. On the client side there are three different categories of technologies being used. Starting with the most specific system, special software programs can be used for user interaction. These stand alone solutions are generally designed to run on one specific platform only. Today web based applications more and more are becoming accepted. In Web-RDE a standard WWW browser is used to present HTML forms that can be filled with trial data or the HTML page just serves as a host for an Java applet that directly connects to the trial database. This last solution is proved to work well in the e-banking field, too, and is supported by most major databases. Minor drawbacks are that there always has to be a special software or a browser for the patient to work with and that you need an online connection (which might be a problem when you travel around internationally). In many cases these systems perform well and offer an accepted platform for multicentre trials.

In some trials however, a better compliance and in consequence better data quality can be reached when using input devices that are always "near" the patient and help him to protocol things when they occur and not when he is possible to access a computer. One application area is the pain diary where the patient lists pain attacks and his treatment (drug, meditation,). Other areas of use are trials that deal with hay fever or allergic reactions. In this case the immediate event logging could lead to better data quality. One approach to better handle the patient interaction deals with using mobile devices as the input technology. Often used in daily live communications and in mobile commerce (mobile e-business), many persons are quite well familiar with this technology.

5.2 Mobile input devices

If talking about mobile devices many people just think of cellular phones that could be seen in daily live. Other could imagine no other solution than that one using a personal digital assistant (PDA), like the Palm Pilot. Indeed the use of cellular phones as point of interaction with the patient seems reasonable, but the same is true for the Palm solution.

As Wagner points out, you can cluster mobile devices into five categories: mobile phones, smart phones, Palmtop PDA, Handheld PDA and derivatives (Sub notebook, Mini-notebook, ...). There are some characteristics of the devices that have to be recognized, when a mobile clinical trial is to be designed. These characteristics will be discussed in the two sections: hardware and software. The more important aspect seems to be that one of the hardware itself (screen size, weight of the system, ...), but as the software systems mature, software solutions will have bigger impact than hardware issues. We don't present a complete list of all aspects to be mentioned rather than point out some issues to show the diversity of aspects.

5.3 Hardware

To compare different devices categories have to be defined. Several approaches could be chosen, in clinical trials the compliance of the patient has a very important role and hence we look at categories that are directly related to the user interaction:
Size of display
Input of data
Access to the device
Life cycle without recharging the accumulator.

The size of the screen is important in all cases when data is output in a different way than by voice/sound. Because most voice processors don't work really well, manual output – and hence display size – plays an important role for the compliance of the user. When considering cellular phones, screens exceeding 100 x 70 pixels hardly can be found. This means the devices just can present 4 to 6 lines of black / white characters. Font modifications like centered, bold, italic or underlined mode vary from device to device. Things look a little bit better when you consider smart phones (like Ericsson R380s or the Nokia 9110 communicator). In the PDA family, the Palm (and related) devices have 160 x 160 pixel, the Compaq (Aero, IPaq) and Casio (Cassiopeia) systems offer even bigger screens (240x 320 pixel). These systems mostly use graphical character representation, so the screen size in number of characters vary. The next category (handheld like HP Jornada, Psion 5mx/ Revo etc.) nearly offers EGA resolution (640x480), so software that is based on a window based operating system can be used. The last group "derivatives" (SimPad, Compaq Aero, Getac CA35) generally offers at least 800x600 Pixels, meaning you have a "small PC screen" to present the questionnaires. In total screen size increases from cellular phones to the derivatives category. Another trial dependent aspect seems to be the abilities of the patients using the device. Especially in trials with elderly people, modern communication systems seem to present a major barrier to acceptance. These people aren't used to use a cellular phone, they may have problems with identifying the characters on the phone screen, may have problems using the multi function keyboard on the cell phone. Here special attention is needed to choose the right interface.

If you consider data input with the cellular phone, the user has to use the button set on the phone. PDA and handheld systems use a data pencil in combination with a touch screen whereas the derivatives also offer the keyboard. Here special patient groups have to be in mind, too, when the trial is to be designed.

These two dimensions show an increase of possibilities from cellular phone via PDA and handheld towards derivatives. Reciprocally the accessibility of the device decreases from the small cellular phones to the bigger derivatives: users of cellular phones and PDAs can access that device at nearly every time, being private life or in business. This is not true for the other systems since the weight and the usage don't allow full mobility and immediate use.

So it has to be decided which system fits best to the needs in the trial.

5.4 The Software

In this paper we want to consider only two aspects of the underlying software: the operating system itself and the possibility to present WAP pages, since these probably will be the base for most trial systems, at least in near future.

There is hardly no impact of the used operating system (OS) for the design of the RDE system. In today's variety of devices only three different known OS are used: Windows CE (Microsoft), EPOC (Symbian consortium) and Palm OS. There is big effort to modify Linux to fit to these devices and it could be also one "player" in future. Since the cellular phone devices cannot use other programs than those built in by the manufacturer, there is only the possibility to use WAP pages for the RDE purpose.

Palm OS and WinCE based systems offer good support for programmers to implement own programs and run them on the device. The derivatives category is designed to work with software that is loaded to the device by the user.

One of the most important aspects of mobile systems is the ability to present WAP [WAP] pages. Since WAP offers the WWW front end to mobile systems, ordinary WWW based RDE systems can easily be extended to work with mobile devices. But because WAP or better said the Wireless Mark-up Language (WML) and WMLScript aren't really in a stable state [WML], the programs can only use limited possibilities of the devices. Many things, that work with the WWW (be it image maps, mixing of Mark-up and Scripts, ...), don't work with the WAP devices. So the program designer has to know the limited capabilities.

Part II: OBESITY TREATMENT BY TELEMEDICINE – THE MOEBIUS OBESITY TRIAL

1. OBESITY AS A HEALTH CARE PROBLEM

Obesity has become a major public health problem in most industrialized countries because of its high prevalence, causal relationship with serious medical diseases and economic impact. In the United States, it is estimated that 300'000 deaths a year are caused by obesity (1) – analogous, reliable figures are not available from Switzerland. Each year, enormous amounts of money is spent on weight loss efforts and the medical complication associated with obesity, but the long-term results of most weight management efforts have been disappointing. In the following sectors, the medical part of obesity is reviewed to provide the background information for the MOEBIUS obesity trial.

2. DEFINITION OF OBESITY

Obesity has increased at an alarming rate in recent years and is now a worldwide public health problem. In addition to suffering from poor health and increased risk of illness such as hypertension and hearth disease, obese people are often stigmatized socially.

Obesity is formally defined as a significant increase above ideal weight, ideal weight being defined as that which maximizes life expectancy. Actuarial tables indicate that life expectancy is reduced when body-mass index, an indicator of adiposity or fatness, is significantly increased above the ideal level. The most intriguing development is an alarming increase in adolescent obesity in recent years. Thus obesity is associated with a significant increase in morbidity and mortality and is a major public health problem. For reasons that are not known, obesity is associated with an increased risk of hypertension, heart disease, diabetes and cancer. Even modest weight loss ameliorates these associated conditions.

In addition to the prospect of diminished health, obese people are often stigmatized both socially and in the workplace. Although the premium on leanness has become especially prominent in late-twentieth-century Western societies, this view is dependent on the cultural context. In many cultures, obesity is considered to be a sign of affluence and prestige, particularly among those cultures where food is less available.

Excessive body fat content is associated with an increased risk for medical illnesses. The identification of patients at increased risk for adiposity-related medical complications is complex. Accurate assessment of body fat is difficult using sophisticated technologies. Data obtained from large-scale epidemiological studies that correlated body mass index (BMI) with clinical outcome have generated guidelines for identifying patients at risk for adiposity-related medical complications and premature mortality. BMI is accepted as the unit for definition of overweight (normal BMI rabge18.5-24.9 kg/m^2; overweight 25-29.9 kg/m^2 and obesity >30 kg/m^2; details in Table 1 (2)). The worldwide epidemic of obesity is signaled by the rise in the percentage of the population with a BMI of >30kg/m^2. BMI is calculated as weight (in kilograms) divided by height (in square meters). BMI correlates strongly with densitometry measurements of fat mass; the main limitation of BMI is that it does not distinguish fat mass from lean mass.

Obesity is not a single disorder but a heterogeneous group of conditions with multiple causes. Body weight is determined by an interaction between genetic, environmental and psychosocial factors acting through the physiological mediators of energy intake and expenditure.

Both genetic and environmental factors modify body weight. Acceptance that the current epidemic of obesity is largely environmental, because of a mismatch between man's ancient genes and his current environment, will help direct preventive and therapeutic strategies. BMI alone, however, is not a sufficient predictor of the detrimental effect of obesity. Central obesity (defined as waist circumference >102 cm in males, >88 cm in females) increases excess mortality and risk of diabetes, coronary heart disease and some forms of cancer. Weight gain after the age of 20 also predicts increased risk. Finally, a sedentary lifestyle by itself increases mortality rates from all causes (3).

BMI is a risk factor like blood pressure and cholesterol in that, as is increases, so does the risk for medical complications and adverse effects (4). Chronic exposure to any of these three factors produces disease – heart failure, stroke and renal failure for raised blood pressure; arteriosclerosis with coronary and cerebrovascular occlusion for hypercholesterolaemia; and diabetes, gallstones, heart failure, and some forms of cancer for overweight (Table 2).

Effects of excess weight on morbidity and mortality have been known since the time of Hyppocrates, who said, "sudden death is more common in those who are naturally fat than in the lean". The epidemic of obesity is, however, recent. As a consequence, health-care expenditure has increased significantly. BMI has been associated with the annual number of hospital days, number and costs of outpatient visits, cost of outpatient pharmacy and laboratory costs in a large health-maintenance organization (5).

The most variable component of energy expenditure is physical activity. If overweight increases risk of mortality, intentional weight loss ought to reduce mortality and morbidity. As to this date, there is no definite studies documenting such an effect, but several studies document that intentional weight loss reduces risk factors other than obesity (6.7): it reduces blood pressure, improves lipid pattern, and reduces the risk of diabetes. These changes are linear and apparent with weight losses of 5-10%.

3. THERAPY

When prevention fails, medical treatment of obesity may become a necessity. Any strategic medicinal development must recognize that obesity is a chronic, stigmatized and costly disease that is increasing in prevalence.

Because obesity can rarely been cured, treatment strategies are effective only as long as they are used, and combined therapy may be more effective than monotherapy.

The cornerstone of obesity therapy is for the patient to eat fewer calories than are expended to consume indigenous fat stores as fuel. There are seven key principles for successful weight management (Table 3):

Obesity is a chronic illness which requires long-term treatment for long-term success;

Modest weight loss (5-10% of initial body weight) has considerable health benefits and improves insulin sensitivity, blood pressure, blood lipid levels and liver abnormalities;

Behavior modification is necessary for long-term lifestyle changes;

Diet education is needed to facilitate energy intake;

Physical activity is a critical component of weight management programs because it is associated with long-term success and may have beneficial cardiovascular and psychological effects;

Pharmacotherapy may support the weight to management efforts; and

Surgery is the most extreme, but effective approach for extremely obese patients who have been unable to maintain weight loss by dieting.

Diet advice should include encouraging patients to eat three meals a day, avoid snacking between meals, avoid energy-dense and high-fat foods, and increase the intake of fruits and vegetables. A diet with a deficit of 500-800 kcal/d is reasonable for the majority of patients (8). Dietary fat intake should be limited to >30% of total calories.

Behavior modification is needed to help patients change eating habits and increase physical activity and can be facilitated by appropriate physician-patient interaction. Physical activity is a key for long-term weight management success and improved health.

The amount of physical activity needed to prevent weight regain is not trivial and requires professional support. Physical activity should be increased slowly over time until the target goal is reached. Studies suggest that 80 minutes of moderate-intensity activity per day (example: brisk walking) or 35 minutes of vigorous activity (fast bicycling or aerobics) is needed for long-term weight maintenance after initial weight loss has been achieved (8).

Pharmacotherapy can support weight loss efforts. For a drug to have a significant impact on body weight, it must ultimately reduce energy intake, increase energy expenditure, or both. As 1958, the year in which thiazides were introduced, was a watershed in the treatment of hypertension, so 1998 may be a watershed for obesity (3). The effects of the weight-losing drug orlistat for the treatment of obesity might be analogous to the effect of thiazides on sodium excretion. As a potent inhibitor of pancreatic lipase, orlistat impairs fat digestion and absorption (9).

One complaint about treatments for obesity is that they "don't work" (10). A more realistic interpretation is that all treatments for obesity can produce weight loss, but overweight is not a curable disease. When treatment is stopped, weight is regained. A major difficulty in conducting long-term weight management programs is the high dropout rate (11). The completion rate of subjects in several behavioral-pharmacologic weight loss studies over 6 months to 2 years ranged from 30-63% (12-14). Keeping patients in a program promotes weight loss, lessens weight regain and improves obesity-related disease risk factors.

With this background in mind, the MOEBIUS weight control program was designed. The treatment included a controlled-energy diet. Energy intake was prescribed for each subject on the basis of estimated daily maintenance energy requirements. Dieticians at the site periodically provided instruction on dietary intake recording procedures as part of a behavior modification program. In addition, individuals were included in a physical activity program. The recommended changes in physical activity throughout the study were not assessed. The physical activity was, however, supported by regular sessions with a physiotherapist. Finally, patients were given orlistat (120 mg t.i.d.) to support the weight management program. orlistat was administered with the subjects three main meals and the controlled-energy diet. Drug intake was not assessed, as this was not an orlistat efficiency trial. Medication was given to the patients to support the weight control efforts.

4. METHODS

4.1. Subjects

Subjects were recruited, evaluated and monitored in the region of Schwyz. Entry criteria included age older than 18 years, BMI of >29 kg/m^2, subjects were excluded if they had a history or presence of substance abuse, excessive intake of alcohol, significant cardiac, renal, hepatic, gastrointestinal (GI), psychiatric or endocrine disorders, drug-treated diabetes mellitus type 2, or the concomitant use of medication that alter appetite or lipid levels.

4.2. Study Design

This was a feasibility study. The hypothesis that interactive communication technology is effective supportive measures for weight management that can be used by patients with obesity was evaluated in an open 4-month observational study.

There are are two distinct classes of telemedicine interactions – those that are pre-recorded (sometimes referred to as store-and-forward or asynchronous telemedicine) and those that occur in real-time (i.e. interactive or synchronous telemedicine). Pre-recorded telemedicine often depends on the use of e-mail or Internet access to online care. Real-time techniques often employ telephone consulting. In the present trial, both techniques were employed and integrated into the trial procedures.

Study procedures:

The initial screening visit included a medical history taking, physical examination, body weight evaluation, electrocardiogram, and clinical chemistry, hematology and urine-analysis laboratory tests. Fasting serum lipid levels were evaluated according to standard procedures. Fasting serum glucose and insulin levels were measured. The patients were then trained in self-assessment of the following parameters: blood pressure measurements and heart rate, blood glucose and cholesterol measurements, weight assessment. Body weight, physical parameters, blood glucose and serum lipid levels were evaluated every 2 weeks. In addition, patients were seen by one of the study physicians every 4 weeks. Each subject provided written informed content before entry into the trial. The study protocol was reviewed and approved by the local institutional review board.

4.3. Statistical Analysis

This was an open, observational study. Therefore no formal efficacy analysis was performed.

5.   RESULTS

5.1. Enrollment

Seventeen patients were initially screened for this pilot trial. Two patients withdrew from the study after the initial teaching session as they considered the study procedures as too complicated. Therefore 15 subjects entered the treatment phase. During the treatment phase 2 patients dropped out of the study due to a lack of effectiveness, according to the opinion of the patients. All patients were white Caucasians. Demographic, anthropometrics, and baseline characteristics are given in Table 4.

5.2. Weight loss

Figure 2 shows the weight change for all 15 patients of the intention-to-treat population. During the treatment period, patients lost a mean 4.9 kg of body weight (-3.9 ± 0.8 kg in female subjects, - 7.7 ± 2.1 kg in males). The percentage of patients losing 5% of baseline weight (measured at visit 1) was also evaluated. This percentage acknowledges that the reduction of bodyweight seen during this short treatment period is a substantial weight loss. 31% of patients experienced a 5% weight loss response. The majority of patients and investigators assessed the effectiveness of treatment as good to very good: 13 patients were willing to continue the programme outside the trial.

5.3. Cardiovascular Risk Factors

The results of the life style changing program are summarized in Table 5. Despite the short duration of treatment and the small number of patients, a significant reduction in blood pressure values was observed (Table 5). In contrast, patients with no weight loss exhibited no change in blood pressure values.

Total cholesterol values were significantly changed between visit 1 and end of study visit, but the decrease did not reach statistical significance. There were similar decreases in LDL-cholesterol values and a non-significant increase in triglyceride levels. The mean LDL/HDL ratio was relatively stable across the 4 months study period (Table 6).

5.4. Glucose and Insulin levels

Fasting plasma glucose and insulin concentrations obtained from measurement at baseline and at the end of the treatment period are given in Table 7. 3 patients exhibited abnormalities in glucose tolerance enabling the diagnosis of diabetes mellitus type 2. The weight loss program did not significantly change plasma glucose concentrations. Weight loss resulted in a decrease in plasma insulin concentration: 9/13 subjects exhibited lower insulin levels after treatment in comparison to pre-treatment plasma concentrations. In parallel, fasting glucose concentrations decreased from 6.1 ± 0.3mmol/l to 5.8 ± 0.2 mmol/l (NS), a 4 percent decrease. HbA1c concentrations were not affected by the treatment.

6. DISCUSSION

The key to successful telemedicine is not the technology but the delivery of care. The technology itself is simply a means to an end. The key issue concerning technology is that it does not operate in isolation, but as part of a system. Peripheral equipment - in the present study a device for measuring heart rate and blood pressure, a second device for measuring blood glucose and cholesterol concentrations - were connected to a hand-hold computer and a mobile telephone for online interactive communication.

This pilot study showed that patients enrolled into a comprehensive weight loosing programme which consists of a diet, an exercise programme, drug treatment (Xenical®) and an interactive communication system will loose weight. Despite the short duration of the trial, the results are consistent with those seen in previous studies lasting for longer periods. As even a moderate weight loss of approximately 5% provides unquestionable benefits for obese patients, the number of patients achieving such a weight loss reflects a possible advantage of our approach. The present study was primarily a feasibility study and not a comparative trial. Extrapolation of the data to larger populations and to other treatment strategies is therefore dangerous, as a possible bias of the results cannot be excluded by the design of the study. Nevertheless, the following statements are valid: 1.The programme provided a marked weight loss in the majority of patients; in addition, there were beneficial changes in several risk factors. 2. Plasma total cholesterol and LDL-cholesterol concentrations fell during the treatment; more important a significant, clinically relevant reduction in systolic and diastolic blood pressure was documented. The results form a valid basis for an integrated mobile health care solution with three key elements: 1. Diagnosis, 2. Treatment, 3. Surveillance of treatment.

As in previous short-term studies (≤ 1 year), the treatment showed improvements in most risk factors after 4 months of treatment. Premature withdrawals were rare and less frequent than expected; although all participants in this study did constitute obese people within a randomly selected population, they were probably highly motivated to loose weight. Despite the selection they were representative of individuals who seek help for their obesity. The results of the present study support the treatment concept for long-term management of obese patients in conjunction with an appropriate diet, an exercise programme and, if necessary, specific drug treatment. The set-up used in the MOEBIUS trial showed that the majority of patients accepted the procedures. More important, the patients felt that the support they were given was extremely helpful. All patients (N=13) who lost weight were willing to continue the therapy outside the trial. The time frame of 4 months is considered too short for assessments of long-term results in a weight-loosing program. This time frame was dictated by the MOEBIUS framework, which did not allow for a longer follow-up period. With these limitations, the current pilot study can be classified as very encouraging. An appropriate controlled clinical study with prolonged fellow-up is warranted to support this statement. It is anticipated that a disease management program can be developed which would include the key elements of the present study resulting in an integrated mobile solution for the treatment of obesity.

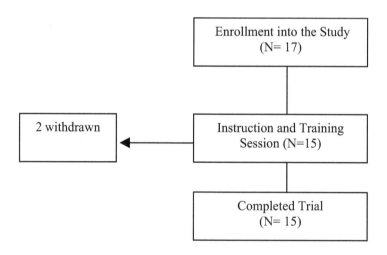

Figure 3. Flow and Disposition of Subjects Entered into the Study

Table 1. Body-weight associated disease risk

Obesity Class		BMI (kg/m²)	Risk
Underweight		>18.5	Increased
Normal		18.5 – 24.9	Normal
Overweight		25.0 – 29.9	Increased
Obesity	I	30.0 – 34.9	High
	II	35.0 – 39.0	Very High
Extreme Obesity	III	> 40	Extremely High

Table 2. Medical Complications associated with obesity

Gastrointestinal	Reflux disease, gallstones, Steatohepatitis, abdominal hernia, pancreatitis
Endocrine/metabolic	Diabetes, dyslipidemia, abnormal menses, infertility
Cardiovascular	Hypertension, coronary heart disease, stroke, deep venous thrombosis, pulmonary embolism
Pulmonary	Obstructive sleep apnoea, abnormal pulmonary function test results
Muscolosceletal	Osteoarthritis, gout, low back pain
Genito-urinary	Urinary stress incontinence
Cancer	Colon, prostate, breast, uterus, cervix, gallbladder
Postoperative events	Atelectasia, pneumonia, deep vein thrombosis, pulmonary embolism

Table 3. Seven principles of weight management (According to Klein (8))

1. Long-term weight loss therapy needed for long-term success
2. Even modest (5-10%) weight loss with considerable health benefits
3. Behavior modification
4. Nutrition education
5. Physical activity
6. Pharmacotherapy
7. Surgical therapy

Table 4. Characteristics of the study population at baseline (4 males, 11 females)

Age (years)	45.0 ± 2.4
Weight (kg)	95 ± 4
Height (cm)	172 ± 2
BMI (kg/m^2)	32 ± 1
Alcohol intake (g/ day)	3
Number of smokers	1

Table 5. Blood pressure measurements before during treatment. Data ± SEM

	Systolic Blood pressure (mm Hg)	Diastolic Blood pressure (mm Hg)
Baseline	130.0 ± 1.5	80.1 ± 0.8
1 month	124.1 ± 1.0	75.0 ± 0.7
2 month	125.7 ± 1.0	76.0 ± 0.7
3 month	122.4 ± 1.2	75.0 ± 1.0
End of treatment	119.8 ± 1.0	74.8 ± 1.1

Table 6. Fasting plasma lipid levels before and after treatment. Data ± SEM

	Baseline	At 4 month	% change
Cholesterol (mmol/l)	5.6 ± 0.2	5.0 ± 0.2	- 10.5 %
HDL (mmol/l)	1.43 ± 0.1	1.35 ± 0.1	- 6.5 %
LDL mmol/l)	3.19 ± 0.2	3.0 ± 0.2	- 5.7 %
Triglycerides (mmol/l)	1.46 ± 0.18	1.45 ± 0.15	- 6.6 %

Table 7. Fasting glucose levels (mmol/l), postprandial glucose levels (mmol/l), fasting insulin concentrations ($\square$V/ml) and HbA1c values before and after treatment. Data are mean ± SEM.

	Baseline	At 4 month	% change
Fasting glucose	6.1 ±0.3	5.8 ± 0.2	- 3.9 %
Postprandial glucose	6.4 ± 0.3	5.9 ± 0.4	- 8.5 %
Insulin	32.2 ± 3.5	21.2 ± 1.4	- 26.0 %
HbA1c	5.6 ± 0.1	5.5 ± 0.1	- 2.1 %

7. REFERENCES

Deloitte & Touche: The Emerging European Health Telematics Industry. Market Analysis. A Health Information Society Technology-Based Industry Study -- Ref C13.25533, 2000.

IST -- Information Society Technologies: IST homepage. Retrieved 2001 at URL: http://www.cordis.lu/ist

Izbicki G, Trachsel D, Rutishauser M, Perruchoud AP, Tamm M: Early detection of exacerbation of lung infections in patients with cystic fibrosis by means of daily spirometry (in German). Schweiz Med Wochenschr 2000;130:1361--1365.

Koschinsky T, Heinemann L: Sensors for glucose monitoring: Technical and clinical aspects. Diabetes Metab Res Rev 2001;17:113--123.

Mansfeld, J.: "Realisierung und Evaluation eines elektronischen Schmerztagebuches für die Kölner Schmerzambulanz" (Realization and evaluation of an electronic pain diary for use in the pain clinic of the university hospital of Cologne, Germany), diploma thesis, University of Heidelberg, 1998.

Rogers MA, Small D, Buchan DA, Butch CA, Stewart CM, Krenzer BE, Husovsky HL: Home monitoring service improves mean arterial pressure in patients with essential hypertension. A randomized, controlled trial. Ann Intern Med 2001;134:1024--1032.

Roine R, Ohinmaa A, Hailey D: Assessing telemedicine: A systematic review of the literature. CMAJ 2001;165:765--771.

SGTM -- Swiss Association of Telemedicine: Definition of Telemedicine. Retrieved 2001 from URL: http://www.sgtm.ch

Smith V: The Strong Extranet White Paper. Thawte Consulting South Africa. Retrieved 2000 from URL: http://www.thawte.com/certs/strong extranet/ whitepaper. html

Standage T: Survey: The mobile internet: The Internet, untethered. The Economist, Oct 13, 2001.

Wagner, M.: "Mobile Endgeräte und Internet – Hardware, Software, Anwendungsbereiche – eine vergleichende Studie", diploma thesis, University of St. Gallen, 2000/2001

Wheeler T: Strategies for delivering tele-home care-provider profiles. Telemed Today 1998;6:37--40.

Wootton R: Recent advances: Telemedicine. BMJ 2001;323:557--560.

Wu J, Kessler DK, Chakko S, Kessler KM: A cost-effectiveness strategy for transtelephonic arrhythmia monitoring. Am J Cardiol 1995;75:184--185.

MOBIHEALTH: MOBILE HEALTH SERVICES BASED ON BODY AREA NETWORKS

Val Jones[*], Aart van Halteren, Ing Widya, Nikolai Dokovsky, George Koprinkov, Richard Bults, Dimitri Konstantas, and Rainer Herzog

1. INTRODUCTION

In this chapter we describe the concept of MobiHealth and the approach developed during the MobiHealth project (MobiHealth, 2002). The concept was to bring together the technologies of Body Area Networks (BANs), wireless broadband communications and wearable medical devices to provide mobile healthcare services for patients and health professionals. These technologies enable remote patient care services such as management of chronic conditions and detection of health emergencies. Because the patient is free to move anywhere whilst wearing the MobiHealth BAN, patient mobility is maximised. The vision is that patients can enjoy enhanced freedom and quality of life through avoidance or reduction of hospital stays. For the health services it means that pressure on overstretched hospital services can be alleviated.

During the project a generic BAN for healthcare and a generic m-health service platform was developed, trialled and evaluated. The MobiHealth BAN incorporates body-worn devices and handles communication amongst those devices. The MobiHealth Service Platform manages (a population of) deployed BANs and handles external communication between the BANs and a remote healthcare location.

During the course of the project the main BAN devices used were medical sensors. Biosignals measured by sensors connected to the BAN were transmitted to a remote healthcare location over 2.5/3G public wireless networks (GPRS and UMTS). The MobiHealth BAN and service platform were trialled in four European countries with a variety of patient groups. The trials focussed on home care, trauma care and outdoor settings. The remote healthcare locations involved were hospitals in Spain, The Netherlands and Sweden and a medical call center in Germany. The remote monitoring

[*] Val Jones, Centre for Telematics and Information Technology/Faculty of Electrical Engineering, Mathematics and Computer Science, University of Twente, PO Box 217, 7500 AE, Enschede, The Netherlands, v.m.jones@ewi.utwente.nl.

services allow patients to follow their normal daily life activities rather than being confined to hospital or to their home during monitoring.

The overall goal of MobiHealth was to evaluate the ability of 2.5 and 3G communication technologies to support innovative mobile health services. The main output of the project therefore is an assessment of the suitability of GPRS and UMTS to support such services. In addition the project delivers an architecture for, and a prototype of, a health BAN and a generic m-health service platform for provision of ubiquitous healthcare services based on Body Area Networks.

The MobiHealth concept arose out of work at the University of Twente since 1999 (Jones et al., 1999; Jones 2001) and was further developed within the work of the Wireless World Research Forum (Jones et al., 2001b; Jones et al., 2001c; WWRF, 2001). The MobiHealth consortium involving 14 European partners was brought together during 2001. The project began on May 1st 2002 and is funded by the European Commission under IST FP5 and has been reported in (Jones et al, 2001a; Konstantas et al., 2002a,b; Widya et al., 2003a, Dokovsky et al., 2003). The project coordinator is Ericsson GmbH with the University of Twente taking the scientific lead.

The MobiHealth System can support not only sensors, but potentially any body worn (medical) device provided it has a communication interface. (An example of another kind of device already used with the BAN is a Bluetooth enabled digital camera.) Hence the system has potentially very many applications in healthcare. Some of these are explored in other chapters in this book. The chapter entitled *Telematic Reequirements for Emergency and Disaster Response* (Widya et al, 2003b) models the extension of the MobiHealth Trauma application to large scale disaster and emergency scenarios. Another chapter entitled *Remote Monitoring for Healthcare and for Safety in Extreme Environments* (Jones et al, 2003) examines the potential of the MobiHealth approach for provision of remote monitoring services to support work and recreation in extreme environments. That chapter focuses on polar and underwater environments as examples.

This remainder of this chapter presents the MobiHealth technology (Section 2), the trials and the evaluation (Section 3), and identifies some of the research issues raised by the project (Section 4).

2. THE MOBIHEALTH TECHNOLOGY

A mobile health BAN and a generic service platform for BAN services for patients and health professionals were developed as part of the work of the MobiHealth project. Remote (patient) monitoring services are just one of the kinds of services that can be provided by such a platform. Here we focus on monitoring as an example of BAN services in order to present the MobiHealth technology.

A wide range of patient monitoring equipment is used in healthcare. Traditionally monitoring equipment was operated by health professionals in hospitals or at other healthcare locations. Some of the equipment is quite bulky and heavy, but for some kinds of monitoring the use of small ambulatory monitors together with telemetry has improved mobility for patients within the hospital.

Nowadays there is an extensive range of miniature wearable monitors available. Some are on sale to the public as consumer electronics products. A wrist worn blood pressure and heart rate monitor can be purchased for as little as 30 Euro. Such products can be used for example by cardiac patients to monitor their own health state and response to medication. Another wrist-worn device allows diabetes patients to monitor their blood glucose non-invasively. Some products provide local processing only (such as readouts, logging of readings, reminders, warnings and alarms). Some also allow upload of data to a PC whilst others combine monitoring and (tele)communication functions. One example is a cardiac monitoring product embedded in a mobile phone, which transmits cardiac signals to a doctor or a to a distant healthcare location over public cellular networks. Some products incorporate positioning to provide positioning information in case of emergency so that the emergency services can locate the casualty easily. There are also many examples of commercially available wearable monitoring systems targeted at non-patients. Examples are heart monitors for use by athletes during training and motion sensors and gas sensors used to protect workers in the chemical industry.

With so many devices available, why do we need to introduce the concept of a Body Area Network? In the next section we explain the concept of Body Area Networks (BANs), and why we apply a BAN approach in MobiHealth.

2.1. Body Area Networks

The concept of the Body Area Network originally came from IBM (Zimmerman, 1999) and was developed further by many other researchers, for example at Philips (van Dam, 2001), at the University of Twente (Jones et al., 2001a; Jones et al., 2001b), and at Fraunhofer (Schmidt, 2001). In the Wireless World Research Forum's *Book of Visions*, we define a BAN as "a collection of (inter) communicating devices which are worn on the body, providing an integrated set of personalised services to the user" (WWRF, 2001). For the purposes of the MobiHealth project a BAN can be thought of simply as a computer network which is worn on the body and which moves around with the person (that is, it is the unit of roaming).

A BAN incorporates a set of devices which perform some specific functions and which also perform communication, perhaps via a central controlling device. Devices may be simple devices such as sensors or actuators, or more complex multimedia devices such as cameras, microphones, audio headsets or media players such as MP3 players. The central controlling device (if there is one) may perform computation, coordination and communication functions. Communication amongst the elements of a BAN is called *intra-BAN communication*. If the BAN communicates externally, i.e. with other networks (which may themselves be BANs), this communication is seen (from the point of view of the BAN itself) as *extra-BAN communication*.

2.1.1. Intra-BAN Communication

IntraBAN communication is carried over a wired or a wireless medium. Wired options include copper wires, of course, also optical fibre and many 'wearable computing' solutions where circuitry is embedded in or woven into or printed onto fabric or incorporated into 'body furniture' such as spectacles or jewelry.

Wireless options include infrared light, microwave, radio and even skin conductivity. Two important standards for short range wireless communication are Bluetooth and Zigbee. Bluetooth (Bluetooth, 2003) is an example of a wireless communication standard (IEEE 802.15) based on radio waves. It supports speeds of up to 721 Kbps and offers a user-friendly way of 'ad hoc' networking providing automatic admission and removal of devices in a network. The range is up to approximately 100m. Although a very attractive technology for BANs, the relatively high power consumption for wireless transmissions make long term unattended use of a Bluetooth BAN problematic. ZigBee (ZigBee Alliance) (also known as RF-lite) is a low cost, low power, two-way wireless communications standard (IEEE 802.15.4) and is particularly promising for BANs. ZigBee supports a relatively low data rate (< 250 Kbps), but combines this with long battery lifetime. The range is approximately 50m.

2.1.2. Extra-BAN Communications

To support convenience and mobility for the BAN user, extra-BAN communication is in general based on wireless technologies. A range of technologies is available including Bluetooth, WLAN, GSM, GPRS and UMTS. The selection of transmission technology will depend largely on range and (transmission) speed requirements. There are other important criteria (e.g. reliability, security) for communication links, but these can be fulfilled by application level protocols. Bluetooth and WLAN (Wireless Local Area Network) are short and medium range communication technologies (Bluetooth 10m and WLAN 100m) and depend on a local infrastructure (within a building for example). GSM, GPRS and UMTS are wide area technologies, offering the possibility for the wearer of the BAN to be truly mobile (i.e. have the freedom to go anywhere without losing connectivity).

GSM (also called 2nd Generation, or 2G, technology) is as a voice network and has national coverage in many countries. Pan-European coverage is made possible by means of international agreements between network operators. GSM can also be used as a data network and offers a minimum guaranteed speed of 9.6/14.4kbps. GPRS (an instance of the so-called 2.5G technologies) is a voice and data network running over the GSM infrastructure. GSM timeslots not assigned to voice connections can be used for data transmission, however voice has priority, creating potential quality of service problems for data transmission. GPRS offers transmission speeds of up to ~50kbps. New technological developments are expected to result in increased speed for GPRS communications. UMTS (a 3G technology) promises wideband communication speeds of up to 2Mbps. At time of writing UMTS rollout is underway but with coverage limited to heavily populated areas (e.g. urban areas). Initial UMTS networks are expected to support a transmission speed of 64kbps, with an option for 144kbps.

2.1.3. Body Area Networks for Healthcare

The *Body Area Network* is a generic concept which can have many applications in various domains. When the devices of a BAN measure physiological signals or perform other actions for health-related purposes we call this kind of specialization of Body Area Networks a *health BAN*. As a further specialisation, a BAN whose devices are biosignal sensors enables patient monitoring. Clearly there can be different kinds of sensors for monitoring different biosignals. Thus we can envisage any number of specific health BANs for specific classes of patients/clinical conditions. Some example applications are: cardiac care, diabetes management, sleep analysis, asthma, epilepsy, EEG/EMG monitoring, SIDS, care of the elderly, assistive technologies for the disabled, movement analysis, feedback on medication (e.g. for Parkinsons disease), monitoring of drug delivery systems (e.g. morphine or insulin pump) and home cancer care.

If we combine health BANs with *extra-BAN communications* this enables remote monitoring of patients. This can be achieved by means of fixed networks, such as a fixed (wired) telephone line or an Internet connection, allowing the patient to be monitored at home instead of in hospital for example. However the fixed line limits the mobility of the patient to a fixed position during monitoring, or at least during regular uploading of locally stored monitor data. Store and upload is only feasible when there are no real-time requirements (such as emergency detection and alarms) and is dependent on storage capacity of the wearable device. Some applications generate only small amounts of biosignal data, but others are very data intensive. For example epilepsy monitoring with 32 channels generates 400-500 MB per day and 128 channel monitoring brings (daily) storage requirements into the Gigabyte range.

Combining BANs with extra-BAN wireless communications enables remote (ambulatory) monitoring and full mobility for the patient with support of real-time applications. This means that the patient can pursue normal life activities, leave the house and even travel anywhere in the world, insofar as their health condition permits. In other words wireless communication services open up the possibility of location freedom (global roaming) for patients who are undergoing monitoring.

In the following section we describe the approach of the MobiHealth Project to the use of health BANs for remote monitoring of patients.

2.2. The MobiHealth Body Area Network

One problem with some of the currently available monitoring devices is that they are closed proprietary (non-interoperable) systems; some even operate in a standalone fashion with no option to upload measurements. Whilst these devices have utility and have clearly found a market, the usefulness of such devices can be enhanced in a number of ways. The vision of the MobiHealth project is to develop an open extensible BAN platform which allows integration of different health functions by means of a (wireless) plug-and-play approach.

The use of the MobiHealth BAN together with 2.5/3G communications means that biosignals measured by the BAN can be transmitted to a remote (healthcare) location. This enables remote management of chronic conditions and detection of health emergencies whilst giving the patient the freedom to move anywhere outside the hospital. At time of writing the MobiHealth system is being tested by patients and health professionals in home care, trauma care and outdoor settings. The MobiHealth BAN also features in a sister project called xMotion where it is integrated into a UMTS based tele-ambulance demonstrator (xMotion, 2002).).

2.2.1. Architecture of the MobiHealth system

The MobiHealth BAN consists of a central unit known as the MBU (Mobile Base Unit) and a set of body-worn devices, which may include sensors, actuators, and various multimedia devices. In principle the BAN may incorporate any wearable or implantable devices. Figure 1 shows the generic architecture of the BAN.

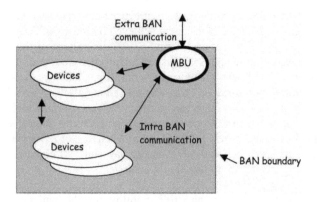

Figure 1. Generic architecture of the BAN.

The MBU can be thought of as a gateway which handles extraBAN communication. It also handles coordination, local computation and communication amongst BAN devices. It can also host local applications such as a local viewer application, or (in future) more complicated functionality such as disease management applications.

In MobiHealth the MBU was implemented on an iPAQ H3870. This device has built-in Bluetooth capabilities and can be extended with a GPRS jacket. Figure 2 shows the MBU running a viewer application to display biosignals measured by a sensor (in this case pulse plethysmogram measured by a pulse oximeter).

Figure 2. MBU platform (iPAQ H3870), displaying output from pulse oximeter

Sensor data is processed by a sensor front-end before being transmitted to the MBU. A range of front-ends can be associated with an MBU, enabling customization of the BAN. Although the MBU currently used in the MobiHealth trials is based on the HP iPAQ platform, future plans include porting to a cell phone platform such as the Sony Ericsson P800. The hardware components of a BAN are shown in Figure 3 which shows (left to right) an MBU, a sensor front-end (known as a "Mobi") and a sensor set consisting of ECG electrodes and a motion sensor.

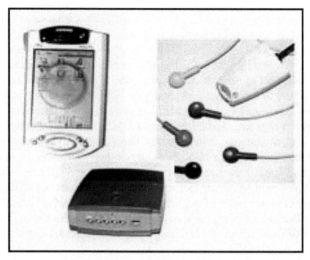

Figure 3. MBU, sensor front-end end and a sensor set

The BAN is only part of the overall MobiHealth system. Figure 4 shows the hardware and software architecture of the MobiHealth system. The lower part of the figure shows the hardware configuration and the upper part shows the software components developed during the course of the project to implement the MobiHealth BAN and its support services. The MobiHealth system includes not only the BANs deployed in the community but also the Back End System (BESys). The BESys is the m-health service runtime environment, which hides the mobile nature of the service provider components (MBUs) and exports the service to the local environment of the healthcare centre. It is composed

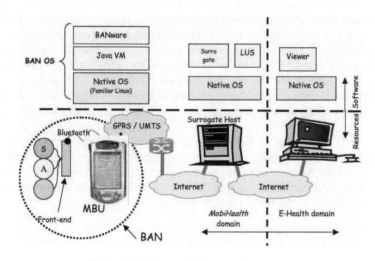

Figure 4. Hardware and software architecture

of several software components which may reside on one or more hardware nodes. Data is exchanged between the BAN and a remote (healthcare) location, which has its own (perhaps legacy) systems in operation. If the remote healthcare location is a hospital then this would be the HIS (Hospital Information System).

The MobiHealth Back End System mediates between the HIS and the BAN. The Back End System may be located at the hospital or at another location entirely. The BAN is shown here with a set of sensors (S), a set of actuators (A) and a set of other devices, all connected (wired) to the sensor front end (FE). The FE communicates with the MBU wirelessly (over Bluetooth).   The MBU software stack contains the BANware implemented on top of a Java virtual machine (J2ME, CDC and Personal profiles) (Sun Microsystems, 2003a) running under Linux. The Back End System runs on a server (possibly at the healthcare location, possibly at another site); it incorporates a Look Up Service (LUS) and hosts a surrogate, which represents and manages the BANs in the field. The end-user application may consist simply of a viewer application or (in future) some much more complex suite of applications interfacing to the clinical systems of the healthcare provider. The software components of the MobiHealth system were developed at the University of Twente and are available as Open Source software.

2.2.2. BAN Devices

The generic BAN is customized with different device sets for different applications. Some of the BAN devices used during the MobiHealth project are electrodes for measuring ECG, pulse oximeters for measuring oxygen saturation, sensors for measuring NIBP (non invasive blood pressure) and motion sensors.

A sensor is responsible for the data acquisition process. A sensor system ensures that a physical phenomenon, such as patient movement, muscle activity or blood flow, is first converted to an electrical signal. This signal is then amplified, conditioned, digitized and communicated within the BAN.

Sensors can be either self-supporting or front-end supported. Self-supporting sensors have their own power supply and facilities for amplification, conditioning, digitization and communication. In case of front-end supported sensors, multiple sensors share a power supply and data acquisition facilities. Consequently, front-end supported sensors typically operate on the same front-end clock and jointly provide multiplexed sensor samples as a single data block. This avoids the need for synchronization between sensors. Self-supporting sensors are independent building blocks of a BAN and ensure a highly configurable BAN. However, each sensor runs at its own internal clock and may have a different sample frequency. Consequently, synchronization between sensors may be needed.

The sensors used in the MobiHealth trials are: electrodes for measuring ExG (where ExG can stand for ECG, EMG or EEG), activity sensors, respiration sensors and pulse oximeter. Other BAN connected devices apart from sensors are: marker/alarm button and digital camera. Readings from external devices (such as NIBP or forced spirometry) can be entered manually on the MBU for transmission.

2.2.3. Intra-BAN Communications

The MobiHealth architecture allows for wired or wireless communications within the BAN, or a combination of the two. The current MobiHealth trials use wired communication from sensors to sensor front end and Bluetooth from the front end to the MBU. However the BAN has actually been implemented using both wired (front-end supported) sensors from TMSI and wireless (self-supporting) sensors from EISlab (Östmark et al, 2003). Both approaches use Bluetooth for intra-BAN communication. The front-end also allows ZigBee as an alternative intra-BAN communication technology.

2.2.4. Extra-BAN Communications

For communication between the BANs and the Back End System (extra-BAN communication), 2.5/3G public networks are used. The MBU communicates via Bluetooth to a GPRS or UMTS handset (cell phone); from there the signals are transmitted to the remote location. In order to abstract from wireless transmission technology, BAN to Internet communication, and vice versa, is based on the IP protocol. National GPRS coverage is available in all the countries where the trials are conducted (Germany, The Netherlands, Spain and Sweden). The first localised UMTS (test) infrastructures became available in Germany and the Netherlands part way through the project (during 2003). At time of writing the Dutch trials were using both GPRS and UMTS for extraBAN communication. In the other trial countries GPRS was being used, with early prospects for UMTS tests. Availability of commercial UMTS services will broaden the range of applications which can be supported by the BAN.

2.3. The MobiHealth m-health Service Platform

The m-health service platform supports the BAN based applications and isolates them from the details of the underlying communications services. It is this service platform which enables genericity and hardware independence. Hardware independence has already been demonstrated by use of sensors from different manufacturers and by the initial experiments with replacing the PDA implementation of the MBU by a cell phone implementation.

2.3.1. Service Platform Architecture

Figure 5 shows the functional architecture of the service platform. The dotted square boxes indicate the physical location where components of the service platform execute. The rounded boxes represent the functional layers of the architecture. The m-health service platform consists of sensor and actuator services, intra-BAN and extra-BAN communication providers and an m-health service layer. The intra-BAN and extra-BAN communication providers represent the communication services offered by intra-BAN communication networks (e.g. Bluetooth), and extra-BAN communication networks (e.g. UMTS), respectively. The m-health service layer integrates and adds value to the intra-BAN and extra-BAN communication providers. The m-health service layer masks applications from specific characteristics of the underlying communication providers, such as the inverted consumer-producer roles.

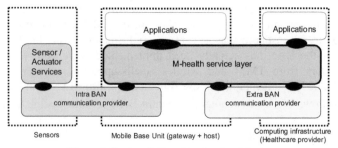

Figure 5. Service platform functional architecture

Applications which run on top of the service platform can either execute on the MBU (for local use by the patient or by a visiting nurse for example) or on the servers or workstations of the healthcare provider, for example at a health call centre or hospital. The services offered by the M-health service platform are:

- BAN registration: the service platform maintains a list of active BANs and allows applications to retrieve the specific configuration of a BAN.
- BAN discovery: applications can subscribe to the platform in order to receive a notification when a BAN becomes active (e.g. when a patient switches on a BAN).
- BAN authorization and authentication: the service platform authenticates BANs and only allows authorized BANs to transfer data.
- BAN data encryption: the platform encrypts data to be transmitted over unsecured networks
- BAN configuration: the service platform allows online configuration and management of the BANs, such as activation and deactivation of specific sensors, or modification of sensor sampling frequency.
- Data acquisition control: the service platform enables applications to start, stop or temporarily suspend the data acquisition process of a BAN.
- Query and modify actuator status: applications can manipulate actuators from a distance.
- BAN data storage: the service platform also offers intermediate storage services for the applications. The application can specify the minimum required lifetime of the stored data.
- BAN data monitoring: the service platform can apply filtering algorithms to BAN data to determine if a significant event has taken place (e.g. a patient has fallen) and report this event to the application layer.

Figure 6 shows a refined view of the m-health service layer. In this figure the m-health service platform user (e.g. a doctor at a hospital) is located remotely from a healthcare provider (in this case a health call center). The arrows indicate the flow of BAN data. The BANip entity is a protocol entity of the BAN interconnect protocol. Peer entities are found on the MBU and on the computing infrastructure (in the 'fixed' network). The BANip entities communicate through a proxy, which authenticates and authorizes the BANs.

The surrogate component uses the BANip protocol to obtain BAN data. It contains a representation of the deployed BANs (i.e. surrogates). The surrogate component is deployed in the fixed network and shields other components in the 'fixed' network from the BANip and direct interaction with the BAN (for example it hides the intermittent availability of the BAN and the use of private IP addresses (Rekhter et al., 1996) for GPRS terminals). Thus it solves the problems arising from the invertion of producer and consumer roles. The surrogate component can be accessed by any application protocol, including Remote Method Invocation (RMI).

The storage entity uses RMI to interact with the surrogate as if it interacts with the remote BAN at the location of the patient, without the responsibility of managing BAN discovery, registration and authentication. The surrogate component is therefore the intermediary to which BAN data from the location of the patient is pushed and from which BAN data is pulled by the application component running in the hospital systems. BAN data may also be pushed by the surrogate to the application component. The storage entity provides a BAN data storage service to the

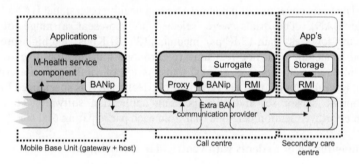

Figure 6. Refined view of the service platform

application layer. Configuration, discovery and monitoring services are offered as separate entities and have a similar structure to the storage entity shown in Figure 6.

Applications which use the m-health service layer can range from simple viewer applications which provide a graphical display of BAN data to complicated applications which analyse and interpret the clinical data.

The Ban Interconnect Protocol (BANip) was implemented using HTTP1.1. The BANip is implemented on the MBU as an HTTP client, which collects a number of samples into the payload of an HTTP POST request and invokes the post on the surrogate. A standard HTTP proxy acts as a security gateway for the surrogate. If the surrogate needs to control the MBU, these control commands are carried as the payload of the HTTP reply.

The surrogate was implemented using the Jini Surrogate architecture (Sun Microsystems, 2003b). Jini provides the implementation for auto-discovery and registration of the BAN. In terms of the Jini architecture the surrogate is a service provider. Other components, such as the BAN data storage component, are service users from the perspective of the surrogate.

Additional security was added by upgrading the BANip protocol from HTTP to HTTPS. This ensures that BAN data is encrypted while it traverses the public wireless network and the Internet.

More details of the BAN service platform implementation may be found in the project deliverables, available at the project website and in (Dokovsky et al, 2003). The following section gives an overview of the MobiHealth trials and evaluation.

3. VALIDATION OF THE MOBIHEALTH APPROACH: TRIALS AND EVALUATION

The overall goal of the MobiHealth project is to test the ability of 2.5 and 3G infrastructures to support value added healthcare services. The trials span four European countries and are targetted at the areas of acute (trauma) care, chronic and high-risk patient monitoring and monitoring of patients in home-care settings. The trials cover a range of conditions including pregnancy, trauma, cardiology (ventricular arrhythmia), rheumatoid arthritis and respiratory insufficiency (chronic obstructive pulmonary disease). The trials cover use of patient BANs and health professional BANs (nurse BAN, paramedic BAN). The trials were selected to represent a range of bandwidth requirements: low (less than 12 Kbps), medium (12 – 24 Kbps) and high (greater than 24 Kbps) and to include both non-real time (e.g. routine transmission of tri-weekly ECG) and real time requirements (e.g. alarms, transmission of vital signs in a critical trauma situation). For each application the generic MobiHealth BAN is specialized by addition of the appropriate sensor set and corresponding application software. To illustrate the validation effort we mention here one trial from each participating trial site.

3.1 Telemonitoring of patients with ventricular arrhythmia

In Germany the MobiHealth system is tested and evaluated with a cohort of patients suffering from ventricular arrhythmias who are undergoing drug therapy for this condition. In such patients, ECG measurements have to be taken regularly to monitor efficacy of drug therapy. In this trial the patient BAN transmits ECG and blood pressure via GPRS to a Medical Service Centre, where vital signs are monitored by physicians and nurses. Instead of staying in hospital during monitoring, the patient can be at home and can even go freely outside the home. The intention is that through remote monitoring developing clinical trends can be identified and corrected before they become serious. From a health economic point of view the intervention is expected to save time and reduce costs (eg by reduction in number of hospitalizations, and by lowering drug costs). For the healthcare/medical evaluation blood pressure and ECG readings during the technology intervention are compared to baseline measurements.

3.2 Integrated Homecare in women with high-risk pregnancies

This Dutch trial evaluates use of the MobiHealth BAN to support Integrated Homecare for women with high-risk pregnancies. Women with high-risk pregnancies are often admitted to hospital because of possible pregnancy-related complications. Admission is necessary for the intensive monitoring of the patient and the unborn child.

Homecare with continuous monitoring of women with high-risk pregnancies, when feasible, is desirable and can postpone hospitalisation and reduce costs. In this trial, pregnant women are monitored from home using the MobiHealth BAN and maternal and foetal signs are transmitted to the hospital. The objective of the trial is to evaluate if such monitoring services can be supported by 2.5-3G communications such that hospitalisation can be postponed and costs reduced. For the healthcare/medical evaluation, main outcome parameters are the number of days in hospital, the number of emergency interventions, patient satisfaction with hospital or telemonitoring and the reliability of the system to monitor the patients and to inform the gynaecologist regarding ECG and EMG data. Although the service, if introduced, is intended for high risk patients, for the trials themselves high risk patients are not included for obvious reasons.

3.3   Support of home-based healthcare services

This Spanish trial involves use of GPRS communication for supporting home-based care for elderly chronically ill patients, including remote assistance if needed. Trial subjects suffer from co-morbidities including Chronic Obstructive Pulmonary Disease. The MobiHealth Nurse BAN is used to perform patient measurements during nurse home visits and the MobiHealth patient BAN is used for continuous home monitoring and also outdoors during patient rehabilitation. It is very important to facilitate patients' access to healthcare professionals without saturating the available resources, and this is one of main expected outcomes of the MobiHealth remote monitoring approach explored in this trial. Physical parameters measured are oxygen saturation, ECG, spirometry, temperature, glucose and blood pressure. The healthcare/medical evaluation focuses on acceptability of the technology solution in this health context and compares quality and process of care received under the technology intervention with that received under current practice. Comparison is in terms of numbers of in-patient hospital stays, ER visits, outpatient visits, primary care physician visits, social support visits, nurse home visits, prescriptions, phone calls (patient-nurse/ nurse-patient) and transports.

3.4 Home care and remote consultation for recently released patients in a rural area

In this Swedish trial the trial subjects are patients with multiple chronic diseases including cardiac failure, diabetes and respiratory disability who have been recently discharged from hospital and who are living at home in a low population density rural area. Measurements (including blood pressure, heart rate, oxygen saturation and blood glucose, as appropriate) are transmitted to a remote physician or a registered district nurse (RDN). The intention is that the remote health professional receives clinical data which offers a better basis for a good and safe decision on whether the patient needs transfer by ambulance to hospital. The expected benefit is that this intervention will reduce the number of cases where the patient is moved to hospital unnecessarily. The healthcare/medical evaluation will focus on qualitative instruments measuring user experiences of the intervention by the involved roles: staff, patients and their next-of-kin.

The trials are evaluated in terms of accuracy and validity of measurements, usability of the GPRS and UMTS networks, business and market potential and also in terms of social and ethical effects.

All trials are evaluated according to four perspectives: Technology, Healthcare/Medical, Market and Social. It cannot be expected that any definitive clinical outcomes can be established with the patient numbers and timescales involved. In fact under the context of the Call the primary evaluation focus must be on the technology, specifically on the potential for value added services over 2.5 and 3G communications. The results of the evaluation will be available in 2004.

4. FUTURE CHALLENGES AND RESEARCH ISSUES RAISED

In the wider context, assessment and evaluation of the value of innovative m-health solutions and the incorporation of worthwhile m-health technologies into existing healthcare systems in terms of structure, organization and working practices presents many challenges. Legal issues and hospital responsibilities such as ownership of the data and legal liability are complicated by the international dimension enabled by the opportunity of seamless access to services for free roaming patients. With the new possibilities come new challenges. These challenges give direction for future research across a range of disciplines.

For our own discipline of computing science, we see (horizontal) research issues arising from MobiHealth in connection with adaptive applications and communications, and vertical research issues such as security, quality of service and scalability. Some of these research issues are listed below:

Mobile service platforms: how to make scalable, platform independent, generic BAN service platforms which can be rapidly configured for different applications and device sets.

Ad hoc networking: investigation of the role of BANs and PANs and the relationship and communication within and between them.

How to provide *seamless vertical handover* between wireless technologies, across the wired/wireless divide and across operator domains.

Adaptive applications and IUIs: how to design intelligent adaptive (dynamically reconfigurable) BAN applications, and how to design Intelligent User Interfaces for the resulting BAN services.

How to ensure *security, privacy, safety, correctness and quality of service* of systems and information based on wireless communications infrastructures, ambient intelligent environments and ad hoc networking.

Scalability: how to support large populations of deployed BANs as part of offered clinical services. This implies high performance distributed computing, broadband wireless communication and automation of clinical analysis and interpretation. For some applications automated analysis will involve complex real time pattern recognition and application of (temporal) abstraction to effect clinical interpretation. This in turn implies that BAN applications interface to large scale and comprehensive clinical knowledge bases and inferencing systems. Emerging technologies such as GRID computing may play a role at the infrastructure level by providing the fast processing, high volume storage and reliable communication resources required.

We expect that in the longer term the current style of implementation of (health) BANs will disappear, replaced by nanoscale devices and contactless sensing from the ambient intelligent environment. We call this vision the Ambient Intelligent BAN or AmI BAN). AmI BAN devices might be implanted rather than worn externally on the body or in clothing. These devices could include for example actuators to control drug delivery systems, nanomanipulators or neural stimulators interfacing directly to motor neurons.

The vision of the future depends on scientific research and technological advances not only in computing science but also from other technical disciplines including electrical engineering, physics and nanoscience. Wearable computing, nanotechnologies, contactless sensing, implantable devices and low-power or self-powered devices are some of the enabling technologies prerequisite to the realization of the vision. Together these multidisciplinary developments can open the way for seamless collaboration between BANs, PANs and the ambient intelligent environment.

Even for the medium term future many problems remain. Apart from the challenges for computing science and other technical disciplines, experience in many application domains indicates that the success factors for take up of technical innovation come from the larger extra-technical context. These factors include organisational and professional issues from the application domain concerning the impact on ways of working, and user acceptance.

In the healthcare domain, first and foremost the technology must gain public acceptance and be perceived as bringing benefits to patients. Of primary concern are potential health benefits, but impact on quality of life for patients and for the patients' families are also part of the very important patient 'utility function'. Clinical effects such as efficacy can be addressed by means of clinical trials. Determining the effects of new health technology on process and outcome of care and subjective factors such as user satisfaction and quality of life are the domain of Health Services Research (HSR) studies. Research in the areas of e-health and e-learning indicates that innovation is embraced by users only when they perceive immediate benefits.

Regarding the organisational aspects of healthcare, exploitation of a technical innovation may imply extensive and radical business process re-engineering of the (national) healthcare system. Such implications have caused many (technically sound) health technology innovations to fail in the past. Unless business process re-engineering is driven by and accepted by the stakeholders, including the patients, the profession, the payers (eg. health insurance companies) and the policy makers, the most promising health technology innovation is likely to fail. Evolutionary change is more likely to succeed than revolutionary change and we believe that provision of m-health services as a supplementary part of the service package from healthcare providers is one possible path in such incremental change in service delivery.

The economic impact of introduction of new technology can also make the difference between success and failure in take-up of the innovation. This is the area of study for health economists. A comprehensive cost benefit analysis should take into account all the direct and indirect financial costs (including cost savings) and set them against the non-financial factors, that is the benefits (such as improved health outcomes or increased user satisfaction) and the disbenefits, for all the stakeholders.

A further set of challenges involves the ethical, legal and regulatory issues associated with the use of BANs and the offering of mobile healthcare services.

5. CONCLUSIONS

At time of writing the MobiHealth project has been running for 17 months (since May 2002). In the rather short duration of the project a great many problems and challenges have been encountered and much progress has been made. The starting point was a vision of ubiquitous mobile health services based on Body Area Networks. During the project we have designed and prototyped a health BAN and a BAN service platform and developed services for different patient groups according to the requirements specified by the clinical partners. Patient trials have begun during which evaluation data will be collected and analysed.

MobiHealth has resulted in an early prototype of the BAN, engineered mainly by integration of existing technologies without focusing on miniaturization or optimisation of power consumption. The main focus has been on the architecture, design and implementation of an m-health service platform. The result is a first version of a service platform whose architecture is comprised of a set of clearly defined components.

Perhaps ironically, the greater the potential of such services for the population at large, the bigger the problem of scalability of solutions becomes. We believe our solution to be scalable. Each of the service components can be replicated on multiple servers and, with implementation of a load balancing strategy, can serve a large user group. We expect the platform to be scalable to a large-scale remote healthcare monitoring situation, defined as 100,000 plus BANs. During the course of the project we believe we have reached a better understanding of the services such a platform should offer.

Prospects of substituting the IPAQ MBU platform with a Sony Ericsson P800 seem promising. As well as allowing substitution of the MBU platform, the generic service platform enables integration of BAN devices from different manufacturers. During the current project sensors from different sources, including front-end supported and self-supporting sensors, have been integrated into the BAN. Thus the objective of platform independence is being fulfilled.

It should not be thought however that all problems have been overcome even with use of current technologies. Ambulatory monitoring is more successful for some biosignals than others, for example some measurements are severely disrupted by movement artifacts. Some monitoring equipment is still too cumbersome for ambulatory use, because of the nature of the equipment or because of power requirements. In the area of wireless (tele)communication technologies (even with 2.5 and 3G) we still suffer from limited bandwidth for some applications, such as those which require serving many simultaneous users with conversational applications requiring high quality two-way audio and video, or generation of 3D animation in real time.

The use of BANs and wireless communications in personal healthcare systems still raises important challenges relating to security, integrity and privacy of data during transmission. This applies to both local transmission (eg. intra-BAN, BAN-PAN) and long range (eg. extra-BAN) communications. End-to-end security and Quality of Service guarantees need to be implemented. Safety of hardware (eg. electrical safety, emissions, interference) and reliability and correctness of applications must also be a priority in deployment of mobile services. Comfort and convenience of sensors or BANs worn long term for continuous monitoring is important for usability and user acceptance. Timeliness of information availability in the face of unreliable performance of underlying network services is another issue. Provision of seamless services across regional and national boundaries multiplies these difficulties. Powering *always on* devices and continuous transmission will continue to raise technical challenges. Business models for healthcare and accounting and billing models for network services need to evolve if technical innovations are to be exploited fully. Standardisation at all levels is essential for open solutions to prevail. At the same time specialization, customisation and personalisation are widely considered to be success criteria for innovative services.

Emerging technologies such as nanotechnology, smart sensors, ambient intelligent environments, grid computing, advanced power management and new implantation techniques open up many future possibilities for innovative mobile health services, however we expect that, along with these advances, issues such as quality of service, safety, security and privacy will pose ever more interesting challenges in the future.

6. ACKNOWLEDGEMENTS

Our thanks go to the European Commission, who fund the MobiHealth project under the IST FP5 programme, and to the MobiHealth partners. Thanks are also due for the support of the University of Twente (Faculty of Electrical Engineering, Mathematics and Computer Science, CTIT, and APS group).

7. REFERENCES

BlueTooth, 2003; http://www.bluetooth.org/

van Dam, K, S. Pitchers and M. Barnard, 'Body Area Networks: Towards a Wearable Future', Proc. WWRF kick off meeting, Munich, Germany, 6-7 March 2001; http://www.wireless-world-research.org/.

Dokovsky, N., A.T. van Halteren, I.A. Widya, "BANip: enabling remote healthcare monitoring with Body Area Networks", In proceedings of IEEE Conference on scientiFic engIneering of Distributed Java applIcations (FIDJI 2003), Luxemburg, November 2003.

Jones, Val, Rob Kleissen, Victor V. Goldman (1999), *Mobile applications in the health sector*, Presentation at Mobile Minded Symposium, 22 September 1999, University of Twente.

Jones, Val, *Mobile applications in future healthcare*, Presented at CTIT Workshop, University of Twente, 8 February 2001.

Jones, V. M., Bults, R. A. G., Konstantas, D., Vierhout, P. A. M., 2001a, Healthcare PANs: Personal Area Networks for trauma care and home care, *Proceedings Fourth International Symposium on Wireless Personal Multimedia Communications* (WPMC), Sept. 9-12, 2001, Aalborg, Denmark, http://wpmc01.org/, ISBN 87-988568-0-4

Jones, V. M., Bults, R. A. G., Konstantas, D., Vierhout, P. A. M,. 2001b, Body Area Networks for Healthcare, *Wireless World Research Forum meeting*, Stockholm, 17-18 September 2001; http://www.wireless-world-research.org/

Jones, Val, Richard Bults, Pieter AM Vierhout, Virtual Trauma Team, 2001c, *Wireless World Research Forum meeting*, Helsinki, 10-11 May 2001; http://www.wireless-world-research.org/

Jones, Val, Nadav Shashar, Oded Ben Shaphrut, Kevin Lavigne, Rienk Rienks, Richard Bults, Dimitri Konstantas, Pieter Vierhout, Jan Peuscher, Aart van Halteren, Rainer Herzog and Ing Widya, Remote Monitoring for Healthcare and for Safety in Extreme Environments, IN *M-Health: Emerging Mobile Health Systems*, Robert H. Istepanian, Swamy Laxminarayan, Constantinos S. Pattichis, Editors, Kluwer Academic/Plenum Publishers.

Konstantas, Dimitri, Val Jones, Richard Bults and Rainer Herzog, 2002a, MobiHealth - Wireless mobile services and applications for healthcare, *International Conference On Telemedicine - Integration of Health Telematics into Medical Practice*, Sept. 22nd-25th, 2002, Regensburg, Germany.

Konstantas, Dimitri, Val Jones, Richard Bults, Rainer Herzog, 2002b, *MobiHealth – innovative 2.5 / 3G mobile services and applications for healthcare*, Thessaloniki, 2002.

MobiHealth, 2002; http://www.mobihealth.org

Östmark, Åke, Linus Svensson, Per Lindgren, Jerker Delsing, "Mobile Medical Applications Made Feasible Through Use of EIS Platforms", *IMTC 2003 – Instrumentation and Measurement Technology Conference*, Vail, CO, USA, 20-22 May 2003.

Rekhter, Y., B. Moskowitz, D. Karrenberg, G. J. de Groot, E. Lear et al., "Address Allocation for Private Internets", RFC 1918, February 1996.

Schmidt, R., 2001, *Patients emit an aura of data*, Fraunhofer-Gesellschaft, www.fraunhofer.de/english/press/md/md2001/md11-2001_t1.html

Sun Microsystems, 2003a. *CDC: An Application Framework for Personal Mobile Devices*, June 2003, http://java.sun.com/j2me

Sun Microsystems, 2003b, *Jini Technology Surrogate Architecture Specification"* July 2001, http://surrogate.jini.org/

Widya, A. van Halteren, V. Jones, R. Bults, D. Konstantas, P. Vierhout, J. Peuscher, 2003a. Telematic Requirements for a Mobile and Wireless Healthcare System derived from Enterprise Models. *Proceedings IEEE ConTel 2003: 7th International Conference on Telecommunications, June 11-13, 2003*, Zagreb, Croatia.

Widya, I., Vierhout, P. A. M., Jones, V. M., Bults, R. A. G., van Halteren, A., Peuscher, J. and Konstantas, D., *Telematic Reequirements for Emergency and Disaster Response derived from Enterprise Models*, 2003b, IN *M-Health: Emerging Mobile Health Systems*, Robert H. Istepanian, Swamy Laxminarayan, Constantinos S. Pattichis, Editors, Kluwer Academic/Plenum Publishers.

Wireless World Research Forum, 2001, *The Book of Visions 2001: Visions of the Wireless World*, Version 1.0, December 2001; http://www.wireless-world-research.org/

xMotion , 2002; http://www.ist-xmotion.org/

ZigBee Alliance, "IEEE 802.15.4, ZigBee standard", http://www.zigbee.org/

Zimmerman, T.G., 1999, 'Wireless networked devices: A new paradigm for computing and communication', *IBM Systems Journal*, Vol. 38, No 4.

MOBIHEALTH: MOBILE SERVICES FOR HEALTH PROFESSIONALS

Val Jones[*], Aart van Halteren, Nikolai Dokovsky, George Koprinkov, Jan Peuscher, Richard Bults, Dimitri Konstantas, Ing Widya, and Rainer Herzog

1. INTRODUCTION

The concept behind the MobiHealth project (MobiHealth, 2002). was to bring together the technologies of Body Area Networks (BANs), wireless broadband communications and wearable devices to provide mobile healthcare services for patients and health professionals. For patients, these technologies enable remote patient care services such as management of chronic conditions and detection of health emergencies. For health professionals the technology offers access to information and communication services from a mobile device, thus enabling mobility for the individual professional and supporting the operation of distributed 'virtual' healthcare teams (*sensu* Pitsillides, 1999).

During the project a generic BAN for healthcare and a generic m-health service platform were developed, trialled and evaluated. The main devices used in patient BANs were medical sensors. Biosignals measured by sensors connected to the BAN were transmitted to a remote healthcare location over 2.5/3G public wireless networks, using commercial GPRS services and test infrastructures for UMTS. The MobiHealth BAN and service platform were trialled in Spain, The Netherlands, Sweden and Germany.

The overall goal of MobiHealth was to evaluate the ability of 2.5 and 3G communication technologies to support innovative mobile health services. The main output of the project therefore is an assessment of the suitability of GPRS and UMTS to support such services.

The MobiHealth concept arose out of earlier work at the University of Twente (Jones et al., 1999; Jones 2001, Jones et al., 2001b; Jones et al., 2001c; WWRF, 2001). The MobiHealth consortium involving 14 European partners was brought together during

[*] Val Jones, Centre for Telematics and Information Technology/Faculty of Electrical Engineering, Mathematics and Computer Science, University of Twente, PO Box 217, 7500 AE, Enschede, The Netherlands, v.m.jones@ewi.utwente.nl.

2001. The project began on May 1st 2002 and was funded by the European Commission under IST FP5 and has been reported in (Jones et al, 2001a; Konstantas et al., 2002a,b; Widya et al., 2003, Dokovsky et al., 2003).

This remainder of this chapter presents the MobiHealth system (Section 2), MobiHealth services from the health professional point of view (Section 3), and some of the issues and future challenges raised by the project (Section 4).

2. THE MOBIHEALTH SYSTEM

The components of the MobiHealth system are presented below. The software was developed at the University of Twente and is available as Open Source software.

2.1. Body Area Networks

The concept of the Body Area Networks has been developed in different directions since the original work at IBM (Zimmerman, 1999), eg. by (van Dam, 2001; Jones et al., 2001a; Jones et al., 2001b; Schmidt, 2001). In our definition, a BAN is "a collection of (inter) communicating devices which are worn on the body, providing an integrated set of personalised services to the user" (WWRF, 2001). A BAN can be thought of as a computer network which is worn on the body and which moves around with the person.

A BAN incorporates a set of devices which perform specific functions and which also perform communication, perhaps via a central controlling device. Communication amongst the elements of a BAN is called *intra-BAN communication*. Intra-BAN communication may be carried over a wired or a wireless medium. If the BAN communicates externally, i.e. with other networks, this communication is referred to as *extra-BAN communication*. Extra-BAN communication is usually mediated over wireless links. A range of technologies is available for extra-BAN communication including Bluetooth, WLAN, GSM, GPRS and UMTS.

When the devices of a BAN measure perform actions for health-related purposes we call this kind of specialization of Body Area Networks a *health BAN*. As a further specialisation, a BAN whose devices are biosignal sensors enables patient monitoring. We can envisage any number of health monitoring BANs for specific clinical conditions. Some example applications are: cardiac care, diabetes management, sleep analysis, asthma monitoring, epilepsy monitoring, SIDS, movement analysis and monitoring and control of implanted drug delivery systems. If we combine health BANs with *extra-BAN communications*, this enables a range of distributed applications including remote monitoring of patients by health professionals.

The MobiHealth BAN consists of a central unit known as the MBU (Mobile Base Unit) and a set of body-worn devices, which may include sensors, actuators, and various multimedia devices. In principle the BAN may incorporate any wearable or implantable devices. Figure 1 shows the MobiHealth BAN and its relation to the m-health service provider and the healthcare service provider.

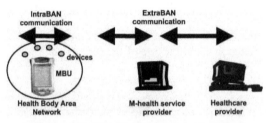

Figure 1. The MobiHealth Body Area Network

The BAN software or *BANware* is implemented on top of a Java virtual machine (J2ME, CDC and Personal profiles) running under Linux. The MBU platform is currently the HP iPAQ 3870. Future plans include porting to other platforms including cell phones (eg. Sony Ericsson P800).

The generic BAN can be customized with different device sets for different applications. Some of the BAN devices used during the MobiHealth project were: electrodes for measuring ECG and EMG, activity sensors, respiration sensors and pulse oximeters. Other BAN devices used were marker/alarms and digital cameras.

2.2 The Back End System

Deployed BANs are supported by a MobiHealth Back End System (BeSys)which mediates between end user applications in the healthcare provider domain (eg HISs) and the BANs. The BeSys may be located at a healthcare location such as a hospital or at another location entirely, eg at an m-health service provider site. The BeSys runs on a server and incorporates a Look Up Service (LUS) and hosts a surrogate which represents and manages the BANs in the field. The end-user application may consist simply of a viewer application or (in future) some much more complex suite of applications interfacing to the clinical systems of the healthcare provider.

2.3. Communications

The BAN has been implemented using both wired (front-end supported) sensors from TMSI and wireless (self-supporting) sensors from EISlab (Östmark et al, 2003). Both approaches use Bluetooth (IEEE 802.15) for intra-BAN communication. The front-end also allows Zigbee (IEEE 802.15.4) as an alternative intra-BAN communication technology. For extra-BAN communication 2.5 and 3G public networks are used. The MBU communicates via Bluetooth to a GPRS or UMTS handset; from there the signals are transmitted to the remote healthcare location. The first UMTS (test) infrastructures became available in the Netherlands during 2003 and the trials were run over commercial GPRS services in all countries and over test UMTS infrastructures in the Netherlands, Spain and Sweden. A commercial UMTS service was launched in the Netherlands shortly after the completion of the trials.

2.4. The m-health Service Platform

Together the components described above comprise an m-health service platform. Figure 2 shows the functional architecture of the service platform.

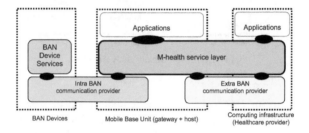

Figure 2. Service platform functional architecture

The m-health service platform supports the BAN based applications and isolates them from the details of the underlying communications services. It is the service platform which enables genericity and hardware independence. The services offered include BAN registration, discovery., security services, online configuration and management of BANs and BAN data storage. BAN data may be pulled by or punched to the remote application components. The BAN Interconnect Protocol (BANip) was implemented using HTTP1.1. The surrogate was implemented using the Jini Surrogate architecture. More details of the implementation may be found in the project deliverables, available at the project website and in (Dokovsky et al, 2003). The following section gives an overview of the MobiHealth services seen from the point of view of the health professional.

3. PROFESSIONAL USE OF THE MOBIHEALTH SYSTEM

The generic BAN concept and the generic m-health service platform permit any number of possibilities for mobile services for health professionals. Here we mention three classes of functionality for use by health professionals which have been explored in the course of the project. They are mobile monitoring systems, remote access to patients' BAN data, and mobile multimedia communication systems. These classes of functionality represent some of the possible configurations for a health professional BAN.

3.1 Mobile monitoring system

A BAN equipped with medical sensors can be carried by mobile health professionals so that they can perform various measurements on patients at the point of care in the community. The measurements can be viewed locally by the professional who applies the BAN to the patient. Furthermore the measurements can be transferred to a remote location over a variety of wireless communication services. At the remote locations(s) the clinical information can be uploaded and stored; it may also be viewed remotely by other members of the medical team.

Technically the health professional mobile monitoring system is very similar to the patient-controlled telemonitoring BAN. The major difference is that in the former case the BAN is part of the mobile equipment carried and used by a mobile health professional. User interface and access control requirements vary between patient-controlled and health professional-controlled use of BAN based monitoring systems. For some patient applications it may be that no local display of biosignals is needed or appropriate. For professional applications local viewing is normally required and the display must be of adequate quality for clinical use in that setting. For some purposes a PDA display will suffice; for others a higher resolution device is required.

In the MobiHealth trials different configurations of the BAN monitoring system were used by visiting nurses and by ambulance paramedics. Two version of the nurse BAN were developed for the trials. The trauma patient BAN is designed to be applied by ambulance paramedics to patients in an emergency setting.

Home monitoring following discharge from hospital. In one of the Swedish trials (*Home care and remote consultation for recently released patients in a rural area*) the target group is patients with multiple chronic diseases including cardiac failure, diabetes and respiratory disability. The intention is to support patients living in remote rural areas following discharge from hospital. If the patient or their family have concerns about the patient's condition they can request a home visit from a nurse. The visiting nurse carries a nurse BAN on home visits. The BAN can be used to measure ECG, heart rate respiration rate and oxygen saturation. Blood glucose, NIBP (blood pressure) and respiration are manually input to the BAN by the nurse. At the same time the measurements are transmitted over UMTS or GPRS to a remote physician or to a registered district nurse (RDN). The intention is to support clinical decision making by distributed teams; here the remotely located health professional views the clinical data during the home visit and determines together with the visiting nurse whether the patient needs transfer by ambulance to hospital. The expected benefit is that the use of the BAN will reduce the number of cases where the patient is moved to hospital unnecessarily.

Vital signs monitoring in emergency medicine. In one of the Dutch trials (the *Trauma trial*), ambulance paramedics make use of two kinds of BAN. The first is known as the trauma patient BAN and is applied by the paramedic to the patient at the scene of the accident. The BAN devices are sensors to measure oxygen saturation, ECG and respiration. If they have time, the paramedics can also make manual inputs (pupil reaction, fluids) to the BAN. During trials information was transmitted from the trauma patient BAN via GPRS and UMTS to the emergency surgical team at the receiving hospital.

The trauma scenario implies much stricter requirements on the health professional user interface and on transmission parameters. In a critical emergency scenario the front-line professional needs to work fast and it is even more important in this kind of setting that the technology does not impede or interfere with their work by demanding they do extra tasks or deal with clumsy equipment. For transmission, the wireless communication with the hospital needs to operate well not only from the static situation (the scene of the incident) but also from the moving ambulance during transfer of the patient. This means maintenance of communication despite a fast moving source which possibly passes through different areas of coverage, generating tougher communication requirements relating to robustness and quality of service and the possible need to perform seamless horizontal and vertical handovers. The expected benefits of transferring clinical information from the scene are improved clinical outcomes for patients due to an earlier informed, and hence better prepared, emergency department at the hospital.

The following section describes another BAN configuration which supports a different set of mobile services for health professionals. The variants of the BAN mentioned so far are all telemonitoring systems and as such are principally *producers* of data. Corresponding to the data producers are the data *consumers* which were originally conceived as applications running at a remote health care location such as a hospital or health call centre within the jurisdiction of the health service provider. Some of these applications will involve exchange of m-health data with the EMR and hospital information systems and as such are outside the scope of the current project. However some of the applications are appropriate to run on a health professional's mobile device, allowing the health professional to be fully mobile and to have ubiquitous access to BAN services and data. Examples explored during the project, and future planned extensions, are discussed below.

3.2 Mobile access to patient BAN data

Within the scope of the current project the end user applications developed for use by healthcare professionals consist of alarm services and applications to process and view data from patient's mobile telemonitoring systems.

Alarm services. If needed a patient BAN can be equipped with an alarm button which the patient can press to summon help. The alarm service is a real time service linking a (mobile) patient with a (mobile) health professional. In the current implementation the health professional receives the alarm notification by SMS message to a mobile phone.

Future developments could include more sophisticated versions of the alarm service. From the health professional point of view the reception of alarms could be one of the services with which the generic health professional BAN could be configured as appropriate. An important evolution would be the automatic detection of (clinical) emergencies and generation of alarms without patient intervention. Automated alarm services could range from simpler cases such as detecting when threshold values are exceeded to intelligent interpretation of multiple data sources such as activity sensors, biosensors and, possibly, environmental sensors together with patient specific information such as the EMR or an emergency dataset.

Viewer applications. A simple viewer application was implemented to run on the MBU. The viewer displays numerical values such as heart rate and SaO2 and visualisations for continuous signals such as ECG and pulse plethysmogram. The intention was that the viewer application should be used to verify that sensors are attached and functioning properly. However the graphical display and the numerical values can be used for some clinical purposes. For other clinical purposes better display capabilities and a more sophisticated viewer application is needed. In the trials the Portilab 2 viewer from TMSI was used. This system can be run on a health professional's laptop to give higher resolution and scalable displays. Portilab 2 running on a laptop with wireless internet access has already been tested as a mobile application for health professionals. Such applications are candidate services for the customizable health professional BAN.

Processing, analysis and interpretation of biosignals. Underlying the viewer applications are functions to process the raw sensor output. Signal processing functions (analogue to digital conversion, filtering and compression) are applied to the raw sensor data by the sensor system. Higher level applications perform further processing and analysis of incoming sensor data. An example is the derivation of calculated parameters from measured parameters (eg heart rate derived from pulse plethysmogram) or the separation of different signals (eg separation of foetal ECG from maternal ECG). These functions are distributed across the sensor system (the sensors themselves and the DSP in the sensor front end) and the Portilab 2 software.

For the future we can postulate a range of clinical interpretation functions involving sophisticated high level clinical inferencing systems applied to multiple biosignal streams sampled over time. Intelligent automated processing of BAN data is a prerequisite for large scale deployment of some telemonitoring services in the population, since continuous monitoring requiring observation of BAN output 24/7 by clinical staff would make impossible resource demands on health services.

3.3 Mobile Multimedia communication system

Earlier we referred to the telemonitoring BAN used in the trauma setting. The other type of BAN used in the Trauma trial is the paramedic BAN. This has different kinds of functionality from the patient BANs and the nurse BANs already described. Rather than being applied to a patient by a health professional this kind of BAN is worn by the health professional and in this case the BAN devices are multimedia devices. The purpose is to provide real time communication between the paramedics and the team at the hospital. The original vision was to equip members of the distributed trauma team with health professional BANs enabling high quality hands free audio and video communication between the paramedics and the hospital staff.

At time of writing the paramedic BAN consists of an MBU and one device: a Bluetooth enabled digital camera; hence visual communication only is supported. Still images from the scene are transmitted from the camera to the MBU via Bluetooth and to the hospital over GPRS or UMTS. When digital video cameras conforming to the Bluetooth video profile become available (expected 2004) the still images will be replaced by motion video. As with the trauma patient BAN, the paramedic BAN exchanges information with the hospital both from the scene and from the moving ambulance.

In future the intention is that by combining information from the paramedic BAN and the trauma patient BAN, the hospital team can participate through telepresence and, by means of augmented reality style presentation, be able to simultaneously view the patient, the scene of the accident and visualisations of the patient's biosignals.

The paramedic BAN and the Nurse BANs each represent a selected set of mobile services for health professional use; that is they are different instantiations of the generic BAN. They are all supported by the generic service platform which can support mobile services customized to the particular needs of individual professionals and which can be used to link multidisciplinary teams of mobile health professionals enabling them to function as a virtual team.

4. FUTURE CHALLENGES

The vision of ubiquitous healthcare poses challenges to many disciplines, both technical and non-technical. Currently the major limiting factor for provision of mobile services seems to be the rather mundane but problematic issue of the limited battery life of mobile devices. Power management strategies can make some inroads here but a breakthrough in power technologies which can make a difference in order of magnitude is needed. Regarding communication technologies, even with UMTS we still suffer from limited bandwidth for some remote monitoring applications, such as full 12-lead ECG or EEG transmission. There are many software engineering issues arising in connection with intelligent adaptive applications and communications, complicated by cross-cutting issues of security, quality of service and scalability.

For health service providers, large scale usage of mobile services implies access to high performance distributed computing and broadband wireless communication resources. More radically it also implies a high degree of automation of clinical analysis and interpretation. For some applications automated analysis will involve complex real time pattern recognition and application of temporal abstraction in order to automate clinical interpretation. This in turn implies that BAN applications interface to large scale and comprehensive clinical knowledge bases and inferencing systems.

We expect that in the longer term the current style of implementation of (health) BANs will disappear, replaced by nanoscale devices and contactless sensing from the ambient intelligent environment. We call this vision the Ambient Intelligent BAN or AmI-BAN). AmI-BAN devices might be implanted rather than worn externally on the body or in clothing. These devices could include for example actuators to control drug delivery systems, nanomanipulators or neural stimulators interfacing directly to motor neurons.

The vision of the future depends on scientific research and technological advances not only in computing science but also from other technical disciplines including electrical engineering, physics and nanoscience. Wearable computing, nanotechnologies, contactless sensing, implantable devices and low-power or self-powered devices are some of the enabling technologies prerequisite to the realization of the vision. Together these multidisciplinary developments can open the way for seamless collaboration between BANs, PANs and the ambient intelligent environment.

Beyond the technical domains, experience indicates that critical success factors for take up of technical innovation include organisational and management issues concerning the impact on ways of working and business processes and trust and acceptance by professional users. Research in the areas of e-health and e-learning indicates that innovation is embraced by busy professional users only when they *perceive immediate* benefits.

Regarding the organisational aspects of healthcare, exploitation of a technical innovation may imply extensive and radical business process re-engineering of the national (or pan-European) healthcare system. Failure to account for this kind of impact has caused many health technology innovations to fail in the past. Unless business process re-engineering is driven by and accepted by the stakeholders, including the patients, the different categories of health professionals and other involved professional groups, the payers (eg. health insurance companies) and the policy makers, the most promising health technology innovation is likely to fail. Even then, evolution goes more smoothly than revolution.

Economic effects can also spell failure for take-up of sound technical innovations. A comprehensive cost benefit analysis is prerequisite, but should take into account all the direct and indirect financial costs (including cost savings) and set them against the non-financial factors, that is the benefits (such as improved health outcomes or increased user satisfaction) and the disbenefits, for all the stakeholders involved. The social and ethical issues, and legal and regulatory factors, also need to be included in the analysis.

5. CONCLUSIONS

The MobiHealth project ran for 22 months from May 2002. The main development focus was on the architecture, design and implementation of an m-health service platform to support mobile services for patients. The objective was to test out the potential of 2.5 and 3G communications to support mobile services. In the relatively short time of the project the phases covered were: user requirements, design and implementation of a generic health BAN including several specialisations of the BAN, and architecting of a generic m-health service platform. In addition trials and evaluation were conducted. At the time of writing the results of the evaluation are being compiled. Regarding mobile services for the professional user, the experience has demonstrated that the generic MobiHealth concept can be applied for the mobile professional user as well as for the patient user.

Further development continues, targeted at a fully generic and evolvable m-health service platform supporting mobility for both patients and professionals and supporting personalised and evolvable services adaptable to the needs of different categories of users operating in dynamically changing contexts.

6. ACKNOWLEDGEMENTS

Our thanks go to the European Commission, who funded the MobiHealth project under the IST FP5 programme, and to the MobiHealth partners. Thanks are also due for the support of the University of Twente (EEMCS, CTIT, and APS group).

7. REFERENCES

van Dam, K, S. Pitchers and M. Barnard, 'Body Area Networks: Towards a Wearable Future', Proc. WWRF kick off meeting, Munich, Germany, 6-7 March 2001; http://www.wireless-world-research.org/.

Dokovsky, N., A.T. van Halteren, I.A. Widya, "BANip: enabling remote healthcare monitoring with Body Area Networks", In proceedings of IEEE Conference on scientiFic engIneering of Distributed Java applIcations (FIDJI 2003), Luxemburg, November 2003.

Jones, Val, Rob Kleissen, Victor V. Goldman (1999), *Mobile applications in the health sector*, Presentation at Mobile Minded Symposium, 22 September 1999, University of Twente.

Jones, Val, *Mobile applications in future healthcare*, Presented at CTIT Workshop, University of Twente, 8 February 2001.

Jones, V. M., Bults, R. A. G., Konstantas, D., Vierhout, P. A. M., 2001a, Healthcare PANs: Personal Area Networks for trauma care and home care, *Proceedings Fourth International Symposium on Wireless Personal Multimedia Communications* (WPMC), Sept. 9-12, 2001, Aalborg, Denmark, http://wpmc01.org/, ISBN 87-988568-0-4.

Jones, V. M., Bults, R. A. G., Konstantas, D., Vierhout, P. A. M,. 2001b, Body Area Networks for Healthcare, *Wireless World Research Forum meeting*, Stockholm, 17-18 September 2001; http://www.wireless-world-research.org/

Jones, Val, Richard Bults, Pieter AM Vierhout, Virtual Trauma Team, 2001c, *Wireless World Research Forum meeting*, Helsinki, 10-11 May 2001; http://www.wireless-world-research.org/

Konstantas, Dimitri, Val Jones, Richard Bults and Rainer Herzog, 2002a, MobiHealth - Wireless mobile services and applications for healthcare, *International Conference On Telemedicine - Integration of Health Telematics into Medical Practice*, Sept. 22nd-25th, 2002, Regensburg, Germany.

Konstantas, Dimitri, Val Jones, Richard Bults, Rainer Herzog, 2002b, *MobiHealth – innovative 2.5 / 3G mobile services and applications for healthcare*, Thessaloniki, 2002.

MobiHealth, 2002; http://www.mobihealth.org

Östmark, Åke, Linus Svensson, Per Lindgren, Jerker Delsing, "Mobile Medical Applications Made Feasible Through Use of EIS Platforms", *IMTC 2003 – Instrumentation and Measurement Technology Conference*, Vail, CO, USA, 20-22 May 2003.

Pitsillides A., Samaras G., Dikaiakos M., Christodoulou E., DITIS: Collaborative Virtual Medical team for home healthcare of cancer patients, Conference on the Information Society and Telematics Applications, Catania, Italy, 16-18 April 1999.

Rekhter, Y., B. Moskowitz, D. Karrenberg, G. J. de Groot, E. Lear et al., "Address Allocation for Private Internets", RFC 1918, February 1996.

Schmidt, R., 2001, *Patients emit an aura of data*, Fraunhofer-Gesellschaft, www.fraunhofer.de/english/press/md/md2001/md11-2001_t1.html

Widya, A. van Halteren, V. Jones, R. Bults, D. Konstantas, P. Vierhout, J. Peuscher, 2003. Telematic Requirements for a Mobile and Wireless Healthcare System derived from Enterprise Models. *Proceedings IEEE ConTel 2003: 7th International Conference on Telecommunications, June 11-13, 2003*, Zagreb, Croatia.

Wireless World Research Forum, 2001, *The Book of Visions 2001: Visions of the Wireless World*, Version 1.0, December 2001; http://www.wireless-world-research.org/

Zimmerman, T.G., 1999, 'Wireless networked devices: A new paradigm for computing and communication', *IBM Systems Journal*, Vol. 38, No 4.

DITIS: A COLLABORATIVE VIRTUAL MEDICAL TEAM FOR HOME HEALTHCARE OF CANCER PATIENTS

Andreas Pitsillides[*], Barbara Pitsillides, George Samaras, Marios Dikaiakos, Eleni Christodoulou, Panayiotis Andreou, and Dimosthenis Georgiadis

1. INTRODUCTION

Complex and chronic illnesses, such as cancer, demand the use of specialised treatment protocols, administered and monitored by a patient centric co-ordinated team of multidisciplinary healthcare professionals. Care of chronic illnesses (e.g. cancer patients) by a team of health care professionals at home is often necessary due to the protracted length of the illness, the differing medical conditions, as well as the different stages of the chronic illness. Most importantly home care can offer comfort for the patient and their family, in the familiar surrounding of their home, and at the same time being cost effective, as compared to the high cost of hospital beds. Hospital based treatment for chronic patients is limited, often demand based for short periods of time, used mainly for acute incidents. As it is not possible for the health care team to be physically present by the patient at all times, or at any time physically together, whilst the patient is undergoing treatment at home (or work), a principal aim is to overcome the difficulty of coordination and communication, through DITIS (ΔΙΤΗΣ, in Greek, stands for: Network for *Medical Collaboration*). DITIS is a system that supports dynamic **Virtual Collaborative HealthCare Teams** dealing with the home-healthcare. It supports the dynamic creation, management and co-ordination of virtual medical teams, for the continuous treatment of the patient at home, and if needed for periodic visits to places of specialised treatment and back home.

DITIS was initiated in 1999, supporting the activities of the home healthcare service of the Cyprus Association of Cancer Patients and Friends (PASYKAF) in the district of

---

[*] Andreas Pitsillides, Department of Computer Science, University of Cyprus, P.O. Box 20537, Nicosia, CY 1678, Cyprus, Andreas.Pitsillides@ucy.ac.cy

Larnaca, using a **patient centric** philosophy, **focused on home-based rather than facility-based care**, receiving governmental and industry grants. It deploys a novel networked system for Tele-collaboration in the area of patient care at the home by a virtual team of medical and paramedical professionals, implemented using existing networking and computing components, initially addressing cancer patients (A.Pitsillides et al., 1997; Dikaiakos et al., 1998; A.Pitsillides et al., 1999; A.Pitsillides et al., 1999; B.Pitsillides., 1999; B.Pitsillides et al., 2003). DITIS was originally developed with a view to address the difficulties of communication and continuity of care between the home health care multidisciplinary team (PASYKAF) and between the team and the oncologist often located over 100kms away. DITIS has through its database and possibility of access via mobile or wire line (computers) offered much more than improved communication. Its flexibility of communication and access to the patient's history and daily record at all times and from anywhere (e.g. home, outpatients, or even during emergency admission) has offered the team a continuous overall assessment and history of each symptom. DITIS has thus offered improved quality of life to the patient, for example by offering the nurses the possibility of immediate authorisation to change prescription via mobile devices and the oncologist the possibility of assessment and symptom control without necessarily having to see the patient. It has offered the home care service the opportunity to plan future services and lobby for funding by offering audit, statistics, and performance evaluation and also the possibility for research. The pilot initial results indicated the usefulness of the proposed system, as well as highlighted practical, at times frustrating, problems. DITIS is currently being extended to island-wide implementation to support the PASYKAF home care service.

DITIS is expected to deliver a product that can improve the quality of the citizen's life. Contrary to today's health processing structure which is, in all practical terms **facility-based care**, this project aims to shift the focus onto **home-based care,** where everything is moving around the patient. The virtual healthcare team will be able to provide dedicated, personalized and private service to the home residing patient on a need based and timely fashion, under the direction of the treating specialist. Thus it is expected that chronic and severe patients, such as cancer patients, can enjoy 'optimum' health service in the comfort of their home (i.e. a focus on **wellness**), feeling safe and secure that in case of a change in their condition the health care team will be (virtually) present to support them.

In section 2 we analyse the justification and need for the proposed system, and in section 3 we provide the identification and analysis of the home healthcare professionals roles and collaboration scenarios. In section 4 we present the system design, and section 5 the pilot implementation in PASYKAF Larnaca. Finally in section 6 we provide some concluding remarks on our experiences, and present the benefits derived from the system, as well as future enhancements, and directions.

2. JUSTIFICATION OF NEEDS AND AIMS OF DITIS

The current context of health and health care is characterized by change and transition associated with health care system reform and restructuring (Cherry et al., 2001). Restructuring initiatives are intended to develop a more results-oriented, integrated and accountable health system that delivers the **right services, to the right people, at the most appropriate time, in the right place, in the most cost-effective**

manner. The aim has been to enhance the health quality of populations by better balancing health promotion and disease prevention, community-based and institutional services. Further technological advances are enabling a greater shift from institutional services to ambulatory and community-based services, as for example home-care.

2.1 Home Health Care

No issue embodies more the meaning of "reform" than the **shift from facility-based care to home-based care**. Reform and restructuring of the health system has seen hospitals closed at a faster pace than the planning and organization of the infrastructure for home and community-based care. Cost containment measures in acute care facilities have given rise to shortened lengths of stay and the shifting of care provision from the inpatient setting to the ambulatory care setting. The **need for home-based care is growing but the capacity to manage that need is still being developed**. This has placed a significant burden of care on individuals, their families and volunteer caregivers.

Home care services are an important and growing component of the health service delivery system. The pressure to expand and enhance home-based services is expected to grow as a result of demographic shifts in the population, changing consumer expectations with respect to service and care options, and technological and scientific advancements in the delivery of health services. Many see home care as a more cost-effective alternative to acute care and/or to long-term institutional care (Cherry et al., 2001).

The **definition** of **home care,** that we adopt, is well captured in (Canadian Institute report, Jun 1999):

"A range of health-related, social and support services received at home. These services enable clients incapacitated, in whole or in part, to live in their home environment. These services help individuals achieve and maintain optimal health, well being and functional ability through a process of assessment, case co-ordination, and/or the provision of services. Service recipients may have one or more chronic health conditions or recently experienced an acute episode of illness or hospitalization. The range of services provided includes prevention, maintenance, rehabilitation, support and palliation."

2.2 The need for DITIS:

Although the context of health reform may vary across countries, the major objectives are similar and include:

- o   a move towards *people-centred* services;
- o   a commitment to *healthy public policy* and a desire to improve the health status of individuals and communities;
- o   increased emphasis on *knowledge/evidence-based decision making* and *efficiency and effectiveness* in service delivery;
- o   a shift from facility-based health services and a focus on illness, to *community-based health services* and a focus on *wellness*;
- o   the *integration* of agencies, programs and services to achieve a seamless continuum of health and health-related services; and
- o   greater *community involvement* in priority setting and decision-making.

DITIS aims to support the above healthcare reform objectives. We focus our analysis on cancer patients, but expect our results to be applicable to home healthcare in general.

For home health care:

- The patient care is provided by a team of more than one healthcare professional, as for example cancer specialists, general medical doctors, nurses, physiotherapists, psychotherapists, dieticians, social workers, and so on. We believe that a key factor that needs to be kept in mind is that the provider is not a monolithic entity but rather consists of a diverse group of people. Thus, the provision of as optimum and effective care as possible demands the *cooperation, communication* and *coordination* among all these professionals, and the formation of a '*team of care*'
- Specialist nurses and other healthcare professionals visit patients regularly at home, offer care, which must be provided in co-operation, and often under the direction by the treating doctors of a hospital (e.g. oncologist).

Providing top quality care necessitates close collaboration, secure, easy and timely exchange of information, and coordination of the team activities. This should be achieved irrespective of the physical presence of the individual members of the team, or even if different doctors treat the patient, for possibly different symptoms at different hospitals, or visit him at home. It is of course obvious, that the *concurrent physical presence* at the *point of care* of all members of the team is rarely possible. This creates serious difficulties for providing the quality care that Cancer patients deserve to obtain in a friendly (to them) home environment. Through DITIS we expect to assist in the delivery of better home-care, by offering the health-care team services that are aimed in achieving a seamless continuum of health and health-related services, despite the structural problems of home-care, as compared to facility based care, as for example the geographic proximity and geographic separation between the team members and the patient.

DITIS aims to overcome these difficulties by maintaining a ***dynamic collaborative virtual healthcare team*** (Figure 1), as well as secure, easy, and timely access to the **unified Electronic Medical Record** database (Figure 2) for the continuous home-treatment of patients. The dynamic virtual healthcare team is created explicitly to satisfy the needs of each particular patient at a point in time.

Figure 1. Patient centric home health care

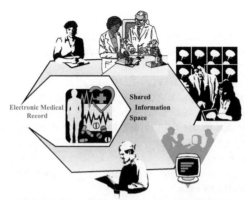

Figure 2. The unified information space of DITIS

As a result the following **clinical objectives** are addressed:

- To provide the presence of the (virtual) team by the patient at any given time, irrespective of locality, or cross country movement.
- To improve communication within the dynamic (virtual) home care team and between the home care team and the hospital (locally, or cross country), thus providing enhanced quality care.
- To provide flexible and secure access and management of healthcare records at any time and from anywhere, to improve continuity of care.
- To improve collection of statistical data for further audit and research within home care setting, enhancing knowledge and offering possibility of evidence-based care.
- Provide continuation of care for chronic illness via Virtual Collaborative Medical Teams, finally leading to a Pan-European scale (to cover the needs of visiting or retiring patients or healthcare professionals in a foreign country).
- To aid in making the dependant role of the home-care nurse legally binding (for example, in the home setting when interacting with a hospital doctor for the prescription of a pain drug in the home).

Given the above are satisfied, the quality of life of chronic and severe patients will improve.

2.3 Prior to DITIS

The case of the home-care provision prior to DITIS is described below:

Home care professionals visit the patient at home. Traditionally, the team of professionals was (loosely) coordinated by weekly meetings, or in case of some urgent event information was exchanged by telephone calls, or face-to-face meetings. Often the same information is requested from the patient, so as each professional can build their own medical and psychosocial history and treatment notes (handwritten). These handwritten notes were filed at the PASYKAF offices, once the health-care professional returned to the office, often at the end of the day.

If a visit to the patient was scheduled, then the file from the office had to be removed and taken with the health-care professional to the patient home. This was inflexible, as there was no possibility of access by another health-care professional at the same time (say the physiotherapist, unless separate records were kept). Also the health-care professional had to make a visit to the office to collect the patient file (even if there was no other business with the office). Additionally, in case of an emergency, even though a health-care professional (not necessarily belonging to the team) may had been able to respond immediately due to his geographic location with respect to the patient, a visit to the office would had been necessary, just to collect the patient file. Of course, for a patient visit to the hospital, especially in emergency, there was no possibility of immediate access to the patient file by the treating doctors. Therefore there was limited possibility for **continuity of care**.

As with every manual system, there was limited possibility for audits and statistics, research was difficult, evidence-based medicine was not supported, dynamic coordination of the team was almost impossible, and communication overheads were very high and costly in human and monetary terms. DITIS was aimed to address these problems in the provision of home-care services by a team of professionals.

2.4 Competing Alternatives

The DITIS system integrates a number of state of the art technologies to provide a sophisticated tele-application in the medical domain. To substantiate the importance of such as system one must differentiate it from other related attempts. Indeed a number of related projects and efforts exist; however, no one, as far as we can ascertain, offers the features and innovation of DITIS within a single integrated application. We can classify these systems (and compare them to DITIS) based on their focus activities.

There are various **projects** covering accurate **diagnosis** and treatment until a patient gets to the hospital or in a home care center. Among them are: Ambulances Health Telematics HC 1001, Emergency-112 Health telematics HC 4027, Vital-Home ISIS 98 502085) that have already developed such telemedicine services over GSM. However these have not addressed the notion of virtual collaborative medical teams for the home care sector. The projects that deal with **home care** can be divided in two types: The first type is closer to research projects concerned with delivery of home care developing new generation of bio-sensors and transmitters (e.g. CHS, CHRONIC, EPIDEMICS, M2DM, TELEMEDICARE, E-REMEDY). The second type improves collaboration between health professionals for better home care delivery (e.g. D-LAB, PHARMA, MTM, MOEBIUS). None of these projects consider mobile technologies to support health-care team collaboration in home care applications.

Similarly a number of European projects are actively focusing on the use of **handheld devices** for health care provision. WARD IN HAND (IST-1999-10479) allows the management of key clinical information while providing decision support to mobile medical staff of a hospital ward, MOBIDEV (IST-2000-26402) promises to provide mobile users with secure access to the Hospital Information System in and outside the hospital, using web interfaces based on Bluetooth technology and GPRS/UMTS networks and also improve user friendliness via voice commands, DOCMEM (IST-2000-25318) and MOMEDA (HC 4015) aim to offer web access to electronic patients records (EMR) via multimedia terminal and possibilities of remote consultation, SMARTIE (IST-2000-25429) aims to develop web tools for multi-platform EMR access and support for medical error prevention, MTM (IST-1999-11100) provides via a local wireless network multimedia medical support to the mobile hospital personnel. These projects are based on a variety of technologies (e.g., GSM, GPRS or local wireless networks) and face the mobile health care problem from a variety of angles. As far as the authors know, none considers collaboration and virtual healthcare teams using commonly available GSM/GPRS based communication channels and handheld devices (e.g. Smart Phones). It is worth mentioning that DITIS, along with some of the above projects, has been invited to participate in a cluster proposal dealing with "Mobile Medical Devices (MEMO)" due to its unique approach in facing home health care provision (i.e., patient-centric virtual collaborative medical teams), and was also invited to demonstrate in the High Level Ministerial Conference: "eHealth 2003: ICT for Health", Brussels, 22-23 May 2003.

3. IDENTIFICATION OF THE HEALTHCARE TEAM, THEIR ROLES AND COLLABORATION SCENARIOS

The healthcare team includes oncologists who are based in the oncology centre, treating doctors who are usually located in the community, home care nurses who visit the patient regularly at home, and a number of other professionals called in as the demand arises, typically physiotherapists, psychologists, and social workers (see Figure 1). The home care nurses spend most of the time with the patient, and thus the analysis has initially focused on them and their interactions with the rest of the healthcare team.

3.1 DITIS functionality

With the following plausible scenario we illustrate aspects of DITIS functionality.

*Nurse Mary wakes up early in the morning. After showering and a quick breakfast she connects (**with her Mobile Computing Unit, MCU**, such as a Smart Phone, Pocket PC etc...) to the central office web server to check in and acquire her day's visit schedule (the nurse does not have to physically check in). Her schedule includes a number of visits. For each patient to visit, Mary downloads on her MCU (**via her personalized agent**) the needed information, plans her route, and updates the Web server of her schedule. At first she visits Mr. P., a nice sweet old man who while enjoying the comfort and love of his home he regularly requires (and receives) attention due to the seriousness of his situation. Arriving at his house and examining Mr P nurse Mary realises an extreme change in his condition. She immediately (**via GSM/GPRS**) connects with the web server and via the MCU's application interface reports the incident. The system records (**into the electronic medical record of the patient in the database**) the new data, and based on the identified/pre-wired scenario triggers and transmits messages (**via the mobile agents**) to the other members of the team alerting them and requesting their services. [Future implementations may include intelligent host interfaces and intelligent event identifiers which identify the implications of such events, and adapt/learn and modify the scenarios dynamically].For each member of the team DITIS extracts (**via the personalized agents**) from the EMR only the needed (and authorised) parts of the information, prepares it and then forwards to them. That is the MCU device of every team member is receiving not only the data of the triggering incident (i.e., patient P. experiences extreme pain in the lower back) but also relevant patient history. (**The virtual team around patient P. has now been set in motion in seconds defying distance***

and geographical barriers.*) The responsible doctor, Dr. A., studies the information submitted to her and prescribes the health care protocol which nurse Mary and other team members (e.g. a prescribed physiotherapy) should administer. Also a new appointment is scheduled for patient P. with the oncologist. Nurse Mary performs her tasks and Mr P. returns back to his smiley and happy self; the virtual medical team made the 'miracle'. His family are thanking nurse Mary who is headed to her next appointment. She is checking her schedule when a message (**via her agent***) comes in informing her that Mrs. D., her next patient, has suffered an extreme crisis and she has been taken to the hospital. Nurse Mary, while sorry for Mrs. D., is headed to the next patient in her list, Mr. X. On route she is informed of a new admission that she can schedule right now. Nurse Mary visits the patient Mr Y. for the first time. DITIS already has his EMR (sent by the treating doctor) to the PASYKAF Web server, and has forwarded (***via her personal agent***) the relevant information to the MCU of Nurse Mary. Nurse Mary examines patient Y. and updates the Web database by performing **on the spot data entry** through the user friendly interface of her MCU. And so the day goes on.*

Efficient and effective nursing care has been provided. All this is made possible by DITIS which provides a **distributed web based database with direct wireless connectivity and mobile agents linking all members of the virtual medical team.**

3.2 DITIS Modelling

Several scenarios were identified and analysed (see Figure 3) in order to implement the collaboration system.

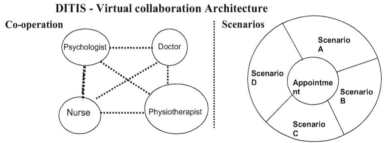

DITIS - Virtual collaboration Architecture

Figure 3. Scenarios for virtual-collaboration.

The UML (Unified Modeling Language) (Lipnack and Stamps, 1997; Kraut et al., 1998) has been used to identify roles and analyse and formalise collaboration scenario between virtual healthcare team members. Using results of the analysis the collaborative system software is developed. Some Common Scenarios include:

  – Referral of a new patient to home-care, Referral to other professionals, and First home-care-visit
  – Home-care virtual team creation / addition of members and Communication with the virtual team members
  – Change of prescription in the community, Bloods taken in community, and Chemotherapy in the community
  – Continuity of care in outpatients, Continuity of care for patients admitted to a hospital, and Continuity of care for staff members on call

To illustrate the process we present a simplified UML class diagram (Figure 4) which identifies a collaborative healthcare team and possible interactions.

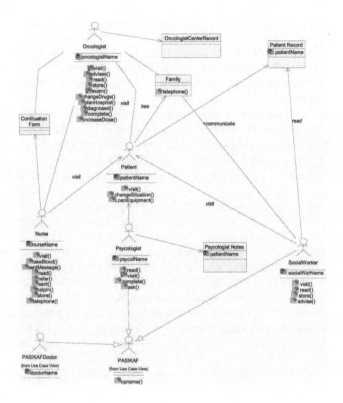

Figure 4. UML Class Diagram identifying a simple collaborative healthcare team and possible interactions.

Note that UML standardises a number of structural, behavioral, and model management diagrams.

3.3 Virtual team creation and management

In this part we discuss virtual teams and the dynamic virtual healthcare team creation for DITIS.

Even though interactivity is often presented as a key characteristic of a computer-mediated communication system, the emphasis is often on the computer-human interaction rather than on human-to-human computer-mediated interaction and trust[9]. The latter is particularly important since virtual teams are effective not only because of technological advancements but also and most importantly because individuals are able to interact and thus constructively engage in knowledge sharing and creation in the increasingly emergent virtual work environments. In particular, we focus on interactivity among the key actors in medical virtual teams. In such virtual teams, where effective and quality patient management care are the expected outcomes, high levels of interactivity often need to be developed quickly and it is important that they last throughout the short duration of the interaction. During the last few years is increasing volume of literature on virtual organisations and virtual teams (Panteli and Dibben, 2001; Jarvenpaa and Leidner, 1998; Lipnack and Stamps, 1997; Kraut et al., 1998; Lewicki and Buncker, 1996). This body of research generally agrees that virtual teams consist of a collection of geographically dispersed individuals who work on a joint project or common tasks and communicate electronically.

The building of a virtual team around a patient will normally commence with the arrival of a referral form for a new patient. Once the primary home-care is assigned, the virtual team can be progressive and dynamically created, following the evolving needs. A simple illustrative UML sequence diagram of the virtual team creation is presented in Figure 5. This simplistic scenario can be expanded to handle various collaboration scenarios

The communication aspects can be illustrated with the following simplified scenario involving a home-care nurse using a wireless (laptop, MCU, or mobile phone) device to connect to the network of a mobile telecom operator and from there to the DITIS system server running at the premises of the home-care organization. The information could be in HTML or WML format, depending on the type of the connection device. A visual representation of the scenario can be illustrated in UML. The scenario while simple can be used to depict the flow of communication among the various partners. In a more complex scenario the mobile agent will communicate, in addition to the database, with other mobile agents that represent other members of the virtual medical team.

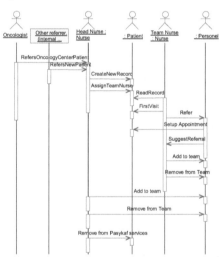

Figure 5. A simple UML sequence diagram for the virtual health-care team creation once a referral for a new patient is received.

4. SYSTEM DESIGN AND IMPLEMENTATION

DITIS is an Internet (web) based Group Collaboration system with fixed and GSM/GPRS mobile connectivity. It employs Mobile Agents, Web Databases with Java Database Connectivity, and web based database for storage and processing of information, including Electronic Medical Record (EMR) pertinent to cancer patients, in accordance with National and International standards (e.g. WHO ICD-10, ICD-O, HL7) (see reference ICD-10, Health Level Seven), software for collaborative work, intelligent interface for uniform access to the common database and the Group Collaboration software from both fixed and mobile computing units. DITIS also supports the integration of new technologies in Telemedicine and the home care service for cancer patients through the use of Mobile Computing Units (e.g. Tablet PC, Pocket PC 2002, Handheld PCs, Smart Phones, PDAs).

In the context of DITIS, a number of research and development issues have been addressed, including: Requirements analysis, Infrastructure development, Design of Electronic Medical Record, Design of collaborative platform, Design of wireless e-services, Design of collaborative software agents, Design and implementation of prototypes, Design of user interface, and Studies of system functionality.

The details of the system architecture are presented below, together with some aspects of the design criteria. These include:

Standards

The development of DITIS was based on the HL7, ICD-0 and ICD-10 standards, with a view toward an open Healthcare Information Infrastructure (Blobel, 2002). In the ISO telecommunication stack model, the 7th layer is the Application layer. HL7 stands for 'Health Level Seven', which is an application-level protocol dedicated to health services. It is meant to be an international communication standard between all digital health services, much like a health-care middleware. It sets 120 standard classes distributed in 22 domains to allow different systems to have a common vocabulary. DITIS is and must be open to the other services, in order to retrieve information from any medical facility such as hospitals. HL7 is tightly bound to UML, because it uses its methodology to help design messages, from use-cases to message definition. The 3rd version of HL7 is now currently under development, and has many concerns about respecting privacy in the storage and transmission of medical data. Another standard that was used was the ICD-10. This stands for International Classification of Diseases. This allows exchange of messages either in electronic health records (EHR) or in HL7, to clearly identify diseases using standard codes. DITIS mainly used its subpart called ICD-0, which deals with Cancer-related diseases, however there are plans to migrate to ICD-10, as the Oncology Centre is migrating to ICD-10. Note that continuous monitoring of international standards is necessary. In particular, in the light of the high priority for electronic records, messaging (e-prescriptions), protecting personal information (PKI and health cards) (Rogers et al., 2002) the use of the following standards is reviewed: the electronic patient record, e.g CEN standard EN 13606, ISO PKI Technical Specification, multipart ISO standard on health cards, CEN standard for electronic prescriptions, and for messaging HL7 Version 3 and use of XML.

Mobile devices

Mobile devices were a necessity since most team members are mobile workers, visiting the patients at home, or need to be accessible from anywhere at anytime. At the time of development high power mobile devices such as Pocket PC 2002 or Tablet PC were not available. Therefore the team turned to existing mobile devices such as the Smart-Phones, Pocket PC, Palm PC and Handheld PC. As a result, DITIS interface was built as simple as possible to support such devices. WAP and HTML technologies were used to address the problems of platform independence and portability. If the device supported a WML or HTML browser then it was supported as a candidate host for the application. An example interface is shown in

Figure 6 for two commonly available mobile devices, which show a number of menu selections for the home-nurses.

Mobile agents and computational model

Mobile agents are a special category of agent that can migrate on any node of a heterogeneous wireless or wire-line network of computers, in order to perform errands that were assigned by the user of whom the agent is dedicated (Samaras et al., 2003; White, 1996; Harrison et al., 1994). The mobile agent can travel from a node to another using the resources on each node it visits. The advantages of using resources on each node that the mobile agent visits include a reduction of the network bandwidth, and use and distribution of processing.

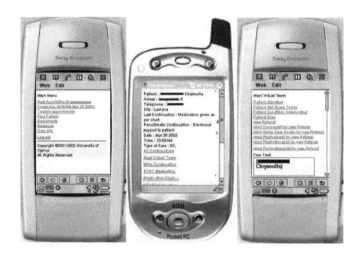

Figure 6. Example collaborative system screen on mobile devices

A simple scenario illustrating the benefits of the migration of the mobile agent is the following:

The application initiates a complex query on a Database server which resides on a different network. The networks are connected using a slow, unreliable, Internet link.

According to the mobile agent paradigm, an agent can move to the node that the Database Server resides, perform the query, filter the results according to the application server preferences and then forward them to the application, which may initiate further queries to the database. Thus, for network reduction concerns, it is cheaper to transport a simple agent on the server that the database resides than retrieve the entire results of a complex query, especially if the device that launches the application is a less powerful device, as for example a Pocket PC or a Smart-Phone. Pocket devices, as mentioned above, lack bandwidth, memory resources and in many cases persistence in connection. These problems led the development team to adopt the use of persistent mobile agents. Each user is assigned explicitly one mobile agent who is personalized to suit his needs, and ensure his/her continuous presence as a member of the virtual team (see Figure 7). To illustrate the personalisation we present the following example for accessing a list of drugs. As the database complies with certain standards such as ICD-10 and HL7, a vast list of drugs is represented covering every medical situation.

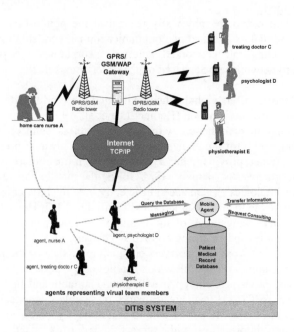

Figure 7. The Client/Agent/Server application model used in DITIS, showing every user represented by a mobile agent.

The agent can be customized to retrieve info on the common drugs that are being used by its user (e.g. pain specialist) and therefore lower unnecessary access to the database and presentation problems in a small device. This scenario promotes the use of mobile agents in terms of network reduction. An additional benefit is to also use intelligence, which can be downloaded into the agent helping him become a better assistant for any user. These features of mobile agents are currently being investigated (Samaras et al., 2003) and planned for implementation. The majority of DITIS users are mainly busy health care professionals, e.g. home-care nurses, oncologists and psychologists. Such users need fast access to medical information, for example, concerning the side effects of drugs. The combination of two drugs may lead to an unfortunate situation for the patient. This is classified as a standard human error, but the built in intelligent of the agents can minimise this problem. The agent can retrieve information about the two drugs which are about to be combined from the database and deliver it to the user when it "feels" there is a need to do so. An additional equally important feature is the ability of the user agent to adapt its interfaces to any format the user device supports. In this case the agent can work as a proxy for transforming the output into the appropriate desired format. For example, in the case of mobile devices that support WML, the agent will reformat the output for the WML browser. In the case of desktop computers the agent will reformat the output for the HTML browser. In this way, it is expected that the system will handle multi-modal devices and provide a better experience to the busy user.

Communication model

Since the team cannot be physically present at the point-of-care at all times, the system creates a virtual team based on communication network solution. The philosophy adopted for modeling the networked solution incorporates the paradigm shift in computing toward network centric, based on the observations that:

1. The Internet offers global connectivity, and the World Wide Web offers access to distributed information (Berners-Lee et al., 1994). The system employs this capability in conjunction with today's wireless connectivity (Samaras and Pitoura, 1998) for the realization of the virtual "always" present medical team. The Internet provides the fixed network infrastructure while the various wireless networks (namely GSM/GPRS) the mobile/nomadic network. These two technologies are enhanced efficiently, intelligently, cooperatively and in a light portable manner (Samaras and Pitsillides, 1997; Housel et al., 1999). The system takes advantage of the evolution of information and communication technologies, and emerging protocols and standards.

2. The fixed network hosts servers containing the data of the patients. Around it an application layer is built which handles cooperation and system intelligence. The virtual team is equipped with small, portable and light handheld PCs or devices (MCU's) that are capable of wireless connectivity. Networked handheld PCs or MCUs are able to seamlessly synchronize, communicate and exchange information with servers. They offer standard communication support, enabling access to the Internet for email and Web browsing. Emerging protocols and standards at that time were investigated, as for example the Wireless Application Protocol, WAP (see reference Wireless Application Protocol).

3. Mobility and Tele-presence come however, with certain cost. This cost is the limitations of wireless link; expensive, low bandwidth, high latency and low reliability (e.g pockets of poor mobile network coverage). The cumulative effect of these limitations can be frequent disconnection and weak connectivity. The use of mobile agents minimises this problem. In general, mobile agents offer autonomy and asynchronous connectivity and can be easily enhanced with intelligence.

Implementation technologies

The current technical architecture of DITIS is shown in Figure 8. The design and development of DITIS, is based on commonly available technology, Internet and GSM/ GPRS connectivity, and includes and integrates (Figure 9).

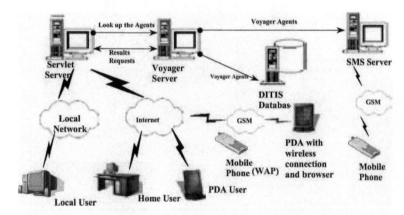

Figure 8. Schematic of DITIS system infrastructure

- Mobile Agents (White, 1996; Harrison et al., 1994) running on the Voyager platform (see reference ObjectSpace Voyager), for the implementation of flexible communication infrastructure (Pitoura and Samaras; 1998) for the support of mobile users.
- Relational Database with Java Database Connectivity (JDBC) (Anuff, 1996; Jepson, 1997; Bradley et al., 1997; Papastavrou et al., 1999) for information storage and processing of Electronic Medical Records and Agents.
- Tele-cooperation system for sharing of information, team communication, coordination of team activities.
- Adaptive intelligent interface (Chess et al., Oct 1993; Beyer and Holtzblatt, 1998) for database access from a variety of access units, such as MCUs with GSM/GPRS Internet connectivity, and Fixed units with Internet Access supporting Tele-cooperation.
- GSM Short Message Service (SMS) to enable push and pull of data and alerts, for example, whenever an agent updates information that affects the virtual medical team. MMS the emerging messaging system is under review.

5. EXPERIENCE, EVALUATION AND FUTURE DIRECTIONS

DITIS, supports the activities of PASYKAF (Cypriot Association of Cancer Patients and Friends), who run a national home-based healthcare service for cancer patients living in the community. It is currently installed in all Cyprus counties with the collaboration system currently operating fully in Larnaca (due to the limited availability of MCUs).

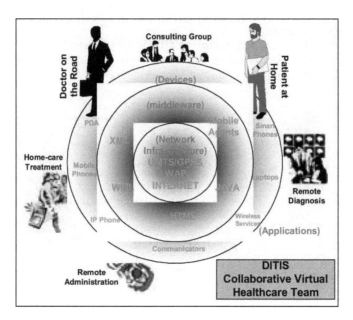

Figure 9. Overview of technologies adopted by DITIS system

Clinical objectives:

The Clinical Objectives addressed include:

- The presence of (Virtual) Collaborative Medical Team by patient at any given time, irrespective of locality, or cross country movement. Continuity of care is supported.
- Improved communication within (virtual) home care team and between home-care team and hospital, thus providing capability to consult within a team of experts (e.g. home nurse with treating doctor or oncologist), without need to move patient from his home to each one of them. This results in reduction of number of visits to health professionals away from patient's home, as well as duration of stay, and reduces burden not only on patient but also his relatives. It further reduces cost of treatment by a reduction in transport costs and time saved by his/hers working relatives. This capability also makes better use of the scarce and expensive medical professionals and scarce hospital beds, irrespective of geographic or organisational barriers.
- Improved and secure, timely access to patient information, in accordance with their authorisation levels, through unified information space centred around patient. As an added benefit, patient need not provide the same history to multiple professionals.
- Improved and flexible collection of statistical data for further audit and research within the home care setting.
- Improved evaluation through the capability to offer audit and research.

- Improved cost effectiveness through improved communications and better planning of services.
- Improved health practices (shift toward evidence-based) and reduction of bureaucratic overhead.
- Assists in promoting the dependant role of the home-nurse legally binding (for example, in the home setting when interacting with a hospital doctor for the prescription of drugs in the home).

As a consequence of meeting the above clinical objectives the system **improves the provision of health care to Cancer patients** in Cyprus, thereby achieving better quality of life, in the warmth of their own home.

Technological objectives:
In addition to the clinical objectives some technological objectives were also met.

- As all the technologies are Internet, GSM/GPRS, and JAVA based there is platform independence. The components collaborate seamlessly to implement the system. The DITIS architecture can be adapted also for other systems, such as a cardiac patients system, insurance agents and any type of collaborative environment requiring only minor adjustments on the customized database and the cooperative scenarios.
- DITIS has developed novelty in the integration of new technologies, in home-care, and the support of home healthcare service through the use of Mobile Computing Units (e.g. Handheld PCs, Smart-Phones).

Efficient and effective nursing care has been supported by DITIS, which provides a centralised web, **based database with direct wireless connectivity linking all members of the virtual medical team**.

Identified difficulties:
Several difficulties, some non-technical, were also reported during the development and deployment phase, including:

- Long delays – often beyond partner's control, research funding limitations, and research associates high initial turnover caused some frustration and mistrust.
- Mismatch between users high expectations and inability of technical team to enlighten users about the complexity and time it would take for development.
- Initial health-care professional's phobia of technology.
- Underestimation of the initial workload to initially populate database, and generally to switch to a new system, which was still being developed.
- Current technology limitations (e.g. WAP over GSM). The migration to new technologies (GPRS and ADSL) is resolving many of the original technical problems i.e. service is always on and bandwidth is much higher.

The Research Program DITIS was successful in attaining the initial goals promised.

Some of the new goals and open issues include:

- The improvement of the secure transaction, process and data archiving with the users able to access only the levels that they are authorized, in a dependable, secure, legal, and trustworthy way.
- Improve TRUST, not only between the (virtual) team members, but also between the team members and the technology.
- Improving robustness of system (to commercial levels suitable for application), whilst ensuring that the open design of DITIS is interoperable and compatible with other European systems, for data exchange.
- The expansion of the system, to be used from all healthcare providers that work on cancer, including the connection with the cancer archive (Cancer Registry).
- Assess fully the national and European legal framework with e-health records, as well as prescription.
- Study on spot entry, as e.g. whether in front of patient data entry is acceptable.
- Minimise system management costs.
- Investigate viability of mobile agent paradigm and also of intelligent mobile agents and intelligent interfaces, able to identify the implications of triggering events and send the right information to the members of the virtual team, or new members.
- Study 3G (UMTS) and new wireless handheld devices and their effect on the computing and networking model selected for DITIS.
- Adapt, customize and validate a sound clinical model and a service delivery model which is financially viable, secure, and legally acceptable, at a Pan-European basis.
- The expansion of the collaborative system for usage in other fields (e.g. cardiac home-care, insurance sales, etc…) and its eventual commercialisation.
- Disseminate/promote at government and national policy-making level.

6. CONCLUSIONS

DITIS uses a number of state of the art technologies which are seamlessly put together, such as collaboration and personalization via mobile agents, access to medical data from anywhere and any time via a variety of mobile devices and a variety of protocols (i.e., WAP, HTML) and continuous connectivity via new communication technologies such as GPRS and soon UMTS. The mobile access technologies are JAVA based therefore there is access device independence.

DITIS, supports home-care by offering wireless health care services for chronic illnesses with emphasis on prevention, assessment and diagnosis. The main service is the dynamic creation, management and co-ordination of virtual collaborative healthcare teams for the continuous treatment of patient at home, independently of physical location of the team's members, or the patient. For each patient a flexible (dynamic) virtual medical team is provided, made up from visiting home-care nurses, doctors, and other health care professionals, responsible for each case. This virtual team is able to provide dedicated, personalized and private service to home residing patient on a need based and timely fashion, under the direction of the treating specialist, without the necessity to move the patient from his home. Also in case of need for hospitalisation better planning can be achieved, so as to minimise expensive hospital stays, as well as better manage scarce resources, by coordinating the admission and discharge with the cooperation of the home-care team. This results in a reduction of number of visits to health professionals away from patient's home, as well as shorter duration of stay in case of hospitalisation. This decreases burden not only on patient but also his relatives. It further reduces cost of treatment by a reduction in transport costs and time saved by his/hers working relatives. This capability also makes better use of scarce and expensive medical professionals and scarce hospital beds, irrespective of geographic or organisational barriers. The system provides secure and improved, timely, access to patient information, in accordance with each user's authorisation level, through a unified information space centred around the patient. Thus, as an added benefit, patient does not have to provide the same history to multiple professionals. With improved and flexible collection of statistical data further audit and research within the home care setting is made possible, which improves evaluation, and better cost effectiveness through improved communications and better planning of services. Thus improved health practices (shift toward **evidence-based**) and reduction of bureaucratic overhead is offered. Furthermore, system assists in promoting the dependant role of the home-nurse legally binding (for example, in the home setting when interacting with a hospital doctor for the prescription of drugs in the home).

DITIS delivers a product that can improve the quality of the citizen's life. Contrary to today's health processing structure which is, in all practical terms **facility-based care**, this project shifts the focus onto **home-based care,** where everything is moving around the patient. Thus chronic patients, such as the cancer patient, can now enjoy 'optimum' health service, with improved quality of life, in the warmth of their own friendly environment, without a degradation in the quality of care provided to them, feeling safe and secure that in case of a change in their condition the health care team will be (virtually) present to support them.

7. ACKNOWLEDGEMENT

This project was funded by the Cyprus Research Promotion Foundation (RPF) (1999-2001 and 2004-2006) and Microsoft Cambridge Research Labs (2003-2004). The total support of the Cyprus Telecommunications Authority (CYTA) with telecommunications infrastructure, NetU as a partner in RPF grant (1989-2001), WinMob Technologies Ltd for wireless technologies, Ericsson (through S.A. Petrides Ltd) for the provision of handsets, Microsoft for the .net framework and XDA devices, and the Cyprus Development Bank (CDB) for financial support are gratefully acknowledged.

8. REFERENCES

Berners-Lee T., Caililiau R., Luotonen A., Nielsen H.F., Secret A., The World Wide Web, Journal Communications of the ACM, Vol. 37, No 8, pp. 76-82, August, 1994.
Beyer H., Holtzblatt K., "Contextual Design", Morgan Kaufmann, 1998.
Blobel B., Analysis, Design and Implementation for Secure and Interoperable Distributed Health Information Systems, Volume 89: Studies in Health Technology and Informatics, 352 pp, ISBN 1-58603-277-1, IOS Press, Amsterdam, 2002.

Bradley F., Burton and Victor W. Marek. Applications of Java programming language to databases management. University of Kentucky, 1997.

Canadian Institute for Health Information. National Consensus Conference on Population Health Indicators Final Report. Ottawa:CIHI, 1999.

Cherry M., Ogilvie L., Paquette D., Evaluation of Information Standards for Home Care Health Transition Fund Final Project Report, Canadian Institute for Health Information, Canada, March 2001.

Chess D., Grosof B., Harrison C., Levine D., Parris C., Tsudik G.,. Itinerant Agents for Mobile Computing. Journal IEEE Personal Communications, Vol. 2, No. 5, October 1993.

Dikaiakos M., Christoyiannis C., Papamichalopoulos A., Pouliou E., Kyprianou T., Nanas S., Tsanakas I., Rasidakis A., Roussos Ch., "Designing and Internet-based Collaborative Environment for Cystic Fibrosis Treatment", Summary Proceedings of Euromednet '98, Nicosia, Cyprus, March 1998.

Ed Anuff. Java sourcebook. Whiley Computer Publishing, 1996.

G. Samaras and E. Pitoura. Computational Models for the Wireless and Mobile Environments. Technical Report TR-98-4, University of Cyprus, Computer Science Department, 1998.

Harrison C.G., Chessm D.M., Kershenbaum A.. Mobile Agents: are they a good idea? Research Report, IBM Research Division 1994.

Health Level Seven Web Site http://www.hl7.org/

Housel B.C., Samaras G., Lindquist D.B, WebExpress: A Client/Intercept Based System for Optimising Web Browsing in a Wireless Environment, Journal of ACM/Baltzer Mobile Networking and Applications (MONET), special issue on Mobile Networking on the Internet, 1999.

ICD-10 International Statistical Classification of Diseases and Related Health Problems (10th Revision). Web site: http://www.mcis.duke.edu/ standards/termcode/ icd10.htm

Jarvenpaa S.L., Leidner D.E., Communication and Trust in Global Virtual Teams, Journal of Computer-Mediated Communication, 3, 4, June 1998.

Jepson B., Java Database Programming. Wiley Computer Publishing, 1997.

Kraut R., Steinfeld C., Butler B. and Hoag A., Coordination and Virtualization: The Role of Electronic Networks and personal Relationships, Journal of Computer Mediated Communication, 3, 4, 1998.

Lewicki R.J., Buncker B.B., Developing and Maintaining Trust in Working Relationships, in R.M. Kramer and Tyler T.R. (eds) Trust in Organizations: Frontiers of Theory and Research, Sage Publications, Thousand Oaks, CA, 1996

Lipnack J., Stamps J. (1997), Virtual Teams: Reaching Across Space, Time, and organizations with Technology, John Wiley & Sons, Inc. NY

Object Space Voyager. http:www.objectspace.com/ voyager/ whitepapers/ VoyagerTechOview.pdf

Panteli N., Dibben M.R., Reflections on Mobile communication systems, Futures, 33/5, 379-391, 2001.

Papastavrou S., Samaras G., Pitoura E., "Mobile Agents for WWW Distributed Database Access", Proc. 15th International Data Engineering Conference, Sydney, Australia, March 1999.

Pitoura E. and Samaras G., "Data Management for Mobile Computing", Kluwer Academic Publishers, ISBN 0-7923-8053-3, 1998.

Pitsillides A., Samaras G. Dikaiakos M., Christodoulou E., DITIS: Collaborative Virtual Medical team for home healthcare of cancer patients, Conference on the Information Society and Telematics Applications, Catania, Italy, 16-18 April 1999. (Invited)

Pitsillides A., Pattichis C., Pitsillides B., Kioupi S., "Tele-homenursing: A cooperative model for patient care in the home", Comprehensive Cancer Care: Focus on cancer pain, Limassol, Cyprus, 28-31 May 1997, pp 48.

Pitsillides A., Samaras G., Dikaiakos M., Christodoulou E., Olympios K., DITIS, Collaborative Virtual Medical team for home healthcare of cancer patients, Re-engineering Cyprus for the digital age, December 1999.

Pitsillides B., Pitsillides A., Samaras G., Georgiades D., Andreou P., Panteli N., DITIS: A collaborative system to support home-care by a virtual multidisciplinary team, 8th Congress of the European Association for Palliative Care, The Hague, Netherlands, April 2003.

Pitsillides, B. DITIS: The user perspective, 4th Hellenic Health Telemetric Forum, Limassol, April 2002.

Rogers R., Sembritzki, J. Village P., CEN/TC 251 Health Informatics, Priorities for application of ICT Standards–A project report for Ethel Thematic working group T1, CEN/TC 251/N02-073, Dec 2002.

Samaras G., Pitsillides A., "Client/Intercept: a Computational Model for Wireless Environments", Proc. 4th International Conference on Telecommunications (ICT'97), Melbourne, Australia, April 1997.

Samaras G., Pitsillides A., Georghiades D., Computational and Wireless Modeling for Collaborative Virtual Medical Teams, Book Chapter, M-Health: Emerging Mobile Health Systems, (R. H. Istepanian, S. Laxminarayan, C. S. Pattichis, Editors), Kluwer Academic/Plenum Publishers, 2003.

White J. E.. Mobile Agents. General Magic White Paper. http://www.genmagic.com/agents, 1996.

Wireless Application Protocol, Web Site: http://www.wapforum.org/docs/ technical.htm

FUTURE CHALLENGES AND RECOMMENDATIONS

Val Jones, Francesca Incardona, Clive Tristram, Salvatore Virtuoso, and Andreas Lymberis[*]

Rapid advances in information technology and telecommunications, and in particular mobile and wireless communications, converge towards the emergence of a new type of "infostructure" that has the potential of supporting a large spectrum of advanced services for healthcare and health. Currently the ICT community produces a great effort to drill down from the vision and the promises of wireless and mobile technologies and provide practical application solutions. Research and development include data gathering and omni-directional transfer of vital information, integration of human machine interface technology into handheld devices and personal applications, security and data interoperability and integration with hospital legacy systems and electronic patient record. The ongoing evolution of wireless technology and mobile device capabilities is changing the way healthcare providers interact with information technologies. The growth and acceptance of mobile information technology at the point of care, coupled with the promise and convenience of data on demand, creates opportunities for enhanced patient care and safety. The developments presented in this section demonstrate clearly the innovation aspects and trends towards user oriented applications.

However, several major challenges still need to be addressed in order to broaden the implementation and use of mobile health devices and services and strengthen the market development.

The main obstacles to the penetration of mobile solutions based on handheld computerised devices and wireless technology in the healthcare sector can be roughly classified into objective and subjective:

Objective obstacles have to do with deficiencies and limitations of the current technology, specifically interoperability, compliance to widely accepted standards and security. Interoperability and standards issues are especially relevant, as they often impede or make a tough job to integrate mobile solutions into the framework of existing systems currently in healthcare institutions for information storage and management

---

[*] Andreas Lymberis, European Commission, Directorate General Information Society, Av. De Beaulieu 31, 1160 Brussels.

purposes. To help overcome this difficulty, the adoption of open standards and cross-platform technology is advised wherever possible. Where this is not possible it is recommended that a layer be constructed on the legacy system with a standards compliant interface to the external world. Access control and security are crucial to the eventual acceptance of mobile solutions by the healthcare community and should not be underestimated. The confidential nature of the data to be accessed obviously requires that access rights be carefully matched against user authentication parameters. Wired network solutions usually approach the problem by adopting security models based on digital certificates and electronic signatures. This may not feasible with the mobile devices currently available on the market. On the contrary, the first results obtained in this respect by some of the projects presented in this section (e.g. Mobi-Dev) show a substantial immaturity of PKI technology for mobile platforms – problematic support for smart card readers, unreliable functioning of available cryptoAPI libraries for digital certificate management, pocket browsers not enabling client authentication in SSL sessions etc. These problems may be overcome when new PDA models and OS versions become available. It remains however a strong recommendation that transmissions be encrypted, authentication be included in non-revocable transactions from the mobile devices to the legacy and new systems and that the data be stored in an encrypted format with access being dependent on certificated keys

Subjective obstacles concern the reluctance of healthcare professionals – often busy people with techno-phobic attitudes and little time left for technical training – to abandon traditional mostly paper and voice based workflows for new practices based on highly technological appliances such as new generation PDAs. This is especially true of applications that require excessively awkward or lengthy procedures. The experience accumulated so far evidences the importance of parameters such as user-friendliness, lightness and functionality in application design, even at the cost of realizing less attractive user interfaces.

A further major obstacle to the adoption of advanced mobile solutions in the healthcare sector is related to the need to comply with usually very restrictive regulations. These are designed to ensure decision traceability in each step of care delivery and guarantee the patients against privacy violations. This is a quite delicate issue that can often only be dealt with by promoting actions aimed at awakening in the legislators the necessary appreciation for the benefits deriving to the healthcare system from the adoption of information technology mature solutions.

Several technical aspects remain to be solved, starting with existing constraints e.g. screen size, resolution and battery life.

Vocal input is a major goal in order to increase usability of mobile devices and applications. It is presently approached in two ways: single word discrete voice recognition, where speech recognition is carried on selecting from long lists; and continuous speech recognition with natural language understanding. This second approach must be included in an architecture in which dictation is entered into the MMD and the recognition is done on a server (this architecture has been adopted in the Mobi-Dev project). Although this is a clear limitation, so as the low quality of the incorporated microphones, there remain considerable opportunities for improved user interfaces enabling vocal input. This is being investigated by the MEMO project as an implementation for SMARTIE and Mobi-Dev an associated project.

Security remains a major issue for authentication of the user as well as encryption of transmitted data. Within the associated to MEMO projects the use of smart cards has been investigated, where the smart card holds encryption codes and authentication certificates. Contacts have been made with other suppliers to identify the feasibility of adding biometric based smart cards to MMDs.

Initial investigations into remote communications architectures and technologies indicate that for many cases GPRS will be an adequate substitute for UMTS based devices for the foreseeable future. However, for other applications we are already beyond the capabilities of UMTS (eg 12-lead ECG transmission or EEG, not to mention VR and AR applications and rendering 3D graphics. Local communications are shown to work well with both wireless LAN and Bluetooth. The difference in usage is dependent upon distance and numbers of devices connected. The main constraint in this area at the present time is the cost of Bluetooth devices and the power of Wireless LAN which is being addressed by the widespread introduction of WiFi systems.

Integration of mobile platforms with existing systems can be pursued by adopting the approaches usually followed in ordinary Enterprise Application Integration (EAI). The integration between different information systems, possibly legacy systems using obsolescent technology, is normally realised by interposing in between a middleware software layer, responsible of mediating between two separate and already existing software platforms, typically DBMSs. EAI may involve developing a new total view of an organisation's business and its applications, seeing how existing applications fit into the new view, and then devising ways to efficiently reuse what already exists while adding new applications and data. This is made possible by using methodologies such as object-oriented programming, distributed, cross-platform program communication using message brokers with Common Object Request Broker Architecture and COM+, organisation-wide content and data distribution using common databases and data standards implemented with the Extensible Markup Language (XML) (both data and voice XML), middleware, message queuing, intelligent agents and other approaches. The common belief that application integration is a short-term patch to apply while awaiting a perfectly interoperable, homogenous application base is a myth. Heterogeneity is an intrinsic characteristic of information systems, and which can hardly be dismissed. Network-oriented development platforms such as Java 2 Enterprise Edition (J2EE) or Microsoft's .NET are not going to solve semantic differences among applications, even though they will narrow down platform variability and provide greater interoperability.

At the level of personal healthcare systems such as the patient BAN there are serious challenges relating to security, integrity and privacy of data during transmission. These issues apply to both local transmission (eg. intra-BAN, BAN-PAN) and long range (e.g. extra-BAN) communications. End-to end security and Quality of Service guarantees need to be implemented. Safety of hardware (eg. electrical safety, emissions, interference) and reliability and correctness of applications must also be a priority. Comfort and convenience of sensors or BANs worn long term for continuous monitoring is important for usability and user acceptance. Timeliness of information availability in the face of unreliable performance of underlying network services is another issue. Provision of seamless services across regional and national boundaries multiplies these difficulties. Powering *always on* devices and continuous transmission will continue to raise technical challenges as will exploitation of the possibilities raised by ambient intelligence and ad hoc networking whilst maintaining security and privacy levels. Business models for healthcare and accounting and billing models for network services need to evolve to incorporate the new possibilities. Perhaps ironically, the bigger the potential of such services for the population at large, the bigger the problem of scalability of solutions becomes. Standardisation at all levels is essential for open solutions to prevail. Nevertheless specialization, customization and personalisation are needed, for instance for some applications BANs will have specific requirements to operate in special conditions (for example requirements to be water-proof, air tight, reusable etc.)

In the wider context, assessment and evaluation of the value of innovative m-health solutions and the incorporation of worthwhile m-health technologies into existing healthcare systems in terms of structure, organization and working practices presents many future challenges. Legal issues and hospital responsibilities such as ownership of the data and legal liability are further complicated by the international dimension enabled by the opportunity of seamless access to services for free roaming patients. With new possibilities come new challenges.

III. SIGNAL, IMAGE, AND VIDEO COMPRESSION FOR M-HEALTH APPLICATIONS

Marios S. Pattichis

Section Editor

SECTION OVERVIEW

Marios S. Pattichis[*]

An essential component of mobile e-Health systems is the effective compression and communication of biosignals, images, and digital videos. The clinical requirements of e-Health systems imply the use of high-rate, high-quality, high SNR compression systems. Thus, lossless and near-lossless performance is usually desired. It is important to note, however that significantly better compression can be achieved using perceptually-lossless and diagnostically-lossless compression criteria. With the acceptance of JPEG2000 and the recent consideration of MPEG-2 by the Digital Imaging and Communications in Medicine (DICOM) committee, it is clear that lossy compression techniques will become an essential component of every mobile e-Health system. There is a variety of biomedical signals that need to be compressed and transmitted. This section reflects this variety. It begins with a discussion of biosignals and compression, the description of a variety of algorithms for a resilient ECG coding system for telecardiology, followed by a discussion of the use of the JPEG2000 image compression standard, the compression of volumetric data, and an overview of digital video compression.

In the chapter on biosignals and compression standards, some of the best known signal compression algorithms are compared for ECG, EEG, surface EMG (SEMG), Lung sounds (LS), Heart Sounds (HS), and Bowel Sounds (BS). Starting from the basic principle that signal compression methods should not sacrifice important diagnostic information, the chapter proceeds to give a careful summary of the origins of each biosignal and how important diagnostic information corresponds to frequency content. This is followed by an overview of signal compression methods, and a presentation of a wide-range of lossy signal compression error measures. The performance of a variety of different algorithms is evaluated using different error measures. The chapter concludes with a telemedical perspective on signal compression, with an example of mobile transmission in emergency systems.

In the chapter on resilient ECG coding, different Wavelet coding algorithms are presented for both constant and variable bit rate coding for real-time telecardiology applications. The algorithms are evaluated using simulated 3G channel conditions.

---

[*] Marios S. Pattichis is with the Department of ECE, Room 229-A, ECE Building, The University of New Mexico, Albuquerque, NM 87131-1356, USA.

Initially, the ECG signal is broken into non-overlapping blocks. Each block is pre-processed and then encoded independently from any other blocks. For error-resilience, the authors discuss the use of Reversible Variable Length Codes (RVLC), using header redundancies, varying the block size, as well as error detection based on sign changes of the sorted Wavelet coefficients. The authors give error measures and also provide a visual analysis of the different error resilience methods. Different algorithms are compared based on the received root mean square error (RMS) for each algorithm versus the channel bit error rate (BER). Then, the authors propose the use of the best algorithm that provides the least RMS error at any given channel bit error rate. Based on this approach, algorithms without any error resilience protection were found to perform best at low bit error rates, while algorithms with error-resilience protection perform best at higher bit error rates. The bit error rates for which different algorithms give the same received RMS are also given.

In the following chapter of this section, a detailed presentation of the JPEG2000 standard is given. We note that JPEG2000 offers many important provisions for mobile health applications. The chapter begins with an overview of the bandwidth and channel characteristics of mobile network systems, and continues with an overview of JPEG2000. In addition to introducing the basic features, the chapter provides brief descriptions of the basic components of a JPEG2000 system, including a description of the Wavelet Transform, quantization methods, and how the code stream is formed. The Wavelet transform discussion includes the analysis and synthesis filters for the Le Gall 5/3 Wavelet system. An expanded section on the code-stream formation gives many details on how and why the JPEG2000 system has adopted the methods that are described.

In section 3, a summary of JPEG2000 features intended for mobile health applications is given. The region of interest (ROI) capability is demonstrated using an X-ray image example. Scalability features of JPEG2000 offer great flexibility for mobile health applications. An important advantage of JPEG2000 over JPEG is that JPEG2000 does not require that an image be compressed multiple times to meet a target bit-rate. Instead, JPEG2000 attempts to achieve the target bit rate using different scalability options. Quality scalability trades lower SNR for requiring a lower rate, while spatial scalability can generate lower to higher resolution image representations that can be used for display on a variety of devices ranging from cellular phones to high-quality, DICOM-compliant, image displays. A combination of quality and spatial scalability allows us to provide spatial resolution within a tolerable SNR. This is followed by sections on error-resilience, lossless compression, and methods for expanding the standard. The chapter also includes references for high-quality, free reference software.

The chapter on the compression of volumetric data in mobile health systems focuses on state of the art compression algorithms for volumetric, medical data sets. The basic result is that intra-band coders outperform current state of the art, inter-band coders. This is first justified theoretically, by examining the mutual information within neighbourhoods of Wavelet coefficients from inter-band, intra-band, and composite (inter-band and intra-band). Under a wide range of distributions of the Wavelet coefficients, the mutual information between the coefficients is less for inter-band, followed by intra-band, and then the composite of the two. Therefore, intra-band coders are expected to perform better than inter-band ones. The chapter also derives the conditions under which spatial block-based coding outperforms zerotree coding for lossless applications. Block-based coding offers a number of advantages over zerotree coding from the perspective of spatial resolution scalability, at a reduced computational cost. The chapter presents comparative, rate-distortion results for JPEG2000, JPEG2000-3D, 3D SPIHT, 3D-DCT, and the new intra-band decoders QT-L and CS-EBCOT for ultrasound images, PET, CT, and MRI. The results are used to establish the better performance of the intra-band coders.

In the last chapter of the section, an overview of medical video compression is given. The chapter gives an overview of video compression algorithms, including the MPEG and the H.26x based standards. The chapter includes an extensive section on motion estimation methods. Motion estimation algorithms are classified as block-based, pixel-based, and mesh-based. The basic underlying equations are given for each motion estimation method, and the methods are illustrated using basic examples. The DCT equations are given and the role of linear transforms is briefly described. Object-based coding and region of interest methods for second generation coding are also presented. The chapter includes an extensive section on communication error control and recovery for real-time video processing applications. The traditional focus on signal recovery from single bit errors is now replaced by a focus on packet-based error recovery methods. The basic problems are introduced and a number of error-control mechanisms are discussed, including algorithms for error resilient encoding and error concealment.

Overall, the section provides an overview of state of the art compression technologies to address the needs of mobile e-Health systems. Scalable compression methods are meant to address the heterogeneous mixture of communications devices ranging from mobile phones and PDAs to laptops, desktops, and Computer Aided Diagnosis (CAD) systems. Efficient compression methods are intended to meet the bandwidth requirements, while robust encoding and decoding methods are intended to deal with the bit error rates of wireless communications systems.

BIOSIGNALS AND COMPRESSION STANDARDS

Leontios J. Hadjileontiadis[*]

1. INTRODUCTION

Electric, mechanical or chemical signals of biological origin delivered by living things can always be of interest for diagnosis, patient monitoring, and biomedical research. Such biological signals, namely *biosignals*, as electrocardiogram (ECG), electroencephalogram (EEG), surface electromyogram (SEMG), bioacoustic signals (lung, heart, and bowel sounds), are usually presented by large amounts of data when digitized for storage and analysis within a signal processing framework. On the contrary, other medical information, i.e., drug description, demographic and anamnestic data, result in substantially smaller data archives. To this end, it is noteworthy the 80 kbytes data equivalence of 10 sec digitized resting ECG of 8/12 leads to 40 pages of text of an encyclopedic dictionary! (Zywietz, 1998). As a result, data reduction and compression processes are of great interest to biosignal analysis, especially when data transmission (e.g. in telemedicine applications) or long-term recordings (e.g. in sleep laboratories, intensive care) are involved. With appropriate processing of biosignals, the redundant data stream could be reduced to the most significant parameters that could efficiently contribute to medical decision making. In this way, high density storage could be achieved. In the same vein, reduction of the number of bits required to describe a biosignal could facilitate the data transmission, but it must be done with great care if it results in a loss of information. Although new emerging data compression techniques with very promising results are seen in the recent years, some problems have not been entirely addressed. In particular, the following items are still under consideration:

1. The lack of widely adopted compression accuracy standards.
2. The limited number of publicly available benchmark-databases, i.e., accurately acquired, pre-categorized, and standardized biosignals recorded from controls

---

[*] Leontios J. Hadjileontiadis, Dept. of Electrical and Computer Engineering, School of Engineering, Aristotle University of Thessaloniki, GR-54124 Thessaloniki, Greece, Tel.: +302310-996340, FAX: +302310-996312, E-mail: leontios@auth.gr.

and patients with various pathologies, for objective testing, evaluation, and performance comparison among the proposed compression techniques.

3. The lack of interoperability of data acquisition and processing equipment of different research groups and/or manufacturers.
4. The difficulty to exchange compressed data between medical databases from different research groups and/or manufacturers because of their incompatibility.

Unfortunately, due to the aforementioned issues, technical barriers to the interconnection of telemedicine centers around the world are set, hampering the use of telemedicine in an economically favorable way, and delaying the development of organizational and healthcare structural adaptations. Standardization in healthcare informatics could contribute to the elimination of these barriers and could provide a dynamic field for organization, coordination, and follow-up of biosignal compression development at a worldwide level. From this perspective, standards and principles that could be applied in the compression process of biosignals, preserving their diagnostic characteristics, are discussed in this chapter.

The rest of the chapter is organized by first covering the categorization and characteristics of the examined biosignals. Section 3 then describes the main compression types. Next, Section 4 refers to the principles and methods used for biosignal compression, while Section 5 gives a telemedical perspective of medical data compression and standardization. Finally, Section 6 concludes the chapter by summarizing the main points of the issues addressed.

2. BIOSIGNALS: CATEGORIZATION AND CHARACTERISTICS

Biosignals, being the acquired output from biological and physical systems, may possess various properties and characteristics that contribute to their diagnostic value. Prior to any analysis, these characteristics must be clearly identified. From the plethora of the available biosignals the following categories and some specific biosignals will be considered in this chapter. In particular:

1. Bioelectric signals, i.e., ECG, EEG, and SEMG.
2. Bioacoustic signals, i.e., lung sounds (LS), heart sounds (HS), and bowel sounds (BS).

These are all noninvasive (recorded from the surface of the human body), one-dimensional biosignals; two-dimensional biosignals, such as X-rays, ultrasound-magnetic resonance-computed tomography images, will not be examined in this chapter. A short description of the characteristics of the examined biosignals follows.

2.1. Characteristics of Bioelectric Signals

2.1.1. Electrocardiogram (ECG)

The ECG is the recording of the electrical activity of the heart, well associated with the mechanical activity of the heart function. In that way, diagnostic analysis of the latter is achieved through the assessment of the ECG. The ECG reflects the temporal changes

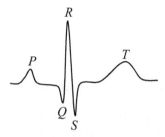

Figure 1. Time domain representation of the important deflections of a typical scalar ECG.

in the electrical potential between pairs of points on the skin surface. In general, its important parts consist of *P, QRS* and *T* waves, shown in Fig. 1. The *P-R* interval is a measure of the time from the beginning of atrial activation to the beginning of ventricular activation, ranging from 0.12 to 0.20 second (Berne and Levy, 1990). The *QRS* complex has duration between 0.06 and 0.10 second and its abnormal prolongation may indicate a block in the normal conduction pathways through the ventricles (Berne and Levy, 1990). The *S-T* interval reflects the depolarization of the entire ventricular myocardium, while the *T* wave its repolarization (Berne and Levy, 1990). The first step in ECG processing is the identification of the *R* wave in order to synchronize consecutive complexes and for *R-R* interval, i.e., heart rhythm, analysis. The latter plays an important role in the heart patient monitoring.

Most of the energy of the ECG is included in the frequency band of 0.05 to 100 Hz (Riggs *et al.*, 1979). Nevertheless, it has been found that the so-called *notches* and *slurs*, superimposed on the slowly varying *QRS* complexes, located in the higher-frequency band of 100 to 1000 Hz, contain additional information (Kim and Tompkins, 1981). In addition, the recording of the electrical field generated by the His and Purkinje activities (Peper *et al.*, 1982) produces a signal in the ECG with an amplitude range of about 1 to 10 μV, useful in the identification of conduction abnormalities. Unfortunately, the low amplitude of this signal makes it comparable to the noise level in the ECG recordings.

2.1.2. Electroencephalogram (EEG)

The EEG is the recording of the electrical activity of the brain, associated with the functioning of various parts of the brain. In routine clinical procedures, the EEG has been used for the diagnosis of epilepsy, head injuries, psychiatric malfunctions, sleep staging/disorders, and others. Usually, it is recorded noninvasively from the scalp by means of surface electrodes. The EEG could reflect both the spontaneous activity of the brain, i.e., the result of the electrical field generated by the brain with no specific task assigned to it, and the evoked potentials, that is, the potentials evoked by the brain as a result of sensory stimulus (e.g. flash lighting and/or audio clicking).

The major power of the EEG is distributed in the range of 0.5 to 60 Hz, although it could be extended into the bandwidth range of DC to 100 Hz, while its amplitude ranges from 2 to 100 μV (Cohen, 1986). The subdivision of the EEG spectrum into fine bands, i.e., the *delta* range (0.5-4 Hz), the *theta* range (4-8 Hz), the *alpha* range (8-13 Hz), and the *beta* range (13-22 Hz), provides useful diagnostic information in many neurological pathologies (Cohen, 1986).

2.1.3. Surface Electromyogram (SEMG)

Noninvasive measurement of the electrical behavior of the muscle is achieved through the surface EMG (SEMG), recorded by means of surface electrodes placed on the skin overlying the muscle. The SEMG amplitudes depend on the muscle under investigation and the electrodes used, ranging (in normal case) from 50 μV to 5 mV (Cohen, 1986). SEMG can record both voluntary and involuntary muscle activity in addition to the action potentials, produced by external stimulation, such as motor evoked potentials after central or peripheral nerve stimulation (Kimura, 1990). Nevertheless, SEMG has limited spatial resolution, is more susceptible to mechanical artifact, and is more likely to show cross-talk between adjacent muscles than the needle EMG (Pullman *et al.*, 2000).

The useful SEMG spectral content for skeletal muscles lies between 2 to 500 Hz, whereas for smooth muscle between 0.01 to 1 Hz (Cohen, 1986). The characteristics of SEMG can be used as diagnostic tool for kinesiologic analysis of movement disorders; for differentiating types of tremors, myoclonus, and dystonia; for evaluating gait and posture disturbances; and for evaluating psychophysical measures of reaction and movement time (Pullman *et al.*, 2000).

2.2. Characteristics of Bioacoustic Signals

2.2.1. Lung Sounds (LS)

The LS refer to the sounds produced by the structures of the lungs during breathing, usually examined by auscultation (listening) with a stethoscope. When listening to the lungs, the categories of findings include *normal* breath sounds and *abnormal* breath sounds (Kraman, 1983).

Normal LS occur in all parts of the chest area, including above the collarbones and as low as the bottom of the rib cage. In particular, they are categorized as: *tracheal* LS (heard over the trachea having a high loudness and a wide frequency band of 0-2kHz); *vesicular* LS (heard over dependent portions of the chest, not immediate proximity to the central airways, within a frequency band of 0-600 Hz); *bronchial* LS (heard in the immediate vicinity of central airways, but principally in the trachea and larynx); *bronchovesicular* LS (resembling the character between vesicular LS and bronchial LS, heard at intermediate locations between the lung and the large airways); and *normal crackles* (inspiratory LS heard over the anterior or the posterior lung bases).

There are several types of abnormal breath sounds: *rales*, *rhonchi*, and *wheezes* are the most common. Wheezing can sometimes be heard without a stethoscope, and other abnormal sounds are sometimes also loud enough to be detected with the unaided ear. Rales (crackles or crepitations) are small clicking, bubbling, or rattling sounds in a portion of the lung. They are believed to occur when air opens closed alveoli (air spaces) and they are further categorized to *fine* and *coarse* crackles (Kraman, 1983). Rales may be further described as moist, dry, fine, and coarse, among other descriptors. Rhonchi are sounds that resemble snoring. They are produced when air movement through the large airways is obstructed or turbulent. Wheezes are high-pitched (200-800 Hz) musical sounds produced by narrowed airways, often occurring during expiration.

For getting the full frequency range and electrical signals that can be processed, microphones, i.e., electronic stethoscopes, are used, with an almost flat response in the frequency range from 20 Hz to 5 kHz. The variety in the categorization of LS implies changes in the acoustic characteristics either of the source and/or the transmission path of the LS inside the lungs, due to the effect of a certain pulmonary pathology; hence, time and frequency characteristics of the LS signals reflect these anatomical changes and serve as diagnostic tools (Gavriely and Cugell, 1994; Hadjileontiadis *et al.*, 2002).

2.2.2. Heart Sounds (HS)

Heart sounds are defined as the repetitive "lub-dub" sounds of the beating of the heart (Gavriely and Cugell, 1994), categorized to: *first heart sound* (or S_1), considered as normal HS, occurs at the beginning of ventricular systole when ventricular volume is maximal, and lasts about 100 to 120 msec (Cohen, 1986); *second heart sound* (or S_2), also considered as normal HS, occurs at the end of ventricular systole; *third heart sound* (or S_3 or "S_3 gallop"), occurs just after the S_2 as a result of decreased ventricular compliance or increased ventricular diastolic volume, it is a low-frequency (20 to 70 Hz) transient with low amplitudes, lasts about 40 to 50 msec (Cohen, 1986), and serves as a sign of congestive heart failure; *fourth heart sound* (or S_4 or "S_4 gallop"), occurs at the time of atrial contraction, is similar to S_3 in duration and bandwidth, and it denotes ventricular stress. Additional heart malfunctions and reflected through the abnormal HS that include changes in intensity of normal HS, splitting of sound components, *ejection clicks* and sounds, *opening snaps*, and *murmurs* (systolic and diastolic) (Cohen, 1986).

Over the 95% of the acoustic energy of S_1 and 99% of the one of S_2 is concentrated under 75 Hz. There is decay in the average spectrum after its peak at 7 Hz containing one or more shallow and wide peaks ending at 150 Hz (Arnott *et al.*, 1984). In a similar vein, the frequency content of most cardiac murmurs is also in the low range (Gavriely and Cugell, 1994), but in some cases is extended, overlapping the low-frequency part of LS.

2.2.3. Bowel Sounds (BS)

Bowel sounds refer to the sounds heard when contractions of the lower intestines propel contents forward (On-line Medical Dictionary, 1997). Unfortunately, there is no reference of what can be considered normal bowel sound activity, thus, only subjective description of the acoustic impression of normal BS exists, employing terms such as "rushes" or "gurgles". Nevertheless, time and frequency domain characteristics of BS could be used as a means for defining normal BS. In particular, BS with frequency content in the range of 100-1000 Hz, with durations within a range of 5-200 msec, and with widely varying amplitudes, could be characterized as *normal* BS (Bray *et al.*, 1997). The "staccato" character seen in many BS of the colon corresponds to frequencies within 500 to 700 Hz and time durations within 5 to 20 msec (Bray *et al.*, 1997). Moreover, differences in the sound-to-sound intervals of BS from different pathologies, i.e., irritable bowel syndrome, Crohn's disease, and from controls, establish a time-domain tool for associating changes in the BS characteristics with bowel pathology (Craine *et al.*, 2001).

Advanced signal processing of BS could reveal more diagnostic features of BS from bowel pathologies (Hadjileontiadis *et al.*, 2002).

3. COMPRESSION TYPES

Compression processes, whether they operate on audio, video, images, or a random collection of files, are categorized into two basic types: *lossless* and *lossy*. In the lossless processes the original data can be *exactly* reconstructed from their compressed form, while in the lossy ones only an *approximation* of the original data can be retrieved. In that way, in the lossless compression all information is saved and the compression is reversible. On the contrary, in lossy compression some information is "thrown away" based on the perceptual response of an observer, hence, the compression is irreversible.

Lossless compression is typically adopted for text compression, while lossy for images and sound where a little bit of resolution loss is often undetectable or, at least, acceptable. It is noteworthy the abstract sense of the term "lossy", since it does not imply random lost data, but instead refers to a loss of a quantity, such as a frequency component, or to the loss of noise. In that way, although lossy compression involves a kind of information loss, it is critical to see if the "authentic meaning", in our case the diagnostic feature of the compressed biosignal, is fully maintained, or even improved, after compression.

The structure of both lossless and lossy compression algorithms always involves two basic components, i.e., the *model* and the *coder*. The model component detects any possible bias on the input data, i.e., any unbalanced probability distribution over the input data, by knowing or discovering something about the structure of the input. In that way, the coder component takes under consideration the probability biases captured by the model in order to effectively generate codes. The latter is achieved by lengthening low-probability data and shortening high-probability ones. There are many different ways to design the model component of compression algorithms with various degrees of sophistication. Nevertheless, the coder components tend to be quite generic, such as, for the case of lossless compression, Huffman (Huffman, 1952), Lempel-Ziv-Welch (LZW) (Welch, 1984) or arithmetic codes (Moffat *et al.*, 1998). The Huffman code maps fixed length symbols to variable length codes and it is optimal only when symbol probabilities are powers of 2. In its adaptive form, an *a priori* estimation of probabilities is not needed. The LZW is a dictionary-based code that maps a variable number of symbols to a fixed length code. The arithmetic code uses a variable number of bits for data encoding according to the probability assigned to the encoded symbol. Low probability symbols use many bits, while high probability symbols use fewer bits. This makes arithmetic coding sound very similar to Huffman coding. However, an arithmetic encoder does not have to use an integral number of bits to encode a symbol. For example, if the optimal number of bits for a symbol is 2.4, a Huffman coder will probably use 2 bits per symbol, whereas the arithmetic encoder my use very close to 2.4. This means that an arithmetic coder can usually encode data using fewer bits.

Table 1. Most common lossless and lossy compression techniques/standards

Lossless	Lossy
Huffman	Scalar Quantization (Uniform/Non-uniform)
Adaptive Huffman	Vector Quantization
Lempel-Ziv 77 (LZ77)/LZSS	Differential or Predictive
LZ-Welch (LZW)	Transform Coding (Discrete Cosine, Polynomial,
LZ-Oberhumer (LZO)	Wavelet)
Run-Length Encoding (RLE)	JPEG
Arithmetic	MPEG
Burrows Wheeler	MP3/AAC
	Fractal Compression
	Model-based Compression

As far as lossy compression is concerned, the most common coding techniques include mapping of regions of a data set onto elements of a smaller one (scalar quantization); mapping of multidimensional space into a smaller set of data (vector quantization); transform of the input into a different form that can then either be compressed better or for which we can more easily drop certain terms without as much qualitative loss in the output (transform coding); association of a family of functions, which have fixed points that may be found in an iterative way without unduly long convergence, with data point values (fractal compression); and characterization of the source data in terms of an underlying model so the entire system is described by only a few bytes of parameter data (model-based compression). JPEG, MPEG and MP3/AAC are widely applied compression standards that are used for still images (JPEG), videos (MPEG), and audio (MP3/AAC) compression. All of these standards combine many compression techniques, including Huffman, arithmetic, residual, run-length, scalar quantization, transform, and model-based (psychoacoustic) coding.

The aforementioned lossless and lossy compression techniques/standards are summarized in Table 1. More details can be found in Moffat's and Turpin's book (2002) and at the Data Compression Info (2002). The comparison of the quality of one versus another compression algorithm depends on the type of the algorithms. Criteria like the compression time, the reconstruction time, the size of the compressed data, and the degree of generality, could be easily adopted in the comparison among lossless compression algorithms (ACT, 2002). Nevertheless, these criteria should be revised under the perspective of the "degree of goodness" of the lossy approximation. In that case, tradeoffs between the amount of compression, the runtime, and the quality of reconstruction are involved, varying in significance according to the application and the signal characteristics. This is clearly described in the following Section.

4. BIOSIGNAL COMPRESSION: PRINCIPLES AND METHODS

Biosignal compression requires deep consideration of the signal properties in order to achieve both successful compression and accurate representation of the signal diagnostic features. To this end, bioelectric and bioacoustic data compression should be handled differently than image or speech data compression, since more constrains are imposed by the former case. This is due to the property of the human eye or ear to act as a smoothing filter, permitting a certain amount of tolerance for distortion with the image and speech data, where with the bioelectric and bioacoustic signals, not only should the overall distortion be low, but, in addition, its essential areas/characteristics (as they are described in Section 2) need to be preserved with the highest possible morphologic fidelity for accurate representation of the diagnostic information. From this perspective, visual inspection of the reconstructed biosignal after compression and/or transmission should also be included in the assessment of the quality of reconstruction, especially when noise contamination is present.

4.1. Principles

The basic principles that apply in the case of biosignal compression are (Zywietz, 1998):

1. Adoption of correct evaluation criteria.

2. Bandwidth limitation and sampling rate reduction.

3. Redundancy reduction.

4. Information reduction

These principles, when followed by efficient signal processing methods, lead to enhanced biosignal compression.

4.1.1. Evaluation Analysis Criteria

The main points in the evaluation analysis of biosignal compression are: *error figures* estimation, computation of *expenditure for encoding/decoding procedures*, consideration of *stability versus transmission errors, archive transportability, scalability, standardization*, and *encryption* (Zywietz, 1998).

Focusing on error figures, the most common ones (listed in Table 2), are based on the differences between the N-sample original biosignal, $S(k)$ and the reconstructed, $\hat{S}(k)$, with $k=1,\dots,N$. Usually, more than one error type should be used in the evaluation of the biosignal compression procedure, since absolute errors may be as misleading as *RMSE* figures (see Table 2) (Zywietz, 1998). For instance, if a large absolute error occurs *between* events of interest, such as successive *PQRST* complexes in ECG, it may result in negligible affection of the reconstructed signal quality, from a diagnostic point of view. Moreover, small *RMS* figures may be misleading when mean squared differences are small but large differences appear in diagnostically relevant short periods of the signal, such as at the location of explosive bioacoustic signals (crackles or sound bursts) or the QRS complex in ECG. According to Zywietz (1998), a *peak amplitude related error* (absolute or in prevent of a maximum amplitude to be measured), and an *RMSE figure* are the two principal error figures that should be used in biosignal compression. Like *RMS* figures, the *SNR* (see Table 2) often smoothes out large local errors; the *PRMSD* (see Table 2) is preferred for the validation of the reconstruction of a zero-mean signal. In any case, error figures have to be handled with caution.

4.1.2. Bandwidth Limitation and Sampling Rate Reduction

Before embarking for biosignal compression it is important to consider issues like bandwidth limitation and sampling rate reduction. According to the Nyquist criterion, the sampling frequency should be at least two times the upper frequency of the signal in order to avoid spectrum aliasing. Nevertheless, this upper frequency does not always correspond to the diagnostic parts of the recorded biosignal, but, instead, to the superimposed noise. In that case, the "useful" bandwidth, i.e., the frequency range within the spectrum of the diagnostically useful signal lies (see biosignals frequency characteristics in Section 2), could be shortened and the sampling frequency could be reduced accordingly (refer to Zywietz *et al.* (1983) for an example in the case of ECG).

Table 2. Most common error figures used in compression evaluation

Name	Definition[a]
Local Absolute Error (LAE)	$LAE(k) = \mid S(k) - \hat{S}(k) \mid$
Max Absolute Error (MAE)	$MAE(k) = \max\{\mid S(k) - \hat{S}(k) \mid\}$
Peak Amplitude Related Error (PARE)	$PARE(k) = [S(k) - \hat{S}(k)] / \max\{S(k)\}$
Normalized Maximum Amplitude Error (NMAE)	$NMAE = \max\{\mid \mathbf{S} - \hat{\mathbf{S}} \mid\} / [\max\{\mathbf{S}\} - \min\{\mathbf{S}\}]$
Root Mean Square Error (RMSE)	$RMSE = \sqrt{\{\sum_{i=1}^{N} [S(k) - \hat{S}(k)]^2\} / N}$
Percentile Root Mean Square Difference (PRMSD)	$PRMSD = 100 \cdot \sqrt{\{\sum_{i=1}^{N} [S(k) - \hat{S}(k)]^2\} / \{\sum_{i=1}^{N} S^2(k)\}}$
Signal-to-Noise Ratio (SNR)	$SNR \cong 10\log\{\sum_{i=1}^{N} [S(k) - \overline{S}]^2 / \sum_{i=1}^{N} [S(k) - \hat{S}(k)]^2\}$

[a] S, $\hat{S}$, and $\overline{S}$ denote the N-sample original biosignal vector, the reconstructed one, and the mean value of S, respectively; $k=1,\ldots,N$.

4.1.3. Redundancy Reduction

Data redundancy in biosignals could be seen in the form of *intersample correlation*, i.e., no statistical independence exists between neighboring samples, and *unequal probability* in the occurrence of the quantized signal amplitudes. The first form of redundancy could be reduced by using first or second order differences between the samples (Zywietz, 1998), resulting in significant shortened digital word lengths per sample. For reducing the second form of redundancy, mostly in the data transmission case, (adaptive) Huffman coding (see Section 3), which assigns shortest and longer codes to digital words with highest and lower probabilities, respectively, or other lossless compression types could be employed (see Table 1).

4.1.4. Information Reduction

Based on the biosignal characteristics, efficient reduction of the "useful" information can be achieved, when appropriate mapping and/or transform is applied to the original signal. For instance, the ECG could be mapped to straight lines (curves are approximated by straight line segments), slopes and plateaus, resulting in an approximation of the original signal with a stepped function that occupies less information. Based on this perspective, lossy ECG compression schemes have been proposed in the literature (Abenstein and Tompkins, 1982; Ishijiama *et al.*, 1983), reporting compression ratios within the range of 2:1-10:1, using predefined accuracy levels. Transform coding reduces the biosignal information by transforming an *N*-sample block into a set of transform coefficients. Characteristic examples of transforms applied in biosignals compression are: *Karhunen-Loeve transform* or *Principal Component Analysis* (PCA), which is based on the eigenvectors derived from the signal characteristics and represent "typical" sequences of the input data set for a given *RMSE* (Elliot and Rao, 1982); *Discrete Cosine transform*, where the signal is represented by a set of cosines; and *Discrete Wavelet transform*, where the signal is represented by a set of wavelets (Gholam *et al.*, 1998).

4.2. Methods

For illustrating the implementation of the aforementioned principles in biosignal compression, a short description of sample compression schemes applied to biosignals follows. The methods are categorized according to the processed biosignal type.

4.2.1. ECG Data Compression

The ECG data compression problem has been widely addressed by many research groups, resulting in many different compression algorithms, mainly categorized to *parameter extraction techniques*, which retain the parameters/characteristics of ECG, *direct time-domain techniques*, such as amplitude zone time epoch coding, delta, and entropy coding, and *transform-domain techniques* (see §4.1.4).

One interesting subcategory of such algorithms employs neural networks (Nazeran and Behbehani, 2000). Although PCA (see §4.1.4) results in uncorrelated transform coefficients (diagonal covariance matrix) and minimizes the total entropy compared with any other transform (Elliot and Rao, 1982), it requires the computation of the eigenvectors of the correlation matrix of the ECG data set, which, usually is a very large matrix. By employing a multiple Hebbian neural network, Al-Hujazi and Al-Nashash (1996), reduced the computational load of the PCA for arbitrary size of the ECG input vector. In this way, they achieved a compression ratio (*CR*) up to 30 with a *PRMSD* of 5%, using data from the MIT/BIH ECG arrhythmia database (MIT/BIH ECG, 2002) and a training set with all expected arrhythmias. By using autoassociative neural network, where the input and output patterns are the same, Hamilton *et al.* (1995), proposed an ECG compression scheme. The underlying idea was the fact that the majority of beats within a given recording ECG segment have the same gross morphology. Consequently, a compression capability could be established by storing an average waveform and compressing the difference only. They achieved a *CR* of 10 with a *PRMSD* of 4.6%, for a network size of 360-30-360 (Hamilton *et al.*, 1995). Vector quantization (see Section 3) involves the creation of a codebook of vectors that best spans the data of interest. A Kohonen neural network-based adaptation of the codebook vectors according to distance measurements and time changes was proposed by McAuliffe (1993). The resulted *CR* ranged from 3:1 to 19:1, according to the heart rate and noise content.

Another significant subcategory includes ECG compression schemes that use interbeat correlation between ECG cycles. In particular, long-term prediction (Nave and Cohen, 1993) and average beat subtraction (Hamilton and Tompkins, 1991) are structured on beat-to-beat correlation. However, several limitations of these methods have been reported by Ramakrishnan and Saha (1997), who used the interbeat correlation combined with period-and-amplitude-normalized and discrete-wavelet-transform truncation, for achieving redundancy-free original ECG data. In this way, *CR*s ranging from 13.4:1 to 19.1:1 with *PRMSD* values ranging from 9.9% to 13.3% were found. An enhancement of this approach was proposed by Istepanian *et al.* (2001), where higher-order statistics-based criteria and modeling were applied to the wavelet-transform domain, resulting in higher *CR*s (11:1-31.5:1) and smaller *PRMSD* values (1.74%-12.9%). A telemedical application of the latter algorithm, combined with adaptive arithmetic coding, is presented in Section 5. In the same vein, Wei *et al.* (2001), have combined the interbeat correlation of ECGs with the quasi-periodic analysis of the truncated singular value decomposition technique to decompose an ECG sequence into a linear combination of a set of basic patterns with associated scaling factors, producing high compression results (Wei *et al.*, 2001).

Furthermore, ECG data compression by parametric modeling of the discrete cosine transformed ECG signal, using separate modeling of low and high frequency regions of the transformed signal, achieves a *CR* of 40:1 with *PRMSD* values of 5% to 8% (Madhukar and Murthy, 1993). Finally, a lossy ECG data compression scheme that uses an adapted LZ77 algorithm (see Section 3) to code detected repetitions in the ECG data results in *CR*s of 9.2:1 to 13.4:1 and *PRMSD* values of 0.44% to 0.64% (Horspool, 1995).

4.2.2. EEG Data Compression

Both lossless and lossy EEG data compression methods have been exploited in the literature. With regard to the lossless one, the most widely known methods employ repetition count, Lempel-Ziv coding, Huffman coding, vector quantization, and techniques based on signal predictors, such as Markov predictor, digital filtering predictor, (adaptive) linear predictor, maximum likelihood, artificial neural networks (Antoniol and Tonella, 1997). Nevertheless, dictionary-based techniques (LZ77, LZ78) do not work well when applied to the EEG data, since their efficiency is based on the exploitation of the frequent reoccurrence of certain exact patterns detected in the data, a principle that does not apply in the nondeterministic nature of the EEG signal. Using the EEG knowledge, compression can be enhanced by removing redundancy, in terms of statistical dependence between samples. Time dependence, exploited by the prediction methods, can lead to the estimation of the next sample from the previous ones by adding some delay inputs to the predictor, while spatial dependence between input EEG channels can be captured by lossless methods based on multivariate time series analysis (Cohen *et al.*, 1995), and vector quantization (Antoniol and Tonella, 1997). In the latter, the input EEG channels (or derivatives) samples are mapped to a code vector and encoding is performed only to the error vector. A detailed comparison of the performance of the lossless EEG compression techniques can be found in the work of Antoniol and Tonella (1997). Although lossy EEG data compression can yield significantly higher compression ratios, while potentially preserving the diagnostic accuracy, it is not usually employed due to legal concerns. Instead, near-lossless EEG compression is preferred, since it gives quantitative bounds on the errors introduced during compression. In that way, the error is controlled and the precious bandwidth is utilized more efficiently. Memon *et al.* (1999), proposed a context-based trellis-searched delta pulse code modulation technique that satisfies a near-lossless error criterion by minimizing the entropy of the quantized prediction error sequence derived from an autoregressive model.

4.2.3. SEMG Data Compression

Unlike ECG and EEG, only a limited number of SEMG compression techniques have been reported in the literature. Among them, Guerrero and Mailhes (1997) compares widely employed compression methods, such as differential pulse code modulation, multi pulse coding, and code excited linear predictor coding, with techniques based on transforms for SEMG data compression. They propose the encoding of the wavelet coefficients using a four-level multiresolution decomposition scheme, as the one that provides the best results (Guerrero and Mailhes, 1997). An extension to this was proposed by Wellig *et al.* (1998), who used embedded zero-tree wavelet encoding for compressing SEMG data rapidly and with little distortion (*PRMSD* values of 1% to 10%).

It is noteworthy that when muscle fatigue occurs, SEMG bandwidth limitation (see §4.1.2) could be achieved, due to decreasing muscle fiber conduction that results in spectral compression of the SEMG signal (Lindstrom and Magnusson, 1977). From this perspective, Lowery *et al.* (2000), proposed a spectral distribution technique that captures the compression of the SEMG power and amplitude spectra by calculating the mean shift in all percentile frequencies throughout the entire spectrum, providing an efficient way to define the "useful" upper frequency that results in sampling frequency reduction (see §4.1.2).

4.2.4. LS, HS, and BS Data Compression

Similarly to the SEMG case, compression of bioacoustic data is limited in the literature. Usually, data reduction-based methods are preferred in order to eliminate the undesired signal, i.e., noise, resulting in less data that need encoding. To this vein, several de-noising algorithms have been proposed that separate LS, HS, and BS from the noisy background, based on their transient (nonstationary) characteristics (see §2.2). A thorough description of these algorithms can be found in Hadjileontiadis *et al.* (2002). Most of them employ wavelet transform, fuzzy logic, and higher-order statistics. An alternative way of data reduction that performs fractal dimension analysis of the bioacoustic signals, capturing their changes in complexity, results in high data reduction percentages in a very fast and efficient way (Hadjileontiadis and Rekanos, 2003). For further modeling and analysis of bioacoustic signals the reader should refer to the ILSA database (ILSA, 2002).

5. FROM A TELEMEDICAL PERSPECTIVE

The introduction of telematics in health care has emphasized the need for organized standardization and for a common use of medical informatics and telematics standards. Regarding compression, the International Telecommunication Union (formerly CCITT) Telecommunications Standardization sector (ITU-T) has the responsibility for compression techniques applied to audio and visual communications, i.e., voice audio and speech codecs, and multimedia terminals (ICU-T, 2002). The ITU-T compression standards recommendation includes the JPEG and MPEG standards for image and video compression, respectively (ICU-T, 2002).

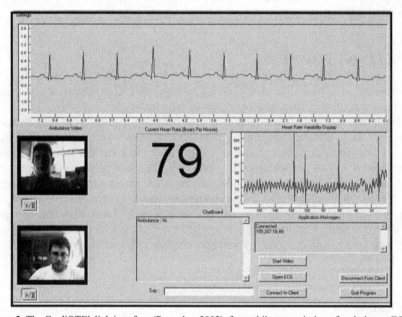

Figure 2. The CardiOTElelink interface (Panoulas, 2002), for mobile transmission of ambulatory ECG.

Regarding the interoperability of the telemedical systems, the Digital Imaging and Communications in Medicine (DICOM) standard, a standard network interface and standard model for imaging devices that can facilitate information systems integration, has been proposed (DICOM, 2002). DICOM is a complex set of documents that provide detailed definition of a rich set of communication services and associated protocols based on an object model that represents an abstraction of certain aspects of the real world. By simply specifying which DICOM functions are supported, different manufacturers could claim conformance, hence, their telemedical products could be applicable to networked environments.

Technological advantages in wireless communications, in recent years, have put into focus wireless biomedical data transmission, giving rise to mobile telemedicine. This perspective appreciates the role of the compression techniques described so far, since bandwidth limitations in mobile telecommunications could be circumvented with efficient compression. An example of a mobile telemedical system for ECG data is presented in Fig. 2. This is a mobile telemedical framework, namely CardiOTElelink (Panoulas, 2002), which uses public wireless cellular data lines (GSM-28.8 kbps up/download) to transmit video and patient biosignals from a moving ambulance to a receiving physician at the basestation-hospital (Panoulas, 2002). The compression scheme adopted by the CardiOTElelink is the one proposed by Istepanian *et al.* (2001). It results in high compression ratios (*CR*s over 20:1 with *PRMSD* values below 2.5%) that allow real-time transmission of ECG data, patient/physician videos (5 frames/sec), heart beat, heart rate variability, full-duplex voice, and text messages (Panoulas, 2002). Other, currently emerging, mobile protocols, such as Bluetooth, Wi-Fi, GPRS, 3GSM, allow wider exploitation of mobile telemedical applications and provide an ample space for new compression techniques.

6. CONCLUSIVE COMMENTS

Compression of biological data remains an important issue, despite the vast increase in storage capacity and transmission speed in communication pathways. This is due to their diagnostic characteristics, which set a common endeavor to all compression approaches, i.e., efficient biosignal data compression yet unaffected diagnostic characteristics. To address this challenge, many customized lossless and lossy techniques have been proposed so far, which reduce data redundancy, employ signal approximation or apply signal transforms towards higher compression ratios and smaller reconstruction errors. Nevertheless, adoption of different data sets and/or evaluation criteria makes their comparison difficult. By reviewing the characteristics of biosignals, the compression types, the main principles, and the most applied methods in biosignals compression, this chapter has tried to enlighten the framework wherein forthcoming compression schemes could blossom.

7. REFERENCES

Abenstein, J., and Tompkins, W., 1982, A new data reduction algorithm for real-time ECG analysis, *IEEE Trans. on Biomed. Eng.* **29**(1):43-48.

ACT: Archive Comparison Test, 2002; http://compression.ca.

Al-Hujazi, E., and Al-Nashash, H., 1996, ECG data compression using Hebbian neural networks, *J. Med. Eng. Technol.* **20**(6):211-218.

Antoniol, G., and Tonella, P., 1997, EEG data compression techniques, *IEEE Trans. on Biomed. Eng.* **44**(2):105-114.

Arnott, P. J., Pfeiffer, G. W., and Tavel, M. E., 1984, Spectral analysis of heart sounds: relationships between some physical characteristics and frequency spectra of first and second heart sounds in normals and hypertensives, *J. Biomed. Eng.* **6**(2):121-128.

Berne, R. M., and Levy, M. N., 1990, Cardiovascular system, in: *Principles of Physiology*, R. M. Berne, and M. N. Levy, eds., Wolfe Publications Ltd, London, pp. 212-213.

Bray, D., Reilly, R. B., Haskin, L., and McCormack, B., 1997, Assessing motility through abdominal sound monitoring, in: *Proc. 19th Int. Conf. IEEE/EMBS*, B. Myklebust, and J. Myklebust, eds., IEEE Press, Chicago, IL, pp. 2398-2400.

Cohen, A., 1986, *Biomedical Signal Processing-Volume II: Compression and Automatic Recognition*, CRC Press, Boca Raton, FL, pp. 113-137.

Craine, B. L., Silpa, M. L., and O'Toole, C. J., 2001, Enterotachogram analysis to distinguish irritable bowel syndrome from Crohn's disease, *Dig. Dis. and Sciences* **46**(9):1974-1979.

Data Compression Info, 2002; http://datacompression.info/Tutorials.shtml.

DICOM: Digital Imaging and Communications in Medicine, 2002; http://medical.nema.org.

Elliot, D. F., and Rao, K. R., 1982, *Fast Transforms, Algorithms, Analysis and Applications*, Academic Press, New York.

Gavriely, N., and Cugell, D. W., 1994, *Breath Sounds Methodology*, CRC Press, Boca Raton, FL.

Gholam, H. H., Nazeran, H., and Moran, B., 1998, ECG compression: evaluation of FFT, DCT, and WT performance, *Australasian Physical & Eng. Sciences in Medicine* 21(4):186-192.

Guerrero, A., and Mailhes, C., 1997, On the choice of an electromyogram data compression method, *Proc. of 19th IEEE EMBS* 4:1558-1561.

Hadjileontiadis, L. J., and Rekanos, I. T., 2003, Detection of explosive lung and bowel sounds by means of fractal dimension, *IEEE Signal Processing Letters* (in press).

Hadjileontiadis, L. J., Tolias, Y. A., and Panas, S. M., 2002, Intelligent system modeling of bioacoustic signals using advanced signal processing techniques, in: *Intelligent Systems: Technologies and Applications – Vol. III: Signal, Image, and Speech Processing*, C. T. Leondes, ed., CRC Press, Boca Raton, FL, chap. 3, pp. III 103-156.

Hamilton, D. J., Thomson, D. C., and Sandham, W. A., 1995, ANN compression of morphologically similar ECG complexes, *Med. Biol. Eng. Comput.* 33(6):841-843.

Hamilton, P. S., and Tompkins, W. J., 1991, Compression of the ambulatory ECG by average beat subtraction and residual differencing, *IEEE Trans. on Biomed. Eng.* 38(3):253-259.

Horspool, R. N., 1995, ECG data compression using Ziv-Lempel techniques, *Computers and Biomedical Research* 28:67-86.

Huffman, D. A., 1952, A method for the construction of minimum-redundancy codes, *Proceedings of the I.R.E.* 40(9):1098-1101.

ILSA: International Lung Sounds Association, 2002; http://www.ilsa.cc.

Ishijiama, M., Castelli, A., Combi, C., Pinciroli, F., 1983, Scan-along polygonal approximation for data compression of electrocardiograms, *IEEE Trans. on Biomed. Eng.* 30(11):723-729.

Istepanian, R. S. H., Hadjileontiadis, L. J., and Panas, S. M., 2001, ECG data compression using wavelets and higher-order statistics, *IEEE Trans. on Information Tech. in Biomedicine* 5(2):108-115.

ITU-T: International Telecommunication Union Telecommunications Standardization sector, 2002; http://www.itu.int/home.

Kim, Y., and Tompkins, W. J., 1981, Forward and inverse high frequency ECG, *Med. Biol. Eng. Comp.* 19:11.

Kimura, J., 1990, *Electrodiagnosis in Diseases of Nerve and Muscle: Principles and Practice*, F. A. Davis, Philadelphia.

Kraman, S. S., 1983, *Lung Sounds: An Introduction to the Interpretation of Auscultatory Findings*, American College of Chest Physicians, Northbrook, IL, pp. 14-21.

Lindstrom, L. H., and Magnusson, R. I., 1977, Interpretation of myoelectric power spectra: a model and its applications, *Proc. IEEE* 65:653-662.

Lowery, M. M., Vaughan, C. L., Nolan, P. J., and O'Malley, M. J., 2000, Spectral compression of the electromyographic signal due to decreasing muscle fiber conduction velocity, *IEEE Trans. on Rehabilitation Eng.* 8(3):353-361.

Madhukar, B., and Murthy, I. S. N., 1993, ECG data compression by modeling, *Computers and Biomedical Research* 26:310-317.

McAuliffe, J. D., 1993, Data compression of the exercise ECG using a Kohonen neural network, *J. Electrocardiol.* 26(Suppl):80-89.

Memon, N., Kong, X., and Cinkler, J., 1999, Context-based lossles and near-lossless compression of EEG signals, *IEEE Trans. on Information Tech. in Biomedicine* 3(3):231-238.

MIT/BIH ECG: Massachusetts Institute of Technology/Beth Israel Hospital ECG Database, 2002; http://ecg.mit.edu.

Moffat, A., and Turpin, A., 2002, *Compression and Coding Algorithms*, Kluwer Academic Publishers, Boston.

Moffat, A., Neal, R., and Witten, I. H., 1998, Arithmetic coding revisited, *ACM Trans. on Information Systems* 16(3):256-294.

Nave, G., and Cohen, A., 1993, ECG compression using long-term prediction, *IEEE Trans. on Biomed. Eng.* 40(9):877-885.

Nazeran, H., and Behbehani, K., 2000, Neural networks in processing and analysis of biomedical signals, in: *Nonlinear Biomedical Signal Processing-Volume I: Fuzzy Logic, Neural Networks, and New Algorithms*, M. Akay, ed., IEEE Press, New York, NY, pp. 89-91.

On-line Medical Dictionary, Academic Medical Publishing & CancerWEB, 1997; http://cancerweb.ncl.ac.uk/cgi-bin/omd?bowel+sounds.

Panoulas, C., 2002, *ECG data processing and compression for transmission through wireless cellular networks*, Diploma Thesis, Dept. of Electrical and Computer Eng., Aristotle Univ. of Thessaloniki, Greece.

Peper, A., Jonges, R., Losekoot, T. G., and Grimbergen, C., 1982, Separation of His-Purkinje potentials from coinciding atrium signals: removal of the P-wave from electrocardiogram, *Med. Biol. Eng. Comp.* **20**:195.

Pullman, S. L., Goodin, D. S., Marquinez, A. I., Tabbal, S., and Rubin, M., 2000, Clinical utility of surface EMG, *Neurology* **55**:171-177.

Ramakrishnan, A. G., and Saha, S., 1997, ECG coding by wavelet-based linear prediction, *IEEE Trans. on Biomed. Eng.* **44**(12):1253-1261.

Riggs T., Isenstein, B., and Thomas, C., 1979, Spectral analysis of the normal ECG in children and adults, *J. Electrocardiol.* **12**(4):377.

Wei, J.-J., Chang, C.-J., Chou, N.-K., and Jan, G.-J., 2001, ECG data compression using truncated singular value decomposition, *IEEE Trans. on Information Tech. in Biomedicine* **5**(4):290-299.

Welch, T. A., 1984, A technique for high performance data compression, *IEEE Computer* **17**(6):8-19.

Wellig, P., Cheng, Z., Semling, M., and Moschytz, G. S., 1998, Electromyogram data compression using single-tree and modified zero-tree wavelet encoding, *Proc. of 20th IEEE EMBS* **3**:1303-1306.

Zywietz, C., 1998, Criteria for data compression of biosignals and methods for data reduction in electrocardiography, in: *Biotelemetry XIV: Proceedings of the XIV International Symposium on Biotelemetry-Chapter 7: ECG Monitoring and Processing*, T. Penzel, S. Salmons, and M. R. Neuman, eds., Tectum Verlag, Marburg, pp. 281-292.

Zywietz, C., Palm, C., Spitzenberger, U., Wetjen, A., 1983, A new approach to determine the sampling rate for ECGs, in: *Computers in Cardiology*, K. Ripley, ed., IEEE Computer Society Press, Los Alamitos, CA, pp. 261-264.

RESILIENT ECG WAVELET CODING FOR WIRELESS REAL-TIME TELECARDIOLOGY APPLICATIONS

Álvaro Alesanco [*], José García, Salvador Olmos, and Robert S.H. Istepanian

1. INTRODUCTION

Mobile telemedicine is a new and evolving area of telemedicine that exploits the recent development in mobile networks for telemedicine applications. The convergence of information and telecommunications infrastructures around telemedicine and mobile telecare systems is fostering a diversity of cost-effective and efficient mobile applications. A comprehensive review of wireless telemedicine applications and more recent advances on m-health systems are presented by Istepanian (1999), Pattichis et al. (2002) and Tachakra et al. (2003).

Telecardiology is one of the most mature and successful areas in telemedicine. The most popular field in telecardiology deals with the transmission of electrocardiographic (ECG) signals over different networks and the inclusion of wireless networks opens a wide range of possibilities for it. It is well known that telecardiology can be carried out in two different ways: pre-recorded (store-and-forward) ECG transmission and real-time ECG transmission.

Pre-recorded ECG transmission constitutes one of the simplest practices in telecardiology since there is no need to arrange a session between a general practitioner (GP) and a cardiologist at the same time. In this case, the GP records patients ECG in a primary assistance centre, and only if the diagnosis is not clear enough, the ECG is sent to the hospital. A cardiologist evaluates this ECG in the hospital, establishes a diagnosis and sends the results back to the GP (Shanit et al., 1996). Thus, the patient only needs to go to the primary assistance centre and does not have to visit the cardiologist at the hospital unless the results suggest that.

---

[*] Álvaro Alesanco, Communication Technologies Group (GTC), Aragón Institute of Engineering Research (I3A), University of Zaragoza, Zaragoza, Spain. email: alesanco@unizar.es

One of the main challenges in telecardiology is found on patient monitoring and early diagnosis in the area of emergency telemedicine, where real-time ECG transmission is required. Wireless networks have evolved from technologies such as Global System for Mobile communication (GSM) or General Packet Radio Service (GPRS) to the recently developed Third Generation (3G) Universal Mobile Telecommunications System (UMTS), thus permitting wider use of new mobile telemedicine services (Tachakra et al., 2003). Emergency telemedicine is synonym of mobile telemedicine since the only way to communicate an ambulance with a hospital is through a wireless channel. One of the most important applications in emergency telemedicine has been reported by Pavlopoulos et al. (1998). Biosignals (ECG among them) were transmitted from an ambulance to a hospital using a mobile channel reducing the time needed to evaluate the patient once he/she arrives to the hospital.

Although wireless telemedicine has been applied successfully in recent years, one of the major challenges that remain for effective and accurate real-time ECG wireless transmission is the selection of the appropriate robust coding method suitable for such critical medical data. For wireless systems these selection is specially critical regarding the resilience of the coding method for issues such as bit error rate (BER) and bandwidth limitations. Although 3G wireless networks increase considerably the available bandwidth, an efficient coding method is still required, especially when the ECG shares the bandwidth with other clinical information and other users within the mobile network. A typical block diagram for a real-time ECG transmission wireless system consists of the basic modules shown in Figure 1.

Working with pre-recorded telecardiology, the errors produced during transmission yield the retransmission of the data until the information in the receiver is exactly the same as in the transmitter. In data-packet networks such as GPRS or 3G systems, the retransmission of information implies the retransmission of the corrupted or lost packets. The higher the BER, the higher the number of packets with errors, therefore leading to a higher number of retransmissions and transfer time. Working in real-time telecardiology, when the conditions of the channel are poor the number of packets suffering transmission errors will be large. If the transmission delay is high, the retransmission of information may not be useful because the retransmitted information may be out-dated when it arrives at the receiver. In this case, an efficient ECG coding strategy that detects these errors and endeavour to recover some information from the corrupted packets may be a better strategy than the retransmission.

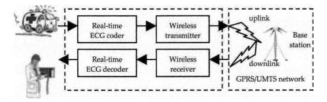

Figure 1. Block diagram of real-time ECG transmission wireless system.

In recent years, the application of wavelet ECG compression has drawn major attention within the biomedical and e-health research. However, the focus of most of this work was on improving the coding for pre-recorded strategies. Most of this research is not reported for recent packetized mobile systems. To the best of our knowledge, no ECG error resilience coding method that can adapt to channel conditions has been reported. Istepanian and Petrosian (2000) presented a wavelet compression method based on the optimal zone compression. The method was used in a telecardiology project transmitting the signal over a GSM wireless network. No error resilience was added to this method. A Vector Quantizer (VQ) in the wavelet domain is proposed by Miaou et al. (2002). A good performance is reported for this method but, it has not been tested in a real-time transmission environment and on the other hand, it does not include an error resilience mechanism. Hilton (1997) presented wavelet and wavelet packet compression algorithms based on Embedded Zerotree Wavelet (EZW) coding for Holter ECG signals. A wavelet ECG data compression method based on the Set Partitioning in Hierarchical Trees (SPIHT) compression algorithm was presented by Lu et al. (2000). Although rate and quality control are possible, no specific results on these issues were reported.

Error resilient data transmission has been studied in the last years especially in video transmission over wireless channels. In this way, various techniques have been proposed to try to palliate the damaging effects of noisy environments. MPEG-4 standard introduces the Reversible Variable Length Codes (RVLCs) to recover data from a corrupted received packet (Talluri, 1998). H263+ and MPEG-4 also use unequal error protection to code important information (Wenger et al., 1998). However, no work is yet reported on the application of these methodologies in wireless telecardiology systems.

The aim of this chapter is two fold. Firstly, we present different ECG wavelet coding strategies suitable for real-time wireless applications based on the transmission of the most significant wavelet coefficients of every segmented ECG block. Secondly, these coding strategies are analyzed through an extensive study carried out under different simulated 3G channel conditions. Real implementation of the presented method in UMTS environment to validate the resilient methodology is also presented.

2. REAL-TIME WAVELET ECG CODING

In this section we present a new wavelet ECG coding approach. The method is based on the estimation and coding of the ECG wavelet transform coefficients. The aim of any transformation being used for compression purposes is to concentrate the signal energy into a small number of transform coefficients. It is this energy concentration what leads to the coding gain (Vetterli and Kovacevic, 1995; Jayant and Noll, 1984). Compression is then achieved by selecting a subset of the most relevant coefficients which afterwards are efficiently coded.

Figure 2. Block diagram of the resilient ECG coding scheme.

To achieve real-time functionality requirements, the ECG signal is segmented into blocks whose coding results are then packetized for transmission. In some network scenarios it is very important to have control over the amount of information transmitted through the channels, that is to say, to control the transmitted bandwidth. To accomplish with this task a Constant Bit Rate (CBR) operational mode has been implemented. However, when the available network bandwidth is large enough, sometimes one would like to guarantee the ECG signal quality instead of the output bit rate. Since the amount of information varies block to block, the bandwidth required to guarantee a reconstruction error is not constant. Hence, this yields a Variable Bit Rate (VBR) transmission mode, which has been also implemented in the proposed method.

In our approach, the ECG coding method is divided into the following stages:

- Pre-processing
- Wavelet transform estimation and coefficients selection
- Coefficients coding
- Packetization

The block diagram of the resilient ECG coding method is show in Figure 2, where the different stages are represented.

2.1 Pre-processing

The ECG data stream is divided into fixed length blocks and every block is coded as an independent entity. Each lead in the block is treated independently through all the coding process. Then the baseline of the acquired ECG signal is removed. As real-time operation is necessary, a fast baseline estimation method is required. Short block length makes the use of cubic splines, technique which has been used in the last years offering accurate baseline identification, not possible. Therefore an Infinite Impulse Response (IIR) low-pass filter was considered. A third-order Butterworth filter, which filters the ECG blocks in both directions to avoid phase distortion, was considered.

2.2 Wavelet transform estimation and coefficients selection

After baseline wandering subtraction, a wavelet transform is applied to the ECG block. A *Coiflet* wavelet was used as mother function (Cooklev and Nishihara, 1999) due to its improved performance on ECG compression (Bradie, 1996). For each block, the wavelet coefficients are estimated. The wavelet coefficients are sorted by decreasing amplitude and the number of coefficients to be transmitted is selected according to a threshold, which depends on the operational mode (CBR or VBR). A typical coefficient distribution and its decreasing pattern are shown in Figure 3.

When operating in VBR mode, the threshold is set up to achieve a fixed Root Mean Square (RMS) reconstruction error by rejecting the smallest coefficients until their accumulated power reaches the threshold value. Hence, in this mode the threshold is defined in μV units. A variable number of coefficients are needed to achieve the fixed RMS error since the coefficient distribution varies from block to block. However, if the operation mode is CBR, then the threshold is set up to achieve a constant bit rate and then it can be defined in bits per second. The number of coefficients is selected (beginning with those of largest energy) to fulfil the desired transmission bit rate considering new bit

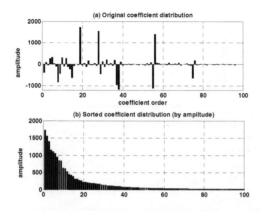

Figure 3. Illustration of the wavelet coefficients for one ECG block (Block length = 512 samples).

allocation strategies, defined in the next section. A general description of commonly used bit allocation strategies is presented in Gersho and Gray (1991).

2.3 Coefficients coding

After threshold operation, a wavelet coefficients set is preserved for transmission. For each preserved coefficient, its order and amplitude need to be coded and transmitted. Since the relevance of one particular coefficient changes block to block, both order and amplitude of the selected coefficient must be coded in every block. Different coding strategies are proposed for order and amplitude information in the next sub-sections.

2.3.1 Order coding

Since the power spectral density of the ECG signal has a smooth low-pass shape, the energy distribution at each scale of the wavelet transform will be similar for different blocks. It also makes that the relevant wavelet coefficients on each ECG block appears very close in the order sequence inside clusters (see Figure 3 (a)). However, since the heart beats are not synchronized with the block window (no QRS detection is performed), there will be some order shifts from block to block. In order to eliminate the order shift and take advantage of the relevant coefficients order proximity, the first difference of the orders sequence for the preserved coefficients is calculated (these difference values are denoted as jumps between the selected orders). The result is that most of the jump values are equal to one. This effect is illustrated in Figure 4 which has been obtained averaging 1000 ECG blocks of 512 samples. Figure 4 (a) shows the average histogram of the jumps when the ECG blocks are coded using 20 coefficients (the 20 coefficients with largest energy). Figure 4 (b) shows the evolution of the probability for jump values ranging from one to four (low jumps with the highest probability values) as the number of coefficients selected to be transmitted increases. As it is shown, most of the probability is concentrated in small jump values. Such *a priori* information may be helpful to design an

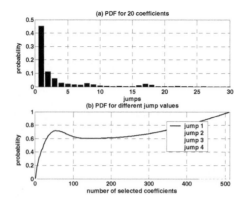

Figure 4. Probability density function (PDF) of the jumps.

efficient coefficient order coding. In this way, a Variable Length Code (VLC) can be used by constructing a Probability Density Function (PDF) that reflects the observed jumps characteristics. Tendencies shown in Figure 4 are similar no matter the ECG signal being studied due to their power spectral density similarity. Although the optimal selection of the PDF depends on the number of selected coefficients, in the VLC implementations only one PDF has been taken into account for simplicity. This PDF is the one shown in Figure 4 (a). The mean number of bits used to code the jumps decreases when the number of transmitted coefficients increases (for 20 coefficients this value is equal to 4 bits). This is the expected behaviour for the PDF showed in Figure 4 since the PDF is concentrating in a few values (especially in value one) as the number of transmitted coefficients increases thus leading to a high compression gain.

2.3.2 Amplitude coding

If the coefficients are sorted by amplitude in absolute value (see Figure 5), we can take advantage of the fact that the amplitude decreases fast from the highest coefficient to the lowest. In this way, an Adaptive Pulse Code Modulation (APCM) can be used. The implementation considered maintains the quantization step and varies the number of bits to code the coefficients amplitudes. For each block, subsets of coefficients are defined as a function of the number of bits they need to be coded. In this way, the first subset of coefficients (1 to n_1, see Figure 5) will be coded with the minimum number of bits (*min*) required to code the maximum amplitude plus one bit per coefficient to code its signum, coefficients ranging from n_1+1 to n_2 will be coded with *min-1* bits and so on until n_m is reached (see Figure 5).

It is worthy to note that in the coefficients coding, amplitude and order coding strategies are incompatible to each other. If coefficients are sorted by amplitude, then the coefficients orders will not be consecutive and the jumps will not present a suitable PDF for compression purposes. Conversely, if coefficients are sorted by order, then the

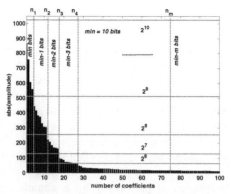

Figure 5. Coefficients sorted by amplitude.

amplitudes will oscillate and the amplitude coding strategy will not be applicable. Therefore, there are two possible coding combinations: a VLC for the coefficient orders with a PCM coder for the amplitudes, or an APCM coder for the coefficient amplitudes with a PCM coder for the orders. These two coding combinations have been studied and compared with two well known different possibilities of implementation for the VLC: a Huffman code and a RVLC.

2.4 ECG coding strategies for error-prone environments

Different coding strategies (that define the bit allocation for the wavelet coefficients) are presented in this section discussing them from the efficiency and error resilience point of view. These strategies are evaluated and their performance is assessed using different mobile conditions.

2.4.1 VLC for jumps and PCM for amplitudes

2.4.1a Huffman code for jumps and PCM for amplitudes: Coding strategy A

The first option to code the order jumps is to use a Huffman code. When the PDF of the source data is known the Huffman code is considered the most efficient VLC (Tanaka, 1987). It is based on codeword assignment as a function of the occurrence probability of the coded value. The higher the probability, the shorter the codeword length. Note that the first order needs to be transmitted in absolute value to reconstruct the original orders from the jump values. Both the coder and the decoder know the PDF used to generate the Huffman code.

When the data is transmitted through a noisy channel, an error introduced in a jump codeword is fatal for this implementation. For instance, an error changing the value of one bit would yield a completely erroneous decoding of the following jumps. Hence, a de-synchronization is produced in the decoder which propagates its effects through the complete coded lead and therefore makes it useless.

2.4.1b RVLC for jumps and PCM for amplitudes: Coding strategy B

An alternative and more efficient code than Huffman regarding error tolerance can be selected especially for error prone channels. A RVLC is a code that can be read in both forward and backward directions. RVLCs have been used in video coding during last years (Talluri, 1998). The RVLC selected follows the approach of Tsai and Wu (2001), consisting on an asymmetrical RVLC generated from the original Huffman coder. When the RVLC decoder detects an error, it goes to the end of the data packet and decodes the bit stream in the backward direction until it encounters an error. Based on the location of the two errors, the decoder can recover some data that would be otherwise discarded.

Due to the special coding of the coefficients order by means of the jump values (the coefficient orders reconstruction is performed by a cumulative addition), it is only possible to use the values decoded from the beginning to the backward error location. Regarding the use of PCM for coefficients amplitude coding in both strategy A and B, since these values are coded as absolute values, a bit channel error will only affect one coefficient amplitude, and the rest of amplitudes in the packet would be correctly recovered.

2.4.2 PCM for orders and APCM for amplitudes: Coding strategy C

In order to apply the APCM coder explained in section 2.3.2, signals from different databases were studied to group the coefficients in sets of amplitude values that could be coded with the same number of bits. Considering groups of ten coefficients this coding technique can be implemented without the risk of saturation in coefficients amplitudes.

This coding strategy has a clear advantage over a VLC in error-prone environments. There is not risk of de-synchronization in the decoder because the length in bits of each coefficient order and amplitude are *a priori* known. Thus, an error either in the amplitude or in the order does not affect the subsequent data. Furthermore, taking into account that coefficient amplitudes are sorted in decreasing order, a transmission error (affecting an amplitude) could be located in the received packet and removed if an amplitude appears breaking the monotonically decreasing tendency.

2.5 Packetization

In this section we describe the protocol data unit (PDU) formats used for the proposed coding strategies in ECG transmission. Each ECG block is coded and then packetized for the transmission into the data field of the transport layer packet.

2.5.1 VLC for jumps and PCM for amplitudes

The proposed PDU format is shown in Figure 6:

PDU length	N of leads (n)	Block length	N of coefficients for lead j (j = 1 to n)
N of bits to code coeffs. amplitude for lead j (j = 1 to n)			Coefficients amplitude for lead j (j = 1 to n)
N of bits to code coefficients order for lead j (j =1 to n)			First coefficient order for lead 1
Jumps for lead 1		...	First coefficient order for lead n
Jumps for lead n			

Figure 6. PDU format for strategies A and B.

where the fields correspond to:

- *N of coefficients for lead j*: the number of coefficients selected to transmit the ECG data for lead *j* after applying the corresponding threshold.
- *N of bits to code coefficients amplitude of lead j*: the number of bits used in the PCM coder to code the amplitudes of lead *j*.
- *Coefficients amplitude for lead j*: the PCM coded amplitudes of the wavelet coefficients for lead *j*. The number of amplitudes is equal to the value in the field *Number of coefficients for lead j*.
- *N of bits to code coefficients order for lead j*: the number of bits to code the first coefficient order and the following jumps for lead *j*.
- *First coefficient order for lead j*: the first coefficient order coded with a PCM coder.
- *Jumps for lead j*: the VLC coded jumps of the wavelet coefficients selected for lead *j*.

2.5.2 PCM for orders and APCM for amplitudes

The proposed PDU format in this case is shown in Figure 7:

PDU length	N of leads (n)	Block length	N of coefficients for lead j (j = 1 to n)
N of bits to code amplitude for lead j (j = 1 to n)			Coefficient amplitude for coeff. .x lead y
Coefficient order for coeff. x lead y			

Figure 7. PDU format for strategy C.

where the fields correspond to:

- *N of bits to code coefficients amplitude in lead j*: the number of bits used in the APCM coder to code the first subset of coefficients amplitudes (*min* in Figure 5) for lead *j*.
- *Coefficient amplitude for coefficient x lead y*: the APCM coded amplitude of the xth wavelet coefficient selected for lead *y*. Note that *x* ranges from 1 to *Number*

of coefficients for lead j (in this case $j = y$) and y ranges from 1 to n, with n the number of leads in the PDU.

- *Coefficient order for coefficient x lead y*: the PCM coded order of the xth wavelet coefficient selected for lead y. Note that x ranges from 1 to *Number of coefficients for lead j* (in this case $j = y$) and y ranges from 1 to n.

2.5.3 Header redundancies

Not all the information within a data packet has the same relevance. Some fields can be essential for the reconstruction of the original block. For example, if an error affects the field *Number of leads*, the rest of information in the packet will be erroneously decoded. Therefore, it could be a good policy to use an error correction methodology (Liu et al., 1997). Although there are better alternatives, a repetition code was considered for simplicity. Hence, each bit is sent n times ($n = 3$ in our implementation) and the decoder decides whether it has received either a one or a zero by majority. In the coding strategies A, B and C, all the information fields are coded with redundancy except for coefficients order (or jump) and amplitude fields. Obviously, these redundancies increase the data packet length for every packet sent, and then the compression performance of the methods decrease. However, the benefit obtained from redundancy might be crucial for a correct reconstruction of the received information in error-prone environments.

Another coding strategy that could be interesting is to use strategy A without header redundancies. This strategy is denoted as A'. This option was only considered with strategy A and not with the others because strategy A is the unique that does not include error detection methodologies.

3. SIMULATION RESULTS AND DISCUSSION

3.1 Compression performance in error-free environments

3.1.1 Comparison between the proposed coding strategies

The MIT-BIH Arrhythmia database (1997) has been selected for performance evaluation of the proposed coding strategies. The sampling rate of the two-lead ECG recordings is 360 Hz and the resolution 11 bits per sample. The Rate Distortion (RD) curves obtained using the four proposed coding strategies for channel 1 of record 100 are shown in Figure 8. Using a RVLC to code the jumps (strategy B) increases the mean codeword length compared with a Huffman coder (strategy A). Nevertheless, the small difference between the mean codeword length in the Huffman and RVLC used for jump coding makes, in this specific implementation, that the Huffman coder presents only a slight improvement compared with the RVLC. Coding strategy A provides a better performance than strategy C except for very low transmission rates, where the number of transmitted coefficients is small thus making the jump coding inefficient compared with PCM coding for orders (the number of bits used to code each jump decreases when the number of transmitted coefficients increases). The best performance is obtained for strategy A', as it was expected since strategy A' does not use header redundancies. These

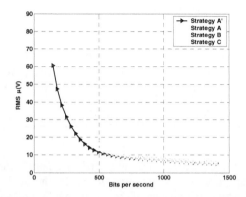

Figure 8. Comparative between the four coding strategies (Block length = 512 samples).

conclusions are independent of the type of the ECG signal being coded because the effect is produced by the coders used and not by the way the coefficients are selected and the jumps are obtained.

3.1.2 Compression results and comparative for the best coding technique

Coding strategy A' was applied to the signals from the MIT-BIH database when no transmission errors are present. Since the method permits two operational modes (VBR and CBR), results for both modes were obtained and some are shown in Table 1 and Table 2, respectively. In VBR mode (see Table 1), the target reconstruction error is achieved with a high level of accuracy (not only in mean but also with a low standard deviation). On the other hand, the data rate achieved presents a large standard deviation indicating that the number of bits needed to guarantee a constant reconstruction error may vary largely block to block due to the different number of transmitted coefficients.

Table 1. *VBR mode compression performance (mean ± sd).*

	Target RMS (μV)					
	10		30		50	
Record	Achieved RMS	Achieved data rate	Achieved RMS	Achieved data rate	Achieved RMS	Achieved data rate
100	10 ± 0.04	586 ± 81	29.4 ± 0.4	259 ± 43	48.6 ± 0.9	173 ± 37
119	10 ± 0.04	898 ± 101	29.4 ± 0.4	404 ± 81	48.7 ± 0.9	300 ± 68

Like in VBR mode, when working in CBR mode the target bit rate is achieved in an accurate way, both in mean and standard deviation (see Table 2). On the other hand, the reconstruction error presents a large standard deviation (compared with the value obtained in VBR), indicating that the reconstruction error may vary largely block to block, depending on the ECG signal block waveform. These conclusions are the same for all the records in the MIT database.

Table 2. *CBR mode compression performance (mean ± sd).*

	Target Data Rate (bps)					
	300		500		700	
Record	Achieved RMS	Achieved data rate	Achieved RMS	Achieved data rate	Achieved RMS	Achieved data rate
100	24 ± 5.3	300 ± 3	12 ± 1.4	500 ± 3	9 ± 0.8	700 ± 3
119	51 ± 18	301 ± 3	22 ± 6.2	501 ± 3	13 ± 2.3	701 ± 3

Although mathematical error indices are used to quantify distortion of the coding methods, a visual inspection is also important to ensure the quality of the reconstructed signal. Figure 9 shows the performance of the coding method on channel 1 record 119. In Figure 9 (a), two blocks of the original signal are shown and the results of applying the coding method in VBR and CBR modes are presented in Figure 9 (b) and (c) respectively. In Figure 9 (b) a RMS target of 30 μV was selected. The reconstruction error adjusts very well to the selected threshold in both blocks but the number of bits required to achieve this goal varies largely from one block to another. This effect is due to a generalized increase in the energy of the coefficients. The higher the coefficients energy, the higher the number of coefficients (and hence the transmitted bits) needed to guarantee a determined reconstruction error. In Figure 9 (c) the threshold was set up to 350 bits per second (bps). The target is accurately achieved but the reconstruction error varies greatly from one block to another. Conversely to VBR mode, if a fixed transmission rate is achieved, the number of transmitted wavelet coefficients would be approximately

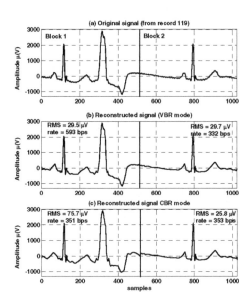

Figure 9. Original and reconstructed signal in VBR and CBR modes. Block length 512 samples.

constant. Thus, the higher the signal energy, the lesser these coefficients represents from the total energy, leading to a high reconstruction error.

3.1.3 Comparison with other wavelet compression coding strategies

Figure 10 shows the RD curve of the coding method (block length = 512 samples) for record 117 compared with the results reported by some authors in the literature. Percentage RMS distortion (PRD %) (Miaou et al., 2002) has been used for comparison purposes. The results of the proposed coding method are not superior from the methods reported by Miaou et al. (2002) and Lu et al. (2000). The method presented by Hilton (1997) presents a worse performance.

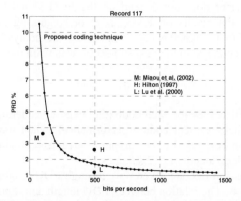

Figure 10. Comparison with other coding techniques (record 117).

In the proposed method real-time operation and error resilience have been given priority. Coders that consider more ECG information (not only a ECG block but the complete recording) would achieve a better performance. In addition, the proposed method includes a header in every coded data block, which would be included only once in a non real-time operation.

3.2 Performance in error-prone environments

Real-time ECG transmission through wireless channels involves bit errors in reception. In order to simulate the performance of the system in different 3G wireless channel conditions, a 3G channel simulator previously developed in the University of Zaragoza has been used. The transmission rate selected for simulation was 64 Kbps and the mobile speed 120 Km/h. Further details of the channel and UMTS simulator can be found in Gállego et al. (2002).

Wireless channels are more challenging than wired ones in data transmission; not only the BER is higher but also the error distribution presents a difficult pattern regarding error correction: burst pattern. Burst errors are concentrated in a few number of PDUs but these erroneous PDUs present a large number of toggled bits thus making error recovery difficult.

3.2.1 Performance analysis measurement of received ECG signals

When receiving a transmitted ECG signal, there are two quantities that contribute to the total RMS error ($RMS_{received}$). There is a reconstruction error due to lossy compression (RMS_{codec}). Additionally there exists an error introduced by the channel ($RMS_{channel}$) due to transmission errors.

In strategy A', packets with transmission errors are discarded. In order to calculate the RMS error index, this block of data is replaced by a constant value equal to zero representing that no signal information is available. In strategies A, B and C, the packet information is tried to be recovered. Nevertheless if an inconsistence is found in the decoding process, then the packet is discarded and a constant value of zero is placed instead.

Since $RMS_{channel}$ error does not depend on the ECG signal being transmitted and general tendencies in the results on RMS_{codec} error (regarding coding strategies) are similar no matter the ECG signal being compressed, the analysis of the performance results presented for record 100 in the next sub-sections is valid to any ECG signal.

3.2.2 Performance analysis in 3G channel environment

The distortion-BER curves for each coding strategy considering only the $RMS_{channel}$ (in order to analyze their resilient coding performance), when transmitting record 100 (first minute) are shown in Figure 11. The graphics have been obtained averaging 40 runs in CBR mode with a block length of 512 samples.

Values of transmission rate are given now through PDU length, which is the length of the transmitted packet. The relation between PDU length and transmission rate is given in Eq. 1,

$$R = PDU_{length} \cdot \frac{f}{N} \tag{1}$$

where R is the transmission rate, f is the sampling frequency (360 samples per second) and N is the number of samples per ECG block (512 in this case). The results correspond to the coding strategies when working in CBR mode (target PDU length 500 and 700 bits). In strategies A and A', the $RMS_{channel}$ error consists only on discarding corrupted packets substituted by zero. However in strategies B and C, many packets will be discarded but some corrupted packets would be reconstructed (with less data than the originally transmitted or even with erroneous data if the detecting error method fails).

Strategies A, A' and B present almost similar performance (see Figure 11). Strategy A presents a slightly better performance than A'. Strategy C presents the best performance because it is able to recover data from erroneous packets, presenting the best error resilience characteristic. On the other hand, a characteristic that affects the performance of all the coding strategies is the PDU length. The larger the PDU length, the higher the error ($RMS_{channel}$) introduced by the UMTS channel (compare Figure 11 (a) and (b)). This is the expected effect: the larger the PDU length, the higher the probability

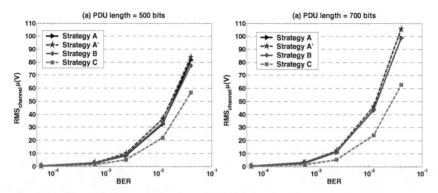

Figure 11. Average $RMS_{channel}$ values showing the performance of the coding methods when using in transmission over simulated UMTS channel conditions (CBR mode. Block length = 512).

to find erroneous bits in the packet. Care must be taken with this result because although the error introduced by the channel ($RMS_{channel}$) is smaller with smaller PDUs, the reconstruction error (RMS_{codec}) is larger (see Figure 8) and there is a minimum reconstruction error below which the reconstructed signal would be clinically useless (Woodward et al., 2001).

Both $RMS_{channel}$ and RMS_{codec} are now analyzed together to take into account not only the error resilience characteristic (through $RMS_{channel}$) but also their compression efficiency (through RMS_{codec}). The expected results yield crossing points in the curves due to different initial RMS_{codec} error values. These crossing points will determine which strategy is better for different BER values. Figure 12 shows the performance of the different coding strategies working in CBR mode. These results are similar to Figure 11 but including the RMS_{codec}. Simulation results from Figure 11 and Figure 12 suggest that strategy B has no significant advantages over strategy A and A' despite the fact that strategy B implements a RVLC. The reason is two fold. On one hand, the use of this code to code the jumps makes it inefficient because once the error is located, only the forward values can be used, since they are differential values. On the other hand, error location is guaranteed only if there is one erroneous bit in the stream. Otherwise (as it usually happens in a bust error channel), error location with a RVLC would be useless.

Strategies A and A' present a similar performance as the BER increases (strategy A slightly better, see Figure 11) but strategy A' is better when the BER is low due to a RMS_{codec} error lower than the one for strategy A. A burst error pattern makes the corrupted packets to present a large number of errors, even in low BER transmissions and the use of header redundancies becomes inefficient if they are not implemented with a robust coding technique to detect errors in the wavelet coefficients because they cannot correct such amount of concentrated errors. Header redundancies make the PDU length larger to maintain the same RMS_{codec} error. The gain obtained using only header redundancies (or combined with RVLCs) is not significant to compensate the drawback of increasing the PDU length.

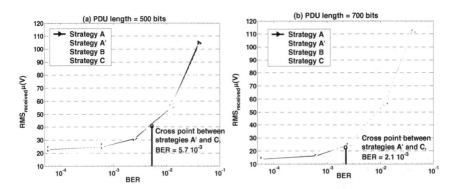

Figure 12. Average RMS$_{received}$ values showing the performance of the coding methods when using in transmission over a simulated UMTS channel (CBR mode. Block length = 512).

Strategy C presents the best performance as the BER increases but it presents the worst efficiency for low BERs (see Figure 12). Header redundancies and robust wavelet coefficient coding make the PDU increase to maintain the same RMS$_{codec}$ error compared with strategy A and A'. The combination of header redundancies, PCM coders (robust in wireless environments) and error amplitude detection makes this strategy the best option when the channel BER increases. The cross point (in BER) between strategy A' and C depends on the PDU length (and thus on the transmission rate). This effect can be explained considering that the differences in RMS$_{codec}$ error between strategy A' and C decrease as the transmission rate increases (see Figure 8). Hence, the approach of RMS$_{codec}$ error values as the PDU length gets higher (see Figure 8) leads to a cross point located in lower BERs values.

RMS$_{channel}$ error decreases with a decrease in the PDU length but on the other hand, RMS$_{codec}$ increases. As commented previously, we cannot decrease the PDU length too much since the increase in the RMS$_{codec}$ error would lead to clinically useless signals. However it is possible to reduce the PDU length without increasing the reconstruction error by reducing the ECG block length. All the results showed until now have been obtained with a block of 512 samples. Figure 13 shows the comparison results for strategies A' and C considering different coding block lengths. For both block lengths in Figure 13 (256 and 512 samples), there is a cross point between the two block sizes and again, this cross point depends on the transmission rate. The conclusion is clear: when the BER increases a more efficient option is to use small block lengths. Obviously, by

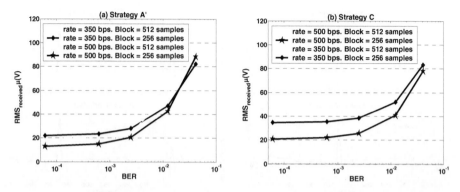

Figure 13. Comparative performance analysis on different ECG block lengths.

reducing the block length the transmission efficiency decreases due to an increase in the ratio between headers and the other data (packetization efficiency). Thus, in order to maintain a constant RMS_{codec} error an increase in the transmission rate will be required. Maintaining the same transmission rate would lead to a higher RMS_{codec} error (see Figure 13). In conclusion, there is a trade off between packetization efficiency and error resilience: the higher the block length, the higher the packetization efficiency but also, the lower the performance in error-prone environment. The optimal choice depends on the channel BER and transmission rate, as it is shown in Figure 13.

In strategies A and A', signal blocks stored after decoding are the same ones that have been sent (after coding) or zero. Strategy A and B are not suitable in a UMTS channel because they present no significant improvement compared to strategy A'. Strategy C presents the best performance as the BER of the channel increases. This result is owed to the error detecting property implemented in strategy C. But in some cases an error would be not detected and thus, the reconstructed block after decoding would yield to erroneous interpretations. Besides this, although no errors were introduced in the reconstructed block, the coefficients discarded due to the transmission errors would yield a low quality reconstructed signal block. Figure 14 shows these effects in the received signal block. Figures 14 (a) and (c) show two original signal blocks. In Figure 14 (b) the effects of not detected erroneous coefficients is shown. Signal peaks are introduced in the reconstructed signal due to erroneous wavelet coefficients. Figure 14 (d) shows the effect of removing a coefficient (after being marked as erroneous) from the coded signal block. The error is proportional to the energy of the removed coefficient (the coefficient removed in Figure 14 (d) is a high energy one). Therefore, care must be taken with these effects because although data is saved compared with strategy A, it may lead to a clinical erroneous interpretation.

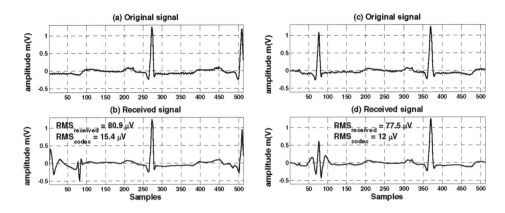

Figure 14. Possible effects of transmission errors in strategy C.

4. REAL-TIME ECG TRANSMISSION APPLICATION

The new ECG wavelet coding method has been implemented in a 3G mobile telemedical environment. Strategy A' has been selected to be adopted in the system. The scenario of use is that of an ambulance communicating with a hospital by means of a 3G mobile communications network. An overview of the biomedical module screen in the telemonitoring system is shown in Figure 15. The ECG signal is represented in the window in the format selected by the ambulance staff (represented leads, time and amplitude scales, etc.) together with other ambulatory biomedical and emergency parameters such as oxygen saturation, heart rate and systolic/diastolic blood pressure. Different transmission rates, which correspond to different compression ratios can be applied to the ECG before transmission.

5. CONCLUSIONS

In this chapter, we described a new resilient ECG coding technique based on the wavelet transform suitable for real-time wireless telecardiology applications. The codec can work on two operational modes: VBR and CBR thus enabling the control of the reconstructed signal error or the transmission rate respectively. We have proposed and tested four coding strategies to code the relevant amplitude and order (or jump) wavelet coefficients: strategy A, which uses Huffman coder for jumps and PCM coder for amplitudes, strategy A' which is similar to A but without header redundancies, strategy B which uses RVLC for jumps and a PCM coder for amplitudes and strategy C which uses a PCM coder for orders and an APCM coder for amplitudes. Strategies A, B and C use header redundancies.

The simulation results on error-free environments suggest that the proposed coding method (strategy A' for transmission in error-free environments) presents a good performance for compressing the ECG signals with high compression ratios and acceptable clinical data for further diagnosis.

The simulations results in 3G wireless channels (UMTS) suggest that strategies A and B are not suitable in these environments, and the selection between A' and C depends on the channel BER and the transmission rate. As the BER increases, strategy C takes advantage over strategy A'. The cross point between both strategies depends on the transmission rate. The higher the transmission rate, the lower the BER of the cross point. On the other hand, for every coding strategy, a reduction of the coding block size has a positive effect when the BER increases although for low BER values, the decrease in header/data ratio makes the performance worse than using larger blocks. Again, the optimum block length selection depends on the channel BER and the transmission rate.

Ongoing work is currently underway to test the system in this real-time transmission 3G wireless environment between selected patient population and specialist medical centre, both in U.K and Spain.

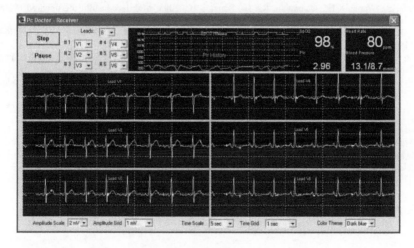

Figure 15. Telemonitoring application (receiver side).

6. REFERENCES

Bradie, B., 1996, Wavelet packet-based compression of single lead ECG, *IEEE Trans. Biomed. Eng.* **43**(5):493-501.

Cooklev, T., Nishihara, A., 1999, Biorthogonal Coiflets, *IEEE Trans. Signal Proc.* **47**(9):2582-2588.

Gállego, J.R., Valdovinos, A., Canales, M., De Mingo, J., 2002, Analysis of closed loop power control modes in UTRA-FDD under time varying multipath channels, *IEEE International Symp. Personal, Indoor and Mobile Radio Comm, (PIMRC)*:1616-1620, Lisboa (Portugal).

Gersho, A., Gray, R.M., 1991, *Vector Quantization and Signal Compression*, Kluwer Academic Publishers, Boston, pp. 225-252.

Hilton, M.L., 1997, Wavelet and wavelet packet compression of electrocardiograms, *IEEE Trans. Biomed. Eng.* **44**(5):394-402.

Istepanian, R.S.H., 1999, Telemedicine in the United Kingdom: current status and future prospects, *IEEE Trans. Inform. Technol. Biomed.* **3**(2):158-159.

Istepanian, R.S.H., Petrosian, A.A., 2000, Optimal zonal wavelet-based ECG data compression for a mobile telecardiology system, *IEEE Trans. Inform. Technol. Biomed.* **4**(3):200-211.

Jayant, N.S., Noll, P., 1984, *Digital Coding of Waveforms: Principles and Applications to Speech and Video*, Prentice Hall. Englewood Cliffs, New Jersey, pp. 510-576.

Liu, H., Ma, H., El Zarki, M., Gupta, S., 1997, Error control schemes for networks: an overview, *Mobile Networks and Applications*, **2**(2):167-182.

Lu, Z., Kim, D.Y., Pearlman, W.A., 2000, Wavelet compression of ECG signals by the set partitioning in hierarchical trees algorithm, *IEEE Trans. Biomed. Eng.* **47**(7):849-856.

Miaou, S.G., Yen, H.L., Lin, C.L., 2002, Wavelet-Based ECG compression using dynamic vector quantization with tree codevectors in single codebook, *IEEE Trans. Biomed. Eng.* **49**(7):671-680.

MIT-BIH arrhythmia database CD-ROM, 1997, Third ed. Cambridge, MI: Harvard-MIT Division of Health Sciences and Technology.

Pattichis, C.S., Kyriacou, E., Voskarides, S., Pattichis, M.S., Istepanian, R.S.H., Chizas, C.N., 2002, Wireless telemedicine systems: an overview, *IEEE Ant. Prop. Mag.* **44**(2):143-153.

Pavlopoulos, S., Kyriacou, E., Berler, A., Dembeyiotis, S., Koutsouris, D., 1998, A novel emergency telemedicine system based on wireless communication technology—AMBULANCE, *IEEE Trans. Inform. Technol. Biomed.* **2**(4):261-267.

Shanit, D., Cheng, A., Greenbaum, R.A., 1996, Telecardiology: supporting the decision making process in general practice, *J. Telemed. Telecare* **2**(1):7-13.

Tachakra, S., Wang, X.H., Istepanian, R.S.H., Song, Y.H., 2003, Mobile e-health: the unwired evolution of telemedicine, *Telemed. J. and e-Health* **9**(3):247-257.

Talluri, R., 1998, Error-resilient video coding in the ISO MPEG-4 standard, *IEEE Comm. Magazine*, **36**(6):112-119.

Tanaka, H., 1987, Data structure of Huffman codes and its application to efficient encoding and decoding, *IEEE Trans. Inform. Theory* **33**(1):154-156.

Tsai, C.W., Wu, J.L., 2001, On constructing the Huffman-code-based Reversible Variable-Length Codes, *IEEE Trans. Comm.* 49(9):1506-1509.

Vetterli, M., Kovacevic, J., 1995, Wavelets and subband coding, *Prentice-Hall*. Englewood Cliffs, New Jersey, pp. 369-451.

Wenger, S., Knorr, G., Ott, J., Kossentini, F., 1998, Error resilience support in H.263+, *IEEE Trans. Cir. Sys. Video Tech.* **8**(7):867-877.

Woodward, B., Istepanian, R.S.H., Richards, C.I., 2001, Design of a telemedicine system using a mobile telephone, *IEEE Trans. Inform. Technol. Biomed.* **5**(1):13-15.

THE JPEG2000 IMAGE COMPRESSION STANDARD IN MOBILE HEALTH

Athanassios N. Skodras[*]

1. INTRODUCTION

Provision of expert advice and consultation, in real time, to rural or remote communities or even in mobile emergency (ambulance) cases has always been a demanding need. The vast expansion of mobile and wireless networks can greatly contribute towards the fulfilment of this need. However, medical image transmission remains the main drawback in such applications. Medical imagery (X-ray, ultrasound, mammograms, MRI, etc) is of high volume and has to be compressed in a lossless (bit preserving) manner in order to maintain its diagnostic value unaffected. For example, digital X-ray images of size 2k x 2k with 12 bits of pixel accuracy have become a commodity. Lossless compression of such an image will result in a reduction of its file size by 2 to 3 times (Adams and Kossentini, 2000), namely it will need approximately 20Mbits of storage instead of 50Mbits. But even such an image size is prohibited for a wireless or mobile network. Bluetooth performs below 1Mbps, which means that more than half a minute is needed for the transmission of the above example image. Existing mobile radio networks (GSM), which provide 9.6kbps connections, require over half an hour! The upcoming third generation (3G) cellular standard (UMTS, Universal Mobile Telecommunication System) will support transfer rates of up to 2Mbps, i.e. 144kbps in rural areas, 384kbps in hotspots, and 2Mbps in indoor scenarios. This is really a remarkable improvement in transfer rate in comparison to the second generation. However, the constraints related to the transmission over wireless networks are still present and stronger that those over connected networks. Apart from the lower transmission rate (bandwidth), there is also the higher channel bit error rate (BER). Recent studies report that the average packet loss rate in a wireless environment is 3.6% and it occurs in a bursty manner (Chang, 1998).

[*] Athanassios N. Skodras, Computer Science, School of Science and Technology, Hellenic Open University, GR-26222, Patras, Greece. ∈

It can be deduced from the above that in order for mobile e-health to become a reality, a combined effort is required. Not only the wireless network should become faster and more robust, but also the coding of images should be more efficient, flexible and feature rich. There should be a balanced trade-off between bandwidth (i.e. transmission rate) and image quality at the receiver at each step of the transmission process, (i.e. quality of service, QoS). These requirements are highly feasible by means of the new image coding standard, the JPEG2000. This standard provides an improved quality at low bit rates, as compared to other coding standards (Skodras et al., 2001). Moreover, it offers all kinds of scalability, lossy up to lossless capability, region of interest (ROI) efficiency, and error resiliency possibility. All these features enable a wide range of QoS strategies, thus justifying its adoption by the DICOM[‡] standard (Oh, 1999; Page, 2003).

In the following sections the JPEG2000 standard is presented. Special emphasis is put on its main features that are necessary for mobile health, namely scalability, lossy to lossless, region of interest, and error resilience. Finally, a brief reference to the upcoming parts of the standard is given.

2. JPEG2000: AN OVERVIEW OF THE STANDARD

The JPEG2000 standard call for contributions was launched in March 1997. The aim was the development of a new image coding system for different types of still images (bi-level, grey-level, colour, multi-component), with different characteristics (natural images, scientific, medical, remote sensing, text, rendered graphics, etc) allowing different imaging models (client/server, real-time transmission, image library archival, limited buffer and bandwidth resources, etc) preferably within a unified system. This coding system should provide low bit-rate operation with rate-distortion and subjective image quality performance superior to existing standards, without sacrificing performance at other points in the rate-distortion spectrum, and at the same time incorporating many new useful features. And all this should be offered on a royalty and fee free basis. The standardization process, coordinated by the JTC1/SC29/WG1 of ISO/IEC[§] has produced, as of January 2001, the International Standard (IS) for the core coding system (Part 1). The whole standard will be comprised of twelve parts (IT, 2000; Taubman and Marcellin, 2002).

The JPEG2000 standard provides a set of features that are of importance to many high-end andemerging applications by taking advantage of new technologies. It addresses areas where existing standards fail to produce the best quality or performance and provides capabilities to markets that currently do not use compression. The markets and

---

[‡] The American College of Radiology – National Electric Manufacturers Association (ACR-NEMA) Digital Imaging and Communications in Medicine (DICOM) standard was formed in 1993 to meet the needs of manufacturers and users of medical imaging equipment for the interconnection of devices on standard networks, and hence, the exchange of meaningful information. DICOM produced a nine-part document which specifies both an ISO compliant profile and an industry standard protocol stack (TCP/IP). It has been designed to offer an integrated approach for the use of high-speed computer networks and data storage systems in health care.

[§] ISO: International Organization for Standardization, IEC: International Electrotechnical Commission, JTC: Joint (i.e. ISO/IEC and ITU-T) Technical Committee on Information Technology, ITU-T: International Telecommunication Union – Telecommunication Standardization Sector, SC: Sub-committee, WG: Working Group

applications better served by the JPEG2000 standard are Internet, colour facsimile, printing, scanning (consumer and pre-press), digital photography, remote sensing, mobile, medical imagery, digital archives and E-commerce. Each application area imposes some requirements that the standard, up to a certain degree, should fulfil. Some of the most important features that this standard possesses are the following:

- ∈ Excellent performance at low bit-rates.
- ∈ Continuous-tone and bi-level compression.
- ∈ Lossless and lossy compression.
- ∈ Progressive transmission by resolution, quality, spatial locality and component.
- ∈ Region-of-Interest (ROI) Coding.
- ∈ Robustness to bit-errors.
- ∈ Random access to the bit-stream.
- ∈ Compressed-domain processing capabilities.

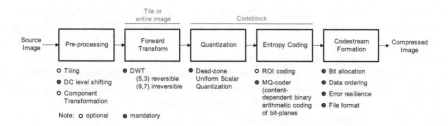

Figure 1. Block diagram of the core JPEG2000 encoder (Part 1).

The JPEG2000 standard (ISO/IEC 15444-1 or ITU-T Recommendation T.800) specifies only the decoder and the code-stream (bit-stream) syntax, thus ensuring compatibility and interoperability among systems developed by different manufacturers. In our presentation, however, the encoder only is discussed, in order to provide a better understanding of the standard. The fundamental building blocks of the JPEG2000 encoder are shown in Fig. 1. The decoder is the reverse of the encoder.

2.1. Image Tiling

Source image components could be decomposed into rectangular tiles. The tile-component is the basic unit of the original or reconstructed image. The term 'tiling' refers to the partition of the source (original) image into rectangular non-overlapping blocks (tiles), which are compressed independently, as though they were entirely distinct images. All operations, including component mixing, wavelet transform, quantization and entropy coding are performed independently on the image tiles. Tiling reduces memory requirements and since they are also reconstructed independently, they can be used for decoding specific parts of the image instead of the whole image. All tiles have exactly the same dimensions, except maybe those at the boundary of the image. Arbitrary tile sizes are allowed, up to and including the entire image (i.e. the whole image is regarded as one tile). Tiling affects image quality by introducing edge artefacts. Special processing is needed for the reduction of the artefacts at tile boundaries (JPEG2000 N1487, 1999; Skodras et al., 2001; Rabbani and Joshi, 2002).

2.2. DC Level Shifting

Prior to computation of the forward discrete wavelet transform (DWT) on each image tile, all samples of the image tile component are DC level shifted by subtracting the same quantity 2^{P-1}, where P is the component's precision. DC level shifting is performed on samples of components that are unsigned only. Level shifting does not affect variances. It actually converts an unsigned representation to a two's complement representation, or vice versa. At the decoder side, inverse DC level shifting is performed on reconstructed samples by adding to them the bias 2^{P-1} after the computation of the inverse component transform (IT, 2000; Taubman and Marcellin, 2002).

2.3. Component Transformations

JPEG2000 supports multiple component images. It can actually support up to 16384 (2^{14}) different components. Component transformations improve compression and allow for visually relevant quantisation. The standard supports two different component transformations, one *irreversible component transformation* (ICT) that can be used for lossy coding and one *reversible component transformation* (RCT) that can be used for lossless or lossy coding. Of course, encoding without any colour transformation is also possible. The RCT is a decorrelating transformation, which is applied to the three first components of an image. Three goals are achieved by this transformation, namely, colour decorrelation for efficient compression, reasonable colour space with respect to the Human Visual System for quantisation, and lossless compression, i.e. exact reconstruction with finite integer precision. For the RGB components, the RCT can be seen as an approximation of a YUV transformation (Skodras et al., 2001; IT, 2000; Taubman and Marcellin, 2002; Rabbani and Joshi, 2002).

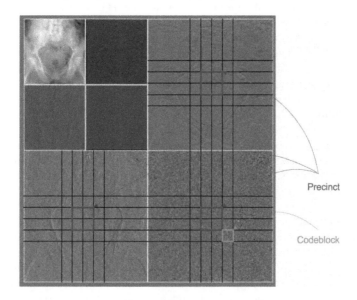

Precinct

Codeblock

Figure 2. Two-level dyadic wavelet decomposition of the test image *X-ray*. A precinct is illustrated in the first level HL (highpass-lowpass), HH, LH subbands. A code-block is also depicted.

2.4. Wavelet Transform

The wavelet transform is used for the analysis of the tile components into different decomposition levels (Vetterli, 2001; Usevitch, 2001). These decomposition levels contain a number of subbands, which consist of coefficients that describe the horizontal, vertical and diagonal spatial frequency characteristics of the original tile component. In Part 1 of the JPEG2000 standard only power of 2 (Mallat) decompositions are allowed in the form of dyadic decomposition as shown in Fig. 2 for the test image *X-ray*.

To perform the forward DWT the standard uses a 1-D subband decomposition of a 1-D set of samples into low-pass samples and high-pass samples. Low-pass samples represent a down-sampled low-resolution version of the original set. High-pass samples represent a down-sampled residual version of the original set, needed for the perfect reconstruction of the original set from the low-pass set. The DWT can be *irreversible* or *reversible*. The default irreversible transform is implemented by means of the Daubechies 9-tap/7-tap filter (Antonini et al., 1992).The default reversible transformation is implemented by means of the Le Gall 5-tap/3-tap filter, the coefficients of which are given in Table 1 (Calderbank et al., 1997).

Table 1. Le Gall 5/3 analysis and synthesis filter coefficients[··]

	Analysis Filter Coefficients		Synthesis Filter Coefficients	
	Lowpass Filter	Highpass Filter	Lowpass Filter	Highpass Filter
i	$h_L(i)$	$h_H(i)$	$g_L(i)$	$g_H(i)$
0	6/8	1	1	6/8
±1	2/8	-1/2	1/2	-2/8
±2	-1/8			-1/8

[··] h_L, h_H and g_L, g_H are the impulse response coefficients of the subband analysis (lowpass and highpass) and synthesis (lowpass and highpass) filters, respectively.

The standard supports two filtering modes: a *convolution-based* and a *lifting-based*. Convolution-based filtering refers to performing a series of dot products between the two filter masks and the extended 1-D signal. Lifting-based filtering refers to a sequence of very simple filtering operations for which alternately odd sample values of the signal are updated with a weighted sum of even sample values, and even sample values are updated with a weighted sum of odd sample values. For the reversible (lossless) case the results are rounded to integer values (Sweldens, 1996; Taubman and Marcellin, 2002; Rabbani and Joshi, 2002). For both modes to be implemented, the signal should first be extended periodically. This periodic symmetric extension is used to ensure that for the filtering operations that take place at both boundaries of the signal, one signal sample exists and spatially corresponds to each coefficient of the filter mask. The number of additional samples required at the boundaries of the signal is filter-length dependent. The symmetric extension of the boundary is of type (1,1), i.e. the first and the last samples appear only once, and can be whole-sample (WS) or half-sample (HS) depending on whether the length of the kernel is odd or even, respectively (IT, 2000; Brislawn, 1996).

2.5. Quantization

After transformation, all coefficients are quantized. Uniform scalar quantisation with dead-zone about the origin is used in Part 1 and Trellis Coded Quantisation (TCQ) in Part 2 of the standard (IT, 2000; Taubman and Marcellin, 2002). Quantisation is the process by which the coefficients are reduced in precision. This operation is lossy, unless the quantisation step is 1 and the coefficients are integers, as produced by the reversible integer 5/3 wavelet.

2.6. Entropy Coding

Entropy coding is achieved by means of an arithmetic coding system that compresses binary symbols relative to an adaptive probability model associated with each of eighteen different coding contexts. The MQ coding algorithm is used to perform this task and to manage the adaptation of the conditional probability models. This algorithm has been selected in part for compatibility reasons with the arithmetic coding engine used by the JBIG2[**] compression standard and every effort has been made to ensure commonality between implementations and surrounding intellectual property issues for JBIG2 and JPEG2000 (Taubman and Marcellin ,2002; Rabbani and Joshi, 2002).

2.7. Code-stream Formation

After quantisation, each subband is divided into rectangular blocks, i.e. non-overlapping rectangles. Three spatially consistent rectangles (one from each subband at each resolution level) comprise a *packet partition location* or *precinct*. Each precinct is further divided into non-overlapping rectangles, called *code-blocks*, which form the input to the entropy coder (Fig. 2). The size of the code-block is typically 64x64, and no less than 32x32. Each dimension of a code-block should be a power of two. When the precinct size in a particular subband is less than the code-clock size, then the code-clock size is set equal to the precinct size (IT, 2000; Taubman and Marcellin ,2002; Rabbani and Joshi, 2002).

Within each subband the code-blocks are visited in raster order. These code blocks are then coded one bit-plane at a time starting with the most significant bit-plane with a non-zero element to the least significant bit-plane. Each code-block is coded entirely independently, without reference to other blocks in the same or other subbands. This independent embedded block coding offers significant benefits, such as spatial random access to the image content, efficient geometric manipulations, error resilience, parallel computations during coding or decoding, etc. The individual bit-planes of the coefficients in a code-block are coded within three coding passes. Each of these coding passes collects contextual information about the bit-plane data.

Each bit-plane of a code-block is scanned in a particular order. Starting from the top left, the first four bits of the first column are scanned. Then the first four bits of the sec ond column, until the width of the code-block is covered. Then the second four bits of the first column are scanned and so on. Each coefficient bit in the bit-plane is coded in only one of the three coding passes, namely the *significance propagation*, the *magnitude*

[**] JBIG (Joint Bi-level Image processing Group) is the standard for the compression of bi-level images.

refinement, and the *cleanup pass*. For each pass, contexts are created which are provided to the arithmetic coder. In addition to the above, a *lazy coding* mode is used to reduce the number of symbols that are arithmetically coded. According to this mode, after the fourth bit-plane is coded, the first and second passes are included as raw (uncompressed data), i.e. the MQ coder is bypassed, while only the third coding pass of each bit-plane employs arithmetic coding. This results in significant speedup for software implementations at high bit-rates. Lazy coding has a negligible effect on compression efficiency for most natural images, but not for compound imagery (Taubman, 2000; Taubman and Marcellin ,2002; Rabbani and Joshi, 2002).

For each code-block, a separate bit-stream is generated. No information from other blocks is utilized during the generation of the bit-stream for a particular block. Rate distortion optimization is used to allocate truncation points to each code block. The bit-stream has the property that it can be truncated to a variety of discrete lengths, and the distortion incurred, when reconstructing from each of these truncated subsets, is estimated and denoted by the mean squared error. During the encoding process, the lengths and the distortions are computed and temporarily stored within the compressed bit-stream itself. The compressed bit-streams from each code-block in a precinct comprise the body of a *packet*. A collection of packets, one from each precinct of each resolution level, comprises the *layer*. A packet could be interpreted as one quality increment for one resolution level at one spatial location, since precincts correspond roughly to spatial locations. Similarly, a layer could be interpreted as one quality increment for the entire full resolution image. Each layer successively and monotonically improves the image quality, so that the decoder is able to decode the code-block contributions contained in each layer in sequence. The final bit-stream is organized as a succession of layers. Each component is coded independently, and the coded data are interleaved on a layer basis. There are four types of progression in the JPEG2000 bit-stream, namely resolution, quality, spatial location and component. Different types of progression are achieved by the appropriate ordering of the packets within the bit-stream (assuming that the image consists of a single tile).

Once the entire image has been compressed, a post-processing operation passes over all the compressed code-blocks. This operation determines the extent to which each code-block's embedded bit-stream should be truncated in order to achieve a particular target bit-rate or distortion. The first, lowest layer (of lowest quality), is formed from the optimally truncated code-block bit-streams in the manner described above. Each subsequent layer is formed by optimally truncating the code-block bit-streams to achieve successively higher target bit-rates (Taubman, 2000; Taubman and Marcellin ,2002).

3. JPEG2000: FEATURES MEETING MOBILE HEALTH

The JPEG2000 standard exhibits a variety of features, the most significant of which, for the mobile e-heath, are the possibility to define Regions-of-Interest (ROI) in an image, the spatial and quality or SNR (signal-to-noise) scalability, the error resilience, and the lossless compression efficiency. All these features are incorporated within a unified algorithm.

3.1. Region-of-Interest

The functionality of ROI is important in applications where certain parts of the image are of higher importance than others. In such a case, these regions need to be encoded at a higher quality than the background. During the transmission of the image, these regions are transmitted first or at a higher priority, as for example in the case of progressive transmission.

The ROI coding scheme in Part 1 of the standard is based on the so-called *maxshift* method (Christopoulos et al., 2000; Grosbois et al., 2001; Askeloef et al., 2002). The *maxshift* method is an extension of the general ROI *scaling-based* coding method. The principle of the general ROI scaling-based method is to scale (shift) coefficients so that the bits associated with the ROI are placed in higher bit-planes than the bits associated with the background. Then, during the embedded coding process, the most significant ROI bit-planes are placed in the bit-stream before any background bit-planes of the image. Thus, the ROI will be decoded, or refined, before the rest of the image. If the bit-stream is truncated, or the encoding process is terminated before the whole image is fully encoded, the ROI will be of higher fidelity than the rest of the image.

According to the *maxshift* method, used in Part 1 of the JPEG2000 standard, the scaling value is computed in such a way that it makes possible to define arbitrary shaped ROI's without the need of transmitting shape information to the decoder. The encoder scans the quantized coefficients and chooses a scaling value a_s such that the minimum coefficient belonging to the ROI is larger than the maximum coefficient of the background (non-ROI area). The decoder receives the bit-stream and starts the decoding process. Every coefficient that is smaller than a_s belongs to the background and is therefore scaled up. The decoder needs only to upscale the received background coefficients.

The advantages of the *maxshift* method, as compared to the scaling-based method, is that encoding of arbitrary shaped ROI's is now possible without the need for shape information at the decoder and without the need for calculating a ROI mask. The encoder is also simpler, since no shape encoding is required. The decoder is almost as simple as a non-ROI capable decoder, while it can still handle ROI's of arbitrary shape.

Experiments have shown that for the lossless coding of images with ROI's, the *maxshift* method increases the bitrate by approximately 1%, in comparison to the lossless coding of the image without ROI (Askeloef et al., 2002). This figure is even smaller compared to the general scaling based method, depending on the scaling value used. This is true for large images (larger than 2k x 2k) and for ROI sizes of about 25% of the image. Such an overhead is indeed small, given the fact that the general scaling based method for arbitrary shaped ROI would require shape information to be transmitted to the decoder, thus increasing the bitrate (in addition to the need of shape encoder/decoder and ROI mask generation at the decoder side). In Fig. 3 the effect of ROI coding is demonstrated. Each column of pictures represents decompressed images at bit-rates of 0.01bpp (bits per pixel), 0.05bpp and 0.125bpp. In the first (second) row a circular (rectangular) ROI was employed. In the third row, no ROI was defined. The test image *X-ray*, which in our experiments is of size 1800x1400 pixels of 8 bits each, is compressed in a reversible way to 2.26bpp using the JPEG2000 standard (Part 1). From this figure the usefulness of the ROI becomes apparent, since at bit-rates as low as 0.01bpp, the diagnostic quality of the image is far better for the ROI cases than that of the non-ROI case. A bit-rate of 0.01bpp corresponds to approximately 3k bytes of information.

ROI coding is a process performed at the encoder. The encoder decides the ROI that is to be coded in better quality than the background. If the ROI's though, are not known to the encoder in advance, there is still possibility for the decoder to receive only the data needed. (New methods for interactive ROI selection have started appearing in the open literature, as for example the one by Rosenbaum and Schumann (2002)). The simplest way of doing so is tiling, but this implies that the image is encoded in tiling mode. Another way is to extract packet partitions from the bit-stream. This can be done easily, since the length information is stored in the header. Due to the filter impulse response lengths, extra care has to be taken to extract all data required to decode the ROI. Fine grain access can be achieved by parsing individual code blocks. As in the case of packet partition precincts, it is necessary to determine which code-blocks affect which pixel locations (since a single pixel can affect four different code blocks within each subband at each resolution of each component). The correct packet affecting these code blocks is determined from the progression order information. The location of the compressed data for the code-blocks can be determined from the packet headers (Taubman and Marcellin ,2002; Rabbani and Joshi, 2002).

3.2. Scalability

The JPEG2000 compression system supports scalability. Scalable image coding involves generating a coded representation (code-stream / bit-stream) in a manner which facilitates the derivation of images of more than one quality and/or resolution by scalable decoding. Bit-stream scalability is the property of a bit-stream that allows decoding of appropriate subsets of the bit-stream to generate complete pictures of quality and/or resolution commensurate with the proportion of the bit-stream decoded. Decoders of different complexities (from low to high) can coexist for a scalable bit-stream. While low performance decoders may decode only small portions of the bit-stream producing basic quality, high performance decoders may decode much more and produce significantly higher quality. The most important types of scalability are *quality or SNR scalability* and *spatial or resolution scalability*. A key advantage of scalable compression is that the target bit-rate or reconstruction resolution need not be known at the time of compression. A related advantage of practical significance is that the image need not be compressed multiple times in order to achieve a target bit-rate, as is common with the JPEG compression standard. An additional advantage of scalability is its ability to provide resilience to transmission errors, as the most important data of the lower layer can be sent over the channel with high error performance, while the less critical enhancement layer data can be sent over the channel with inferior error performance. Both types of scalability are very important for mobile, wireless, Internet and database access applications. The quality and spatial scalability types include the progressive and hierarchical coding modes defined in the JPEG (Pennebaker and Mitcell, 1993), but they are more general.

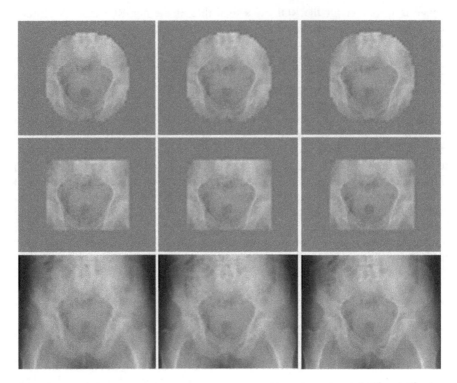

Figure 3. Coding of the *X-ray* image at a rate of 0.01bpp, 0.05bpp and 0.125bpp (1st, 2nd, and 3rd columns of images, respectively) using a circular ROI (1st row), a rectangular ROI (2nd row) or employing no ROI at all (3rd row). (VM8.6 software was used in the experiments).

3.2.1. Quality Scalability

Quality scalability involves generating at least two image layers of the same spatial resolution, but different qualities, out of a single image source. The lower layer is coded by itself to provide the basic image quality and the enhancement layers are coded to enhance the lower layer. An enhancement layer, when added back to the lower layer, regenerates a higher quality reproduction of the input image. Quality scalability is illustrated in the third row of images in Fig. 3. The test image, which was first reversibly compressed, is gradually decompressed at 0.01bpp, 0.05bpp, and 0.125bpp. It is seen that the quality improves rapidly as more information becomes available, i.e. as we move from 0.01bpp or 3149 bytes to 0.05bpp or 15750 bytes. In fact, this particular image is almost indistinguishable from the original one after the bit-rate of 0.05bpp.

3.2.2. Spatial scalability

Spatial scalability involves generating at least two spatial resolution layers out of a single source such that the lower layer is coded by itself to provide the basic spatial resolution and the enhancement layer employs the spatially interpolated lower layer and carries the full spatial resolution of the input image source. Spatial scalability is useful for fast database access as well as for delivering different resolutions to devices (terminals) of different capabilities in terms of display and bandwidth (Fig. 4).

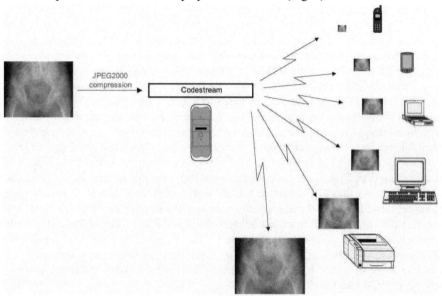

Figure 4. Spatial scalability example: different resolutions can be extracted out of the same code-stream in accordance with the user requirements.

JPEG2000 also supports a combination of spatial and quality scalability. It is possible therefore to progress by spatial scalability at a given (resolution) level and then change the progression by quality at a higher level. This order in progression allows a thumbnail to be displayed first, then a screen resolution image and then an image suitable for the resolution of the printer. It is evident that quality scalability at each resolution allows the best possible image to be displayed at each resolution.

Notice that the bit-stream contains markers that identify the progression type of the bit-stream. The data stored in packets are identical regardless of the type of scalability used. Therefore it is trivial to change the progression type or to extract any required data from the bit-stream. In order to change the progression from quality to progression by resolution, a parser can read the markers, change the type of progression in the markers and then write the new markers in the new order. In this manner, fast transcoding of the bit-stream can be achieved by a server or gateway, without requiring the use of image decoding and re-encoding, not even the employment of the MQ-coder. The corresponding complexity is almost equal to that of a copy operation (Colyer and Clark, 2003).

In a similar fashion, applications that require the use of a grey scale version of a colour compressed image, as for example printing a colour image to a grey-scale printer, do not need to receive all colour components (component scalability). A parser can read the markers from the colour components and write the markers for one of the components (discarding the packets that contain the colour information).

3.3. Error Resilience

Error resilience is one of the most desirable properties in mobile, wireless, and Internet applications (JPEG2000 N2696, 2002, Martina et al., 2002). Entropy coding is known to be prone to channel or transmission errors. A bit error results in loss of synchronization at the entropy decoder and the reconstructed image can be severely degraded. To improve the performance of transmitting compressed images over error prone channels, error resilient bit-stream syntax and tools are included in the standard. Actually, Part 11 of the standard is dedicated to wireless applications. The error resilience tools deal with channel errors using the following approaches: data partitioning and resynchronization, error detection and concealment, and QoS transmission based on priority. Error resilience is achieved at the entropy coding level and at the packet level.

Entropy coding of the quantized coefficients is performed within code-blocks. Since encoding and decoding of the code-blocks are independent processes, bit errors in the bit-stream of a code-block will be restricted within that code-block. To increase error resilience, termination of the arithmetic coder is allowed after every coding pass and the contexts may be reset after each coding pass. This allows the arithmetic decoder to continue the decoding process even if an error has occurred. At the packet level, a packet with a resynchronization marker allows spatial partitioning and resynchronization. This is placed in front of every packet in a tile with a sequence number starting at zero and incremented with each packet.

The lazy coding mode is also useful for error resilience. This relates to the optional arithmetic coding bypass, in which bits are used as raw bits into the bit-stream without arithmetic coding. This prevents the error propagation types to which variable length coding is susceptible.

3.4. Lossless Compression

The lossless compression efficiency of the reversible JPEG2000 (JP2), JPEG-LS, lossless JPEG (L-JPEG) and PNG (Portable Network Graphics) for a variety of the JPEG2000 test images, is reported in Table 2 (Skodras et al., 2001; Santa-Cruz et al., 2002). It is seen that JPEG2000 performs equivalently to JPEG-LS in the case of the natural images, with the added benefit of scalability. JPEG-LS, however, is advantageous in the case of the compound image. Taking into account that JPEG-LS is significantly less complex than JPEG2000, it is reasonable to use JPEG-LS for lossless compression. In such a case though, the generality of JPEG2000 and the rich set of the offered features are sacrificed.

4. JPEG2000: EXPANDING ITS POSSIBILITIES

Part 1 of the JPEG2000 standard was described above. However, by the year 2004 that the full JPEG2000 standard will be completed, twelve Parts will have been released (Table 3). *Part 1* describes the core coding system, consisting of a limited number of possible coding algorithms, in order to provide maximum interchange. *Part 2* consists of optional technologies not required for all implementations. Evidently, images encoded

Table 2. Lossless compression ratios (in bpp)

	JP2	JPEG-LS	L-JPEG	PNG
aerial2	1.47	1.51	1.43	1.48
bike	1.77	1.84	1.61	1.66
cafe	1.49	1.57	1.36	1.44
chart	2.60	2.82	2.00	2.41
cmpnd1	3.77	6.44	3.23	6.02
target	3.76	3.66	2.59	8.70
ultra sound	2.63	3.04	2.41	2.94
average	**2.50**	**2.98**	**2.09**	**3.52**

with Part 2 technology will not be able to be decoded with Part 1 decoders. As an example, Part 2 includes Trellis Coded Quantization, user defined wavelets, wavelet packets and other decompositions, general scaling-based ROI coding, mixed fixed-length coding and variable-length coding, and advanced error resilience schemes. It is actually a toolbox with technologies useful for various specialized applications. *Part 3* defines motion JPEG2000 (MJ2 or MJP2) and is based on Part 1 of JPEG2000. MJ2 will be used in many different areas, as for example in applications where it is desired to have a single codec for both still pictures and motion sequences (which is a common feature of digital still cameras), or in applications where very high quality motion pictures are required (e.g. medical imaging and motion picture production), or in video applications in error prone environments (e.g. wireless and the Internet). JPEG2000 compressed image sequences, synchronized audio and metadata can be stored in the MJ2 file format. Motion JPEG2000 is also targeting interoperability with the JPEG2000 file format (JP2) and the MPEG-4 file format (MP4). A new ad hoc group was created in Oct. 2002 (the Motion JPEG2000 for Medical Imaging) for the promotion of the JPEG2000 standards in the medical community (JPEG2000 Medical, 2004). *Part 4* of the standard defines the conformance testing. *Part 5* defines the reference software (high quality free software). Two reference software implementations do exist, namely, the `JJ2000` software in Java and the `JasPer` software in C (Skodras et al., 2001). The `Kakadu` software is also available (Taubman and Marcellin ,2002). *Part 6* defines the compound image file format. *Part 7*, originally intended to produce a technical report with guidelines of minimum support function of Part 1, has been withdrawn. *Part 8* will address security issues such as authentication, data integrity, protection of copyright and intellectual property, privacy, and conditional access. *Part 9* will define interactivity tools, APIs and protocols for the transmission of JPEG2000 images over a client/server environment. *Part 10* will extent the JPEG2000 algorithm to 3D image compression (something extremely important for medical applications) and floating point data. *Part 11* will define a file format for image compression and transmission in wireless environments such as mobile, wirelessLAN and Radio. Finally, *Part 12* will define the common file format for MPEG-4 and Motion JPEG2000. The work plan of these parts is shown in Table 3 (Colyer and Clark, 2003; JPEG Home, 2004).

Table 3. JPEG2000 standardisation schedule

Part	Description	CFP[□]	Status[#]
1	JPEG 2000 Image Coding System: Core Coding System	Mar-97	IS
2	JPEG 2000 Image Coding System: Extensions	Mar-97	IS
3	Motion JPEG 2000	Dec-99	IS
4	Conformance Testing	Dec-99	IS
5	Reference Software	Dec-99	IS
6	Compound Image File Format	Mar-97	IS[+]
7	Technical Report: Guideline of minimum support function of Part 1	Withdrawn	
8	JPSEC: Secure JPEG 2000	Mar-02	WD
9	JPIP: Interactivity tools, APIs and protocols	Mar-02	FCD
10	JP3D:3-D and floating point data	Mar-02	WD
11	JPWL: Wireless	Jul-02	WD
12	ISO Base Media File Format	Oct-02	FDIS

[□] CFP: Call for Proposals   [#] IS: International Standard   WD: Working Draft   FCD: Final Committee Draft
FDIS: Final Draft International Standard   IS[+]: IS awaiting publication

5. CONCLUSIONS

JPEG2000, the new standard for still image coding, provides a new framework and an integrated toolbox to better address increasing needs for compression. It offers a wide range of functionalities that are of great benefit for mobile health. Lossless and lossy coding, embedded lossy to lossless, progression by resolution and quality, high compression efficiency, error resilience and lossless color transformations are some of its characteristics. Comparative results have shown that JPEG2000 is indeed superior to established image compression standards. Overall, the JPEG2000 standard offers the richest set of features in a very efficient way and within a unified algorithm. The price of its additional complexity in comparison to JPEG and JPEG-LS should not be perceived as a disadvantage, as the technology evolves rapidly.

6. ACKNOWLEDGMENTS

The author would like to thank C. Christopoulos, Editor of the Verification Model of the JPEG2000, for providing an updated list of the JPEG2000 Parts, and T. Ebrahimi, Editor of the JPEG2000 Requirements, for useful discussions on the medical image requirements. The help of E.B. Christopoulou and V. Fotopoulos in the manipulation of some of the medical images is also acknowledged.

7. REFERENCES

Adams, M.D., and Kossentini, F., 2000, Reversible integer-to-integer wavelet transforms for image compression: performance evaluation and analysis, *IEEE Trans. Image Proc.*, 9(6), pp. 1010-1024.

Antonini, M., Barlaud, M., Mathieu P., and Daubechies, I., 1992, Image coding using the wavelet transform, *IEEE Trans. Image Proc*, pp. 205-220.

Askeloef, J., Larsson Carlander, M., and Christopoulos, C., 2002, Region of interest in JPEG2000, *Signal Processing Image Communication*, 17(1).

Brislawn, C.M., 1996, Classification of nonexpansive symmetric extension transforms for multirate filter banks, *Appl.Comput. Harmonic Analysis*, 3, pp. 337-357.

Calderbank, A.R., Daubechies, I., Sweldens, W., and Yeo, B.-L., 1997, Lossless image compression using integer to integer wavelet transforms, *Proc. IEEE Int. Conf. Image Processing*, Santa Barbara, CA, Vol. 1, pp. 596-599.

Chang, E., 1998, An image coding and reconstruction scheme for mobile computing, *Proc 5th IDMS* (Springer-Verlag LNCS 1483), Oslo, Norway, pp. 137-148.

Christopoulos, C.A., Askelof, J., and Larsson, M., 2000, Efficient methods for encoding regions of interest in the upcoming JPEG2000 still image coding standard, *IEEE Signal Processing Letters*, 7(9), pp. 247-249.

Colyer, G., and Clark, R., 2003, *Guide to the Practical Implementation of JPEG2000*, BSI Published Document PD 6777:2003.

Grosbois, R., Santa-Cruz, D., and Ebrahimi, T., 2001, New approach to JPEG2000 compliant region of interest coding, *Proc. of the SPIE 46th Annual Meeting*, Applications of Digital Image Processing XXIV, San Diego, USA.

IT, 2000, JPEG2000 Image Coding System, ISO/IEC International Standard 15444-1, ITU Recommendation T.800.

JPEG Home Page, 2004; http://www.jpeg.org

JPEG2000 Medical Imaging Applications, 2004; http://groups.yahoo.com/group/j2k-medical/

JPEG2000 N1487, 1999, ISO/IEC JTC1/SC29/WG1 N1487, Post-processing approach to tile boundary removal.

JPEG2000 N2696, 2002, ISO/IEC JTC1/SC29/WG1 N2696, JPWL Scope and Requirements.

Martina, M., Masera, G., Piccinini, G., Vacca, F., and Zamboni, M., 2002, System architecture for error-resilient embedded JPEG2000 wireless delivery, *Proc. 2002 14th Int. Conf. Digital Signal Processing* (DSP 2002), Santorini, Greece, 2-4 July 2002.

Oh, J., 1999, A review on medical image compression and storage options, Educational Technology Technical Report Series, The University of Birmingham, ETTR1, ISSN 1463-9424.

Page, D., 2003, JPEG2000 stirs new interest in radiology, in *PACSweb*, http://www.diagnosticimaging.com/pacsweb/ stories/news05290201.shtml

Pennebaker, W.B., and Mitcell, J. L., 1993, *JPEG: Still Image Data Compression Standard*, Van Nostrand Reinhold.

Rabbani, M., and Joshi, R., 2002, An overview of the JPEG2000 still image compression standard, *Signal Processing Image Communication*, 17(1)1.

Rosebaum, R., and Schumann, H., 2002, Flexible, dynamic and compliant region of interest coding in JPEG2000, *Proc.9th IEEE Int. Conf. Image Processing* (ICIP-2002), Rochester, New York, USA.

Santa-Cruz, D., Grosbois, R., and Ebrahimi, T., 2002, JPEG2000 performance evaluation and assessment, *Signal Processing Image Communication*, 17(1).

Skodras, A.N., Christopoulos, C.A., and Ebrahimi, T., 2001, The JPEG2000 still image compression standard, *IEEE Signal Proc. Magazine*, 18(5), pp. 36-58.

Sweldens, W., 1996, The lifting scheme, a custom-design construction of biorthogonal wavelets, *Appl. Comput. Harmonic Analysis*, 3(2), pp. 186-200.

Taubman, D.S., 2000, High performance scalable image compression with EBCOT, *IEEE Trans. Image Proc.*, 9(7), pp. 1158-1170.

Taubman, D.S., and Marcellin, M.W., 2002, *JPEG2000: Image Compression Fundamentals, Standards and Practice*, Kluwer Academic Publishers.

Usevitch, B.E., 2001, A tutorial on modern lossy wavelet image compression: foundations of JPEG2000, *IEEE Signal Proc. Magazine*, 18(5), pp. 22-35.

Vetterli, M., 2001, Wavelets, approximation and compression, *IEEE Signal Proc. Magazine*, 18(5), pp. 59-73.

COMPRESSION OF VOLUMETRIC DATA IN MOBILE HEALTH SYSTEMS

Adrian Munteanu[*], Peter Schelkens, and Jan Cornelis

1. INTRODUCTION

An ever-increasing fraction of the visual content contributing to medical diagnosis has a volumetric character. A plethora of volumetric imaging devices is available, such as Magnetic Resonance Imaging (MRI), Computer-assisted Tomography (CT), Ultrasound (US), Single Photon Emission Computed Tomography (SPECT) and Positron Emission Tomography (PET). Several diagnostic and emergency scenarios exist that rely on the exploitation in mobile stations of on-line available volumetric data or on data generated by mobile imaging systems. Consequently, there is an increasing demand for compact data representations used for storage and transmission purposes, which simultaneously support the functionalities needed to ensure fast interactivity over wireless channels.

The main problem of handling volumetric datasets in a mobile environment is their size. The storage problem in fixed workstations can be considered solved, but efficiently sending this content over wireless channels (even broadband) still poses theoretical and technological challenges. Hence, mobile health systems need to be equipped with the adequate mechanisms for efficient data handling. A natural component of these mechanisms is a compression tool that exploits the redundancy of the volumetric data along all axes, which is not the case in classical slice-based (two-dimensional) coding technologies. Additionally, fast interactivity and random access need to be supported as well: the medical staff has to be able to access parts-of-interest within the dataset without having to upload it entirely. Moreover, progressive data transmission, implementing functionalities like quality and resolution scalability is mandatory in this context. Especially in mobile health systems, the data representation should support flexible adaptation to a wide range of channel capacities and screen resolutions. Finally, lossy-to-lossless compression is required to ensure that medically important image details are not overlooked, and to secure the legal aspects of the medical diagnosis or intervention.

---

[*]Adrian Munteanu, Vrije Universiteit Brussel, Electronics and Information Processing Department, Pleinlaan 2, B-1050 Brussels, Belgium.

Embedded wavelet-based coding algorithms are natural candidates to meet these functionality requirements. In the context of mobile health applications, the compression performance of the employed coding algorithm and its compliance to the aforementioned functionality constraints are of particular importance.

This chapter makes an analysis of the underlying models used in embedded wavelet-based coding, and gives an insight of the achievable performance obtained on volumetric medical datasets with several state-of-the-art 3D coding techniques. The analysis focuses on two main classes of algorithms: the inter-band and intra-band codecs. The chapter summarizes recent theoretical findings, indicating that (1) intra-band models should be favored over inter-band models, and (2) wavelet-based coders exploiting the intra-band dependencies improve the lossless coding performance in comparison to the algorithms that make use of the inter-band dependencies. An additional advantage of intra-band coding is that it inherently provides support for quality and resolution scalability and Region-Of-Interest (ROI) coding.

The coding performance of several state-of-the-art 2D/3D codecs is evaluated on five volumetric data sets. The test set of algorithms includes the embedded block coding by optimized truncation (EBCOT) codec (Taubman 2000) used in the JPEG2000 standard (ISO/IEC 2000, Christopoulos 2000, Skodras 2001, Taubman 2002), JPEG2000 equipped with a 3D wavelet transform (JPEG2K-3D), 3D set-partitioning into hierarchical trees (3D SPIHT) (Kim 2000), and finally, three other algorithms, including 3D QuadTree-Limited coding (3D QT-L) (Munteanu 2003), Cube-Splitting EBCOT (CS-EBCOT) (Schelkens 2001), and a 3D version of the JPEG standard, all recently summarized in an overview paper (Schelkens 2003). The experimental results show that 3D QT-L and CS-EBCOT (Schelkens 2001) provide the best lossless coding performance. JPEG2000 typically yields the worst results of all. In lossy coding, 3D QT-L provides the best overall lossy coding results for an integer wavelet transform.

The experiments demonstrate that the intra-band volumetric coding algorithms provide excellent compression performance, reduce the storage and bandwidth needs, and improve the rate-distortion performance especially at low and medium bit-rates. The latter benefit is important in environments with limited channel capacities, since it enables early noteworthy visualization in a progressive signaling set-up, allowing the user for on-the-fly definition of several ROI's that have to be prioritized for decoding.

2. INTER-BAND AND INTRA-BAND WAVELET CODING

The embedded wavelet-based coding techniques can be classified into three main classes: inter-band, intra-band, and composite wavelet codecs. In these classes, the underlying models that drive the coding process capture the inter-band, intra-band, and composite dependencies respectively between the wavelet coefficients.

Embedded zerotree coding of the wavelet coefficients (EZW) (Shapiro 1993), SPIHT (Said 1996), and the algorithms falling in the same category (Lewis 1992, Zandi 1995) are zerotree-based coders, and their efficiency comes from exploiting the inter-band dependencies between the wavelet coefficients. The underlying model used in these algorithms is based on the zerotree hypothesis (Shapiro 1993), which assumes that if a wavelet coefficient w at a certain scale is not significant with respect to a given threshold T, i.e. $|w| < T$, then all the coefficients of the same orientation in the same spatial location at finer scales are also not significant with respect to the threshold T.

These algorithms apply successive approximation quantization (SAQ) to provide a multi-precision representation of the wavelet coefficients and to facilitate the embedded coding. The significance of the coefficients in the wavelet image W with respect to a monotonically decreasing set of thresholds T_p of the form $T_p = 2^p$, $p_{\max} \geq p \geq 0$ is recorded into a set of the significance (binary) maps W_p. The zero regions in the significance maps W_p are efficiently encoded by grouping the non-significant coefficients in trees spanning exponentially across the scales, and by coding them with "zerotree" symbols (Shapiro 1993). The main disadvantage of this approach is the fact that the zero regions in the significance maps are approximated as a highly constrained set of tree structured regions. As a consequence, certain zero regions that are not aligned with the tree structure may be expensive to code, and some portions of zero regions may not be included in zerotrees at all (Munteanu 1999b).

In contrast to these zerotree-based coders, the recently emerged family of intra-band coding algorithms exploits the intra-band dependencies between the wavelet coefficients. Basically, these techniques employ either a fixed-size, block-based decomposition of the significance maps (e.g. the wavelet lattice partitioning coding (Munteanu 1999a) and the JPEG2000's EBCOT algorithm (Taubman 2000)), either a variable-size block-based decomposition, i.e. a quadtree decomposition (e.g. the square-partitioning (SQP) codec (Munteanu 1999c)), either a combination of the two (e.g. the QT-L codec (Munteanu 2003) or the nested quadtree coding (NQS) algorithm (Chui 1998)). These intra-band coding algorithms abandon the cross-subband tree-structure grouping of the wavelet coefficients of its predecessors, which prevents them from exploiting the inter-band dependencies. However, intra-band dependencies are exploited to a larger extent. The zero regions in the significance maps are represented by a union of independent rectangular zero regions of fixed or variable sizes. This simple model accounts for the clustering property of the wavelet transform: the significant coefficients are clustered in certain areas in the wavelet domain, corresponding to the edges and textural regions in the image. Therefore, by using this model, one can isolate interesting non-zero details in the significance maps by immediately eliminating large insignificant regions from further consideration (Munteanu 1999b, Munteanu 1999c). The coding gain resulting from using this model compensates for the losses incurred by not explicitly exploiting the inter-band dependencies.

2.1. Information-Theoretic Analysis

The information-theoretic analysis of inter-band and intra-band dependencies between the wavelet coefficients (Liu 2001, Galca 2001, Munteanu 2003) indicates that intra-band models should be favoured over the inter-band models in wavelet image coding. These theoretical findings are briefly presented in the following. Define X as a random variable representing the value of an arbitrary wavelet coefficient, $\mathcal{N}X$ as a predefined neighbourhood of X (excluding X) and $\mathcal{P}X$ as the parent of X defined in the sense of Shapiro (Shapiro 1993). Denote by $I(X,Y)$ the mutual information (Cover 1991) between two random variables X,Y, and by Ψ the summarizing function:

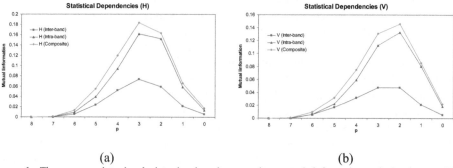

(a)   (b)

Figure 1. The average inter-band, intra-band and composite mutual information calculated over 10 photographic images, for the (a) horizontal and (b) vertical finest details. The results are obtained using the Daubechies 4-tap filters.

$$\Psi = \sum_i X_i^2, \ X_i \ \square \ \mathcal{N}X \ . \tag{1}$$

The inter-band and intra-band dependencies between the wavelet coefficients are quantified using the mutual information. The results in (Liu 2001) and extensive experimental results reported in (Galca 2001) on a variety of images for several types of wavelet filters show that $\hat{I}(X;\mathcal{P}X)$, $\hat{I}(X;\Psi)$, $\hat{I}(X;\Psi;\mathcal{P}X)$, i.e. the estimated inter-band, intra-band and composite mutual information values respectively, satisfy:

$$\hat{I}(X;\mathcal{P}X) < \hat{I}(X;\Psi) < \hat{I}(X;\Psi;\mathcal{P}X) \tag{2}$$

This inequality is constantly satisfied (Liu 2001, Galca 2001) for different non-parametric density estimators including the log-scale histogram method, the adaptive partitioning method (Darbellay 1999), and the wavelet shrinkage method (Donoho 1992). Furthermore, inequality (2) is also satisfied if one defines X as a binary random variable $\tilde{X}$, representing the significance of an arbitrary wavelet coefficient with respect to the applied SAQ thresholds $T_p = 2^p$. In this case, the summarising function returns binary values, and can be defined as (Galca 2001):

$$\tilde{\Psi} = \left\lfloor \frac{1}{2} + \frac{1}{N} \sum_i \tilde{X}_i \right\rfloor, \ \tilde{X}_i \ \square \ \mathcal{N}\tilde{X} \ , \tag{3}$$

where N is the number of binary values in $\mathcal{N}\tilde{X}$, and $\lfloor x \rfloor$ is the integer part of x. The average mutual information values $\hat{I}(\tilde{X};\mathcal{P}\tilde{X}), \hat{I}(\tilde{X};\tilde{\Psi}), \hat{I}(\tilde{X};\tilde{\Psi};\mathcal{P}\tilde{X})$ calculated over a set of ten photographic images for the horizontal and vertical finest details are depicted in Figure 1 for different values of p. This numerical example shows that similar to (2), one has indeed that: $\hat{I}(\tilde{X};\mathcal{P}\tilde{X}) < \hat{I}(\tilde{X};\tilde{\Psi}) < \hat{I}(\tilde{X};\tilde{\Psi};\mathcal{P}\tilde{X})$.

Additionally, extensive experiments (Liu 2001, Galca 2001) show that $\hat{I}(X;\Psi;\mathcal{P}X)$ is always significantly larger than $\hat{I}(X;\mathcal{P}X)$ and slightly larger than $\hat{I}(X;\Psi)$. This observation indicates that the intra-band models capture most of the dependencies between the wavelet coefficients, and that only minor gains can be obtained with composite models that capture both types of dependencies (Liu 2001, Galca 2001). It is important to notice though that quantization further reduces the mutual information, and this might explain why in practice, in comparison to the coding results obtained by exploiting the inter-band models, the coding gains attained by using intra-band models are not as large as such an information-theoretic analysis might indicate (Liu 2001).

2.2. Lattice-Partitioning Coding versus Zerotree Coding

A second argument for favouring intra-band coding over inter-band coding was given in (Munteanu 1999a). This paper introduced the lattice partitioning coding (or the block-based coding) of the wavelet coefficients, and defined the conditions wherein in lossless coding, the block-based coding approach outperforms the zerotree coding.

Within the class of intra-band wavelet codecs, the lattice partitioning coding of the wavelet coefficients plays an important role. The basic concept of the wavelet lattice-partitioning (WLP) coding (Munteanu 1999a) is simple, and it consists of dividing the wavelet image W into a rectangular lattice, each lattice-cell delineating sets of wavelet coefficients that are jointly coded, with each of these sets being independently coded of each other. This concept has been employed in the WLP codec (Munteanu 1999a, Munteanu 2003) and in the EBCOT codec (Taubman 2000) used in the JPEG2000 standard. The extension of the block-based coding concept towards quadtree decomposition and quadtree coding of the significance maps generated a whole family of quadtree codecs, e.g. the SQP codec (Munteanu 1999c), the WQT (Wavelet QuadTree) codec (Munteanu 1999b), the NSQ codec (Chui 1998), the SB-SPECK (Subband-Block Set Partitioned Embedded BloCK) codec (Wheeler 2000), the EZBC (Embedded ZeroBlocks coding of wavelet coefficients and Context modeling) codec (Hsiang 2000), the QT-L codec (Munteanu 2003), etc.

The viability of the lattice-partitioning coding approach resides in its benefits: this approach provides a certain degree of spatial random access to the coded bit-stream, allows for resolution scalability, reduces the memory requirements in both hardware and software implementations (Taubman 2002), and provides coding results competitive to the state-of-the-art, even if the employed entropy coding technique is simple, such as the one employed in the WLP codec (Munteanu 1999a). Furthermore, the lattice-partitioning coding concept coupled with sophisticated context-based entropy coding yields state-of-the-art 2D lossy and lossless coding performance, as it is the case of, for instance, the EBCOT codec used in the JPEG2000 standard (Taubman 2000).

The competitive lossless coding performance of these algorithms stems from the joint coding of the wavelet coefficients using a block-based coding concept. With this respect, recent theoretical findings (Munteanu 1999a, Munteanu 2003) explain the conditions wherein in lossless coding, the block-based coding approach outperforms the zerotree coding. Specifically, the bit rate needed to entropy code the zero regions in the significance maps with a block-based method is lower than the rate needed to entropy code the zerotree symbols, for block sizes confined to some theoretical bounds (Munteanu 1999a, Munteanu 2003). These theoretical findings are summarized in the following.

The formulations are generalised for n-dimensional images (Munteanu 2003). The number of wavelet decomposition levels is denoted by J; an arbitrary wavelet subband of any level k is denoted as W_k, $1 \le k \le J$, and the lowest-frequency subband is denoted by A_J. A lattice-cell of index i located in the subband $W_k, 1 \le k \le J$ is denoted by w_k^i, and the lattice-cell size (in each dimension) is denoted by v.

Successive approximation quantization (SAQ) is used to quantize the wavelet coefficients. The starting SAQ threshold is denoted by $T_{p_{max}} = 2^{p_{max}}$, with $p_{max} = \lfloor \log_2 |w_M| \rfloor$, and with $|w_M|$ denoting the highest absolute magnitude of the wavelet coefficients in the wavelet representation W.

For any lattice cell w_k^i located in the subbands A_J, $W_k, 1 \le k \le J$ one associates the *logarithmic threshold* T_k^i defined as (Munteanu 1999a):

$$T_k^i = \lfloor \log_2 w_{max} \rfloor \tag{4}$$

where w_{max} is the maximum absolute value of the wavelet coefficients delimited by w_k^i. The values T_k^i indicate that (1) all the coefficients in w_k^i are not significant with respect to the whole set of SAQ thresholds 2^T, $T_k^i < T \le p_{max}$, and (2) there exists at least one coefficient within w_k^i that becomes significant with respect to $2^{T_k^i}$.

The *logarithmic thresholds representation* (Munteanu 1999a) given by the T_k^i values encodes jointly the significance of the wavelet coefficients delimited by the lattice-cells w_k^i. Thus, this representation encodes the zero-regions in the significance maps and represents an alternative to the classical zerotree coding. Notice finally that the significance of the coefficients from w_k^i with respect to the lower SAQ thresholds 2^T, $0 \le T \le T_k^i$ is entropy coded in a bit-plane-by-bit-plane fashion (Munteanu 1999a).

Assuming, as in the case of the zerotree-based coders that the wavelet coefficients satisfy the zerotree hypothesis (Shapiro 1993, Said 1996), one can determine sufficient conditions on the lattice-cell size so that coding of the zero-regions in the significance maps using the logarithmic thresholds representation is more efficient than the classical zerotree coding, when both representations are used in lossless coding mode (Munteanu 1999a).

Consider a "parent cell" w_J^1 located in the subband W_J, $W_J \ne A_J$. By using the logarithmic thresholds representation, it can be shown that the total number of bits needed to code the zero-regions covered by the v^n trees of wavelet coefficients having a root in w_J^1 is (Munteanu 1999a, Munteanu 2003):

$$N = \beta \sum_{k=1}^{J} 2^{n(J-k)} = \beta \frac{2^{nJ} - 1}{2^n - 1} \le \lfloor \log_2 (p_{max} + 1) \rfloor \frac{2^{nJ} - 1}{2^n - 1}, \tag{5}$$

where β denotes the average number of bits needed to entropy code the logarithmic thresholds. Equivalently, by using the zerotree representation, the total number of bits needed to code the zero-regions covered by the v^n trees is (Munteanu 2003):

$$\rho N_{ZTR} = \rho v^n \cdot \left[\left(p_{max} - T_J^1 \right) + \sum_{k=1}^{J-1} \sum_{c(i)=1}^{2^{n(J-k)}} \left(T_{k+1}^i - T_k^{c(i)} \right) \right]^{notation} = \rho v^n \cdot A, \tag{6}$$

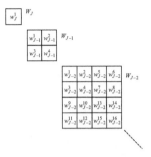

Figure 2. Indexing of the lattice cells belonging to subbands of different decomposition levels (2D case).

where N_{ZTR} is the number of zerotrees in the considered set of trees, ρ denotes the average number of bits needed to entropy code the zerotree symbols, and i, $c(i)$ are the indices of the lattice cells w_{k+1}^{i} and $w_{k}^{c(i)}$, $1 \le k < J$ delimiting the parents and the corresponding descendants in the subbands W_{k+1} and W_{k} respectively.

The joint coding of the wavelet coefficients via the logarithmic thresholds representation is more efficient than the zerotree coding, if N (given by (5)) is smaller than the number of bits needed to code the zerotree symbols. Thus, the logarithmic thresholds representation yields a better coding efficiency than zerotree coding if the following necessary and sufficient condition is satisfied (Munteanu 2003):

$$\rho \cdot N_{ZTR} > N \Leftrightarrow v > n\sqrt{\frac{\beta\left(2^{nJ}-1\right)}{\rho\left(2^{n}-1\right)A}}. \tag{7}$$

Expression (6) is further evaluated by considering an indexing of the entire set of lattice cells delimiting the v^{n} trees of coefficients with a root in w_{J}^{1}. An example of such indexing is given for the two-dimensional case in Figure 2. Equation (6) is equivalent to:

$$\rho N_{ZTR} = \rho v^{n} \cdot \left[P_{max} + \left(2^{n}-1\right)\left(T_{J}^{1} + \sum_{l=1}^{2^{n}} T_{J-1}^{l} + \sum_{l=1}^{2^{2n}} T_{J-2}^{l} + ... + \sum_{l=1}^{2^{n(J-2)}} T_{2}^{l} \right) - \sum_{l=1}^{2^{n(J-1)}} T_{1}^{l} \right]. \tag{8}$$

According to the zerotree hypothesis, it can be shown (Munteanu 2003) that the logarithmic threshold differences that occur in (8) and have the form:

$$I_{k}^{s_{J},s_{J-1},...,s_{3},s_{2}} = \left(T_{k}^{2^{n(J-k)}+s_{J}2^{n(J-k-1)}+...+s_{k+2}2^{n}+s_{k+1}} - T_{1}^{2^{n(J-1)}+s_{J}2^{n(J-2)}+...+s_{k}2^{n(k-2)}+...+s_{3}2^{n}+s_{2}} \right),$$

satisfy:

$$P_{max} \ge I_{J}^{s_{J},s_{J-1},...,s_{2}} \ge I_{J-1}^{s_{J},s_{J-1},...,s_{2}} \ge .... \ge I_{k}^{s_{J},s_{J-1},...,s_{2}} \ge ... \ge I_{2}^{s_{J},s_{J-1},...,s_{2}} \ge 0, \tag{9}$$

for any $s_{J}, s_{J-1}, ..., s_{3}, s_{2}$ in the interval $\left[-2^{n}+1, 0\right]$. Denote by:

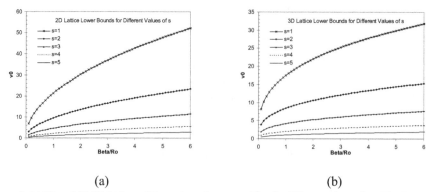

(a)   (b)

Figure 3. (a) 2D and (b) 3D lattice-cell lower bounds versus β/ρ for different values of s. The factor β/ρ varies in the range 0.1–6, and s takes all the values between 1 and $J-1$. In this figure one has chosen $J = 6$.

$$I_k^{min} = \min_{s_J, s_{J-1}, \ldots, s_k} \left(I_k^{s_J, s_{J-1}, \ldots, s_k, 0, \ldots, 0} \right).$$

It can be shown that (Munteanu 2003):

$$p_{max} \geq I_J^{min} \geq I_{J-1}^{min} \geq \ldots \geq I_k^{min} \geq \ldots \geq I_3^{min} \geq I_2^{min} \geq 0. \tag{10}$$

A lower bound for (8) is then given by:

$$\rho N_{ZTR} \geq \rho v^n \left[\left(p_{max} - T_J^1 + I_J^{min} \right) + \left(2^n - 1 \right) \left(I_J^{min} + \ldots + 2^{n(J-k)} I_k^{min} + \ldots + 2^{n(J-2)} I_2^{min} \right) \right]. \tag{11}$$

Following (10), it can be assumed that in the most general case, the s largest values I_k^{min}, $J \geq k \geq J-s+1$ ($1 \leq s \leq J-1$) are different from zero, and that $I_k^{min} = 0$ for any $J - s \geq k \geq 2$. Then, a sufficient (but not necessary) condition so that (7) is satisfied is:

$$v > v_0 = \left[\frac{2^{nJ} - 1}{\left(2^n - 1 \right) \left(2^{ns} - 1 \right)} \right]^{1/n} \cdot \left(\frac{\beta}{\rho} \right)^{1/n}. \tag{12}$$

Figure 3 depicts the lattice-cell lower bound v_0 as a function of the ratio β/ρ, for different values of the model parameter s. The curves correspond to the 2D and 3D cases ($n = 2, 3$). We chose a typical value of $J = 6$ for the number of decomposition levels, while the β/ρ factor varies in the range $[0.1, 6]$. From Figure 3 one may conclude that even if the ratio β/ρ varies in a very broad range, that is, even if little can be said about the relative entropy-coding efficiency of the zerotree symbols and logarithmic thresholds, the lower bound v_0 is reasonably small (i.e. $v_0 < 12$ in 2D and $v_0 < 8$ in 3D) for large values of s ($J - 3 \leq s \leq J - 1$). This brief analysis indicates that in lossless compression, the block-based coding is indeed a viable alternative to zerotree coding.

The joint coding of the wavelet coefficients using the logarithmic thresholds representation has been successfully used in various block-based and quadtree-based wavelet codecs providing lossless coding performance competitive to the state-of-the-art. For instance, the WLP codec (Munteanu 1999a) uses this representation and entropy codes the logarithmic thresholds by using a simple high-order arithmetic entropy coder. In the EBCOT codec (Taubman 2000), the logarithmic thresholds (referred to as the number of magnitude bit-planes used to represent the coefficients in every code-block) are encoded using the tag-tree concept (Taubman 2002). Other algorithms encode the logarithmic thresholds using quadtree coding and adaptive arithmetic entropy coding or context-based entropy coding – see for example (Chui 1998, Munteanu 1999b, Munteanu 1999c, Hsiang 2000, Schelkens 2003).

Finally, the excellent rate-distortion performance provided by EBCOT (Taubman 2000), and by other codecs falling into the same family of 2D intra-band coding algorithms – see e.g. (Munteanu 1999a, Munteanu 1999b, Chui 1998, Hsiang 2000) constitutes an additional argument for favouring intra-band models over inter-band models in embedded wavelet-based image coding.

Triggered by these set of observations, we focus our attention on several representative codecs that exploit the intra-band dependencies between the wavelet coefficients, and evaluate their compression performance on a set of medical volumetric data.

3. WAVELET-BASED CODING OF VOLUMETRIC DATA

The first two codecs included in the test-bench of 3D algorithms are 3D QT-L (Munteanu 2003) and CS-EBCOT (Schelkens 2001), both recently summarized in an overview paper (Schelkens 2003). The QT-L codec (Munteanu 2003) is an enhanced version of the SQP and WQT algorithms (Munteanu 1999b, Munteanu 1999c). Basically, 3D QT-L employs octtree-decompositions of the significance maps and context-based entropy coding. The octtree-partitioning process is limited, such that, once the volume of a node becomes smaller than a predefined threshold, the splitting process is stopped, and the entropy coding of the coefficients within the significant leaf-nodes is activated. For each bit-plane, the coding process consists of the non-significance, significance and refinement passes. The non-significance pass identifies new significant coefficients located in the immediate neighbourhood of coefficients already encoded as significant. The significance pass identifies new significant coefficients within the non-significant leaf nodes. The magnitudes of the coefficients already identified as significant with respect to the higher SAQ thresholds are refined during the refinement pass. Finally, context-based entropy coding of the symbols generated by this algorithm is applied, using four different sets of models; for a detailed description of the algorithm, the reader is referred to (Munteanu 2003, Schelkens 2003).

The CS-EBCOT algorithm (Schelkens 2001) combines the octtree-decomposition of the significance maps with a 3D instantiation of EBCOT (Taubman 2000). The coding module consists of two main units, the Tier 1 and Tier 2 parts. The Tier 1 is a hybrid module combining the octtree-decomposition of the significance maps and a fractional bit-plane coder using context-based arithmetic coding. The wavelet image is partitioned into separate, equally sized cubes, called code-blocks. The octtree-decomposition is applied on the individual code-blocks in order to isolate smaller entities, i.e. sub-cubes, possibly containing significant wavelet coefficients. The fractional bit-plane coder operates only on those leaf nodes that have been identified as significant during the octtree-decomposition. The fractional bit-plane coding is a 3D extension of the 2D version implemented in the EBCOT coder (Taubman 2000); also, the post-compression rate-distortion optimization procedure implemented in the Tier 2 module is identical to the original EBCOT implementation.

A third algorithm introduced in the 3D test-bench is a JPEG-alike, 3D DCT-based codec (Schelkens 2001, Schelkens 2003), which has been considered in order to have a good reference for DCT-based systems. The algorithm is composed of a discrete cosine transform, followed by a uniform scalar quantizer and a combination of run-length coding and adaptive arithmetic encoding. The entropy coding system follows its 2D JPEG counter-part, but Huffman coding is replaced by adaptive arithmetic coding.

Finally, the 3D SPIHT codec (Kim 2000) has been also included in the test set of wavelet codecs, in order to provide a reference of the achievable compression performance obtained with a state-of-the-art inter-band coding algorithm. The 3D SPIHT codec is similar to its 2D version, with the exception that the states of the tree nodes are encoded with a context-based arithmetic coder (Kim 2000).

4. PERFORMANCE EVALUATION

4.1. Lossless Coding

The lossy and lossless compression performance of these 3D algorithms is evaluated on a set of volumetric data obtained with different imaging modalities, including: positron-emission tomography – PET (128x128x39x15bits), computerised tomography – CT1 (512x512x100x12bits), CT2 (512x512x44x12bits), ultrasound – US (256x256x256x8bpp), and MRI data – MRI (256x256x200x12bpp).

Lossless coding results are reported for the techniques mentioned above: 3D QT-L, 3D SPIHT, CS-EBCOT and JPEG2000. We have also included the results obtained with 3D SB-SPECK (only for the CT2 and MRI data, based on results reported by its author in (Wheeler 2000)). Additionally, the results obtained with the JPEG2000 codec equipped with a 3D wavelet transform (JPEG2K-3D) are given. The latter corresponds to one of the functionalities provided by the JPEG2000 Verification Model software (from V7.0 on), which was developed to support multi-spectral image coding. For all the tests performed in lossless coding (as well as for the lossy coding experiments later), a 5-level wavelet transform with a lossless 5x3 lifting kernel was applied in all spatial dimensions, except for the low-resolution PET image, where only 4 levels were used. To allow a fair comparison with the 3D SPIHT algorithm, the same number of decomposition levels was used in all dimensions. This is because the SPIHT implementation used in this study employs balanced 3D spatial orientation trees. Hence, a non-balanced decomposition would lead to the destruction of the parent-child dependencies within the spatial-orientation trees of the 3D SPIHT codec. The other algorithms are not limited by this drawback, but to ensure a fair comparison, we applied the same restriction for them too.

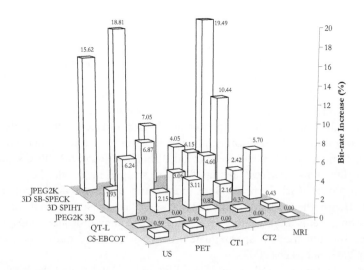

Figure 4. The increase of the lossless bit-rate in terms of percentage, with the reference technique taken as the algorithm yielding the best coding results for each test volume. All techniques use a lossless 5x3-lifting kernel.

Figure 4 shows the increase in terms of percentage of the bit-rate achieved in lossless compression, with the reference technique taken as the algorithm yielding the best coding results for each test volume. We notice that for the US and PET volumes, the 3D QT-L codec provides the best coding performance, while for the other three volumetric data, the CS-EBCOT performs better. If one refers to the average increase in percentage taking the CS-EBCOT codec as the reference, then one can notice that the 3D QT-L codec yields similar performance, since the average difference between the two is only 0.1%.

The 3D SPIHT and the JPEG2K-3D codecs provide similar results, with an average difference of 3.56% and 3.65% respectively. Finally, the average difference increases up to 7.07% and 12.78% for the 3D SB-SPECK and JPEG2K codecs.

One notices also from Figure 4 that the relative performance of the several techniques is heavily dependant on the data set involved. For example, 3D SPIHT provides excellent results for the US, CT1 and MRI sets, while for the other ones the performance is relatively poor. JPEG2000 yields the worst coding results for all, except for the CT2 image, which has a low axial resolution. One notices also that activating the 3D wavelet transform feature of JPEG2000 boosts the lossless coding performance of the JPEG2000 codec (except again for CT2). The results of 3D SB-SPECK were reported only for the CT2 and MRI data sets (Wheeler 2000) and are situated in between JPEG2K and 3D SPIHT for CT2 and in between JPEG2K and JPEG2K-3D for MRI.

In summary, these results lead to the following observations for lossless coding:

- ☐ CS-EBCOT and 3D QT-L yield the best lossless coding results on all images;
- ☐ The 3D wavelet transform as such significantly boosts the coding performance;
- ☐ As spatial resolution and consequently inter-slice dependency diminishes, the benefit of using a 3D transform and implicitly a 3D coding system decreases.

4.2. Lossy Coding

Lossy coding experiments were carried out on the five volumetric datasets for the aforementioned algorithms; in addition, the 3D DCT-based coding engine is included as well. The peak signal-to-noise ratios (PSNRs) are measured at seven different bit-rates: 2, 1, 0.5, 0.25, 0.125, 0.0625 and 0.03125 bpp. This section does not contain all the obtained results, and we refer to (Schelkens 2002) for a complete report.

Similar to the lossless coding experiments, the performance of the wavelet codecs is evaluated using a wavelet transform employing a lossless 5x3 filter kernel. To simulate a unitary transform, the CS-EBCOT, JPEG2000, JPEG2K-3D and SB-SPECK intrinsically compensate the non-unitariness by applying a scaling factor in the rate-distortion optimization process (Schelkens 2003). Unfortunately, this approach is not feasible for 3D SPIHT. Hence, the reported results for this algorithm are sub-optimal, since the high frequency components are overemphasized during encoding. For the 3D QT-L codec, a sub-optimal solution used to approximate (roughly) a unitary transform was adopted (Munteanu 2003, Schelkens 2003).

Figure 5 reports the PSNR values as a function of bit-rate and the lossless coding rates obtained on the five datasets. Since the rates obtained with the JPEG2K, JPEG2K-3D, CS-EBCOT and 3D DCT do not match exactly the target dyadic rates, the best rate-distortion points can be identified following a graphical performance comparison; the best points resulting from such a comparison are indicated in bold letters in the tables.

This set of experiments yields surprising results. If in a first evaluation one excludes the results for CT2 and PET data, we observe that despite its sub-optimal approximation of unitariness, the 3D QT-L codec outperforms all the other wavelet codecs in the whole range of rates. Additionally, the results obtained on the PET volumetric data indicate that at rates below 0.25 bpp, the 3D QT-L codec outperforms all the other codecs, while at higher rates JPEG2K-3D performs better.

The 3D DCT coder provides good rate-distortion performance but only at high rates. For instance, the PSNR figures at 2bpp on the US, CT1, and MRI datasets are higher than those obtained with 3D QT-L; however, this situation changes on the PET data, for which both JPEG2K-3D and 3D QT-L are better. At low bit-rates the performance of 3D DCT tends to decrease fast, a characteristic that is also observed for its 2D counterpart.

If one refers to JPEG2000, one notices that this codec typically produces poor results on the PET, US and MRI volumes; on CT1 it yields good coding results at rates higher than 0.25bpp, but the results are modest at lower rates. However, JPEG2000 is the best codec for CT2 at high bit-rates, and is only beaten by JPEG2K-3D and SB-SPECK at low rates. At that moment JPEG2000 equals the performance of CS-EBCOT and 3D QT-L. Once again this illustrates the effect of the inter-slice correlation (and consequently resolution).

Finally, if one refers to the 3D SPIHT algorithm, one notices that on the US, CT1, and MRI data sets it is constantly performing worse than the 3D QT-L, JPEG2K-3D and CS-EBCOT algorithms at all rates. We expect that these poor results are partially caused by the fact that 3D SPIHT was equipped with a non-unitary wavelet transform. This aspect has been assessed experimentally in (Schelkens 2002) by equipping the CS-EBCOT, 3D SPIHT and JPEG2000 algorithms with a lossy 9x7 lifting-based wavelet transform (L_2 -normalized).

US	JPEG2K		JPEG2K 3D		QT-L		3D SPIHT		CS-EBCOT		3D DCT	
	Bit-rate	PSNR	Bit-rate	PSNR	Bit-rate	PSNR	Bit-rate	PSNR	Bit-rate	PSNR	Bit-rate	PSNR
Lossless	4.2800	Infinite	3.9328	Infinite	3.7019	Infinite	3.7735	Infinite	3.7238	Infinite		
2	1.9996	42.19	1.9994	43.14	2.0000	44.02	2.0000	40.78	1.9989	42.57	2.1663	45.74
1	0.9982	36.35	0.9986	38.25	1.0000	38.75	1.0000	36.80	0.9992	37.32		
0.5	0.4997	32.36	0.4998	34.81	0.5000	34.93	0.5000	32.53	0.4972	33.80	0.8285	37.32
0.25	0.2498	29.68	0.2497	32.17	0.2500	32.29	0.2500	28.26	0.2499	31.44	0.2618	31.34
0.125	0.1250	27.71	0.1250	30.04	0.1250	30.16	0.1250	27.02	0.1249	29.79	0.1261	28.77
0.0625	0.0624	26.10	0.0625	28.43	0.0625	28.76	0.0625	25.85	0.0623	28.16	0.0641	26.93
0.03125					0.0313	27.49			0.0309	26.70	0.0315	25.15

PET	JPEG2K		JPEG2K 3D		QT-L		3D SPIHT		CS-EBCOT		3D DCT	
	Bit-rate	PSNR	Bit-rate	PSNR	Bit-rate	PSNR	Bit-rate	PSNR	Bit-rate	PSNR	Bit-rate	PSNR
Lossless	10.1787	Infinite	8.7516	Infinite	8.56750	Infinite	9.1558	Infinite	8.6099	Infinite		
2	1.9977	52.70	1.9992	60.38	2.00000	60.05	2.0000	57.96	1.9680	56.57	2.0255	60.11
1	0.9988	46.67	0.9993	52.84	1.00000	52.32	1.0000	50.31	0.9905	50.55	1.0134	52.10
0.5	0.5000	43.10	0.4976	47.45	0.50000	47.18	0.5000	45.11	0.4880	45.56	0.5172	46.85
0.25	0.2371	40.69	0.2356	43.66	0.25000	44.02	0.2500	42.95	0.2456	42.53	0.2523	42.91
0.125	0.1248	39.23	0.1239	41.48	0.12500	42.21	0.1250	39.88	0.1075	40.58	0.1284	40.34
0.0625	0.0625	37.96	0.0623	39.74	0.06250	40.59	0.0625	39.09	0.0539	39.42	0.0647	38.28
0.03125					0.03125	39.13			0.0244	38.00		

CT1	JPEG2K		JPEG2K 3D		QT-L		3D SPIHT		CS-EBCOT		3D DCT	
	Bit-rate	PSNR	Bit-rate	PSNR	Bit-rate	PSNR	Bit-rate	PSNR	Bit-rate	PSNR	Bit-rate	PSNR
Lossless	4.0808	Infinite	3.9304	Infinite	3.8432	Infinite	3.9286	Infinite	3.8120	Infinite		
2	1.9998	66.96	1.9998	66.84	2.0000	66.85	2.0000	64.80	1.9990	65.32	2.0722	68.00
1	0.9999	59.73	0.9999	60.30	1.0000	60.57	1.0000	58.49	0.9996	59.34	0.9524	59.04
0.5	0.5000	54.44	0.4995	55.61	0.5000	55.72	0.5000	53.62	0.4997	54.32	0.4975	54.06
0.25	0.2495	49.36	0.2500	51.24	0.2500	51.48	0.2500	48.41	0.2491	49.82	0.2554	49.19
0.125	0.1250	44.37	0.1250	46.68	0.1250	47.21	0.1250	43.12	0.1250	46.70	0.1268	43.74
0.0625	0.0625	39.48	0.0625	42.64	0.0625	43.27	0.0625	38.26	0.0625	42.31	0.0630	38.47
0.03125	0.0312	35.03	0.0312	38.97	0.0313	39.54			0.0311	39.04	0.0327	33.17

CT2	JPEG2K		JPEG2K 3D		QT-L		3D SPIHT		CS-EBCOT		3D DCT		SB-SPECK	
	Bit-rate	PSNR	Bit-rate	PSNR	Bit-rate	PSNR	Bit-rate	PSNR	Bit-rate	PSNR	Bit-rate	PSNR	Bit-rate	PSNR
Lossless	5.1677	Infinite	5.0741	Infinite	4.9852	Infinite	5.1951	Infinite	4.9668	Infinite			5.1730	Infinite
2	1.9996	59.72	1.9998	59.43	2.0000	59.45	2.0000	57.38	1.9960	57.13	1.9449	57.71		
1	0.9999	53.24	0.9991	53.23	1.0000	52.94	1.0000	51.55	1.0000	51.61	1.0259	51.79		
0.5	0.4996	49.30	0.4994	49.65	0.5000	48.68	0.5000	47.32	0.4993	48.17	0.5073	47.69	0.5000	49.55
0.25	0.2500	45.85	0.2500	47.00	0.2500	45.02	0.2500	43.10	0.2490	45.11	0.2546	43.96	0.2500	46.53
0.125	0.1249	41.84	0.1250	44.51	0.1250	41.26	0.1250	38.59	0.1236	41.89	0.1310	39.93	0.1250	43.32
0.0625	0.0625	38.41	0.0624	41.73	0.0625	37.83	0.0625	34.49	0.0624	38.38	0.0641	35.84	0.0625	39.70
0.03125					0.0313	35.84			0.0312	36.13	0.0318	31.82		

MRI	JPEG2K		JPEG2K 3D		QT-L		3D SPIHT		CS-EBCOT		3D DCT		SB-SPECK	
	Bit-rate	PSNR	Bit-rate	PSNR	Bit-rate	PSNR	Bit-rate	PSNR	Bit-rate	PSNR	Bit-rate	PSNR	Bit-rate	PSNR
Lossless	4.7283	Infinite	4.1825	Infinite	3.9742	Infinite	4.0529	Infinite	3.9571	Infinite			4.3700	Infinite
2	1.9999	63.24	1.9985	65.81	2.0000	66.43	2.0000	64.26	1.9996	65.12	1.9487	67.16		
1	0.9999	56.98	0.9997	59.60	1.0000	60.26	1.0000	58.21	0.9985	58.57	0.9530	59.46		
0.5	0.4996	52.29	0.4994	55.13	0.5000	55.74	0.5000	53.23	0.4987	54.46	0.4854	54.49	0.5000	54.07
0.25	0.2499	48.32	0.2498	51.61	0.2500	52.01	0.2500	49.11	0.2491	51.17	0.2526	50.55	0.2500	50.70
0.125	0.1250	45.13	0.1250	48.67	0.1250	49.09	0.1250	47.00	0.1239	48.55	0.1265	47.00	0.1250	47.55
0.0625	0.0624	42.16	0.0625	46.22	0.0625	46.75	0.0625	41.59	0.0625	46.52	0.0654	44.06	0.0625	44.60
0.03125	0.0156	37.13	0.0156	41.73	0.0313	44.63			0.0312	44.24	0.0322	40.92		

Figure 5. The five volumetric datasets compressed with JPEG2000, JPEG2K-3D, 3D DCT, CS-EBCOT, 3D QT-L and 3D SPIHT. All wavelet-based algorithms use a 5-level, 5x3 integer wavelet-transform. Except for 3D SPIHT, the transforms were unitary, or approximately unitary (3D QT-L).

The rate-distortion figures given in (Schelkens 2002) show that, for this transform, 3D SPIHT is superior to CS-EBCOT and JPEG2000 for the CT1 and MRI volumes. The performance of JPEG2000 was poorer, especially for MRI, having a high inter-slice correlation. JPEG2000 provided the best coding performance for CT2, as it was the case for the experiments obtained with the lossless 5x3 lifting kernel. It is noticeable also that the techniques evaluated with the lossy 9x7 lifting-based kernel do not outperform JPEG2000 at low bit-rates (Schelkens 2002).

5. DISCUSSION AND CONCLUSIONS

Mobile health systems need to be equipped with the adequate mechanisms for efficient data handling. A critical component of these mechanisms is a compression tool that exploits the redundancy of the volumetric data along all axes. This chapter makes an analysis of the inter-band and intra-band models used in wavelet-based coding and gives an insight of the achievable coding performance obtained on volumetric medical datasets with several state-of-the-art 3D coding techniques. The theoretical findings summarized in this chapter indicate that (1) intra-band models should be favored over inter-band models, and (2) wavelet-based coders exploiting the intra-band dependencies improve the lossless coding performance in comparison to the algorithms that make use of the inter-band dependencies. Furthermore, the intra-band coding schemes inherently provide support for quality and resolution scalability and region-of-interest coding.

The experimental results obtained on a test-bench of five volumetric data sets show that 3D QT-L and 3D EBCOT yield in all cases the best lossless coding performance. JPEG2000 typically provides the worst results of all. However, the 3D techniques are sensitive to a reduced spatial resolution along the slice axis. A reduction of this resolution works to the advantage of the classical 2D techniques (JPEG2000). In lossy coding, 3D QT-L yields the best overall coding performance for an integer wavelet transform.

One can conclude that volumetric intra-band coding algorithms provide excellent compression performance, reduce the storage and bandwidth needs, and improve the rate-distortion performance especially at low and medium bit-rates. The latter benefit is important in environments with limited channel capacities, since it enables early noteworthy visualization in a progressive signaling set-up, allowing the user for on-the-fly definition of several ROI's that have to be prioritized for decoding.

An important aspect that needs to be further investigated is the reduction of the implementation complexity and computational load of such 3D codecs. The computational complexity of the 3D techniques is significantly higher than that of their 2D counterparts when applied on the same data. Typical bottlenecks are the required memory bandwidths and memory sizes. A major problem is the coincident access of axially related data, causing huge jumps in the memory and consequently cache failures, resulting in long execution times. Fortunately, data transfer and storage optimised versions of their 2D counterparts have been proposed in the past (e.g. (Schelkens 2001, Martucci 1997)) of which the issued principles can be transferred to 3D. For example, the realization of a 3D version of the local wavelet transform (Lafruit 1999) will significantly improve the memory access and consumption behaviour of such 3D codecs. As a final remark, one can intuitively state that, compared to the inter-band codecs, the intra-band coding algorithms (QT-L, CS-EBCOT, etc.) should be favoured again, since their coding process is localised, allowing for the development of memory-efficient implementations.

6. ACKNOWLEDGEMENTS

This work was partly supported by the IAP–Mobile project funded by DWTC. A. Munteanu and P. Schelkens have post-doctoral fellowships with the Fund for Scientific Research – Flanders (FWO), Egmontstraat 5, B-1000 Brussels, Belgium.

7. REFERENCES

Christopoulos, C., Skodras, A., and Ebrahimi, T., 2000, The JPEG2000 still image coding system: an overview, *IEEE Transactions on Consumer Electronics*, **46**:1103-1127.

Chui, C. K. and Yi, R., 1998, System and method for nested split coding of sparse data sets, Patent US:005748116A, Teralogic Inc., Menlo Park, California, pp. 25.

Cover, T. M. and Thomas, J. A., 1991, *Elements of Information Theory*, Wiley, New York, USA.

Darbellay, G. A. and Vajda, I., 1999, Estimation of the information by an adaptive partitioning of the observation space, *IEEE Transactions on Information Theory*, **45**:1315-1321.

Donoho, D. L. and Johnstone, I., 1992, Minimax estimation via wavelet shrinkage, Technical Report, Department of Statistics, Stanford University, USA.

Galca, M., 2001, *Quantifying Dependencies in the Wavelet Domain and New Developments in 2D/3D Intraband Wavelet Image Coding*, MSc Thesis, Department ETRO, Vrije Universiteit Brussel, Brussel.

Hsiang, S.-T. and Woods, J. W., 2000, Embedded image coding using zeroblocks of subband/wavelet coefficients and context modeling, Proceedings of IEEE International Symposium on Circuits and Systems (ISCAS 2000), Geneva, Switzerland, vol. 3, pp. 662-665.

ISO/IEC, 2000, JPEG 2000 Image Coding System, ISO/IEC JTC1/SC29/WG1, International Standard 15444-1.

Kim, Y. and Pearlman, W. A., 2000, Stripe-based SPIHT lossy compression of volumetric medical images for low memory usage and uniform reconstruction quality, Proceedings of IEEE International Conference on Image Processing (ICIP 2000), Vancouver, Canada, vol. III, pp. 652-655.

Lafruit, G., Nachtergaele, L., Bormans, J., Engels, M., and Bolsens, I., 1999, Optimal memory organization for scalable texture codecs in MPEG-4, *IEEE Transactions on Circuits and Systems for Video Technology*, **9**:218-243.

Lewis, A. S. and Knowles, G., 1992, Image compression using the 2-D wavelet transform, *IEEE Transactions on Image Processing*, **1**:244-250.

Liu, J. and Moulin, P., 2001, Information-theoretic analysis of interscale and intrascale dependencies between image wavelet coefficients, *IEEE Transactions on Image Processing*, **10**:1647-1658.

Martucci, S. A., Sodagar, I., Chiang, T., and Zhang, Y., 1997, A zerotree wavelet video coder, *IEEE Transactions on Circuits and Systems for Video Technology*, **7**:109-118.

Munteanu, A., 2003, *Wavelet Image Coding and Multiscale Edge Detection: Algorithms and Applications*, PhD Thesis, Electronics and Information Processing Department, Vrije Universiteit Brussel, Brussels.

Munteanu, A., Cornelis, J., and Cristea, P., 1999a, Wavelet-based lossless compression of coronary angiographic images, *IEEE Transactions on Medical Imaging*, **18**:272-281.

Munteanu, A., Cornelis, J., Van der Auwera, G., and Cristea, P., 1999b, Wavelet image compression - the quadtree coding approach, *IEEE Transactions on Information Technology in Biomedicine*, **3**:176-185.

Munteanu, A., Cornelis, J., Van der Auwera, G., and Cristea, P., 1999c, Wavelet-based lossless compression scheme with progressive transmission capability, *International Journal of Imaging Systems and Technology, special issue on "Image and Video Coding,"* Edited by J. Robinson and R. D. Dony, **10**:76-85.

Said, A. and Pearlman, W., 1996, A new fast and efficient image codec based on set partitioning in hierarchical trees, *IEEE Transactions on Circuits and Systems for Video Technology*, **6**:243-250.

Schelkens, P., 2001, *Multidimensional Wavelet Coding - Algorithms and Implementations*, PhD Thesis, Department of Electronics and Information Processing, Vrije Universiteit Brussel, Brussel.

Schelkens, P. and Munteanu, A., 2002, An overview of volumetric coding technologies, ISO/IEC JTC1/SC29/WG1, Boston, Massachussets, USA, Report WG1N2613.

Schelkens, P., Munteanu, A., Barbarien, J., Galca, M., Giro-Nieto, X., and Cornelis, J., 2003, Wavelet coding of volumetric medical datasets, *IEEE Transactions on Medical Imaging, special issue on "Wavelets in Medical Imaging,"* Edited by M. Unser, A. Aldroubi, and A. Laine, **22**:441-458.

Shapiro, J. M., 1993, Embedded image coding using zerotrees of wavelet coefficients, *IEEE Transactions on Signal Processing*, **41**:3445-3462.

Skodras, A., Christopoulos, C., and Ebrahimi, T., 2001, The JPEG 2000 still image compression standard, *IEEE Signal Processing Magazine*, **18**:36-58.

Taubman, D., 2000, High performance scalable image compression with EBCOT, *IEEE Transactions on Image Processing*, **9**:1158-1170.

Taubman, D. and Marcelin, M. W., 2002, *JPEG2000: Image Compression Fundamentals, Standards, and Practice*, Kluwer Academic Publishers, Norwell, Massachusetts.

Wheeler, F. W., 2000, *Trellis Source Coding and Memory Constrained Image Coding*, PhD thesis, Department of Electrical, Computer and Systems Engineering, Renselaer Polytechnic Institute, Troy, New York.

Zandi, A., Allen, J. D., Schwartz, E. L., and Boliek, M., 1995, CREW: Compression with Reversible Embedded Wavelets, Proceedings of Data Compression Conference (DCC 1995), pp. 212-221.

AN OVERVIEW OF DIGITAL VIDEO COMPRESSION FOR MOBILE HEALTH SYSTEMS

Marios S. Pattichis[*], Songhe Cai, Constantinos S. Pattichis, and Rony Abdallah

1. INTRODUCTION

Few technologies have had a more positive and profound effect on Medicine than Medical Imaging. Imaging procedures have altered the practice of Medicine not only from the diagnostic perspective, but also from the therapeutic one with the development of Interventional radiology. Such impact is probably due to the fact that imaging is a non-invasive, usually painless and quick procedure. While static imaging approaches (X-ray, CT, etc) have been the norm, the advent of powerful video compression systems now allows medical video to supplement earlier imaging techniques.

Medical video is now used from echocardiography and laparoscopy to Otoscopy and Retinoscopy. Video imaging in medicine is important not only because it allows the Physician to review his/her procedure and reevaluate the initial diagnosis, but also because of its application in Medical education. With the development of the relatively new "hands on" and problem-based system of Medical Education, which includes more clinical and practical exposure and less didactic lectures, the importance of Medical Video Imaging can be felt across the medical education spectrum. Medical students and residents now have the opportunity to watch the whole imaging procedure, while they are in the classroom or even at home, instead of just reading about it in books. In addition, colored videos offer a great diagnostic advantage. By delineating the color of the area under study as well as by contrasting its color to the background, the physician can tell if it corresponds to normal, cancerous, or fibrous tissue. Another advantage of Medical Video imaging is the possibility of having multiple views. The multiple views allow a 3-Dimensional reconstruction of the lesion that is studied, which is especially important when differentiating benign lesions from malignant lesions whether used in Endoscopy, Colonoscopy, or Hysteroscopy.

---

[*] Marios S. Pattichis is with the department of Electrical and Computer Engineering of the University of New Mexico, Albuquerque, NM. 87131. He is currently visiting the department of Computer Science of the University of Cyprus.

Wireless video communication systems consist of a large number of independent components aimed at providing efficient video compression at an acceptable distortion level, with reasonable performance degradation over noisy channels. To introduce some of the basic algorithms of video compression technology, we first provide a summary of the features offered by the most popular standards. MPEG-2 offers (Haskell et al., (1996)):

- $\in$ random access, allowing the user to access any video frame
- $\in$ trick modes, allowing VCR-type controls: slow-motion, freeze-frame, etc.
- $\in$ multicast to many terminal types, broadcast to many different terminal types over a variety of communication channels. This requirement leads to the requirement that the video bitstream be *scalable*.
- $\in$ multiple audio and video, allowing the use of many languages, widescreen video, etc.
- $\in$ compatible stereoscopic 3D, allowing viewing of 3D programs in 3D displays, while providing regular display for monocular displays.

The success of MPEG-2 has been felt in the introduction of DVD players and digital TV.

Recently, MPEG-4 was introduced in order to address the convergence of TV, film, entertainment with computing and telecommunications (Pereira and Ebrahimi, (2002)). The functionalities offered by MPEG-4 include:

- $\in$ content-based interactivity, allowing multimedia data access, manipulation, and coding,
- $\in$ compression efficiency, including improved coding efficiency, and coding of multiple concurrent data streams, and
- $\in$ universal access, including robustness in error-prone environments, and content-based scalability.

Potential applications of MPEG-4 includes DVD, digital TV, as well as mobile multimedia, real-time communications, and *streaming video* over the Internet/Intranet (Pereira and Ebrahimi, (2002)). Here the term streaming video refers to applications that decode and play digital video without having to first download the entire video stream. In terms of bandwidth requirements, MPEG-4 was originally intended for very low bitrate video coding. However, the current standard addresses applications from as low as 5 Kbps to as high as 1 Gbits/s (Pereira and Ebrahimi, (2002)). It is important to note that despite the standardization of MPEG-4, many of the suggested functionalities are under intense development, and are thus not currently available. Progress in the implementation of MPEG-4 is primarily hindered due to the lack of effective video segmentation tools (Delp, (2000)).

The ITU video coding standards began with H.261 (ITU-T, (1993)). H.261 was the first practical digital video coding standard and was widely used in video teleconferencing. The H.263 standard provided a significant enhancement of the coding efficiency over H.261 (Cherriman and Hazno 1998; Cote et al., 1998; Liu et al. 2001; ITU-T, 1998) and was used to form the efficient backbone of the MPEG-4 design. The H.263 and H.263++ standards incorporate numerous enhancements over the previous technologies. They were designed for low data rate and internet applications. The next generation standard being developed is H.26L. H.26L includes a new structure for video coding development, separating the codec design into a video coding layer (VCL) and a network adaptation layer (NAL). VCL represents the video content, while NAL is used for customizing and packetizing the content for delivery over a particular network environment.

The H.320-series governs the basic video-telephony concepts of audio, video and graphical communications by specifying requirements for processing audio and video information, providing common formats for compatible audio/video inputs and outputs, and protocols that allow a multimedia terminal to utilize the communication links and synchronization of audio and video signals. It was extensively used in telemedicine applications (IMTC, 2004). A summary of the most popular video coding standards is given in Table 1.

Assessing the performance of medical video compression is still an active research topic. Research in video image quality assessment is grouped into three broad categories: (i) quantitative methods based on visual perception (Pappas and Safranek, 2000), (ii) quantitative methods based on rate-distortion theory (R-D curves), and (iii) quantitative methods based on the effect on diagnostic performance, such as receiver operating characteristic curves (ROC curves) (Cosman et al., 2000(1); Cosman et al., 2000(2); Cosman et al., 2000(3)).

Table 1. Video Coding Standards

Type	Sub-type	Typical BDR	Usage
MPEG	MPEG-1	1.5 Mbps	Optical Storage Media
	MPEG-2	4~20 Mbps	DVD and Digital TV
	MPEG-4	5kbps~1Gbps	Wireless, WEB
	MPEG-7	Ongoing	Content description of audiovisual documents
H.26x	H.261	Px64 kbps	ISDN applications
	H.263	Less than 64kbps	PSTN Applications
	H.263+		Low-Bit Rate PSTN Applications
	H.26L	Ongoing	ISDN, PSTN, low bitrate PSTN

The rest of this Chapter is organized into eight Sections. In Section 2, we review digital color video standards. Basic concepts used in video image sequence compression are reviewed in Section 3, while motion estimation algorithms used for computing motion between images is summarized in Section 4. This is followed by Section 5, where a discussion on the use of linear transforms for video compression is given. A brief Section on Region of Interest (ROI) methods is given in Section 6. In Section 7, new methods for error control in video communications are discussed. Concluding remarks are made in Section 8.

2. DIGITAL COLOR VIDEO STANDARDS

Typically, color video images are provided as three separate color images through time: the red $R(x,y,t)$, the green $G(x,y,t)$, and the blue $B(x,y,t)$. Yet, when the video images are displayed in either NTSC or PAL systems, they have to be converted to the corresponding CCIR-601 standards. We next explain some of the fundamental concepts involved in these conversions to develop an appreciation of what is being displayed versus what is represented in digital format.

First, it is important to distinguish between actual image intensity and displayed image intensity. The relationship is that (Haskell et al., 1996) $I = cv^{\gamma} + b$, where I denotes the actual light intensity, v denotes the applied voltage, c is a gain factor, b is a constant intensity correction, and *gamma* typically ranges as $1 \leq \gamma \leq 3$. We say that an image is gamma-corrected if applying the voltage to be the same as the gamma-corrected intensity will result in the correct displayed intensity I.

The CCIR-601 digital video standards are given in terms of gamma-corrected R', G', B' inputs. The CCCIR-601 digital video standard is defined in terms of the luminance Y and chrominance components Cr, Cb as given by (Haskell et al., 1996):

$$Y = 0.257R' + 0.504G' + 0.098B' + 16$$
$$Cr = 0.439R' - 0.368G' - 0.071B' + 128$$
$$Cb = -0.148R' - 0.291G' + 0.439B' + 128$$

where R', G', B' are integer-valued between zero and 255, while the resulting luminance and chrominance values vary between 16 and 240 (centered at 128). To convert back, we use

$$R' = 1.164(Y - 16) + 1.596(Cr - 128)$$
$$G' = 1.164(Y - 16) - 0.813(Cr - 128) - 0.392(Cb - 128)$$
$$B' = 1.164(Y - 16) + 2.017(Cb - 128)$$

where the resulting values are constrained to be integers between zero and 255.

The display and thus perception of digital video varies significantly among standards. In the 4:2:0 format, for every four Y-samples, we have one Cr and one Cb sample. In the 4:2:2 format, for every four Y-samples, we have two Cr and two Cb samples. In the 4:4:4 format, we have an equal number of Y, Cr and Cb samples. The most popular interlaced size for MPEG-2 uses the 4:2:0 format, with image sizes of 720×480 (720 pixels by 480 horizontal lines of pixels) at 30 frames per second, or 720×576 at 25 frames per second. The *Source Input Format* (SIF) is a non-interlaced format with either 352×240 pixels at 30 frames per second, or 352×288 at 25 frames per second. The SIF format is used for MPEG-1 video coding. More details can be found in Chapter 5 of Haskell et al., (1996). For the H.261 and H.263 standards, we have support for the Common-Intermediate Format standard (CIF) with 352×288 pixels, and low-bitrate CIF (QCIF) with 176×144 pixels per frame (both using 4:2:0 format). For the CIF and QCIF standards, we have support for 7.5, 10, 15, and 30 frames per second. More details can be found in Chapter 13 of Wang et al., (2002).

3. VIDEO IMAGE SEQUENCE COMPRESSION: BASIC CONCEPTS

Effective video compression can be achieved using prediction between video images. The basic idea is to predict a video frame from its neighboring frames, with the expectation that the residual frames (formed from the difference between the original and predicted frames) can be coded more effectively than the original frames (Haskell et al., 1996; Cote et al., 1998; Liu et al., 2001).

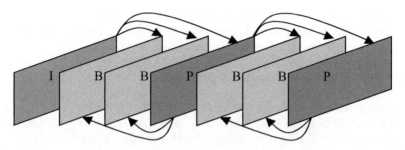

Figure 1. General structure of motion prediction frame

The process of predicting the change in a video frame image is called *motion estimation*. In motion estimation, *motion vectors* are computed for describing the change. The motion vectors are applied to the video frame(s) used for the prediction in order to come up with a predicted image. The process of computing the predicted image is called *motion compensation*.

To understand some of the issues associated with motion estimation and motion compensation, we turn to Figure 1. In Figure 1, we show an MPEG-1 "group of pictures". The images are grouped together because their encoding is dependent upon each other, as depicted by the unidirectional prediction arrows that connect them, while their encoding is not affected, neither does it affect, any images outside the group. An arrow from frame A to frame B denotes prediction of B from A.

There are three types of pictures: I, B, and P. I-pictures are intracoded-pictures that are encoded as still images. In wireless communications, transmission errors from other frames do not propagate to I-pictures. On the other hand, it is clear from Figure 5, that any transmission error in an I-picture will propagate to the entire group, except in the special case, when the encoder does not use motion compensation based on the corrupted pixels.

P-pictures are predictive-coded pictures, and they are predicted from either a previous I-picture, or a previous P-picture using forward prediction. Together, I and P-pictures are used for the prediction of B-pictures. B-pictures are bidirectional-coded pictures. Usually, the average between two predicted images is taken as the outcome of the combined prediction. Alternatively, it may be more efficient to use only one of the two predictions. This is clearly preferable when there is occlusion, as in the case where a video camera is imaging multiple moving objects.

Assuming effective motion compensation, the coding efficiency is proportional to the number of images used in the group of images. This will typically result in highly-compressed B-pictures, less-compressed P-pictures, and lesser-compressed I-pictures. Since the number of B-pictures and P-pictures may be varied, it is possible to increase the number of pictures in the group and expect greater compression. In general though, motion-estimation and motion-compensation for large groups of pictures requires significant processing power that is unavailable in low-power, low-speed processors found in many wireless devices. Furthermore, the use of larger groups of pictures will result in unacceptable transmission delays.

4. MOTION ESTIMATION METHODS

We discuss three types of methods used in motion estimation: (i) block-matching based methods, (ii) pixel-based methods, and (iii) mesh-based methods. The most popular motion estimation methods are based on block matching.

4.1 Block Matching Motion Estimation Algorithms

Block-based motion estimation has been adopted by all video coding standards (H.261, H.263, MPEG-1, MPEG-2, MPEG-4, etc.). The simplicity of block-based methods allows for very fast implementations in both software and hardware.

The basic principle of block matching motion estimation algorithms is to partition each video frame into non-overlapping small blocks of pixels, then approximate the rigid motion between frames by searching the best matching block in the previous frame for each block of the current frame, where a given matching criterion is used. Matching criteria include the minimum mean square error (MSE), the minimum sum of absolute difference (SAD) and the maximum matching pel count (MPC). Here, I_t denotes the intensity of a pixel at position (x, y) in the current frame t and I_{t+1} denotes the intensity of a pixel in the candidate block at position $(x+u, y+v)$ in the reference frame $t+1$. Definitions of these matching criteria are as follows:

$$MSE_{(x,y)}(u,v) = \min_{(u,v)} \left\{ \sum_{i=0}^{M-1} \sum_{j=0}^{M-1} (I_t(x+i, y+j) - I_{t+1}(x+u+i, y+v+j))^2 \right\}$$

$$SAD_{(x,y)}(u,v) = \min_{(u,v)} \left\{ \sum_{i=0}^{M-1} \sum_{j=0}^{M-1} |I_t(x+i, y+j) - I_{t+1}(x+u+i, y+v+j)| \right\}$$

$$MPC_{(x,y)}(u,v) = \max_{(u,v)} \left\{ \sum_{i=0}^{M-1} \sum_{j=0}^{M-1} \varphi(u,v) \right\}$$

Here, $\varphi(u,v) = \begin{cases} 1, & |I_t(x+i, y+j) - I_{t+1}(x+u+i, y+v+j)| \le t \\ 0, & |I_t(x+i, y+j) - I_{t+1}(x+u+i, y+v+j)| > t \end{cases}$.

In all methods, the image block size is $M \times M$. Figure 2 illustrates the general idea of block matching. The motion of object I between frame t and $t+1$ is obtained by considering an $M \times M$ block of pixels containing this object, and searching in the neighbor frame $t+1$ for the best matching block of the same size. The search range is an $R \times R$ square of pixels. The overall computational cost is independent of the block size (Wang et al., 2002; Tekalp (1995)). However, when the frame size of the video sequences is large, the computational cost of an exhaustive full search algorithm is very high. Thus there are many fast block matching algorithms that have been developed to solve this problem of high computational complexity.

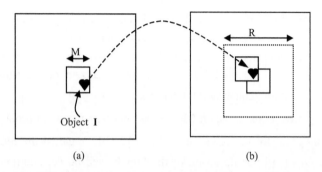

(a)   (b)

Figure 2. Illustration of full search algorithm (a) frame t, (b) frame $t+1$

One popular fast algorithm is the three-step hierarchical search method. Although the video quality obtained using this method is degraded as compared to the results obtained from performing the full search, there are at least two advantages when compared against the full search method. One advantage is that the computational time can be reduced to about one-tenth of the time required for the full search. In three-step search, the initial search step size is equal to, or slightly larger than, half of the maximum search range. But after each step, the search step size is halved and the search center is moved to the best matching point of the result of the previous search. Another advantage is that its block size is adaptable due to its hierarchical structure. In the hierarchical framework, the video sequences are represented at multiple resolution levels from coarse to fine. The block-matching estimation begins at the coarsest level, and then moves to the higher resolution level. The estimation result of the lower level is used as the initial estimation for the higher level. So, estimation at a higher level can begin with a better initial estimation and use a relatively smaller search range. Other algorithms include the 2-D log search (Jain and Jain, 1981), diamond search (Zhu and Ma, 2000), spiral search (Zahariadis and Kalivas, 1996), and predictive zonal search (Tourapis et al., 2002), center-biased adaptive search (Christopoulos and Cornelis, 2000).

4.2. Pixel-Based Motion Estimation

Pixel based methods allow us to estimate the motion of individual video pixels through time. The most popular pixel-based method is based on *optical flow* (Horn and Schunk, 1981). A comparison of the performance of a number of optical flow algorithms can be found in (Barron et al., 1994). Applications in digital video coding are given in (Krishnamurthy et al., 1997; Krishnamurthy and Woods, 1995).

To introduce optical flow techniques, consider an expanding object shown in Figures 3(a) and 3(b). The image in Figure 3(a) is represented by $I(x,y,t)$, and the image in Figure 3(b) is represented by $I(x,y,t+\delta t)$. In optical flow, we assume that each pixel in the second image has been derived from the first image. A velocity is associated with each image pixel given by the vector field $\left(U(x,y),V(x,y)\right)$ assumed to be constant between the two frames. In order to predict the motion, we must estimate the motion vector field $(u(x,y), v(x,y))$. In the optical flow solution, we assume that there is no brightness variation between the frames (intensity constancy assumption), giving that $I(x+u\delta t, y+v\delta t, t+\delta t) = I(x,y,t)$. Using the intensity constancy assumption, and also assuming small deformations between the two video frames, we write a Taylor Series expansion for $I(x+u\delta t, y+v\delta t, t+\delta t)$, drop the second order terms to arrive at the optical flow constraint equation (OFCE): $I_x u + I_y v + I_t = 0$.

The first two terms in the OFCE can be interpreted as a dot product $\nabla I (u,v)$. Any motion orthogonal to ∇I is lost in $\nabla I (u,v)$. This is called the *aperture problem*. It makes the OFCE ill-posed. To make the OFCE well-posed, additional constraints are usually imposed. The original constrain reformulates the OFCE as (Horn and Schunk, 1981):

$$\min_{u,v} \{ \iint ((\frac{\partial I}{\partial x}u + \frac{\partial I}{\partial y}v + \frac{\partial I}{\partial t})^2 + \alpha(\|\nabla u\|^2 + \|\nabla v\|^2)) dx dy \} . \tag{1}$$

Here, α is the weighting coefficient for the velocity smoothness constraint. Derivatives are often estimated using finite differences, and the solution is computed using iterative methods. Returning to Figure 3, the optical flow estimates are shown in Figure 3(c).

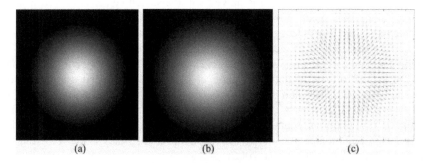

Figure 3. Motion estimation using optical flow: (a) 2D synthetic object at time t, (b) 2D synthetic object at time $t + \delta t$, (c) 2-D optical flow estimate. Synthetic images were constructed using isotropic Gaussians. The pixels that were more than two standard deviations from the center were set to zero.

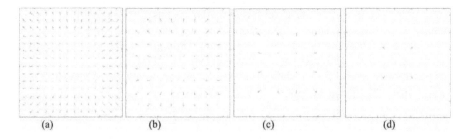

Figure 4. 2-D multi-scale optical flow estimation (a), scale =2, (b) scale =3, (c) scale=4, (d) scale=5.

Multi-scale algorithms were proposed by Martin et al., (2001) and Moulin and Krishnamurthy, (1997). These methods are less constrained in the pixel intensity of image sequences. They produce superior motion estimation results at reduced computational complexity. Figure 4 shows the multi-scale motion estimation results of image sequences in 3(a) and 3(b). Here the Deslauries-Dubuc wavelets were used with four vanishing moments (with integer filter coefficients).

4.3. Mesh-Based Motion Estimation Methods

Mesh based methods have been investigated by many researchers (Nostratinia, 2001; Yeung et al., 1998). This method provides better video quality than block-based approaches and alleviates the block artifacts even at very low bitrates, just because it can also support rotation and scaling, in addition to translation, which is the only type of displacement that is supported by block-based methods.

A procedure generating a shape adaptive meshes is described in (Tekalp et al., 1998; Altunbasak and Tekalp, 1994; Celasun and Tekalp, 2000). Meshes are built using nodes, defined as points of interest on the video images. The combination of three nodes defines a triangle. The collection of all the triangles together defines the mesh (see Figure 5). There are several questions to consider, including (i) how the node points are created, (ii) how they are combined to form a mesh, and then (iii) how they are tracked through time.

We begin by describing the mesh generation process. In general, we would like the triangular meshes to be aligned with intensity edges. The density in the number of nodes is a function of the estimated local motion, local "cornerness", and local "textureness" measures (Celasun and Tekalp, 2000). To develop a multiscale representation, we begin with the generation of the boundary of a detected video object. Originally, the number of node points on the boundary is chosen so that the average distance between the points corresponds to the average distance in the interior. The number of nodes used for describing the boundary is then minimized so that a distance measure between the estimated actual and the approximate boundaries does not exceed some threshold. This is accomplished by deleting node after node, while keeping some vertical deviation distance small.

A multiscale representation of the mesh is then defined in terms of the boundary points (Celasun and Tekalp, 2000). The first layer is defined to be defined in terms of the boundary points. The second layer is then defined in terms of the points that are found to be adjacent to the first layer. Similarly, all the layers are defined in terms of the previous ones. Some internal nodes are deleted from the layers, but the ones that are highly connected, or correspond to edgy, corner nodes are kept.

Once the mesh points and the layers have been identified, the nodes are connected together using a Delaunay triangulation algorithm. The details of the Delaunay triangulation algorithm can be found in Berg et al., (2000).

The problem of computing the corresponding nodes between video frames is again a motion estimation problem. A textbook account of how to estimate the motion so as to minimize some error function is given in Wang et al., (2002). By interpolating the estimated motion vectors from the neighboring nodes, we obtain a continuous-space estimate of the motion vectors. A crossover can occur when a motion vector for a node significantly exceeds those of its neighboring nodes. When a crossover is detected, the nodes in the affected regions are re-computed to avoid the crossover.

Figure 5 shows an example using the shape adaptive mesh-based methods to estimate myocardial motion. The experiment uses two cardiac ultrasound video frames. Here we extract an ROI from the ultrasound frame firstly. The ROI is cropped by simply using a rectangular box and saved as a new image.

Wang and Ostermann (1998) compared the mesh-based and the full search block-matching based motion estimation and compensation algorithms in H.263-like coders. Two mesh-based algorithms, a hierarchical mesh-based matching algorithm (HMMA) and a hierarchical block-based matching algorithm (HBMA), were developed. In both algorithms, the hierarchical search method was used to find nodal motion vectors in order

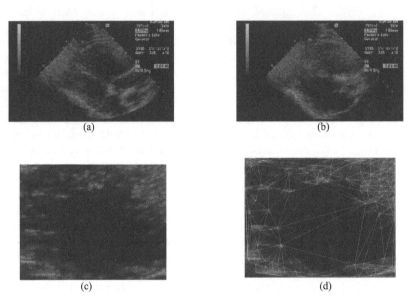

(a)   (b)

(c)   (d)

Figure 5. (a) Cardiac ultrasound video frame k, (b) Cardiac ultrasound video frame k+1
(c) Generated nodes using shape adaptive algorithm, (d) Generated mesh

to speed up the computation and obtain global minima. A pyramid representation was generated for each video sequence at the beginning of the motion computation. Then, a minimum mean square error based full search algorithm was applied at each pyramid level from coarse to fine. The search result of the coarse level was used to initialize the computation at the finer levels. HBMA treated the center of blocks as nodes as opposed to vertices of each quadrilateral mesh in HMMA. Experimental results demonstrated that the mesh-based methods performed better than block-based methods in prediction accuracy, particularly in the presence of non-translational motions, such as rotation and zooming.

5. LINEAR TRANSFORMS IN VIDEO CODING

An essential component in every video compression standard is the use of linear transforms such as the Discrete Cosine Transform or the Discrete Wavelet Transform. Image transforms are applied to a group of pixels directly, or to the difference between the predicted and actual pixel values. In the case of the Discrete Cosine Transform (DCT), the transform coefficients $\hat{I}(u, v)$ are computed from the input $I(x, y)$ using

$$\hat{I}(u, v) = \frac{4C(u)C(v)}{N^2} \sum_{x=0}^{N-1}\sum_{y=0}^{N-1} I(x, y)\cos\left[\frac{(2x+1)u\pi}{2N}\right]\cos\left[\frac{(2y+1)v\pi}{2N}\right]$$

where $C(0) = 1/\sqrt{2}, C(u) = 0$ for $u \neq 0$. The inverse DCT is computed using

$$I(x, y) = \sum_{u=0}^{N-1}\sum_{v=0}^{N-1} \hat{I}(u, v)C(u)C(v)\cos\left[\frac{(2x+1)u\pi}{2N}\right]\cos\left[\frac{(2y+1)v\pi}{2N}\right].$$

The transformed coefficients must be quantized to integral values prior to encoding. In *scalar quantization*, each transform coefficient is quantized independently of the others, through the use of a transform-domain quantization table $Q(u, v)$ using

$$\hat{I}_Q(u, v) = \text{INT}\left[\hat{I}(u, v)/Q(u, v) + 0.5\right],$$

where $\text{INT}[\cdot]$ denotes the integral function that truncates a floating point number into an integral value. The most significant compression gains are due to a stream of zeros resulting from the quantization operation. The zeros are either due to the large values of $Q(u, v)$ for large frequencies, the small values of $\hat{I}(u, v)$ due to effective prediction, or due to both. It is important to use separate quantization tables for transform coefficients resulting from predicted values as opposed to coefficients resulting from still image pixels (Lin and Gray, 1999). Alternatively, in *vector quantization*, the entire array of transformed coefficients $\hat{I}(u, v)$ is mapped to a *code vector*.

It is important to note that in addition to spatial pixel prediction, we need to consider methods for predicting the transformed coefficients. For example, for the DCT coefficients, the DC coefficient $\hat{I}(0, 0)$ is typically predicted from the previous block, leaving only the predicted residual for encoding. More sophisticated transform-domain prediction methods have been investigated for the Discrete Wavelet Transform (DWT).

Woods et al., (2000) give an introduction of spatiotemporal methods for Wavelet video compression. For still images, the original set partitioning in hierarchal trees (SPIHT) is given by Said and Pearlman, (1996). An important 3-D extension of the SPIHT for video compression, also addressing issues of video transmission for mobile systems can be found in Cho and Pearlman, (2002). Essentially though, the Wavelet coefficients are coded within independent spatiotemporal blocks, so as to allow independent decoding. Alternatively, spatial-prediction of the Wavelet coefficients through a quadtree approach has proven useful in Region of Interest (ROI) coding and coding efficiency (Munteanu et al., 1999; Munteanu et al., 2005).

For computing the quantization tables for minimizing the mean-square error, we can use the Lloyd-Max algorithm (Gersho and Gray, 1992). For fast Vector Quantization (VQ), popular algorithms include the classified VQ (Cziho, 1998), the tree-structure VQ (Tong and Gray, 2000; Guo et al., 2001; Ordonez et al., 2001), and the multistage VQ (Khalil and Rose, 2001). Khalil and Rose (2001) built predictive multistage vector quantizers for low bit rate video coding.

Following quantization, the quantized coefficients are *entropy encoded* to minimize the total number of bits required for encoding. For the encoding, a *scanning order* is used to map the two-dimensional quantized coefficient array into a sequential order

$\hat{I}_Q(0, 0), \hat{I}_Q(u_1, v_1), \hat{I}_Q(u_2, v_2), \_$. An important characteristic of the scanning order is that high-frequency coefficients are typically grouped together. This allows the use of *run-length encoding* for effective encoding of the zeros. In run-length encoding, instead of coding individual coefficients, the number of zero coefficients is encoded. Furthermore, given the fact that most high frequency coefficients are zero, a single EOB character is used to signify that the rest of the transformed coefficients are all zero.

For entropy encoding, the most popular algorithms include Huffman coding and Arithmetic coding. Unfortunately, these variable-length coding algorithms are unsuitable for wireless communications. As we discuss later, a transmission error resulting in a single bit error will not permit further decoding.

5.1. Medical Video Compression Using Wavelet-based methods: literature review

Most medical video compression standards operate in a "near-lossless" mode. Fractal and Wavelets transform based algorithms, such as embedded zerotree wavelets coding (EZW) (Munteanu et al., 1999; Marpe and Cycon, 1999; Hang et al., 2000; Chiu et al, 2001), set partitioning in hierarchal trees (SPIHT) (Xiong et al., 1998; Cho and Pearlman, 2002; Youngseop and Pearlman 2000), and joint space-frequency segmentation (JSFS) (Chiu et al., 2001), were used in general video compression research. Some researchers also introduced those methods into medical video compression. Hang et al., (2000) and Youngseop and Pearlman, (2000), SPIHT based methods are used in compressing MRI and volumetric echocardiographic data separately. Hang et al., (2000) showed that a 40:1 compression ratio preserved image quality with bandwidth not exceeding 400kbps.

Chiu et al., (2001), discussed ultrasound compression results using both JSFS and SPIHT based methods. The results claim better performance of JSFS than SPIHT methods. For example, on a specific 0.5bpp bit rate, the PSNR for JSFS was 32.99dB, which is 0.7dB higher than that for the corresponding SPIHT. Marpe and Cycon, (1999), designed a new video coding system based on the discrete wavelet transform (DWT). They compared their system with a standard MPEG-4 DCT based method using some typical MPEG-4 test sequences, such as *News, Foreman* and *Akiyo* in QCIF format with 30 frames/s bit rates. Their experimental results also demonstrated this DWT based coding system achieved better performance than an MPEG-4 DCT based method both for intraframe coding and residual frames coding.

6. OBJECT-BASED CODING AND REGION OF INTEREST (ROI) METHODS: A SUMMARY

For object-based and region of interest (ROI) methods, the video viewed as a collection of objects or regions. For coding purposes, each object or region may have its own codec. The basic idea is to allow different codecs that may be efficient for different objects, while no codec needs to be efficient for all objects. The basic challenge with these methods lies in the development of effective segmentation methods that can be used to extract the objects or regions of interest (Delp, 2000). Nevertheless, for mobile health systems, there are obvious advantages to identifying different objects of diagnostic interest (Rodriguez et al., 2005). Furthermore, it is widely accepted that for very low bit rate coding, there is a need for model-based and object-based methods (see Aizawa and Huang, 1995).

An early introduction to object based methods can be found in the book by Torres and Kunt, (1996). An overview of object-based methods is given by Ebrahimi and Kunt, (2000). Much of the research in object-based methods has focused on mesh methods for tracking facial expressions. A recent account is given by Siu et al., (2001). Doulamis et. al., (1998), introduced a two level neural network method for extracting the ROI in the video frames. Akgul et al. (1998), also described how to extract the tongue surface automatically from ultrasound sequences using a snake-based technique.

An extension of model-based and object-based coding leads to content-based coding, which allows us to process video objects in their proper content. Several future applications in mobile health systems are possible, provided that the video segmentation problems are properly addressed (see some related work by Meier, T. and Ngan, K. N., 1999). Dumitras and Haskell (2004), consider content-based coding for textured regions.

7. COMMUNICATIONS ERROR CONTROL AND RECOVERY IN REAL-TIME VIDEO PROCESSING APPLICATIONS

The use of specialized error control mechanisms in video communications is due to the need for real-time video processing. There is no need for error control in "store and forward" applications, because classical error-control methods can be used to design the source and channel coders separately, so as to deliver error-free video, as long as the source bitstream is encoded below the channel capacity. Unfortunately though, error-free delivery can only be achieved by allowing long (in theory infinite) delays in the video decoding process. In general, in video streaming applications, where the video must be decoded prior to receiving the entire video bitstream, long delays are unacceptable.

Three of the largest sources of communication error are due to the relative significance of different parts of the video bitstream, predictive coding, and variable length coding (Wang et al., 2002). For example, headers provide critical format information, such as image size and type that will have to be communicated without error. On the other hand, we can tolerate some decoder error that relates to a single-block of pixels, and then try to reduce its effect. As we have discussed, in MPEG-based video encoding, much of the video compression performance is due to the prediction of video images from a prior video image (P-pictures predicted from I-pictures) or prediction from both prior and future video images (B-pictures predicted from both I and P-pictures). Clearly, decoding error propagates when a block of pixels (a macroblock) is predicted based on corrupted pixels or a corrupt block of pixels. Usually, the error is detected. The basic problem becomes one of having to minimize the effect of a detected error. A third significant source of communication errors comes from the use of variable-length coding techniques, such as Huffman coding. Again, assuming that the error has been detected, the use of variable-length coding makes it very difficult to isolate its effect, since it is not clear how to resume decoding.

We will discuss four types of general methods for control and recovery from transmission error: *transport-level error control, error-resilient encoding, error concealment,* and *interactive error control methods* (Wang et al., 2002). In transport level error control, we group methods that operate at the transport level of communication systems. In *error resilient encoding*, the source coder is designed to localize and reduce the decoder error by using redundancy in the bitstream. In error concealment, instead of the encoder, the decoder attempts to recover from transmission errors where it is assumed that little or no redundant information is added to the bitstream. Algorithms for both technologies are rapidly evolving. In *interactive error control* methods, we assume that in addition to the communication channel from the encoder to the decoder, we also have a communication channel from the decoder, back to the encoder. The communication channel from the decoder to the encoder is then used to adapt the encoder to the characteristics of the channel, or more specifically, to adjust the encoding process for information lost during transmission.

In *transport-level error control*, we have unequal error protection, interleaved packetization, the classical single bit error detection and correction mechanisms, and delay-constrained retransmission. We will only discuss the first two methods. In unequal error protection, network priority can be assigned to important header and or base-layer information, by using the priority bit in ATM cell headers, using TCP for transmission of sensitive data (Civanlar and Cash, 1996), or simply duplicating the information. In interleaved packetization, the spatial continuity of the encoded bitstream is broken by packetizing data from different spatial locations together, in the same packet. Then, even if a continuous part of the video bitstream is lost, it is very likely that neighboring information is preserved, and can thus be used to interpolate the lost values.

In error-resilient encoding, we discuss techniques for Multiple-Description Transform Coding (MDTC), layered coding with unequal error protection, resynchronization, and robust encoding. The basic idea in MDTC is to use several bitstreams (descriptors) each transmitted over separate physical, independent channels, or virtual channels that approximate the condition of being independent. Since the communication channels are assumed to be independent, we assume that there is a low probability that all of them will fail to deliver the same descriptor. The decoder can then decode the video and deliver acceptable performance by using any one of the successfully transmitted descriptors. The descriptors are not identical. Instead, they are designed so that the decoder performance is proportional to the number of descriptors used during decoding. The use of layered coding with unequal error protection is very popular for wireless networks (Zhang et al., 1994; Hazno, 1998; Khansari et al., 1996; Ghavari and Alamouti, 1999). The basic idea is to protect the so-called base-layer of the video bitstream by using a high-priority channel, while using lower-priority for the enhancement-layers of the video bitstream. In this encoding scheme, the prediction in the enhancement layer maybe modified to be based on the base-layer, thus resulting in coding redundancy. In any case, the basic idea is to provide acceptable performance through the protected transmission of the base layer, while improving performance through the enhancement layers. The idea of breaking the video bitstream into a base-layer and an enhancement-layer comes from the desire to provide *scalable coding* that is appropriate for a variety of viewing conditions.

In resynchronization methods, a resynchronization marker is used to provide sufficient information to resume coding. The basic assumption is that the location of a transmitted error has been detected, and that a resynchronization marker can still be detected after the error. All information between the location of the error and the resynchronization marker will have to be discarded. Through data partitioning, it maybe possible to further localize the error through the use of secondary markers.

In robust encoding methods, error propagation due to the use of variable length coding is minimized. In Reversible Variable Length Coding (RVLC) (Takishima et al.,

1995), after detecting an error and its following resynchronization marker, we decode the bitstream backwards, from the resynchronization marker towards the detected error location. Thus, we can recover most of the information between the error and the resynchronization marker. In Error-Resilient Entropy Coding (EREC) (Redmill and Kingsbury, 1996), fixed-size slots are used for storing bits from a collection of video blocks. When an error occurs, after discarding the remaining slots, we regain synchronization and restart decoding at the start of each block.

In decoder error concealment, spatial, temporal, or combined spatio-temporal interpolation is used to minimize error propagation through the decoded bitstream. For example, spatial interpolation can be used to recover blocks of pixels by interpolating their values from neighboring pixels. An effective spatial interpolation method is based on Projection onto Convex Sets methods (POCS) (Yu et al., 1998; Sun and Kwok, 1995), but requires fast implementation for real-time application. In temporal interpolation, we estimate lost motion vectors by filtering neighboring vectors (Lam et al., 1993; Hong et al., 1999). In combined spatio-temporal interpolation, both pixel intensities and motion vectors are estimated.

In interactive error control mechanisms, the decoder can inform the encoder about the channel error rate or even the location of lost blocks or macroblocks. If the channel error rate can be estimated, then the encoder target bitrate can either be increased or reduced to match the estimated error (Dapeng et al., 2000). Furthermore, if the decoder communicates the location of lost blocks and/or macroblocks to the encoder, the encoder may then keep track of which parts of the video are damaged. Then, the encoder can limit the decoder error propagation by avoiding prediction based on unknown or incorrect pixel intensities (Girod and Farber, 1999; Steinbach et al., 1997; Wada, 1989).

8. CONCLUDING REMARKS

Medical video communication over wireless communication channels is still a very active area of research. In real-time video communication applications, error control methods must be used to minimize the decoded error. For very low bitrate video coding, object-based methods such as those proposed by MPEG-4 are necessary. However, the lack of effective video segmentation algorithms limits the effectiveness of such methods (Delp, 2000). With the introduction of more advanced hardware, error control mechanisms will limit decoded error while allowing for even more effective encoding at higher compression ratios. In addition, the constant increase in wireless communications bandwidth is likely to limit the need for very low bitrate video coding.

9. REFERENCES

Aizawa, K., and Huang, T.S., 1995, Model-based image coding advanced video coding techniques for very low bit-rate applications, *Proc. of the IEEE*, **83(2)**:259-271.

Akgul, Y. S., Kambhamettu, C., and Stone, M. 1998, Extraction and tracking of the tongue surface from ultrasound image sequences, *IEEE Conference on Computer Vision and Pattern Recognition*, pp. 298-303.

Altunbasak, Y., and Tekalp, A. M., 1994, Occlusion-adaptive, content-based mesh design and forward tracking, *IEEE Transactions on Image Processing*, **6**:1270-1280.

Barron, J. L., Fleet, D. J., and Beauchemin, S. S., 1994, Performance of optical flow techniques, *The international journal of computer vision*, **12**:43-77.

Berg, M. D., Kreveld, M. V., Overmars, and M., Schwarzkopf, O., 2000, *Computational Geometry, 2nd ed.*, Springer-Verlag, New York.

Bescós, J., Movilla, A., and Cisneros, G., 2002, Real time temporal segmentation of MPEG video, *IEEE international conference on Image Processing*, **2**:II-409-II-412.

Celasun, I. and Tekalp, M., 2000, Optimal 2-d hierarchical content-based mesh design and update for object-based video, *IEEE Trans. on Circuits and Systems for Video Technology*, **10**(7):1135-1153.

Chang-Su, K., Rin-Chul, K., and Sang-Uk, L., 1998, Fractal coding of video sequence using circular prediction mapping and noncontractive interframe mapping, *IEEE Transactions on image processing*, **7**:601-605.

Chen, W. H. J. and Hwang, J. N., 2001, The CBERC: a content-based error-resilient coding technique for packet video communications, *IEEE Transactions on Circuits and Systems for Video Technology*, **11**:974-980.

Cherriman, P. and Hazno, L., 1998, Programmable H.263-based wireless video transceivers for interference-limited environments, *IEEE Transcation on Circuits and Systems for Video Technology.*, **8**:275-286.

Chiu, E., Vaisey, J., and Atkins, M. S., 2001, Wavelet-based space-frequency compression of ultrasound images, *IEEE Transactions on Information Technology in Biomedicine.*, **5**:300-310.

Cho, S. and Pearlman, W. A., 2002, A full-featured, error resilient, scalable wavelet video codec based on the set partitioning in hierarchical trees (SPIHT) algorithm, *IEEE Transactions on Circuits and Systems for Video Technology*, **12**:151-170.

Choi, M. H., Park, H. J., Han, J. M., Jeong, D. U., and Park, K. S., 1998, A study of the integration and synchronization of video image using H.261 in polysomnography., *Proceedings of the 20th Annual EMBS International Conference.*, **4**:1960-1963.

Christopoulos, V. and Cornelis, J., 2000, A center-biased adaptive search algorithms for block motion estimation, *IEEE Transactions on Circuits and Systems for Video Technology*, **10**(3):423-426.

Chung, D. M. and Wang Y., 2002, Lapped orthogonal transforms designed for error-resilient image coding, *IEEE Transactions on Circuits and Systems for Video Technology.*, **12**:752-764.

Civanlar, M. R., Cash, G. L., 1996, A practical system for MPEG-2 based video-on-command over ATM packet networks and the WWW, *Signal Processing: Image Communications*, **8**:221-227.

Cosman, P., Gray, R., and Olshen, R., 2000(1), Quality Evaluation for Compressed Medical Images: Fundamentals, in: *Handbook of medical imaging processing and analysis*, I. N. Bankman, Ed.,Academic Press, New York, pp. 803-820.

Cosman, P., Gray, R., and Olshen, R., 2000(2), Quality evaluation for compressed medical images: diagnostic accuracy, in: *Handbook of medical imaging processing and analysis*, I. N. Bankman, Ed., Academic Press, New York, pp. 821-840.

Cosman, P., Gray, R., and Olshen, R., 2000(3), Quality evaluation for compressed medical images:statistical issues, in: *Handbook of medical imaging processing and analysis*, I. N. Bankman, Ed., Academic Press, New York, pp. 841-850.

Cote, G., Erol, B., Callant, M., and Kossentini, F., 1998, H.263+: video coding at low bit rates, *IEEE Transactions on Circuits and Systems for Video Technology.*, **8**:849-866.

Cziho, A., 1998, Medical image compression using ROI vector quantization., *Proceedings of the 20th Annual EMBS International Conference.*, **20**:1277-1280.

Dapeng, W., Hou, Y. T., Wenwu, Z., Hung-Ju, Tihao, C., Ya-Qin, Z., Chao, H. J., 2000, On end-to-end architecture for transporting MPEG-4 video over the internet, *IEEE transactions on circuits and systems for video technology*, **10**(6):923-941.

Delp, E., 2000, Personal Communication.

Doulamis, N., Doulamis, A., Kalogeras, A. and Kollias, S., 1998, Low bit-rate coding of image sequences using adaptive regions of interest, *IEEE transactions on circuits and systems for video technology*, **8**:928-934.

Dumitras, A. and Haskell, B. G., 2004, An encoder-decoder texture replacement method with application to content-based movie coding, *IEEE Trans. on Circuits and Systems fro Video Technology*, **14**(6):825-840.

Ebrahimi, T., and Kunt, M., 2000, Object-based video coding, chapter in: *Handbook of Image and Video Processing*, A. C. Bovik, Ed.: Academic Press, New York, pp. 585-595.

Frossard, P. and Verscheure, O., 2001, AMISP: a complete content-based MPEG-2 error-resilient scheme, *IEEE Transactions on Circuits and Systems for Video Technology*, **11**:989-998.

Gersho, A., and Gray, R. M., 1992, *Vector Quantization and Signal Compression*. Kluwer Academic Publishers, Boston.

Ghavari, H., and Alamouti, S. M., 1999, Multipriority video transmission for third generation wireless communication systems, *Proceedings of the IEEE*, **87**: 1751-1763.

Girod, B., and Farber, N., 1999, Feedback-based error control for mobile video transmission, *Proceedings of the IEEE,Special issue on video for Mobile Multimedia*, **87**:1707-1723.

Guo, L., Umbaugh, S., and Cheng, Y., 2001, Compression of color skin tumor images with vector quantization, *IEEE Engineering in Medicine and Biology Magazine.*, **20**(6):152-164.

Hang, X., Greenberg, N., L., Shiota, T., Firstenberg, M. S., and Thomas, J. D., 2000, Compression of real time volumetric echocardiographic data using modified SPIHT based on the three-dimensional wavelet, *Computers in Cardiology*, **27**:123-126.

Haskell, B. G., Puri, A., and Netravali, A. N., 1996, *Digital Video: An Introduction to MPEG-2*, Chapman & Hall, Boca Raton, Florida, U.S.A.

Hazno, L., 1998, Bandwidth efficient wireless multimedia communications, *Proceedings of the IEEE*, **86**:1342-1382.

Hong, M. C., Kondi, L., Scwab, H., and Katsaggelos, A. K., 1999, Video error concealment techniques, *Signal Processing: Image Communications, Special Issue on Error Resilient Video*, **14**:437-492.

Horn, B. and Schunk, B., 1981, Determining optical flow, *Artificial Intelligence*, **17**:185-203. IMTC 2004, (2004); www.imtc.org/h320.htm.

ISO/IEC., 1993, IS 11172: Information technology-coding of moving pictures and associated audio for digital storage media at up to about 1.5M bits/s (MPEG-1).

ISO/IEC., 1995, IS 13818-2:Information technology-generic coding of moving pictures and associated audio information: video (MPEG-2 video).

ISO/IEC., 1995, IS 13818-1:Information technology-generic coding of moving pictures and associated audio information: Systems (MPEG-2 Systems).

ISO/IEC., 1999, IS 14496-X:Information technology-coding of audio-visual objects (MPEG-4).

ISO/IEC., 1999, IS 14496-1:Information technology-coding of audio-visual objects-part 1:Systems (MPEG-4 Systems).

ISO/IEC., 1999, IS 14496-2:Information technology-coding of audio-visual objects-part 2:Visual (MPEG-4 Video).

ITU-T, 1993, Recommendation H.261: Video codec for audiovisual services at px64k bit/s.

ITU-T, 1998, Recommendation H.263:Video coding for low bit rate communication.

Jain, J. R., and Jain, A. K., 1981, Displacement measurement and its application in interframe image coding, *IEEE Trans. Commun.*, **29**:1799–1808.

Khansari, M., Jalali, A., Dubois, E., and Mermelstein, P., 1996, Low bit rate video transmission over fading channels for wireless microcellular systems, *IEEE Transactions on Circuits and Systems forVideo Technology.*, **6**:1-11.

Khalil, H. and Rose, K., 2001, Predictive multistage vector quantizer design using asymptotic closed-loop optimization, *IEEE Transactions on image processing*, **10**:1765-1770.

Khludov, S., Vorwerk, L., and Meinel, C., 2000, Internet-orientated medical information system for DICOM-data transfer, visualization and revision., *13th IEEE CBMS 2000*, pp. 293-296.

Krishnamurthy, R., Moulin, P., and Woods, J. W., 1997, Compactly-encoded optical flow fields for video coding and bidirectional prediction, *Visual Communications and Image Processing*.

Krishnamurthy, R. and Woods, J. W., 1995, Optical flow techniques applied to video coding, *Proc. IEEE International Conference Image Processing*, **I**:570-573.

Lam, W. M., Reibman, A. R., and Liu, B., 1993, Recovery of lost or erroneously received motion vectors, *IEEE Int. Conf. Acoustics, Speech, Signal Proc.(ICASSP' 93), Minneapolis*, **5**:417-420.

Lau, C., Cabral, J. E., A. H. R. Jr., and Kim, Y., 2000, MPEG-4 Coding of Ultrasound Sequences, *Proceedings of SPIE Medical Imaging 2000*, **3976**:573-579.

Lin, K. K., and Gray, R. M., 1999, Vector quantization of video with two codebooks, *Proc. Data Compression Conference 1999*, pp. 537.

Liu, H., Ravindran, A., Miller, D. J., Kavehrad, M., Doherty, J. F., Agoren, I., and Lackpoor, A., 2001, Error-resilient H.263 video coding for wideband CDMA systems, *the 35th Asilomar Conference on Signals, Systems,and Computers, Asilomar Conference Grounds*, **1**:684-688.

Man, H., Queiroz, E.L.D., and Smith, M. J. T., 2002, Three-dimensional subband coding techniques for wireless video communications, *IEEE Transactions on Circuits and Systems for VideoTechnology.*, **12**:386-397.

Marpe, D. and Cycon, H. L., 1999, Very low bit-rate video coding using wavelet-based techniques, *IEEE Transactions on Circuits and Systems for Video Technology.*, **9**:85-94.

Martin, M. A., Alberola, C., Sanz de Acedo, J., and Ruiz-Alzola, J., 2001, Multiresolution compression schemes for medical image and volume data, *Proceedings of the 23rd Annual EMBS International Conference.*, **3**:2437-2440.

Meier, T. and Ngan, K. N., 1999, Video segmentation for content-based coding, *IEEE Trans. on Circuits and Systems for Video Technology*, **9(8)**:1190-1203.

Moulin, P. and Krishnamurthy, R., 1997, Multiscale modeling and estimation of motion fields for video coding, *IEEE Transactions on image processing*, **6**:1606-1620.

Munteanu, A., Cornelis, J., Van Der Auwera, G., Cristea, P., 1999, Wavelet image compression - the quadtree coding approach, *IEEE Transactions on Information Technology in Biomedicine.*, **3(3)**:176-185.

Munteanu, A., Schelkens, P., and Cornelis, J., 2005, Compression of volumetric data in mobile health systems, chapter in: *M-Health: Emerging Mobile Health Systems*, R.H. Istepanian, S. Laxminarayan, and C.S. Pattichis, eds., Kluwer Academic/Plenum, New York.

Nosratinia, A., 2001, New kernels for fast mesh-based motion estimation, *IEEE transactions on circuits and systems for video technology*, **11**:40-51.

Ordonez, J. R., Gazuguel, G., Puentes, G., Solaiman, B., Cauvin, J. M., and Roux, C., 2001, Medical image indexing and compression based on vector quantization: image retrieval efficiency evaluation, *Proceedings of the 23rd Annual EMBS International Conference.*, **3**:2465-2468

Pappas, T. N., and Safranek, R. J., 2000, Perceptual Criteria for Image Quality Evaluation, chapter in: *Handbook of Image and Video Processing*, A. C. Bovik, ed.: Academic Press, New York, pp. 669-684.

Pereira, F. and Ebrahimi, T., 2002, *The MPEG-4 Book*: IMSC Press, Pearson Education, Upper Saddle River, New Jersey.

Puentes, J., Feuvre, J., Roux, C., Solaiman, B., Cauvin, J. M., Robaskiewicz, M., and Debon, R., 2000, Gastroenterology patient record consultation based on MPEG-4 scene composition, *Proceedings of the 22nd Annual EMBS International Conference.*, **3**:2337-2338.

Redmill, D. W., and Kingsbury, N. G., 1996, The EREC: An error resilient technique for coding variable-length blocks of data, *IEEE Transcations on Image Processing*, **5**:565-574.

Rodriguez, V., P., Pattichis, M. S., Pattichis, C. S., Abdallah, R., and Goens, M.B., 2005, Object-based ultrasound video processing for wireless transmission in cardiology, chapter in: *M-Health: Emerging Mobile Health Systems*, R.H. Istepanian, S. Laxminarayan, and C.S. Pattichis, eds., Kluwer Academic/Plenum, New York.

Said, A., and Pearlman, W. A., 1996, A new, fast, and efficient image codec based on set partitioning in hierarchical trees, *IEEE Transactions on Circuits and Systems for Video Technology*, **6(3)**:243-250.

Siu, M., Chan, Y.H., and Siu, 2001, A robust model generation technique for model-based video coding, *IEEE Transactions on Circuits and Systems for Video Technology*, **11(11)**:1188-1192.

Song, X., Chiang, T., and Lee, X., 2000, New fast binary pyramid motion estimation for MPEG2 and HDTV encoding, *IEEE Transactions on Circuits and Systems for Video Technology.*, **10**:1015-1028.

Steinbach, E., Farber, N., and Girod, B., 1997, Standard compatible extension of H.263 for robust video transmission in mobile environments, *IEEE transactions on circuits and systems for video technology*, **7**:872-881.

Sun, H., and Kwok, W., 1995, Concealment of damaged block transform coded images using projections onto convex sets, *IEEE Transactions on Image Processing*, **4**:470-477.

Takishima, Y., Wada, M., and Murakami, H., 1995, Reversible variable length codes, *IEEE Trans Commun.*, **43**:158-162.

Tekalp, A. M., 1995, *Digital Video Processing*: Prentice Hall PTR, Englewood Cliffs, New Jersey. Tekalp, A. M., Beek, P. V., and Toklu, C., 1998, Two-dimensional mesh-based visual-object representation for interactive synthetic/natural digital video, *Proc. IEEE*, **86**:1029-1051.

Tong, X., and Gray, R. M., 2000, Coding of multi-view images for immersive viewing, *ICASSP 2000*, 2000.

Tourapis, A. M., Au, O. C., and Liou, M. L., 2002, Highly efficient predictive zonal algorithms for fast block-matching motion estimation, *IEEE transactions on circuits and systems for video technology*, **12**:934-947.

Turaga, D. S. and Chen, T., 2002, Model-based error concealment for wireless video, *IEEE Transactions on Circuits and Systems for Video Technology.*, **12**:483-495.

Torres, L. and Kunt, M., 1996, *Video Coding: The Second Generation Approach*, Kluwer Academic Publishers, Boston.

Wada, M., 1989, Selective recovery of video packet loss using error concealment, *IEEE J. Select. Areas Commun.*, **7**:807-814.

Wang, Y., and Ostermann, J., 1998, Evaluation of mesh-based motion estimation in H.263-like coders, *IEEE Transcations on Circuits and Systems for Video Technology.*, **8**:243-252.

Wang, Y., Ostermann, J., and Zhang, Y. Q., 2002, *Video Processing and Communications*: Prentice Hall, Englewood Cliffs, New Jersey.

Wang Y and Lin, S., 2002, Error-resilient video coding using multiple description motion compensation, *IEEE Transactions on Circuits and Systems for Video Technology*, **12(6)**:438-452.

Woods, J. W., Han, S., Hsiang, S., and Naveen, T., 2000, Spatiotemporal subband/wavelet video compression, Object-based video coding, chapter in *Handbook of Image and Video Processing*, A. C. Bovik, Ed.: Academic Press, New York, pp. 575-584.

Xiong, Z., Kim, B. J., and Pearlman, W. A., 1998, Progressive video coding for noisy channels, *IEEE Int. Conf. On Image Processing (ICIP 1998)*, **1**:334-337.

Yan, Z., Kumar, S., and Kuo, C.C.J., 2001, Error-resilient coding of 3-d graphic models via adaptive mesh segmentation, *IEEE Transactions on Circuits and Systems for Video Technology*, **11**:860-873.

Yeung, F., Levinson, S. F., Fu, D., and Parker, K. J., 1998, Feature-adaptive motion tracking of ultrasound image sequences using a deformable mesh, *IEEE Transactions on Medical Imaging.*, **17**:945-956.

Youngseop, K., and Pearlman, W. A., 2000, Stripe-Based SPIHT lossy compression of volumetric medical images for low memory usage and uniform reconstruction quality, *IEEE Int. Conf. on ImageProcessing (ICIP 2000)*, **3**:652-655.

Yu, G. S., Liu, M. M., and Marcellin, M. W., 1998, POCS-based error concealment for packet video using multiframe overlap information, *IEEE transactions on circuits and systems for video technology*, **8**:422-434.

Zahariadis, T. and Kalivas, D., 1996, A spiral search algorithm for fast estimation of block motion vectors, *Proc. EUSIPCO '96*, **2**:1079–1082.

Zhang, Y. Q., Liu, Y. J., and Pickholtz, R. L., 1994, Layered image transmission over cellular radio channels, *IEEE Trans. Vehicular Tech.*, **43(3)**:786-796.

Zhu, S., and Ma, K. K., 2000, A new diamond search algorithm for fast block-matching motion estimation, *IEEE Transcations on Image Processing*, **9**:287-290.

FUTURE CHALLENGES AND RECOMMENDATIONS

Marios S. Pattichis[*]

The future needs of signal and image processing applications in e-Health will involve a multitude of different signals, ranging from one-dimensional signals such as the ECG to real-time color video signals. There will also continue to be strong demand to move more and more services to smaller, low-power, compact computing devices. The challenge to eHealth signal and image processing systems is to deliver the highest possible quality while minimizing the computational power and bandwidth requirements. We begin with a general discussion of basic signal compression challenges associated with single signal and image processing modalities, and then consider challenges associated with the joint compression of a collection of different modalities. We discuss future challenges not only from the perspective of the development of real-time, collaborative systems, but also from their intended uses in Computer Aided Diagnosis (CAD) systems.

There has been substantial progress made in the processing and analysis of one-dimensional biomedical signals (see Alesanco et al., 2005). Their bandwidth requirements can usually be met, and their strong diagnostic value can make them an essential part of most future collaborative systems. They are essential for continuous, real-time monitoring. Clearly, if a medical condition can be detected using a one-dimensional signal, such as the ECG, then we should avoid using images and/or video to accomplish the same task. In general, lower dimensional signals require significantly less bandwidth and computational power. In joint-processing, it is important to consider the use of one-dimensional signals to reduce bandwidth and computation of higher-dimensional signals. Furthermore, we should consider scalable coding systems, where one-dimensional biomedical signals belong to the base-layer with strong protection from transmission error.

There are still substantial challenges associated with the representation of medical images (see Chiu et al., 2001; Skodras, 2005; Munteanu et al., 2005). A general observation that can be made that applies to most medical imaging modalities is that they represent two-dimensional slices through moving or deforming 3D objects. This leads to

---

[*] Marios S. Pattichis is with the Department of ECE, Room 229-A, ECE Building, The University of New Mexico, Albuquerque, NM 87131-1356, USA.

motion artifacts in CT and MRI, as well as resolution issues on the distances between the slices. There is also a need to standardize image acquisition parameters such that pixel intensity values become meaningful. In general, device independent representations should be preferred. We offer three examples:

  (i)   device independent representations imply color images that have been gamma corrected,
  (ii) to be able to reproduce ultrasound images, we need to record and standardize gain settings in ultrasound systems, and
  (iii) the non-uniform illumination correction algorithm should be listed for MRI.

For intended use with Computer Aided Diagnosis (CAD) systems, image standardization and calibration methods can be used to reduce the variance in the computed features. This will lead to the development of large, standardized, medical imaging libraries, which will in turn lead to wider application of CAD systems.

Due to the high-bandwidth requirements, image and video compression methods will continue to play an important role in future, real-time collaborative environments (see Vieyres et al., 2005; Kyriacou and Pavlopoulos, 2005). At the most basic level, medical image quality assessment is an important area of future research. Traditional mean-square error measurements do not necessarily correspond to perceptual quality, and may correspond even less to diagnostic quality. As an example, image quality assessment over the near-regions of ultrasound video is not useful, while in general, the users expect the highest quality in regions that are in-focus of the ultrasound beam. In addition, there will continue to be strong interest in region of interest (ROI) and object-based coding methods. The challenges associated with applying these methods require the development of effective segmentation methods.

Scalable image and video coding will see continual development (Cho and Pearlman, 2002). For medical imaging applications, there are many challenges in defining diagnostically relevant scalable methods. There are obvious applications in object-based scalability, not only where the object is of diagnostic interest, but also in defining the base layer and enhancement layers so that the base layer is of diagnostic significance. In transmitting video images of the patient, on-going and future research on facial image coding will be important in determining the patient's feelings and reactions during medical exams. There is also a need to consider new image compression models that correspond more closely with the structured texture and image acquisition characteristics of medical video.

As an example for object-based coding, consider the acquisition and display of motion-mode (M-mode) ultrasound. In M-mode ultrasound, a single image line can be used to capture all the changes in the video signal. A specialized M-mode codec will likely be significantly better than a general-purpose decoder. Similarly, a specialized text codec for video images will also allow future, content-based functionalities for the compressed video content.

From the wireless communications perspective, video image compression research will continue in areas such as error-resilience and error-concealment (Wang et al., 2002). Especially for error-resilience methods, there will be strong interest in new encoding schemes that allow for robust decoding. For medical imaging applications, a small percentage of errors could be tolerated and their effects minimized through error-concealment. Future research in error-concealment methods should take into account the complex nature of biomedical images. Over a single video, different interpolation schemes should be employed for text objects, background, and texture objects. We should also consider the development of new interpolation methods that are a function of well-established imaging parameters such as the use of different interpolation methods for near-field and in-focus regions in ultrasound video. For in-focus regions, we can consider accurate, yet computationally expensive methods. On the other hand, we can use fast and somewhat less accurate methods for near-field regions.

Many signal and image processing challenges lie in the joint processing and transmission of biomedical signals, images, and (see Kyriacou and Pavlopoulos, 2005). We list challenges in three areas: (i) multi-modality signal synchronization, (ii) joint signal, image and video compression, and (iii) interactive collaborative environments. The basic application is the development of high-quality collaborative environments, where a variety of biomedical signal and images are exchanged.

For joint decoding, there are significant synchronization issues. Clearly, the one-dimensional biomedical signals should correspond to the video images. As an example, we note the synchronization of the ECG, respiratory, 2D, and Doppler signals in ultrasound systems. In addition, we note that two-way voice communications must be synchronized to all clinical signals, as well as to real-time video images of the patient and the doctor. Re-synchronization in the presence of wireless communication errors will require innovative error-resilience methods.

For well-synchronized signals and images, we can develop methods for joint signal, image and video compression. We offer an example in video image compression. In cardiac imaging using ultrasound, the ECG signal can be used to predict changes in the video imaging signal. Thus, we can appropriately adapt the rate requirements, so as to allocate more rate to capture significant motion, while allocating less rate for anticipated less motion.

Scalable coding for jointly coded signals, images, and video also presents significant challenges. It is important to decide how to jointly break the one-dimensional biomedical signals with the voice, images and video to form independently encoded blocks. Then, the base and enhancement layers in such layers should be decided based on diagnostic measures, as well as with regards to available bandwidth. Clearly, we will require spatiotemporal scalability, as well as object-based (or region-based) quality scalability. In addition, in the future, we want to consider content-based access for the jointly encoded signals.

There are many challenges associated with the use of interactive, collaborative environments. As an example, all the MPEG-2/MPEG-4 functionalities (see Haskell et al., 1996; Pereira and Ebrahimi, 2002) need to be re-thought of, in the context of synchronized, jointly-compressed signals. The users may be reviewing a particular signal, asking to see the corresponding signals (images or video) from other modalities. Such an interactive preview capability requires the development of fast joint-decoding methods. For real-time collaborative work, the heterogeneity of the networks, computing systems and image displays, will be best served by innovative, scalable, network-aware systems. In conclusion, we note that the high quality requirements of e-Health systems will only be met by addressing particular clinical needs.

REFERENCES

Alesanco, A., Garcia, J., Olmos, S. and Istepanian, R. S. H., 2005, Resilient ECG wavelet coding for wireless real-time telecardiology applications, chapter in: *M-Health: Emerging Mobile Health Systems*, R.H. Istepanian, S. Laxminarayan, and C.S. Pattichis, eds., Springer Science.

Chiu, E., Vaisey, J., and Atkins, M. S., 2001, Wavelet-based space-frequency compression of ultrasound images, *IEEE Transactions on Information Technology in Biomedicine.*, **5**:300-310.

Cho, S. and Pearlman, W. A., 2002, A full-featured, error resilient, scalable wavelet video codec based on the set partitioning in hierarchical trees (SPIHT) algorithm, *IEEE Transactions on Circuits andSystems for Video Technology*, **12**:151-170.

Haskell, B. G., Puri, A., and Netravali, A. N., 1996, *Digital Video: An Introduction to MPEG-2*, Chapman & Hall, Boca Raton, Florida, U.S.A..

Kyriacou, E. and Pavlopoulos, E., 2005, Emergency health care systems and services: future challenges, chapter in: *M-Health: Emerging Mobile Health Systems*, R.H. Istepanian, S. Laxminarayan, and C.S. Pattichis, eds., Springer Science.

Munteanu, A., Schelkens, P., and Cornelis, J., 2005, Compression of volumetric data in mobile health systems, chapter in M-Health: Emerging Mobile Health Systems, R.H. Istepanian, S. Laxminarayan, and C.S. Pattichis, eds., Springer Science.

Pereira, F. and Ebrahimi, T., 2002, *The MPEG-4 Book*: IMSC Press, Pearson Education, Upper Saddle River, New Jersey.

Skodras, A. N., 2005, The JPEG2000 image compression standard in mobile health, chapter in: *M-Health:Emerging Mobile Health Systems*, R.H. Istepanian, S. Laxminarayan, and C.S. Pattichis, eds., Springer Science.

Vieyres, P., Poisson, G., Triantafyllidis, G., Pattichis, M., and Sakas, G., 2005, Future challenges and recommendations on echography systems and services, chapter in: *M-Health: Emerging Mobile Health Systems*, R.H. Istepanian, S. Laxminarayan, and C.S. Pattichis, eds., Springer Science.

Wang, Y., Ostermann, J., and Zhang, Y. Q., 2002, *Video Processing and Communications*: Prentice Hall, Englewood Cliffs, New Jersey.

IV. EMERGENCY HEALTH CARE SYSTEMS AND SERVICES

Sotiris Pavlopoulos

Section Editor

SECTION OVERVIEW

Sotiris A. Pavlopoulos[*]

The main goal of modern health care systems is the improvement of health of the populations they serve. At present, there is a trend for decentralization of health care institutions and for a move from institutional health care to primary and home care and from impatient care to outpatient setting. This results in an increased need for specialized care in outpatient setting and out-hospital environments. Recent advances in telecommunications and informatics technologies has created a significant impact in healthcare service provision and the Emergency Health services have been one of the divisions that has significantly benefited from these advances. In this chapter we present current and emerging applications of mobile technologies in the fields of Emergency Health care systems and services. A brief review of the spectrum of these applications and potential benefits are included in this overview.

Emergency Healthcare Telematics can be defined as the use of modern telecommunication and informatics technologies to support Emergency medical services including clinical practice provision, administration, research and education activities. The ultimate goal is to exploit modern technologies to provide cost-efficient high quality services to the community. Current implementations of telematics applications for Emergency Medicine include Computer Assisted Dispatch applications (with the use of Geographical Information Systems – GIS), computerized patient records, telemetry systems as well Human Resource Management systems, cost-analysis systems, etc.

In a report prepared by the American College of Emergency Physicians (ACEP) in 1998 (The future and Emergency Medicine, 1998) there are specific references to current and future applications of emerging technologies in the practice of Emergency Medicine (Case, 1998). More specifically, it is anticipated that wireless communications technologies will greatly impact Emergency care provision in the near future. It is also evident that wireless telemedicine systems and services are expected to enhance traditional emergency care provision not only within the Emergency Department but also in a variety of pre-hospital Emergency care situations where geographically remote consultation and monitoring can be implemented. The existence of high-bandwidth

[*] Sotiris A. Pavlopoulos, School of Electrical and Computer Engineering, National Technical University of Athens, Greece.

mobile communication links can ensure use of telemedicine resources to facilitate acute and non-acute care provided in the field and to provide a direct link between field personnel and medical direction (Krohmer, 1998).

Many studies have demonstrated the potential benefits, on patient survival, of early and specialized pre-hospital patient management (Guidelines for the Early Management of Patient with Myocardial Infarction, 1994). For example many studies worldwide have proven that a rapid response time in pre-hospital settings resulting from treatment of acute cardiac events decreases mortality and improves patient outcomes dramatically (Canto et al., 1997; Stults et al, 1984; Weaver et al., 1986; Cummins et al., 1987; Schrading et al., 1993; Shuster and Keller, 1993). In addition, other studies have shown that 12-lead ECG performed during transportation increase available time to perform thrombolytic therapy effectively, thus preventing death and maintaining heart muscle functionSedgewick et al., 1993. The reduction of all those high death rates is definitely achievable through strategies and measures, which improve access to care, administration of pre-hospital care and patient monitoring techniques. Along that direction, a number of systems have been developed to allow for specialized pre-hospital care with the use of modern telematic technologies. As early as 1996, researchers at the National Technical University of Athens have successfully demonstrated real time transmission of ECG data from a moving ambulance, using GSM data links (Pavlopoulos et al., 1996; Pavlopoulos et al., 1997; Pavlopoulos et al., 1998; Kyriacou et al., 1999). In 1997, a team from the University of Maryland Hospital has also implemented with success a wireless Telemedicine system for stroke victims (Gagliano, 1998). Furthermore, researchers from the school of medicine of the Kapodistrian University of Athens elaborated an experimental setup to transmit 12-lead ECGs from an ambulance (Giovas et al., 1998) based on a store-and-forward approach. A real-time emergency telemonitoring system has been developed and validated within the Emergency-112 EU funded project (Pavlopoulos, 1998) in 1998 and demonstrated initial patient benefits. Other research groups (McKee et al., 1996; Schimizu, 1999) have performed similar work towards the implementation of emergency telemedicine using mobile communication. From all relevant studies it is evident that mobile communication can provide extensive support in Emergency Healthcare provision both in terms of remote supervision and in terms of teleconsultation. Many of the organizational and technical problems encountered in the experiments preformed can be resolved by implementing appropriate strategies.

In this section, a number of telematics applications for pre-hospital care support are presented. The selected applications demonstrate the usefulness of mobile communication techniques and technologies in supporting emergency care provision. Applications span from pre-hospital collaboration to real-time telemonitoring and remote supervision of patients in emergency care provision.

The paper by S. Walderhaug et al. presents an information system for improvement and support of pre-hospital collaboration. The system presented allows:

- $\in$ the continuous exchange and synchronization of shared information between medics, hospital, transport and other supporting units through ad-hoc wireless communication and passive storage devices.
- $\in$ context and role awareness through explicitly represented plans, tasks and procedures.
- $\in$ collaboration support by exchange of information about performed, ongoing and planned tasks for each patient.

• ∈ distributed and replicated information storage in "Patient Information Carriers" (PICs), PDAs, and centralized record systems.

The paper by E. Kyriacou et al presents the experience from an implementation of a portable emergency telemedicine system based on wireless (mobile) technology. System allows the real-time transmission of vital biosignals (3/12-lead ECG, SpO$_2$, NIBP, Temp) and still images (snap-shots) of the patient. Both technical and clinical issues from the implementation of the system are discussed

The paper by K. Shimizu also deals with mobile emergency telemedicine systems and the concept of emergency care provision in a moving vehicle. The practicality of such a system is investigated through technical considerations required to realize mobile telemedicine. Paper addresses problems such as channel capacity, trade-off between reliability and real-time operation and electromagnetic interference. The results of experimental work on the application era are presented and discussed.

The paper by C. Nugent et al. deals with the application area of telecardiology and presents its evolution in time with emphasis in the present and the future issues related to technological and clinical aspects.

REFERENCES

Auer N.J., Ed., *The Future of Emergency Medicine* (1998).

Canto J.G., Rogers W.J., Bowlby L.J., et al., The prehospital electrocardiogram in acute myocardial infarction: is its full potential being recognized?, *J Am College Cardiol*. **29**, 498-505 (1997).

Case R.B., The role of emerging technologies in the practice of Emergency Medicine, in: *The future of Emergency Medicine*, ed. N. J. Auer, 25-27 (1998).

Cummins R.O., Eisenberg M.S, Litwin P.E., Graves J.R., Hearne T.R., Hallstrom A.P., Automatic external defibrillators used by emergency medical technicians: a controlled clinical trial, *JAMA* **257**(12), 1605-1610 (1987).

Gagliano D., Wireless ambulance telemedicine may lessen stroke morbidity, *Telemedicine Today* **6**(1), 21(1998).

Giovas P., Papadogiannis D., Thomakos D., et al, Transmission of electrocardiograms from a moving ambulance, *Journal of Telemedicine and Telecare*, **4** (Sup. 1), 5-7 (1998).

Krohmer J.R., Emergency Medical Services of the future, in: *The future of Emergency Medicine*, ed. N. J. Auer, 5-9 (1998).

Kyriacou E., Pavlopoulos S., Bourka A., Berler A., Koutsouris D., Telemedicine in emergency care, in: *Proceedings of the VI International Conference on Medical Physics* (Patras, Greece, 1999).

McKee J.J., Evans N.E. and Owens F.J., Digital transmission of 12-lead electrocardiograms and duplex speech in the telephone bandwidth, *Journal of Telemedicine and Telecare* **2**, 42-49 (1996).

Pavlopoulos S., Dembeyiotis S., Konnis G., and Koutsouris D., AMBULANCE - Mobile unit for health care provision via telematics support, in: *Proceedings of the IEEE Engineering in Medicine & Biology*, Amsterdam, The Netherlands, October 1996.

Pavlopoulos S., Dembeyiotis S., Koutsouris D., An augmented reality system for health care provision via telematics support, in: *Proceedings of the MMVR-97 Conference*, San Diego, USA, 1997.

Pavlopoulos S., Emergency-112 demonstrator functional specs-architecture design, *Emergency-112 Project HC4027 Deliverable D3.1*, (1998).

Pavlopoulos S., Kyriakou E., Berler A., Dembeyiotis S., Koutsouris D., A novel emergency telemedicine system based on wireless communication technology – AMBULANCE, *IEEE Trans. Inform. Tech. Biomed. - Special Issue on Emerging Health Telematics Applications in Europe* **2**(4), 261-267(1998).

Schimizu K., Telemedicine by mobile communication, *IEEE Eng. Med. Biol.*, **18**, 32-44 (1999).

Schradi ng W.A., Stein S., Eitel D.R., Grove L., Horner L., Stechkert G., Sabulsky N.K., Ogden C.S., Hess D.R., An evaluation of automated defibrillation and manual defibrillation by emergency medical technicians in a rural setting, *Am J Emerg Med*, **11**(2), 125-130 (1993).

Sedgewick M.L., Dalziel K., Watson J., Carrington D.J., Cobbe S.M., Performance of an established system of first responder out-of-hospital defibrillation. The results of the second year of the heartstart Scotland project in the utstein style, *Resuscitation* **26**, 75-88 (1993).

Shuster M., Keller J.L., Effect of fire department first-responder automated defibrillation, *Annals Emerg Med* **22**(4), 721-727 (1993).

Stults K.R., Brown D.D., Schug V.L., J.A. Bean, Prehospital defibrillation performed by emergency medical technicians in rural communities, *NEJM* **310,** 219-223 (1984).

Weaver W.D., Copass M.K., Hill D.L., Fahrenbruch C., Hallstrom A.P., Cobb L.A., Cardiac arrest treated with a new automatic external defibrillator by out-of-hospital first responders, *Am J Cardiol* **57**, 1017-1021 (1986).

Weston C.F., Penny W.J., Julian D.G., Guidelines for the Early Management of Patient with Myocardial Infarction, *BMJ* **308**, 767-771 (1994).

ECG TELECARE:
PAST, PRESENT AND FUTURE

Chris D. Nugent[*], Haiying Wang, Norman D. Black, Dewar D. Finlay, and Frank J. Owens

1. INTRODUCTION

Current advancements in computing power, digital signal processing techniques and data communications have helped to usher in a new era in the delivery of medicine and patient management. The coming together of these technologies into the discipline of telemedicine facilitates the remote diagnosis and care management of patients, irrespective of location, by providing the opportunity to link remote clinical expertise with any primary care, secondary care or other health care centre. The increasing quality and fidelity of transmitted clinical data along with high levels of interaction means that telemedicine systems allow the possibility for examinations on a level similar to that of conventional examinations. Telemedicine now offers the opportunity for examination and management of patients by a specialist in a remote location almost as if both were co-located.

As communication systems continue to develop and move from being fixed or terrestrial in nature to being mobile, we can expect the flexibility and application of telemedicine based care to expand. At present, some systems allow for remote diagnosis by using a combination of mobile and fixed communication systems. In the future we shall see an increased reliance on mobile technology and with it the potential to offer additional services such as home-based care models, remote and personal consultations and others not yet fully realised.

One area of medicine which is heavily embroiled in telemedicine developments, for obvious reasons, is that of cardiology. Given the name 'Telecardiology' this application has seen a number of very positive developments over the years which has had a major impact on both the treatment of the condition and the mortality rates. Advances in both communication and medical device technologies can now provide flexible telecardiology

[*]Chris D. Nugent, University of Ulster, Jordanstown, Northern Ireland, BT37 0QB. Email: cd.nugent@ulster.ac.uk.

solutions which have significant potential in the prevention of serious cardiac disorder. This chapter examines the developments in telecardiology and looks forward to how they might appear in the future.

1.1. Cardiology

Heart disease still remains one of the major causes of premature deaths. With careful clinical evaluation, some of the causes of heart disease can be foreseen and prevented. Many factors contribute to the likelihood of developing a heart disease; cholesterol levels, blood pressure levels, family history etc. and hence, in certain conditions, these must be carefully monitored. Various clinical tests may be performed to evaluate the status of the heart. One of the most commonly employed non-invasive tests is that of the Electrocardiogram (ECG) which records the electrical activity of the heart. Careful consideration of this electrical information provides sufficient detail to identify and diagnose a number of cardiac abnormalities which may induce untimely deaths.

The ECG has established itself as a standard tool in modern clinical medicine. It is a non-invasive technique, painless to the patient, inexpensive, simple to use and the most popular practical means of recording the cardiac activity, in electrical terms (Wellens, 1990). The major advantage of the ECG is its relationship to physiology. It is possible to correlate the recorded electrical activity of the ECG with the fundamental behaviour of the heart. Hence, through careful analysis, it is possible to establish relationships between electrophysiological events and measured signals.

From a diagnostic point of view, the ECG offers two perspectives. Firstly, by analysis of wave shapes it is possible to describe the condition of the heart's working muscle masses. Secondly, consideration of the rate of the cardiac cycle provides rhythm statements which give additional diagnostic information. In normal subjects, the ECG remains reasonably constant, whereas, under pathological conditions, several pertinent differences appear. Not all cardiac abnormalities, however, are identifiable by the ECG. Nevertheless, when used in conjunction with other clinical techniques e.g. angiography and echocardiography, an accurate 'picture' of the heart may be obtained for diagnostic purposes.

The ECG can be taken in several ways. Single lead recordings employ two electrodes on the patient's body and produce a single ECG trace. These are useful for continuous or ambulatory methods (Holter recordings) of rhythm monitoring. The 12-lead ECG records differences in voltages between 10 electrodes placed on the patient's body (see Figure 1). The 12 leads are subdivided into 2 groups: the limb leads measuring the cardiac activity in the frontal plane of the body and the chest leads or precordial leads measuring the cardiac activity in the horizontal plane of the body. As the number of leads are increased so too is the information attained. Hence, the 12-lead ECG provides a fuller consistent diagnosis in comparison with single (or configurations of single) lead recordings. A Body Surface Mapping approach may also be utilised whereby many recording sites on the patient's body are used (for example >80). With this increase in the number of sensors, in comparison to the 12-lead approach, a more complete picture of the electrical activity of the heart may be obtained. The potentials recorded from the Body Surface Map are reconstructed as contour maps for each instantaneous time period. The Body Surface Mapping approach has been widely accepted as mainly a research tool, however, from an electrical point of view, the 12-lead

ECG is the most commonly performed procedure in the standard routine daily examination of cardiac function (Cox et al., 1972).

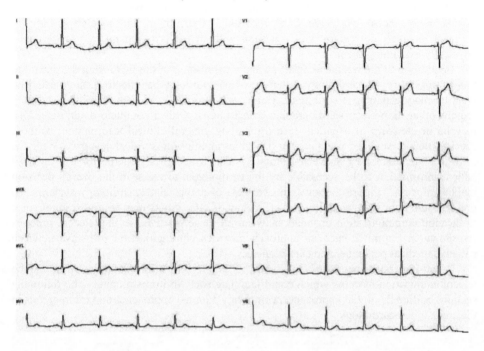

Figure 1. 12-Lead ECG Recording indicating electrical activity in both frontal and horizontal planes of the body.

1.2. Telecardiology

Telemedicine, which may be broadly defined as the use of communication infrastructures to provide medical information and healthcare delivery at a distance is an emerging field in healthcare. Its general aim is to improve the quality of healthcare by at least maintaining patient outcome and offering increased flexibility while reducing delivery costs. The wholesale application of telemedicine is not yet possible but, for selected medical modalities such as telecardiology, there are significant benefits to be achieved in its application.

A telecardiology system generally consists of a centralised communications facility, or hub, to which are attached satellite devices or nodes. The central location, may be a primary or secondary care organisation while the satellite nodes could be GP surgeries and/or (home based) medical devices. Telecardiology in general relates to the transmission of any form of cardiac information for example ECG recordings, echo images, cath images etc. For the purpose of the current study we have focused on the transmission of ECG recordings. ECG recordings can be generated at the nodes and the information subsequently transmitted to the hub for viewing or further processing. Following analysis and possible archival, the diagnosis and potential recommendation for treatment may be relayed back to the remote site (or nodes) i.e. patient and physician/career.   This provides an infrastructure whereby patients with artificial pacemakers, suspected cardiac abnormalities, patients in rural areas, patients with post-operative or post myocardial infarct conditions or general emergency situations (e.g. cardiac ambulances) can have ECG recordings transmitted and receive a form of feedback on their current cardiac condition without the requirement of a cardiologist physically examining them.

In addition to the remote acquisition and transmission of the ECG from the patient to the control centre, telecardiology systems provide additional benefits to cardiologists who wish to analyse patient recordings.   Along with the raw ECG data, cardiologists often require other, non-electrocardiographic data to assist with a complete diagnosis.   This may be in the form of a patient record detailing general clinical information, or more specific details such as previous cardiac abnormalities or surgical history.   With the infrastructure offered by a telecardiology system, it becomes possible for this array of patient information to be accessible by the cardiologist to assist in the overall decision making process.   Figure 2 shows an overview of a typical telecardiology system.   To preserve and improve the quality and efficiency of care, it becomes of utmost importance to develop communication channels between all those involved in the telecare process. As shown in Figure 2, such an approach provides communication pathways between patients, medical personnel and cardiologists.

What has been neglected in Figure 2 is an indication or description of the telecommunication systems which could facilitate such an infrastructure.   The following sections outline the initial approaches, commonly adopted approaches and future trends in the arena of telecardiology.

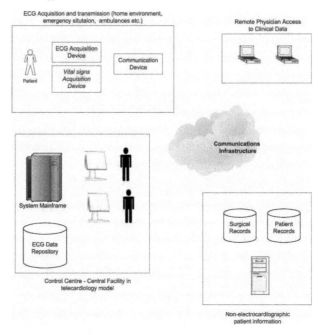

Figure 2. Telecardiology system overview providing communication between patients, medical personnel and cardiologists.

2. ECG TELECARE: THE PAST

Einthoven's work published in 1906 (Einthoven, 1906) is considered to be the first organised presentation of normal and abnormal ECG recordings using the string galvanometer. Although at the time focus was placed mainly on the phenomena associated with the clinical information acquired and presented and additionally the ability to present differences between normal and abnormal patients, this study can be considered to be one of the first documented instances of telecardiology.

Due to logistical problems of obtaining ECGs from hospital inpatients some distance away from his laboratory where the non-portable string galvanometer was located, part of Einthoven's experimental apparatus included a telecommunications link. This took the form of connecting the patient in the hospital to telephone wires which carried the electric current from the patient to the laboratory where the galvanometer traced the ECG. Results from this initial study indicated that a recording taken in this manner and subsequently compared with a recording taken with the patient in the laboratory connected directly to the galvanometer were identical, hence proving the principles of telecardiology.

Although technology has advanced since Einthoven's first transmission, the general principle has remained the same. Telephone systems, or more specifically the public switched telephone network (PSTN) has been used to transfer an ECG from a remote location to a central hospital, control centre or primary care organisation. Techniques during the early stages of telecardiology were largely based on analogue signal processing approaches and frequency modulation (FM) or frequency division multiplexing (FDM). Many single lead systems used the FM approach (Cooper and Caceres, 1965), whereas FDM, allowing the available bandwidth to be divided up, was used mostly to support transmission of a number of leads in multi-lead systems.

Following Einthhoven's initial study, the next subsequent study in ECG transmission was reported by Rahm et al. (Cooper and Caceres, 1965). In this approach ECGs were transmitted over telephone lines through the use of acoustically coupled analogue FM equipment. As with Einthoven's approach, Rahm et al. used specially conditioned telephone lines and it was not until 1953 that the PSTN was used without any special arrangements being made (Cooper and Caceres, 1965). The process of telecardiology in these first 50 years was streamlined from an analogue PSTN perspective. From an ECG perspective, the advances made during this period in nomenclature, wave definition and electrode positioning were significant and have changed little since their inception/definition. The next major developmental step following the digital era and the introduction of computers to medicine, was the digitisation of telecardiology.

3. ECG TELECARE: THE PRESENT

Single or multi-lead transmissions of ECGs over the PSTN using analogue techniques are now well established. Analogue approaches, however, lead to corruption of the ECG signal due to poor quality analogue communication lines. Invariably such transmissions suffer from distortion due to the inherent channel imperfections of the various media i.e. noise and non-ideal frequency response. In addition, associated with the PSTN are bandwidth limitations (300-3400Hz) and an inability to transmit signals down to 0Hz. Analogue approaches are well suited to instances of non-critical monitoring tasks where the added noise and distortion of the PSTN do not have a significant impact on the observation of rhythm changes. However, for more precise clinical diagnosis, the integrity is not acceptable.

A natural progression has been made to support telecardiology recordings via digital communications. In comparison with analogue approaches McKee et al. (1996) indicate the following advantages:

- ∈ Low cost – digital technology, for example in VLSI form, is much cheaper than complex analogue circuitry.
- ∈ Data integrity – channel noise and interference can be addressed.
- ∈ Utilization – Time Division Multiplexing (TDM) can provide better bandwidth than FDM.
- ∈ Integration – diverse signals can be carried in a common format.

Through the use of modems, it is possible to convey digital information over analogue PSTN lines. Many attempts have been made to adopt a digital approach to the transmission of ECGs over the PSTN. Berson in 1963 (Coopers and Caceres, 1965) reported one of the first attempts which involved the use of an acoustically coupled modem. Unfortunately, the bit rate of any such transmissions are constrained by PSTN limits. For example, a single lead digitised ECG at a rate of 250 samples per second with 16 bit resolution generates 4000 bits per second. With modern modems, supporting 56 kilo bits per second or higher, real-time transmission of multi-lead ECGs becomes difficult, especially if other data is also required. Hence to decrease transmission times, and move towards the reality of real-time telecardiology systems various attempts have been made and put in place to compress the ECG prior to transmission (McKee et al., 1994; Berbari, 1995). This has now become a necessity in digital approaches.

It has been recognised that to assist further in the process of telecardiology there is a requirement to transfer additional information in the form of speech. This can be useful to assist with call-setup, transfer of patient details, interrogation of the person making the recording and a means to provide patient diagnosis or feedback. Systems have been developed that provide a means of switchable ECG-Voice transmission (Hagan, 1965) and also simultaneous ECG and voice transmission (McKee, 1996).

3.1. Mobile Telecardiology

Although the approach of transmitting the ECG via the PSTN (via both analogue and digital means) is well established, it was recognised that restrictions of these systems were present due to the logistical presence of PSTN lines. Nevertheless, many manufacturers have integrated digital PSTN ECG transmission facilities into their products. To overcome these limitations it is necessary to introduce a mobile aspect into the telecardiology telecommunications framework. The facilitation of this mobile aspect supports the recording and transmitting of ECGs from, for example, ambulances, accident or emergency scenes and general rural areas not readily supported by the PSTN.

The first instances of mobile telecardiology were reported in the 1970's. These provided remote means of the acquisition and transmission of ECGs using conventional radio equipment. From the 1980's onwards, the popularity of commercial cellular networks were exploited to provide ECG transfer facilities.

One of the first studies carried out employed the transceivers of ambulances and used FM (Uhley, 1970). In this system, when ECGs were not being transmitted, the normal voice channel of the transceivers were available. Lambrew (1973) used a self-contained radio based system to transmit from either ambulances or helicopters either the ECG information or voice. Pozen (1977) and Cayten (1985) also employed mobile telecardiology systems from ambulances. Both systems permitted either voice or ECG information to be transmitted.

Grim et al. (1987) investigated the possibility of using cellular networks for the transmission of the 12-lead ECG. A portable recording machine was used and connected to a cellphone. Prior to transmission, the recorded ECG signal was digitised and compressed. Although, in principle, it was possible to demonstrate a mobile telecardiology approach in this manner, a problem at that time was found with the cellular network's capacity i.e. at certain times, due to network peaks, a transmission was not possible, however, when compared with approaches using ambulance transceivers this was not the case as proprietary radio communications could reserve emergency channels.

To further extend on the aspect of mobile telecardiology, systems have been made available which provide mobile transmission of both ECG and voice simultaneously. This has advantages of no instances of breaks in communications between the two users of the system. Shimizu (1999) provided a system whereby ECG and voice could be transmitted via a single vocal communication channel of a mobile phone. The approach offered real time transmissions and the facility of duplex communications. McKee et al. (1996) presented a system employing GSM mobile and conventional telephone technologies. The system facilitated a 5 second 12-lead ECG recording and supported simultaneous communication of digitised, full duplex speech and ECG signals.

With digital cellular mobile communications and compact acquisition machines employing efficient compression algorithms, mobile telecardiology is at present a reality. The challenges now lie in the standardisation of the approach and tailoring of the solution to all stakeholders involved.

3.2. ECG Transmission Standards

With the proliferation of computer based instrumentation and systems in medicine, the need to exchange information between heterogeneous computers has been recognised. This exchange of information may be ECG data, patient database records, physiological recordings or images. A number of interchange protocols to facilitate the structured transmission, storage and interchange of ECG information have been suggested, however, proprietary standards have tended to predominate. Various manufacturers use different techniques for not only measurement and interpretation, but also for the transmission and storage of data. As a result manufacture monopolisation looms (Willems, 1990).

A flat file structure is a commonly adopted approach within biomedical data transmission, however, it offers a very limited solution due to a lack of referencing, typed values, vocabulary control and constraints. Addressing this issue, was one of the goals of the Common Standards for Quantitative Electrocardiography project (CSE). The aim was to produce/provide a set of common standards for transmission, encoding and storage of digital data (Willems et al., 1990). What was produced was a standard communication protocol referred to as SCP-ECG supported by CENTC/251. Characteristics for this communication standard are its high flexibility which allows not only the transmission of the identification data and interpretation results but also the complete ECG recording. An additional standard provided by the International Society for Holter and non-invasive Electrocardiography provides a standard output file format intended to facilitate the exchange of ECG data in the field of Holter analysis.

As telecardiology systems have become established and recognised, much effort is currently being directed towards the standardisation of their exchange of information.

4. ECG TELECARE: THE FUTURE

As advances have taken place in the communications arena these have been equally matched with the advances in the development of medical devices. Modern ECG machines are no longer required to be devices in their own right, but now are miniature machines, smaller and lighter than standard ECG machines used in conventional hospital settings (Daja et al., 2001). It is no longer necessary to connect cumbersome ECG cables and apply electrodes to the patient, as new acquisition devices can have, for example, inbuilt sealed metal contacts in place of the use of conventional electrodes. These may be simply placed against the chest of the patient and the ECG signal recorded. Such an approach permits the use of the device by the patient and hence reduces the need of clinical input/assistance during ECG acquisition. This can be seen as a significant advantage in the progression towards improved patient empowerment. With ECG devices now being considered as miniature and the reality of mobile telecommunications supporting telecardiology, patients can carry on their person integrated medical and mobile devices and when an irregularity of the heart rhythm is experienced a recording can be made and relayed to a control centre for analysis. This provides the users with a degree of geographic freedom, eliminating the necessity to be in the home environment to facilitate telecare. Many patients who currently exploit home care monitoring systems in the cardiology domain feel uneasy to leave their home environment incase the need to contact, or transmit cardiac information to the control centre arises. With mobility and provision of geographic independence for the patient, telecardiology overcomes these constraints and has now become appealing to the patient.

Examples of systems combining the development of new miniature ECG acquisition devices and digital mobile communications have been presented in the form of wallets/purses and hybrid mobile phone devices. With these systems, the inclusion of miniature electronics permits the production of inconspicuous devices which users may carry on their person without any possible discomfort or embarrassment. It is possible to inconspicuously embed an ECG device into a wallet which at the appropriate time can be placed on the chest to acquire a recording. With this product, the recorded ECG signal can be acoustically coupled to a phone handset (mobile or fixed) and the subsequent recording relayed to a control centre. Figure 3 shows a device developed by Meridian Medical Technologies, which is currently being supported by the Sahal monitoring centre in Israel.

Another possibility is to include the ECG device within the housing of a standard mobile phone. Metal contacts can be incorporated on the back of the phone which can be connected to the acquisition circuitry. By placing the contacts onto the chest of the patient, the ECG can be recorded, subsequently compressed, digitised and transmitted via mobile communications modules all within the handset. These devices also offer the

Figure 3. Wallet ECG based system.

potential to incorporate GPS (Global Positioning System) in the handset to permit the exact location of the patient to be located in instances of emergency. Istepanian et al. (2001) have developed an approach involving a mobile telephone and a miniature ECG acquisition device. For modularity purposes, the mobile telephone and the ECG acquisition device were not integrated. Instead, the digitised and compressed ECG signals could be transferred from the acquisition device to the mobile telephone via an infra-red link. The protocol used was compliant with IrDa standards and hence provides the opportunity to use a number of different handsets to make the remote transmission. The device itself was designed as a waist mounted holster with a cradle for the mobile phone. When not acquiring ECGs, the mobile phone could be operated as normal. As can be seen from these new approaches, trends are moving towards personalised and tailored solutions for the patient, to produce inconspicuous and mobile solutions which strive to place healthcare more in the hands of the patients themselves and alleviate the impositions posed by non portable monitoring equipment.

4.1. Cardiologist Remote Access of Information

Not only is it necessary to provide a means of remote transfer of the acquired ECG data from the patient to the control centre, but it is a necessity to provide Cardiologists with the ability to remotely access patient information from the control centre. This may be necessary in instances were cardiologists are not available at the control centre and when second opinions are required (Shanit et al., 1996), or the general case of access to patient information/history during remote patient examinations.

Vassanyi et al. (2001) addressed this by providing a WAP (Wireless Application Protocol) interface to a cardiology repository via the Internet. Via the WAP interface on a mobile phone limited information could be retrieved and displayed providing the ability to remotely access the information. This provides a platform to address the issues mentioned above where cardiologists' input is not available at the control centre, however, it may be accessed remotely. Black et al. (1999) used a fax routine incorporated into the control centre's software to transmit a 12-lead ECG recording to a fax compatible mobile phone (see Figure 4 (a)). A potentially more interactive approach to this is the use of PDAs and mobile phones (or smartphones). The ability to send a fax to a mobile device is useful, but it only provides a simplex communications channel. WAP systems are also useful, but provide limited functionality in terms of interactivity with the user interface and have limited display content. Through the inclusion of appropriate web browsing software it is possible to access the Internet via a PDA with the communications channel being provided by a mobile phone (or conventional PSTN). PDAs and mobile phones can now be easily connected via infra red or Bluetooth links. Figure 4 (b) shows an example of a PDA displaying cardiac information. With such an approach, the user of the PDA can dictate, in a similar manner when using a standard web interface, the information content to be displayed/accessed. Barro et al. (1999) have developed an approach whereby an expert can retrieve cardiac information from a central location via the use of a PDA and modem. The information details can be selected by the user of the PDA and ECG information can be visually displayed through customised software hosted on the PDA. In a similar manner to providing tailored and personalised solutions to the patient, the aforementioned have provided the same potential benefits to cardiologists and medical personnel.

4.2. Teleconsultation and Decision Support

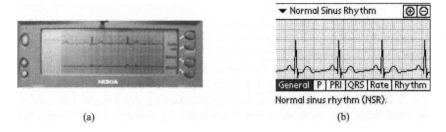

(a)   (b)

Figure 4. (a) 12-lead ECG transmission to a fax compatible mobile phone (b) PDA screen shot.

To date, the majority of efforts relating to telecardiology have been focused on the developments of the telecommunications infrastructure. These have focused on providing a reliable means to transmit the ECG recordings from a patient in a remote location to a central facility by both fixed and mobile telecommunication channels. As described in section 1.2 this forms a key aspect of a telecardiology system, but to provide a holistic approach to patient analysis and actively involve all stakeholders in the process and support their roles, trends have moved towards assisting the cardiologist in making the final patient diagnosis through assistive computerised means. Future trends in mobile telecardiology are currently addressing 2 issues:

1. Provision of interactive platforms between the physician/patient making the recording and the cardiologist.
2. Automation of the ECG classification through applications of Artificial Intelligence (AI).

4.2.1 Interactive Platforms

In use since 1984, computer-supported cooperative work (CSCW) (Reinhard, 1994), and particularly synchronous cooperation, which allows geographically distant users to work together, in realtime and on the same data, opens up new opportunities for telecardiology and remote diagnostic support. The recent advances in mobile telecommunication technologies have allowed telecardiology over mobile communication systems to be a reality. However, building an interactive teleconsultation system over the low-bandwidth mobile channels still remains a technical challenge.

It is evident that an integration of a CSCW-based platform with mobile telecardiology applications will have an important role in creating versatile and cost effective alternatives to health care delivery. Wang *et al.* (2001) reported a promising way to build a CSCW-based mobile telecardiology system. The system consists of a mobile unit and central consultation station, linked by a voice communication channel, which could be the PSTN or GSM. In the presented work, the mobile system was a notebook computer equipped with a PC-based 12-lead sample unit and a Nokia Card Phone 2.0 GSM modem. The central consultation station is used as a processing terminal in the hands of the acting expert doctor who is presented with a number of customised data collection screens and can record the patient details in an orderly fashion during transmission of ECG signals. Fast and error free transmission is achieved through the use of data compression algorithms and error correcting protocols. The interactivity of the GSM-based telecardiology system is enhanced by a series of system-independent tools: (a) a shared pointer, keeping the mouse moving positions consistent on both local and remote systems; (b) interactive window leveling, including visualization window, processing selection window and result window; (c) a bulletin board, allowing communication in plain text format; (d) telemarking, providing the capability of drawing curves, lines, and rectangles on any region of the shared images to indicate areas of interest. Based on Wang's work, it is indicative that a CSCW distributed architecture, comprised by a collaborative toolkit, can improve the quality of consultation in highly interactive user interfaces.

4.2.2 Automated Decision Support

Decision tasks, such as ECG classification—to determine whether the patient under examination is 'normal' or exhibits any cardiac abnormalities requiring treatment, are key components to the successful introduction of reliable and practical telecare systems. Over the past decades, research techniques based on AI, ranging from decision trees and fuzzy logic to advanced neural networks, have produced decision support systems which are becoming more attractive in the field (Nugent et al., 1999). What is important to note, however, is that the role of such systems is not to replace the human cardiologist in their decision making process, but to assist them in the overall patient diagnosis. Decision support systems have the ability to rapidly consider vast amounts of information and produce consistent results, with lower levels of intra-observer variability in comparison with humans. Recent investigations have shown that the computerised approach has the ability to produce comparable diagnostic results with cardiologists (MacFarlane, 1992). It is likely that the role of such systems will be used to analyse certain aspects of the recorded electrocardiographic data and where appropriate take into account other non-ECG data, hence providing indications to operators during potential situations of concern.

Clinical acceptance of medical decision support systems are now gaining popularity as formalisation of evaluation of their potential are being established (Smith et al., 2002). This has the potential for a larger scale uptake of their use.

4.3. Future Standards

In the field of standardization of exchange and management of the ECG, various alternatives still exist. There is a growing need to harmonize transmission, encoding and storage of digital ECG data from the full spectrum of ECG devices and different manufactures (i.e., standard 12-lead ECGs, Holter monitors, transtelephonic monitors, and implanted devices) along with annotations for events (e.g., standard ECG interval measurements, arrhythmic events). Since its adoption as a W3C recommendation in 1998, XML (1998) presents an attractive alternative to the representation and exchange of medical data. As a result, a number of projects such as the KMEHR (De Clercq et al., 2000), I-Med (2002) and standardization organizations in healthcare such as CDISC (Clinical Data Interchange Standards Consortium (2002), HL7 (Health Level Seven), ASTM (American Society for Testing and Materials), are currently working on the development of XML as a standard data interchange format for clinical data. In the ECG domain, a proposal for encoding ECG signals in an XML format was proposed during 2001 (Stockbridge, 2001). A group, consisting of members from government, ECG equipment manufacturers, pharmaceutical companies, clinical research organizations, and academic research laboratories, are refining these initial proposals. The most recent document specifying the XML data format for ECG was circulated mid, 2002 (Brown et al., 2002). To harmonize with existing standards, the format reuses some concepts, nomenclature, data structure, etc., where appropriate from HL7 v3.0, CDISC, and DICOM 3.0. Currently, these efforts have been supported by many manufacturers of ECG equipment. At present, few telecare system vendors have adopted XML for storage and manipulation of ECG data and other relevant medical information within computer systems, however, there is a strong industry drive to adopt XML as the preferred format. It is clear that XML has interesting features presenting great promise for allowing exchange and processing of medical information.

5. CONCLUSIONS

The use of telecardiology has many advantages compared with a classical examination of a patient. It provides the ability to facilitate many patients at the same time, cover large geographical areas, provide (quasi) real-time cardiac information, faster diagnosis and faster therapy provision and reduce effects of patient illness due to cardiac dysfunction. Advanced interactive systems can now be considered to provide diagnostic quality equal to that of a traditional approach. The provision of an infrastructure which facilitates patients, medical personnel and cardiologists has increased the communication pathways between all those involved during and following a cardiac examination, hence the quality and efficiency of the entire process can benefit.

In virtually 100 years of telecardiology the supporting equipment has progressed from devices which would have initially filled a small office, to small, pocket-sized mobile devices, offering the same capabilities. This has permitted the development of mobile, from both a physical and telecommunication perspective, ECG acquisition and transmission devices. Significant progression has been made with respect to telecommunications since the inception of telecardiology, however, bandwidth limitations of GSM still pose restrictions on systems development. Intricate signal processing and data compression techniques have alleviated upon these restrictions but the inevitable solution to the problem is hoped to be resolved with the proposed introduction of the 3rd generation wireless network and associated data rates. To be successful in this evolving arena of telecommunications and telecardiology, foresight of these advances must be considered during present developments to provide open and extensible infrastructures. At national levels, many governments have recognised the potential for telemedicine in general and its potential to revolutionise the way healthcare is delivered and as such have made the issue of e-health a key and strategic item on their healthcare agenda over the next decade. Through effective harnessing of available technologies it is now possible to enable the outreach of medical expertise beyond the primary care setting. With many studies proving the significance and value of telecardiology systems it is to be expected that their usage will prevail in the near future.

6. REFERENCES

Barro, S., Presedo, J., Castro, D., Fernandez-Delgado, M., Fraga, S., Lama, M. and Vila, J., 1999, Intelligent telemonitoring of critical-care patients, *IEEE EMBS Magazine* **18**(4):80-88.
Berbari, E.J., 1995, in: *The Biomedical Engineering Handbook*, J.D. Bronzino, ed., CRC Press, Florida.
Black, N.D. and Nugent, C.D., 1999, Telecardiology: Remote monitoring of the 12-lead ECG, *Proceedings of the British Council Trauma Masterclass Symposium*, 1-2.
Brown, B., Kohls, M., and Stockbridge, N., 2002, FDA XML Data Format Design Specification, April, 2002, http://www.cdisc.org/discussions/EGC/FDA%20_XML_Data_Format_Design_Specification_DRAFT_C.pdf.
Cayten, C.G., Oler, J., Walker, K., Murphy, J, Morganroth, J. and Staroscik, R., 1985, The effect of telemetry on urban prehospital cardiac care, *Annals of Emer. Med.* **14**:976-985.
CDISC, Clinical Data Interchange Standards Consortium (2002); http://www.cdisc.org/.
Cooper, J.K. and Caceres, C.A., 1965, in: *Biomedical Telemetry*, C.A. Caceres, ed., Academic Press.
Cox, T.R., Nolle, F.M. and Arthur, R.M., 1972, Digital analysis of the electrocencephalogram, the blood pressure wave and the electrocardiogram, *Proc. IEEE* **60**(10):1137-1164.
Daja, N., Reljin, I. and Reljin, B., 2001, Telemonitoring in cardiology – ECG transmission by mobile phone, *Annals of the Academy of Studenica* **4**:63-66.
De Clercq, E., Fiers, T. and Bangels, M., 2000, KMEHR: Kindly Marked-up Electronic Healthcare Record, Usage of XML for concrete implementation of messages for EHCR exchange, *Belgian Society for Medical Informatics Bulletin* **7-1**:44-48.
Einthoven, W., 1906, Le Telecardiogramme, *Arch. Internat. Physiol.* **4**:132.
Grim, P., Feldam, T., Martin, M., Donovan, R., Nevins, V. and Childers, R.W., 1987, Cellular telephone transmission of 12-lead electrocardiograms from ambulance to hospital, *Am. J. Cardiol.* **60**(8):715-720.
Hagan, W.K., 1965, Telephone applications, in: *Biomedical Telemetry*, C.A. Caceres, ed, Academic Press.
Imed project (2002); http://www.hnbe.com/healthweb/imedpub.
Istepanian, R.S.H., Woodward, B. and Richards, C.I., 2001, Advances in telemedicine using mobile communications, *Proc IEEE EMBS* CD-Rom publication.
Lambrew, C.T., Schuchman, W.L. and Cannon, T.H., 1973, Emergency medical transport systems: use of ECG telemetry, *Chest* **63**(4):447.
McKee, J.J., Evans, N.E. and Owens, F.J., 1994, Efficient Implementation of FAN/SAPA-2 ecg compression methods using fixed point arithmetic, *Automedica* **16**:109-117.
McKee, J.J., Evans, N.E. and Owens, F.J., 1996, Digital transmission of 12-lead electrocardiograms and duplex speech in the telephone bandwidth, *Journal of Telemedicine and Telecare* **2**:42-49.
MacFarlane, P., 1992, Recent developments in computer analysis of ecgs, Clinical Physiology **12**,313-317.
Nugent, C.D., Webb, J.A.C., Black, N.D. and Wright, G.T.H., 1999, Electrocardiogram 2: Classification, *Automedia* **17**:281-306.

Pozen, M.W., Fried, D.D., Smith, S., Lindsay, L.V. and Voigt, G.C., 1977, Studies of ambulance patients with ischemic heart disease, *Amer. J. Public Health* **67**(6): 527-531.

Reinhard, W., 1994, CSCW tools: concepts and architectures, *Computer* **27**:28-36.

Shanit, D., Cheng, A. and Greenbaum, R.A, 1996, Telecardiology: supporting the decision-making process in general practice, *J. Telemedicine and Telecare* **2**:7-13.

Shimizu, K., 1999, Telemedicine by mobile communication, *IEEE EMBS Magazine* **18**(4):32-44.

Smith, A.E., Nugent, C.D. and McClean, S.I., 2002, Implementation of intelligent decision support systems in healthcare, *European Journal of Management in Medicine* **16**(2/3):206-218.

Stockbridge, N., 2001, FDA Proposed ECG Interchange Standards, November 19th, 2001, http://www.fda.gov/cder/regulatory/ersr/ECGdata.html.

Uhley, H.N., 1970, Electrocardiographic telemetry from ambulances. A practical approach to mobile coronary care units, *Amer. Heart J.*, **80**(6):838-842.

Vassanyi, I., Mogor, E., Szakolczai and Tarnay, K., 2001, Mobile access to a cardiology database, *Proc. MEDINFO 2001*, 881.

Wang, H., Black, N.D. and Nugent, C.D., 2001, Design of an interactive GSM-based mobile telecardiology system, *Physica Medica*, 184.

Wellens, H.J.J., 1990, Electrocardiography past, present and future, *Annals of the New York Academy of Sciences* **601**:305-311.

Willems, J.L., 1990, Quantitative electrocardiography standardization and performance evaluation, *Annals of the New York Academy of Sciences* **601**:329-343.

Willems, J.L., Arnaud, P., van Bemmel, J.H., Degani, R., MacFarlane, P.W. and Zwietz, C., 1990, Common standards for quantitative electrocardiography: goals and main results, *Meth. Inform. Med* **29**:263-271.

XML, Extensible Markup language(1998); http://www.w3.org/xml/ .

MEDICAL ASPECTS OF PREHOSPITAL CARDIAC TELECARE

Periklis Giovas[*], Demetrios Thomakos, Ourania Papazachou, and Demetrios Papadoyannis

1. INTRODUCTION

Each year 900.000 people experience acute myocardial infarction (AMI) in the United States. Roughly 225.000 of them die, including 125,000 who die "in the field" before obtaining medical care. Because rapid identification and early treatment of patients with AMI reduce morbidity and mortality, the main effort has to be focused on minimizing delays.

This is especially important today that a constellation of therapies in the management of AMI has been introduced and therapy is not limited just to the widespread use of thrombolytic agents, but also includes percutaneous transluminal coronary angioplasty (PTCA) and emergency coronary artery bypass graft surgery in suitable patients. The reperfusion era has also embraced the extensive use of aspirin and other antiplatelet agents, adrenoceptor blocking agents, vasodilator therapy, and ACE inhibitors. Providing that the combined use of all these therapies has resulted in an impressive reduction in early and 1-year mortality for patients with AMI, it is essential to deliver them in time.

The components of delay in delivering therapy in AMI are (fig1):
- ■❑ The time elapsed from the patient side to seek medical care (i.e. failure to recognize the seriousness of the chest pain)
- ■❑ pre-hospital evaluation and transport time
- ■❑ time required for diagnosis and treatment in-hospital.

As patient-related delay is a issue of prevention, education and disease management, the main efforts after an incident have to be focused on the other two components. Prehospital 12 lead-ECG, obtained by ambulance staff, could be part of the solution. American College of Cardiology (ACC) and American Heart Association (AHA) in the latest (Ryan et al, 1999) guidelines for the management of AMI, suggested that

[*] Periklis Giovas, Cardiology Laboratory, 1st Department of Propaedeutic Medicine, Laikon Hospital, Athens University, Athens, Greece. E-mail: periklisgiovas@ath.forthnet.gr.

transmission of prehospital ECG to the emergency department physician for interpretation and instructions, when available, should be included in prehospital policy.

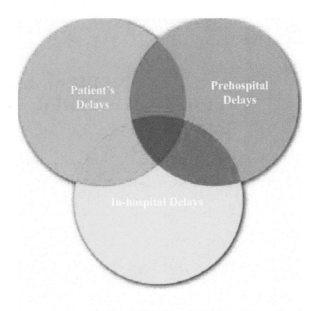

Figure 1. Components of delay

2. HISTORY OF ECG TELEMETRY IN AMBULANCES

Electrocardiograms (ECG) transmission is one of the oldest techniques in telemedicine. The brilliant Hollander Wilhelm Einthoven first investigated on ECG transmission over telephone lines, connecting his laboratory with the University Hospital though a distance of 1.5 kilometer. He suggested that this technique could be also implemented between cities in Holland and described his experience with about one hundred ECG transmissions at his article "Le telecardiogramme", appeared in the "Archives Internationales Physiologie" (Einthoven, 1906).

Prehospital cardiac care was first established in Belfast, Northern Ireland in 1966. It was a revolutionary step towards a new approach: Belfast program "moved" the coronary care unit into the community by treating the early complications of AMI. The program demonstrated that patients with out-of-hospital sudden cardiac arrest could be resuscitated. The idea of prehospital cardiac care spread to other countries after the Belfast experience. Next year, prehospital one-lead telemetry was presented in Miami. With ECG telemetry, firefighters with special training could be authorized by the physician to administer needed drugs and perform procedures before arriving in the emergency department. In this way, a paramedic at the scene was considered a legal extension of a physician and telemetry was the key to extending treatment outside the hospital. Herman Uhley (1970) from San Francisco published his experience with a low cost one-lead ECG telemetry system that permitted wireless transmission to the ambulance base station and from there to any hospital with a demodulator and ECG machine via telephone line.

Prehospital telemetry for cardiac care evolved significantly during the next decades and 12-lead electrocardiographic telemetry was introduced. This was a significant step, as single-lead telemetry could provide information for patients experiencing arrhythmias or sudden cardiac death, but not for those with angina pectoris and acute myocardial infarction (AMI). Pamela Grim (1987) and her colleagues at Chicago University Hospital used cellular network and accomplish transmission of 12-lead ECGs. The system used the standard data link control transmission protocol to produce error-free transmission. This protocol labeled each digital data package with an identifying tag. Any error occurred during the transmission caused a disparity between the data packet and the identifying tag. The receiving ECG machine was checking for this disparity, discarding packets in which errors had occurred and requesting retransmission. In this way of data transfer, errors produced by even the worst transmission pathway could be overcome. Grim's study proved that an ambulance 12-lead ECG transmission is feasible from a clinical perspective.

3. "CARDIOLOGY ON THE MOVE" PROJECT

Our experience on ambulance telemedicine is derived from "Cardiology on the move" project (Giovas et al, 1998) organized by the 1st Department of Propaedeutic Medicine, Athens University. The project started in February 1996 and the first phase began in January 1997. The aim was to determine whether 12-lead ECG transmission from a moving ambulance is feasible and to assess time saving in Athens emergency system.

One ambulance was equipped with a portable digital ECG recorder (Cardio Perfect), which was connected via an optical fiber to a notebook computer, coupled with a mobile phone for wireless transmission over cellular telephone network (fig2).

Figure 2. Ambulance equipment

A custom made software solution ("SendECG") was developed by Proton Labs Ltd, a Greek manufacturer of telemedicine systems. SendECG used data compression algorithms and error correcting protocols for fast, error free, store-and-forward transmission over GSM network. The receiving station was also running SendECG and ECG analysis software (fig 3).

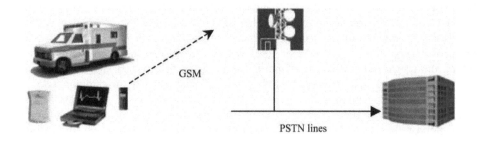

Figure 3. ECG transmission over GSM cellular network and PSTN lines to Laikon Hospital

Ambulance staff was trained to record and transmit the ECG. In every patient with chest pain or relevant history, an ECG was accomplished and transmitted to Laikon Hospital (fig4).

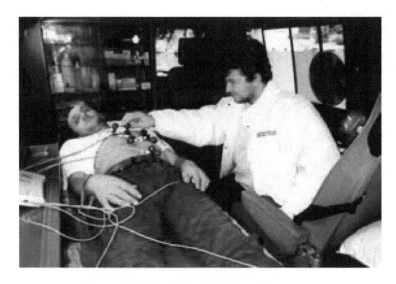

Figure 4. ECG acquisition inside the ambulance

During the first phase (January-May 1997), a total number of 72 patients with suspected cardiac problem was transported with the equipped ambulance. In most cases,

ECG was recorded on the way to the hospital. A moving ambulance is a difficult site for such acquisition and ambulance attendances were adequately trained to record a diagnostic quality ECG.

Although there were previous references that in-field ECG causes negligible delays in overall paramedic time (Karagounis et al, Salt Lake City, 1990), we decided to record the ECG on the move, because greek emergency system cope chest pain patients with the "scoop and run" approach.

The average ECG acquisition time was 2 minutes, depending on patient's co-operation and vehicle's speed. The most difficult point concerning en route recording was the placing of the precordial leads. A practical and ergonomic solution has been introduced, consisted of a silicon belt with all precodial leads integrated (fig5). Our experience showed that its use reduced the time needed for ECG acquisition approximately to the half.

Figure 5. ECG belt with all the precordial leads integrated for immediate acquisition

On the other hand, in-field ECG (recorded prior the transportation on site), might cause a slight delay on the length of paramedic scene time, but provides important information to the cardiologist when key decisions are to be made that could have been affected by the ECG data. For example, in one of the 72 cases, a patient without cardiac history described irrelevant symptoms, but paramedics suspected angina. The first 12-lead ECG was recorded and transmitted from the field, but it was non-diagnostic and the cardiologist requested a new in-field ECG after a few minutes. The second ECG defined unstable angina and the cardiologist decided hospitalization. An additional ECG was recorded and transmitted on the way to the hospital.

Transmission duration via GSM at 9600 bps was 34±14 seconds. In 65 of the 72 cases (90,3%) successful transmission was achieved during the first attempt. Common reasons for retransmission were ambulance's speed, cellular "dead spaces", network's congestion, bad weather and interference of large structures (fig 6).

In order to assess time saving in our study, two groups were taken into account: One with the patients transported by the equipped ambulance and another (control group) with patients arrived with ordinary ambulances.

Figure 6. Interference of large structures en route to the hospital

Average time needed for transportation was 18 (±6.5) minutes, so the ambulance arrived to Laikon Hospital 15.5 (±6.5) minutes after ECG transmission. For control patients with chest pain, it was approximately needed 9.5 (±3.5) minutes from arrival to ECG diagnosis at the Emergency Department. Thus, ECG diagnosis was made 25 minutes (±7.5) earlier for the patients carried by the telemedicine ambulance, in comparison with those transported by conventional ambulances (Fig 7).

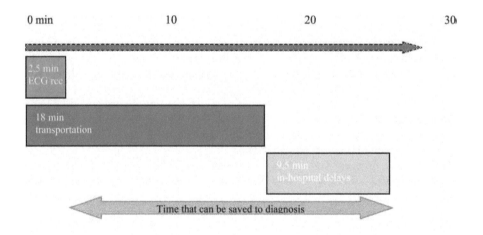

Figure 7. Benefits of prehospital transmitted ECG

4. BENEFITS OF PREHOSPITAL TRANSMITTED ECG

4.1. Reduction of hospital delays

As time is the enemy for patient with AMI, prehospital identification appears to lead to an important reduction of in-hospital delays. Possible sources of delay at the Emergency Department include the period waiting for ECG, X-Ray, routine blood and consultation with the patient's physician. Reducing hospital delay could be the single most important in shortening the time until initiation of treatment and coronary recanalization.

It has now been recognized that the only prehospital strategy that appeared to significantly reduce hospital time delay to treatment is EMS transport with transmission of the 12-lead ECG to the destination emergency department. In most studies to date, it is indicated that prehospital 12-lead ECG can reduce in-hospital delays an average of between 20 and 60 minutes, depending on the emergency medical system and transportation time (urban, suburban or rural area). This finding could be considered at public education programs aimed at encouraging patients with chest pain to utilize prehospital services rather than self-transportation.

4.2. Better Triage

The cardiologist has also the opportunity to triage patients and admit them directly to the Coronary Care Unit for thrombolytic therapy, without delays at the Emergency Department.

In addition, in some cases the cardiologist could decide to divert the ambulance itinerary to a center with more advanced facilities (PTCA or emergency coronary artery bypass graft surgery).

4.3. Continuous Monitoring

The evolution of mobile telecommunications has allowed continuous ECG monitoring during the transportation, rather than a "snap" 12-lead ECG with a duration of 10 or 20 seconds. This is very important because patients with AMI are at relatively high risk of sudden death during the first hour and paramedics -in certain circumstances and under cardiologist instructions- can administrate for instance atropine intravenously in severe bradycardia or xylocaine in ventricular tachycardia. They can also provide pain relief and hemodynamic support under the virtual presence of a physician.

4.4. ECG data accessible for comparison

An additional benefit of prehospital ECG may be in detecting transient prehospital ischemia that may resolve by the time of hospital arrival. Documenting prehospital electrocardiographic ischemia that would otherwise remain undetected to the receiving doctor improves hospital-based evaluation.

It also provides a 12-lead ECG for comparison with the first emergency department ECG and has a potential to more accurately differentiate prehospital cardiac arrhythmias.

4.5. Cardiologist vs EMS and computer analysis

With ambulance transmitted 12-lead ECG to the destination hospital, the cardiologist on duty has the opportunity to prepare treatment before ambulance's arrival, based on his own interpretation and not on the information given by ambulance crew. This approach has been proven to be optimal, in comparison with paramedic or computer-based diagnosis.

A study at Wisconsin (Aufderheide et al, 1990) showed that such telemedicine applications had significantly higher specificity (99,2% vs 70,9% P<.001) and significantly higher positive predictive value (92,9% vs32,7% P<.001) than the paramedic clinical diagnosis, which is the standard practice for initial evaluation in the US.

Computer-interpreted ECG could be useful in enhancing selection accuracy and could help to minimize interpretation errors at patients without evidence of acute infarction. Although the sensitivity of the computer-interpreted algorithm is imperfect in identifying all patients with AMI and ST elevation, its high specificity seems useful to non-experts in excluding patients with non-diagnstics ECG for thrombolytic treatment, and thus the use of computer analysis should be considered in situation in which expert consultation is not immediately available. The accuracy of computer-interpretive algorithm can vary considerably and each algorithm should be carefully validated before being used for this purpose.

4.6. Prehospital Therapy in eligible subjects with AMI

As far as prehospital thrombolytic therapy is concerned, it would be rational to expect that a greater number of lives could be saved if started in-field or during transportation. Randomized controlled trials with fibrinolytic therapy agents have demonstrated the benefit of initiating thrombolytic therapy as early as possible after onset of ischemic-type chest discomfort. The value of reducing delay until treatment depends not only on the amount of time saved but when it occurs. Available data suggest that time saved within the first 1 to 2 hours ("golden hour") has greater biological importance than time saved during the later stages of AMI. Acquisition of prehospital ECG and use of a chest-pain checklist lead to more rapid prehospital and hospital care.

Although none of the individual trials showed a reduction in mortality with prehospital-initiated thrombolytic therapy, a meta-analysis of all available trials demonstrated a 17% relative improvement in outcome associated with prehospital therapy. The greatest improvement is observed when treatment can be initiated in the field 60 to 90 minutes earlier than in the hospital.

The latest AHA/ACC guidelines suggests that although prehospital-initiated thrombolytic therapy results in earlier treatment, the time savings can be offset in most cases by an improved hospital triage with resultant "door-to-needle time" reduced to 30 minutes or less. In addition, prehospital thrombolysis has medical, legal and economic implications. For all these reasons, prehospital initiation of thrombolytic therapy cannot currently be advocated as a general policy. It is suggested that prehospital systems should be focused on early diagnosis instead of therapy, especially in urban areas where transportation time is often short. However, in special settings in which physician is present in the ambulance or prehospital transport times are 90 minutes or longer, prehospital therapeutic strategy should be considered.

Sweden has one of the longest experience on telemedicine-supported prehospital thrombolysis. A reason for this is Sweden's low population density, which leads to longer transportation time. Two thirds of the hospitals in Sweden have some form of prehospital thrombolysis in use today. Data from Sweden (Benger et al, 2002) indicate that the median "pain to needle" time has been reduced by 45 minutes with a concurrent reduction in complications from 50% to 25% (p=0.018).

In our study in Laikon Hospital, a special checklist was used to identify patients eligible for thrombolytic therapy in case of an AMI. This checklist with inclusion and exclusion criteria could lead to a further reduction of in-hospital delays, used by paramedics in-field or on route (fig 8).

Checklist for Thrombolytic Therapy Screening	Yes	No
Ongoing chest pain (>20 min and unrelieved with nitroglycerin)		
Age <75 years		
Systolic Blood Pressure >80 and <200 mm Hg		
Diastolic Blood Pressure <100 mm Hg		
Systolic Blood Pressure of right arm minus left arm <20 mm Hg		
If all answers are YES, then proceed		
History of stroke, brain surgery, aneurysm		
Known bleeding disorder		
Active internal bleeding in the last 2 weeks		
Surgery or trauma in the last 2 weeks		
Terminal illness		
Jaundice, hepatitis, kidney failure *If all answers are NO then inform Laikon Hospital*		

Figure 8. Checklist for identifying eligible patients for thrombolytic therapy

5. CURRENT RESEARCH AND DEVELOPMENT

Prehospital telemedicine for cardiac care is an emerging field and efforts for further implementation is currently underway.

In the future, this technology is expected to be more refined and significantly easier to handle. Telecommunication capacities are expected to be integrated in many medical devices. This means that more vital minutes can be saved providing the ability to utilize mobile information equipment nearer to the accident scene, while third generation (3G) cellular networks will allow real time transmission of biosignals and video, when needed.

Besides, the whole ambulance design and ergonomics are being reconsidered, in accordance with the growing need for constant contact with the hospital.

In addition to ECG and vital signs transmission, other medical devices are being tested for cardiac emergency. Echocardiography is a key diagnostic tool in evaluating patients with cardiac emergencies. The lack of qualified real-time interpretation and high-bandwidth mobile connections limits its use by emergency first responders. Early diagnosis of cardiac emergencies has the potential to facilitate triage and medical intervention to improve outcomes. A team from Brooke Army Medical Center, Houston (Garrett et al, 2003) investigated the feasibility of remote, real-time interpretation of echocardiograms during patient transport. Echocardiograms using a hand-carried ultrasound device were transmitted from an ambulance in transit to a tertiary care facility using a distributed mobile local area network. Echocardiographic images were successfully transmitted greater than 88% of transport time. The evaluation of left-ventricular size and function, and presence of pericardial effusion were greater than 90% concordant, but only 66% of all echocardiographic features were concordant. Most transmission losses were brief (<or=10 seconds) with little impact on interpretability.

Patients with symptoms of acute myocardial infarction or unstable angina pectoris, are very common in emergency departments (EDs). Because of diagnostic difficulties, a certain "overadmission" to in-hospital care for suspected acute cardiac syndromes is usually accepted. The size of this overadmission is generally unknown, despite its negative influence on hospital efficiency and resource utilization. In order to reduce unnecessary admission, to optimize triage and reduce cost, chest pain units with risk-based diagnostic protocols and decision support systems have been established in many hospitals. The adoption of decision making support systems in clinical environments could further exploit ambulance data and biosignals, for patients' classification and triage. For example, a decision making support software for thrombolysis -a software program designed to aid the physician's decision making process by using medication, patient age, gender and ECG features- provides the predicted probability of acute cardiac ischemia or acute coronary syndrome. This software identifies ECG characteristics (abnormal Q-waves, ST segment elevation or depression and T-wave elevation or inversion) and uses them to calculate the predicted probability of acute cardiac ischemia. This probability is available real-time in the emergency department and could be a valuable tool for immediate clinical decision.

Finally, mobile units for telecare at high-risk patients' homes will also be able to give important information about the status before ambulance's arrival. In this way, more time will be saved and the mortality rate may be further reduced.

6. REFERENCES

Aufderheide T.P., et al., 1990, *The diagnostic impact of prehospital 12-lead electrocardiography* Ann Emerg Med. Nov;19(11):1280-7.

Benger J.R., Karlsten R. and Eriksson B., 2002, *Prehospital thrombolysis: lessons from Sweden and their application to the United Kingdom* Emerg Med J.

Einthoven, W., 1906, *Le telecardiogramme* Archives Internationales Physiologie IV:132-164.

Eisenberg, M.S., et al, 1996, *The revolution and evolution of prehospital cardiac care* Arch Intern Med. Aug 12-26;156(15):1611-9.

Garrett, P.D., et al, 2003 *Feasibility of real-time echocardiographic evaluation during patient transport* J Am Soc EchocardiogrMar;16(3):197-201.

Giovas P et al, 1998, *Transmission of electrocardiograms from a moving ambulance*, Journal of Telemedicine and Telecare Volume 4 Supplement 1, 5-7.

Giovas P., et al, 1999, *Ambulance telemetry for emergency cardiac care*, European Telemedicine 98-99 RSM & Kensington Publications, pp. 111-113.

Grim, P., et al, 1987 *Cellular telephone transmission of 12-lead electrocardiograms from ambulance to hospital* Am J Cardiol. Sep 15;60(8):715-20.

Karagounis L., et al, 1990 *Impact of field-transmitted electrocardiography on time to in-hospital thrombolytic therapy in acute myocardial infarction* American Journal of Cardiology; 66:786-791

Ryan, T., et al, 1999 ACC/AHA *Guidelines for the management of patients with acute myocardial infarction,* Circulation.;100:1016-1030.

Uhley, H.N., 1970 *Electrocardiographic telemetry from ambulances. A practical approach to mobile coronary care units* Am Heart J. Dec;80(6):838-42.

AN EMERGENCY TELEMEDICINE SYSTEM BASED ON WIRELESS COMMUNICATION TECHNOLOGY: A CASE STUDY

Efthyvoulos Kyriacou[*], Sotiris Pavlopoulos, and Dimitris Koutsouris

1. INTRODUCTION

Telemedicine can be defined as the delivery of health care and sharing of medical knowledge over distance using telecommunication means. The aim of telemedicine is to provide expert-based health care to understaffed remote sites through modern telecommunication (wireless communications) and information technologies. The availability of prompt and expert medical care can meaningfully improve health care services at understaffed rural or remote areas. The provision of effective emergency telemedicine is the major field of interest of this Chapter. During emergency cases such as coronary artery diseases and cardiac arrests, survival of patients is related to the "call to needle" time, where thrombolysis might require in less than 60 minutes and has to be done during transportation to a main hospital. Furthermore in cases of serious injuries of the head, the spinal cord and internal organs the way of moving the patient from the accident scene and generally the way of providing care before and during transportation is crucial for the future of the patient.

In order to support emergency telemedicine a portable medical device that allows telediagnosis, long distance support and teleconsultation of mobile health care providers by expert physicians has been created. The device is always located near the patient and allows the collection and transmission of vital biosignals and still images of the patient to a consultation site which is located at a main hospital where the expert doctors are. The transmition is performed through wireless communications GSM and GPRS mobile telephony networks, Satellite networks or wired communications where available. The device can telematically "bring" an expert specialist doctor at the site of the medical emergency; thus allowing him evaluate patient status and direct emergency personnel on treatment procedures until the patient is brought to hospital. All data collected through the system are stored and managed using a multimedia database located at the consultation site.

---
[*] Efthyvoulos Kyriacou Department of Computer Science, University of Cyprus, 75 Kallipoleos str, P.O.Box 20537, 1678, Nicosia, Cyprus, e-mail:ekyriac@ucy.ac.cy.

2. BACKGROUND

The availability of prompt and expert medical care can meaningfully improve health care services at understaffed rural or remote areas. The provision of effective emergency Telemedicine is the major field of interest discussed in this chapter. There is a wide variety of examples where those fields are crucial. Ambulance vehicles, Rural Health Centers or other remote health location, Ships navigating in wide seas are common examples of possible emergency sites. In emergency cases where immediate medical treatment is the issue, studies conclude that early and specialized pre-hospital patient management contributes to the patient's survival (Weston et al., 1994). Especially in cases of serious head injuries, spinal cord or internal organs trauma, the way the patients are displaced from the incident place to the ambulance vehicle, treatment on the scene and during transportation is crucial for the future well being of the patients.

A quick look to past car accident statistics points out clearly the issue: During 1997, 6753500 incidents were reported in the United States (Cerreli, E.C., 1997) from which about 42000 people lost their lives, 2182660 drivers and 1125890 passengers were injured. In Europe during the same period 50000 people died resulting of car crash injuries and about half a million were severely injured. Furthermore, studies completed in 1997 in Greece (Hell. Nat. Stat. Service, 1997), a country with the world's third highest death rate due to car crashes, show that 77,4 % of the 2500 fatal injuries in accidents were injured far away from any competent healthcare institution, thus resulting in long response times. In addition, the same studies reported that 66% of deceased people passed away during the first 24 hours.

Coronary artery diseases are another common example of high death rates in emergency cases since still two thirds of all patients die before reaching a central hospital. In a study (Evans, T., 1998) in the UK in 1998, it is sobering to see that among patients above 55 years old, who die from cardiac arrest, and 91% do so outside hospital, due to a lack of immediate treatment. In cases where thrombolysis is required, survival is related to the "call to needle" time, which should be less than 60 minutes (Sandler, D.A., 1999). Thus, time is the enemy in the acute treatment of heart attack or sudden cardiac death (SCD).

It is common knowledge that people that handles emergency situations (rural doctors, emergency medical personnel, nurses) do not always have the required advanced theoretical background and experience to manage all cases properly. Emergency Telemedicine can solve this problem by enabling experienced neurosurgeons, cardiologists, orthopedics and other skilled people to be virtually present in the emergency medical site. This can be done through transmission of vital biosignals and on scene images of the patient to the experienced doctor. Several systems that could cover emergency cases (Hoffman, B., 1995 – Kyriacou, E. et al. 2001) have been presented over the years.

Developments in communications and especially in wireless communications enable the creation of reliable, high quality and efficient telemedicine systems using wireless communication means (Bai, J. et al., 1999 – EMERGENCY-112, 1998).

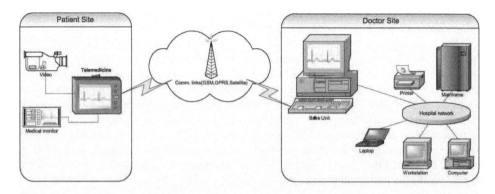

Figure 1. Emergency telemedicine system architecture

Following these different growing demands we created a combined real-time and store and forward facility that consists of a base (doctor's) unit and a telemedicine (patient's) unit where this integrated system can be used in order to handle emergency cases :

- $\in$ Being treated in an ambulance vehicle
- $\in$ Take place in a Rural Health Center (RHC)
- $\in$ Take place on a navigating ship

by using a Telemedicine unit at the patient / emergency site and the expert's medical consulting at the base unit.

The design and implementation of the system was based on a detailed user requirements analysis. The experience was mainly gained from two EU funded Telemedicine projects named AMBULANCE (TAP #HC1001 1996-1998) and Emergency 112 (TAP #HC4027 1998-2000), where functional prototypes of a device with emergency Telemedicine functionalities were built and extensively evaluated. Through these projects we had phased the need to implement a telemedicine device which would facilitate a more flexible architecture and could be used in several emergency cases that have similar needs of information transmition.

3. TECHNICAL APPROACH

The system developed is a "Multi-purpose" telemedicine system consisting of two major parts:

1. Telemedicine unit (located near the patient)
2. Base unit (located at a Central Hospital)

An overview of the system architecture and its basic components is shown in Figure 1.

The Telemedicine unit is responsible for collecting and transmitting biosignals and still images of the patient from the incident place to the Doctor's location while the Base unit is responsible for receiving and displaying incoming data. The Doctor might be using the system either in an Emergency case or when monitoring a patient from a remote place.

The software design and implementation follows the client server model; it was done using Borland Delphi 5 for Microsoft Windows 95/98/NT/2000 platform; the Telemedicine unit site is the client while the Base unit site is the server. Communication between the two parts is achieved using TCP/IP as network protocol, which ensures safe data transmission and interoperability over different telecommunication means (GSM, GPRS, Satellite and Plain Old Telephony System (POTS). System communications are based on a predefined communication protocol for data interchange, which is used to control and maintain connection between the two sites, thus ensuring portability, interoperability and security of the transmitted data. The protocol was based on the experience gained from the creation of an extended codification scheme based on the "Vital" and "DICOM" standards which were developed (Anagnostaki, A. et al. 2002) during the design and implementation phase of the system.

3.1. Telemedicine Unit:

The Telemedicine unit mainly consists of four modules:

1.   The biosignal acquisition module

2.   The image capturing module

3.   The main module

4.   The communication module.

All modules are a combination of hardware and software components.

3.1.1   Biosignal acquisition module:

This module is responsible for the collection of crucial biosignals from the patient. Commercial biosignal monitors are used for the acquisition of biosignals; while the control of the monitors and transmition of biosignals to the main module is performed via a custom developed software component. This was designed to operate with some of the most common portable biosignal monitors used in emergency cases or in intensive care units. The supported monitors are a) CRITIKON DINAMAP PLUS Models 8700/9700 family of monitors (Dinamap-Plus, 1994), b) PROTOCOL - Welch Allyn Propaq 1xx Vital Signs Monitor (Propaq 1xx, 1994), c) PROTOCOL - Welch Allyn Propaq Encore 2xx Vital Signs Monitor (Propaq Encore 2xx, 1996).

The biosignals collected by the patient (and then transmitted to the Base Unit) are: ECG up to 12 leads, depending on the monitor used, Oxygen Saturation (SpO_2), Capnography (CO_2), Heart Rate (HR), Non-Invasive Blood Pressure (NIBP), Invasive blood Pressure (IP), Temperature (Temp), Respiration (Resp).

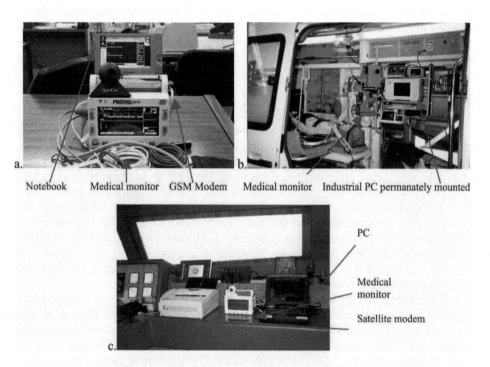

Figure 2. Pictures from different versions of telemedicine units a) Portable unit, Propaq 2xx monitor and Toshiba libretto sub-notebook are used b) Ambulance permanent unit, CRITIKON monitor and a pc are mounted in an Ambulance vehicle, the communication is performed with GSM modem permanently installed in the vehicle c) Yacht unit, Propaq 1xx monitor and a notebook pc are installed in the cockpit of a yacht, communication is performed with satellite modem permanently installed on the yacht

3.1.2   Image capturing module:

This module is responsible for capturing still images of a patient. It consists from a developed software module; which is responsible for the connection to a digital camera and collection of still images. The other part is a commercial digital camera. The module can support any digital camera that is compatible with Microsoft Video for Windows library.

3.1.3   Main module

This is the heart of the telemedicine unit. It consists of a standard commercial PC (portable or not portable depending on the case) (Figure 2) and the main developed software module. This is responsible for synchronizing controlling and maintaining the accurate operation and communication of all other modules. It is also responsible for the local display and storage of all data captured and transmitted during the telemedicine sessions.

Several types of PCs are used depending on the type and role of the telemedicine application / unit. a) In cases where the autonomy and small size of the system are important (mainly ambulances), a sub notebook like Toshiba libretto 100ct portable PC or pen tablet PC is used. b) in cases where we need some autonomy (navigating ships) but size is not considered as an important element a typical portable PC is used and c) in cases where we do not necessarily need autonomy, portability and small system size(Rural Health Centers), a typical Pentium Desktop PC is used.

3.1.4   Communication module

This module is responsible for the accurate and safe communication of telemedicine and base unit. It consists of a communication device / modem and a developed software component which is responsible for the control of the device and the transmition of data to the base unit. Data interchange is done using the TCP/IP network protocol, which allows operation over several communication means. The communication with the base unit is based on a predefined communication protocol which contains all the necessary commands, data compression (using the Huffman algorithm for ECG signal compression (Sayood, K., 1996)) and data encryption scheme (using the blowfish algorithm (Schneier, B., 1996)) needed to ensure portability, interoperability and security between the two sites. The security of the Telemedicine Unit was designed according to the directive 97/66/EC, concerning processing of medical data in the telecommunication sector.

In order to communicate with the base unit the telemedicine unit PC is equipped with the proper modem for each case, i.e. GSM, GPRS, Satellite or POTS. The design was done for standard Hayes modems. The system supports ETSI - AT command set for GSM modem, for Satellite modems and for Standard POTS modems. Several modems types were used for testing: a) a NOKIA card phone 2.0 GSM 900/1800 modem pcmcia card and an Option FirstFone GSM 900 modem pcmcia card is used for GSM communication, b) an Option GlobeTrotter GPRS/GSM pcmcia card is used for GPRS network c) a Micronet pcmcia POTS modem 56K and a US-Robotics 56K external modem are used for POTS communication, d) a mini m terminal for ships "Thrane &Thrane TT-3064A CAPSAT Inmarsat Maritime Phone" is used for satellite communication.

The control of the Telemedicine unit is fully automatic. The only thing the telemedicine unit user has to do is connect the biosignal monitor to the patient and turn on the PC. The PC then performs the connection to the base unit automatically. Although the base unit basically controls the overall system operation, the Telemedicine unit user can also execute a number of commands. This option is useful when the system is used in a distance health center or in a ship and a conversation between the two sites takes place. Pictures from several versions of telemedicine unit are shown in Figure 2.

3.2. Base Unit

The base unit pc is responsible for displaying incoming signals from the Telemedicine unit and the expert doctor uses it as a processing terminal (Figure 1). Through the base unit, user has the full control of the telemedicine session. The user is able to monitor the connection with a client (telemedicine unit), send commands to the

ECG Lead commands   Alarms indication (type, color, sound)

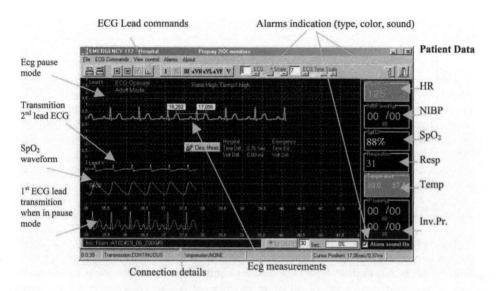

Figure 3. Biosignal transmission – Base unit, continuos mode

telemedicine unit such as the operation mode (biosignals or images). In cases were the base unit is installed in a central hospital that supports several telemedicine units, the user of the base unit is able to choose and connect to anyone of the telemedicine units connected on the network.

The user can monitor incoming biosignals or still images, thus keeping a continuous online communication with the patient site. This unit has the full control of the telemedicine session. The doctor (user) can send all possible commands concerning both still image transmission and biosignals transmission. Figure 3 presents a typical biosignal-receiving window (continuous operation).

When the system operates on still image mode, the doctor can draw-annotate on the image and send the annotations back to the telemedicine unit; the user can also annotate on the freezed image and annotations will then again be transferred to the base unit.

When operating on biosignal mode, the transmission of vital biosignals can be done in two ways, continuous or store and forward way, depending on the ECG waveform channels which are transmitted and the telecommunication channel data transfer rate. In continuous operation, the Base Unit user can send commands to the Telemedicine Unit monitor, such as lead change or blood pressure determination. Furthermore the user can pause incoming ECG, move it forward or backward and perform some measurements on the waveform.

The base unit mainly consists of a dedicated PC equipped with an appropriate modem responsible for data interchange. One base unit can accept several calls from different telemedicine devices.

The base unit consists of three modules,

1. The communication module
2. The main module

3.   The data archiving module

All modules are a combination of hardware and software components.

3.2.1   Communication module

This module is responsible for the communication with the telemedicine units. It consists of a hardware communication component (GSM or GPRS modem, POTS modem or Satellite modem) and a software component which was developed for the system purposes. Data interchange is done using the TCP/IP protocol and the predefined protocol mentioned above (section 3.1.4). This module receives all transmitted data biosignals or images from the telemedicine unit(s) and forwards them to the main module. It is also responsible for getting information such as commands from the main module and transmits them to the telemedicine unit(s).

3.2.2   Main Module

This is the coordinating module of the base unit. It consists of a standard commercial PC (portable or not portable depending on the case) and the main developed software module. This is responsible for synchronizing controlling and maintaining the accurate operation and communication of all other modules. It is also responsible for the local display of all data transmitted from the telemedicine units during the telemedicine sessions.

3.2.3   Data archiving module

In order to record information being exchanged during emergency case handling a hospital database unit has been integrated in the system. For each emergency case, information which includes incident's number, date, time, initial diagnosis, final diagnosis and telemedicine files are being recorded. This information is compliant to the directive *"Standard Guide for View of Emergency Medical Care in the Computerized Patient Record"* (designation E-1744-95) of the American Society for Testing and Materials (ASTM). When the system is used in an Ambulance Emergency Medical Service, the database unit is also responsible for accepting and recording emergency calls, as well as managing Ambulance vehicle fleet.

In cases where a Hospital Information System (HIS) is already available at the base unit site (Hospital), the doctor (base unit user) can retrieve information (using the data archiving unit) concerning the patient's medical history. When HIS is not available, the hospital database unit can handle the patient's medical record by itself.

The database was integrated in the system in order to help users keep all necessary information concerning medical emergencies. The information can be used in order to have a complete record about each emergency case which can be very useful for the treatment of the patients. Furthermore the database is used in order to get statistical results about the operation of emergency medical service departments. Finally for legal purposes all interchanged data has to be recorded in case they might be requested.

The database was designed for Microsoft SQL server and Paradox 7 and was equipped with graphical user interface features built in Borland Delphi 5 for increased user friendliness. All parts of the database are in compatibility with Microsoft Windows 98/NT/2000. For security reasons, according to the directive 95/46/EC, the database is fully protected against unauthorized access and is password protected and encrypted, whereas the whole application is password protected with several access levels depending on user groups.

4. SYSTEM EVALUATION

The final result is a "Multi-purpose" Telemedicine system, which facilitates a flexible architecture that can be adopted in several different application fields. The system was extensively technically and clinically validated.

4.1. Technical evaluation

In order to have the best possible technical quality, before using it in daily clinical practice The system was thoroughly tested and validated (Kyriacou, E. et al., 2003) for a variety of medical devices and telecommunication means in order to cover all user needs with the best possible and reliable way. The tests where performed on each one of the modules of the system as well as on the complete system.

4.1.1   Real time biosignals acquisition

The selection of the biosignal acquisition devices was performed in order to have the best quality of signals. Commercial monitors from two of the most commonly used monitors for ambulatory monitoring where selected. The monitors have digital signal output and satisfy several standards and certifications (Dinamap-Plus, 1994 – Propaq Encore 2xx, 1996). The communication with the monitors is performed through RS232 serial port. It was 100% accurate when following the predefined digital communication protocol provided by the companies. The information acquired from the monitors included biosignals (ECG, SpO_2, CO_2, Resp, Temp, NIBP, HR), monitor information and alarms. The ECG waveform and SpO_2 or CO_2 Waveform (where available) are the continuous signals transmitted, trends are transmitted for the rest of data. ECG data are sampled at a rate of 200 samples/sec by 10 bits/sample or 12 bits/sample, for all monitors used, thus resulting in a generation of 2000 bits/sec and 2400 bits/sec for one ECG channel. SpO_2 and CO_2 waveforms are sampled at a rate of 100 samples/sec by 10 bit/sample; thus resulting in a generation of 1000 bits/sec for one channel. Trends for SpO_2, HR, NIBP, BP, Temp, Resp and monitor data are updated with a refresh rate of one per second, thus adding a small fraction of data to be transmitted approximately up to 200 bits/sec.

4.1.2   Image acquisition

Image acquisition is performed using a digital camera compatible with Microsoft video for windows library. During the development we had tested several cameras which could support video for windows. The devices tested where commercial models of Creative, Zoom, QuickCam, Intel and Logitech companies. All models responded according to their specifications and the program was able to communicate and grab image with success from all cameras. Captured images have resolution 320x240 pixels and are compressed using the JPEG compression algorithm; the resulting dataset is approximately 5-6 Kbits depending on the compression rate or the algorithm.

4.1.3   Biosignals, images and commands transmition

The communication between the two sites is performed using the TCP/IP protocol. The exchanged data and commands were exchanged according to a predefined protocol designed during the development of the project which was based on the "Dicom" and "Vital" standards. TCP/IP was selected in order to ensure safe data transmition and interoperability over several communication means. In order to support / optimize the correct operation of TCP/IP especially in cases where the bit rate is low (GSM, Satellite), the telemedicine unit program is transmitting data packets that have a certain size. This is because of the way TCP/IP operates (Stevens, R., 1994). When sending a certain amount of data through the TCP/IP and the amount is bigger than a certain size called Maximum Transfer Unit (MTU), the data are fragmented into smaller packets. In cases when the amount of data is smaller than MTU they are transmitted without fragmentation. A header is being added on each data packet transmitted from one site to the other. Headers are being used in order to maintain the correct transmition of packets. The problem is that when sending data packets of smaller size than that of the MTU the protocol adds an amount of extra bytes per packet thus resulting to a significant expansion of the transmitted information. When using this practice over communication links of high bit rate, no problem is being appeared, but when doing it over communication links with low communication bit rate, such as GSM, it results to the transmition of unnecessary information thus delaying communication and sometimes causes line failure. In order to avoid this effect several tests where performed and an optimum data package of approximately 430 bytes (Kyriacou, E. et al., 2003) was selected for the transmition.

The medical monitors used in this study can provide three to twelve leads waveform of ECG and numeric data from biosignals such as HR, SpO_2, CO_2, NIBP, IP, Temp, and Resp.

The first group of the monitors used, CRITIKON DINAMAP PLUS Monitor has a digital output of a continuous one channel ECG plus biosignals such as NIBP, SpO_2, HR, IP and data concerning monitor, alarms etc.; all the above information can be transferred using up to 2200 bps. For this reason, the continuous transmission of signals from this monitor can be done when using GSM and POTS and 2400 BPS satellite links.

The second group of the monitors used, PROTOCOL Propaq Monitor has a digital output of a continuous one (model 1xx) or two (model 2xx) channels of ECG, plus another waveform such as SpO_2 or CO_2; plus biosignals trends such as NIBP, SpO_2, HR, IP and data concerning monitor, alarms etc. All above information can be transferred using up to 2400 BPS for one channel ECG, up to 4400 for two channels of ECG or up to 5400 for two channels of ECG plus another waveform (SpO_2 or CO_2). For this reason, the continuous transmission of signals from this monitor can be done when using GSM and POTS but only one lead ECG when using 2400 bps satellite links.

4.1.4   Telecommunications

The signals transmission is done using GSM, Satellite and POTS links. For the time being, the GSM network that the system was technically tested on; allows transmission of data up to 9600 bps (when operating on the normal mode) and is able to reach up to 43200 bps when using the HSDC (High Speed Circuit Switched Data). The satellite link transmission rate depends on the equipment and the satellite system used in each case; its range varies from 2400 bps and upwards. The use of different satellite systems can increase the cost of equipment and cost of use; in our case we had used an INMARSAT-phone Mini-m system which can transmit data only up to 2400 bps, but has low equipment and use cost. Plain Old Telephony System (POTS) allows the transmission of data using a rate up to 56000 bps, thus enabling the continuous and fast information transmission.

The practical maximum data transfer rate over telecommunication means is never as high as the theoretical data transfer rate. Practical data rates depend on the time and the area where the system is used. Biosignals data transmission can be done in two ways: real time transmission where a continuous signal is transmitted from client to server or store and forward transmission where signals of a predefined period of time are stored in the client and transmitted as files to server. It mainly depends on the maximum data transfer rate of the telecommunication link used and the digital data output that the biosignal monitor has in each case.

The main telecommunication means used where tested in real time operation so as to have a complete idea about their performance.

In order to measure the performance of the GSM network using the optimum buffer size and the equipment mentioned above several measurements from a moving vehicle were performed (Kyriacou, E. et al., 2003). A NOKIA card phone 2.0 GSM 900/1800 modem pcmcia card was used during the tests. Measurements were performed using a moving vehicle in some of the main streets of the city of Athens (Greece). The vehicle was moving with a speed of approximately 60 km/h. The tests were performed in such a way and using the relevant equipment in order to simulate the data calls from an Ambulance vehicle. The transmission of ECG signals was done using continuous mode, while at the same time images of several sizes were transmitted. Tests were performed for one month using different routes and for different periods of a day.

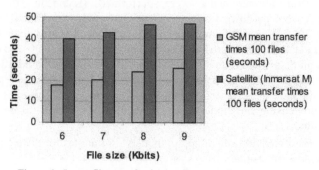

Figure 4. Image files transfer times, GSM and Satellite links

Having performed 40 tests of approximately 30 minutes we had the following results:

- ∈ In order to establish the connection between the telemedicine unit and the base unit an average time of 28 seconds was required.

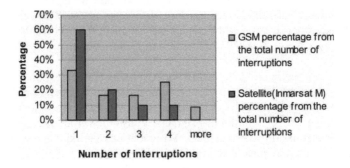

Figure 5. Interruptions percentage for GSM and Satellite connections

- •∈ 10 images per test were successfully transmitted. The average transmission time for several image files was from 18 to 26 seconds. Around 93% of image transmissions were achieved within the first attempt; the rest 7% was transmitted using a second attempt because we had a line failure (Figure 4).
- •∈ The transmition of two ECG lead waveforms and pulse oxymetry waveform was performed in real time. The connection was interrupted once for at least 20 % of all cases. In some cases we had more than one interruption; reconnection of telemedicine unit to the base unit was performed successfully in all cases of interruption (Figure 5).

In order to measure the performance of the system over the Inmarsat satellite network, several measurements were performed (Kyriacou, E. et al., 2003). Measurements were performed on a yacht using a mini m terminal for ships "Thrane &Thrane TT-3064A CAPSAT Inmarsat Maritime Phone" for the telemedicine unit, and a POTS modem US robotics sportster voice 56 kbps for the base unit. The mini m device is able to transmit data with a rate up to 2400 bps. The GSM optimum buffer size was used in this case too. The tests were performed in such a way and using the relevant equipment in order to simulate the emergency data calls from a yacht. The transmission of ECG signals was done using continuous mode, while at the same time images of several sizes were transmitted.

Having performed 40 tests of approximately 30 minutes for different periods of day, we had the following results:

- •∈ In order to establish the connection between the telemedicine unit and the base unit an average time of 40 seconds was required.
- •∈ 10 images per tests were successfully transmitted. The average transmition time for several image files was from 40 to 47 seconds. Around 90% of image transmissions were achieved within the first attempt; the rest 10% was transmitted using a second attempt because we had a line failure (Figure 4).
- •∈ The transmition of one ECG lead waveform was performed in real time. The connection was interrupted once for at least 15 % of all cases, in some cases we had more than one interruptions; reconnection of telemedicine unit

to the base unit was performed successfully in all cases of interruption (Figure 5).

When using standard POTS lines the connection interruptions where rare, this was dependent on the quality of the lines used. When using lines with no interfering noise the connections where operating with no interruptions.

4.2. Clinical evaluation

The success of such systems is always dependent on the way users accept it and use it. In order to be able to create a system that would have the best possible results from the clinical point of view, we had selected a group of users and the collaboration with them started from the beginning of the project / requirements phase and continued until the end / clinical evaluation. A number of distinct demonstration sites across Europe participated in the evaluation. More specifically, the use of the developed system for emergency cases handling in ambulances has been extensively demonstrated in Greece (Athens Medical Centre), Cyprus (Nicosia General Hospital) and Italy (Azienda Ospedaliera Pisa).

The initial demonstration of the system for ambulance emergency cases was performed on 100 (not severe) emergency cases for each hospital. The results of this phase were very promising. The system was able to improve, the percentage of emergency incidents that initial diagnosis did not matched final diagnosis. For 100 cases without the system use, 13% of the initial diagnosis did not matched final diagnosis; while in 100 cases with the system use 8% of initial diagnosis did not matched the final diagnosis.

The system is currently installed and being used in two different countries, Greece (ambulance vehicles, rural health centers, ship) (Figure 6) and Cyprus (ambulance vehicle, rural health center) (Figure 7).

5. DISCUSSION

The final result is an "all-weather" Telemedicine system, with a flexible architecture that can be adopted in several different application fields. The system has been tested and validated for a variety of medical devices and telecommunication means. The system is currently installed and being used in daily basis in two different countries, Greece and Cyprus. We have developed a medical device for telemedicine applications. The device uses GSM and GPRS mobile telephony links, Satellite links or POTS links and allows the collection and transmission of vital biosignals, still images of the patient and bi-directional telepointing capability. The advance man-machine interface enhances the system functionality by allowing the users to operate in hands-free mode while receiving data and communicating with specialists. Initial results and conclusions are very promising. We intend to improve the system by adding several other functions such as the connection to other medical devices and the application in other cases such as airplanes in flight or trains. Further work is being carried out in integrating the system to hospital HIS/PACS networks. Finally since the involvement of medical experts is definitely the key to the success of similar systems, we intent to enhance the training of the involved personnel in order to have the best results out of the use of the system and modify the system so as to become a perfect part of the medical procedures.

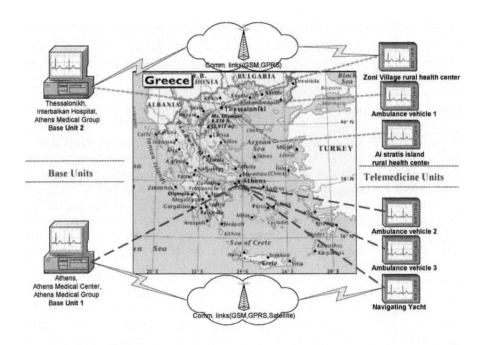

Figure 6. Emergency telemedicine network – GREECE

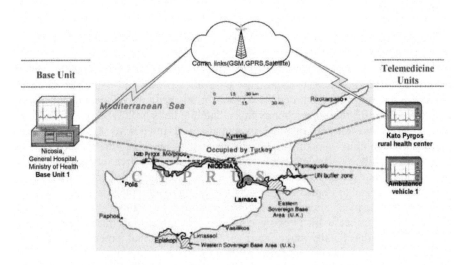

Figure 7. Emergency telemedicine network – CYPRUS

6. ACKNOWLEDGMENTS

The general framework for the above system was developed under EU funded TAP (Telematics Applications Programme) projects. The Emergency-112 project (HC 4027) and the Ambulance (HC1001). Partners in the projects were ICCS-NTUA (project coordinator), Athens Medical Center, Medical Diagnosis and Treatment, Panafon and Epsilon Software from Greece, R&S Informatica and CPR/Pisa Hospital from Italy, Eurotechnology and Malmo University Hospital in Sweden, University of Cyprus and Nicosia general hospital in Cyprus. We would like to express our thankfulness to all participants in the project for their significant contribution and collaboration.

7. REFERENCES

Anagnostaki, A.P., Pavlopoulos, S., Kyriacou, E., Koutsouris, D., A Novel Codification Scheme Based on the "VITAL" and "DICOM" Standards for Telemedicine, *IEEE Transactions on Biomedical Engineering*, December 2002, **49**(12):399-1411.

Bai, J., Zhang, Y., Shen, D., Wen, L., Ding, C., Cui, Z., Tian, F., Yu, B., Dai, B., Dang, J., A Portable ECG and Blood Pressure Telemonitoring System, *IEEE EMB Mag*, Jul/Aug 1999, **18**(4): 63-70.

Cerreli, E. C., 1997 Traffic Crashes Injuries And Fatalities, Preliminary Report, *National Highway Traffic Safety Administration*, U.S. Dept of transportation, 1997.

DINAMAP PLUS monitor host communications, CRITIKON INC., 776591B, USA, 1994.

EMERGENCY 112 project (TAP #HC4027) 1998; http://www.biomed.ntua.gr/emergency112.

Evans, T.,: Cardiac Arrests Outside Hospital, *BMJ*, Apr 4, 1998, **316**:1031-1032.

Gagliano, D., Wireless Ambulance Telemedicine May Lessen Stroke Morbidity, *Telemedicine Today*, Feb 1998, **6**(1): 21.

Giovas, P., Papadogiannis, D., Thomakos, D., et al, Transmission of Electrocardiograms From A Moving Ambulance, *Journal of Telemedicine and Telecare*, 1998, **4**(1): 5-7.

Graschew, G., Roelofs, T.A., Rakowsky, S., Schlag, P.M., Interactive telemedical applications in OP 2000 via satellite, *Biomed Tech*, Berlin, Germany 2002, **47**(1):330-3.

Hofmann, B., General Purpose Telemetry for Analog Biomedical Signals, *IEEE Engineering in Medicine and Biology Magazine*, Nov-Dec 1995, pp.772-775.

Hellenic National Statistics Service, 1997 Report.

Kontaxakis, G., Walter, S., and Sakas, G., EU-TeleInViVo: An integrated portable telemedicine workstation featuring acquisition, processing and transmission over low-bandwidth lines of 3D ultrasound volume images, *Conf. Record, 2000 IEEE EMBS Intl Conf. Inf. Tech. Appl. Biomedicine (ITAB 2000)*, Arlington, VA, USA, November 2000, pp.158-163 .

Kyriacou, E., Pavlopoulos, S., Koutsouris, D., Andreou, A., Pattichis, C., Schizas, C., Multipurpose Health Care Telemedicine System, *Proceedings of the 23rd Annual International Conference of the IEEE/EMBS*, Istanbul, Turkey, October 2001, 8.1.2-3.

Kyriacou, E., Pavlopoulos, S., Berler, A., Neophytou, M., Bourka, A., Georgoulas, A., Anagnostaki, A., Karayiannis, D., Schizas, C., Pattichis,C., Andreou, A., Koutsouris, D., Multi-purpose HealthCare Telemedicine Systems with mobile communication link support. *BioMedical Engineering OnLine*, 2003, **2**:7 .

Murakami, H., Shimizu, K., Yamamoto, K., Mikami, T., Hishimiya, N., Kondo, K., Telemedicine Using Mobile Satellite Communication, *IEEE Trans on Biom. Eng.*, May 1994, **41**(5): 488-497.

Morlion, B., Verbandt, Y., Paiva, M., Estenne, M., Michils, A., Sandron, P., Bawin, C., Assis-Arantes, P., A Telemanagement System for Home Follow-up of Respiratory Patients, *IEEE EMB Mag*, Jul/Aug 1999, **18**(4): 71-79.

Propaq 1xx communications option data interface guide, PROTOCOL SYSTEMS INC., Revision A, 810-0547-00, USA, February 1994.

Propaq Encore 2xx External developers guide, PROTOCOL SYSTEMS INC., Revision A, 810-0791-00, USA, July 1996.

Pavlopoulos, S., Kyriacou, E., Berler, A., Dembeyiotis, S., Koutsouris, D., A Novel Emergency Telemedicine System Based on Wireless Communication Technology – AMBULANCE, *IEEE Trans. Inform. Tech. Biomed. - Special Issue on Emerging Health Telematics Applications in Europe*, 1998, **2**(4): 261-267.

Pattichis, C.S., Kyriacou, E., Voskarides, S.,   Pattichis, M.S., Istepanian, R., Schizas, C.N., Wireless Telemedicine Systems: An Overview, *IEEE Antennas & Propagation Magazine*, 2002, **44**(2):143-153.

Stevens, R., *TCP/IP Illustrated, Volume 1*. Addison-Wesley, USA, 1994.

Sayood, K., Introduction to Data Compression, *Morgan Kaufmann Publishers, Inc*, 1996, pp. 27-54.

Schneier, B., Applied Cryptography, *John Wiley & Sons,* 1996, pp.336-339.

Sandler, D.A., Call to Needle Times After Acute Myocardial Infarction, *BMJ*, Jun 5, 1999, **318**: 1553.

Sachpazidis, I., @Home: A modular telemedicine system, *Conf. Record, 2nd Workshop Mobile Computing in Medicine*, Heidelberg, Germany, April 2002.
   http://www.mocomed.org/workshop2002/beitraege/Sachpazidis.pdf .

Weston, C.F., Penny,W.J., Julian, D.G., Guidelines For The Early Management Of Patient With Myocardial Infarction, *BMJ*, Mar 1994, **308**:767-771.

APPLICATION OF MOBILE COMMUNICATION TECHNIQUES FOR EMERGENCY TELEMEDICINE

Koichi Shimizu*

1. INTRODUCTION

Many studies for the provision of expert medical care to isolated areas have been presented over the years. Application of communication modes such as a telephone network is one of the technological solutions. It is called "telemedicine". To accommodate the following changes in social trends, a new type of telemedicine has to be devised. On the one hand, the population of senior citizens has been increasing, and our living condition has become more stressful than ever. This has resulted in the rapid increase of acute cardiovascular or cerebrovascular crises. On the other hand, people are spending longer periods of time in moving vehicles; for example in aircraft on an international route and in ships on the ocean. Thus, the number of emergency cases in such situations has steadily increased. However, at present, there are no effective means for coping with emergency cases on moving vehicles. In such emergency cases, accurate diagnosis is one of the most important factors for the survival and aftercare of the patient.

Some projects have been conducted in which the techniques of mobile communication and satellite communication were applied to telemedicine. Early experiments were conducted on the transmission of an ECG from an ambulance (Uhley, 1970; Grim et al., 1987), and on the medical consultation service transmitted by a communication satellite (Moller, 1988). Although their feasibility has been verified, the transmission of such medical information has not been widely used in a daily practice.

In this chapter, we define the application of the mobile communication to telemedicine as "mobile telemedicine." We have proposed this mobile telemedicine technique and have verified its practical feasibility in different experiments (Murakami et al., 1989;

*Koichi Shimizu, Graduate School of Engineering, Hokkaido University, Sapporo, 060-8628 Japan, E-mail:shimizu@bme.eng.hokudai.ac.jp

Murakami et al., 1994; Shimizu et al., 1995a; Shimizu, 1999). This paper presents the technical considerations required to realize a practical mobile telemedicine system, techniques developed for multiple medical data transmission, and the satisfactory results of their applications to emergency medicine.

2. TECHNICAL CONSIDERATIONS FOR MOBILE TELEMEDICINE

We have proposed the application of biotelemetry techniques to the mobile telemedicine (Murakami et al., 1994; Shimizu, 1999). The mobile communication links used were a radio communication link and a satellite communication link. Fig.1 illustrates the proposed technique of mobile telemedicine. In a moving vehicle, color images, audio signals and physiological signals such as ECG and blood pressure are obtained from the patient. These images and signals are multiplexed and transmitted to a fixed station. In the fixed station, the signals received are demultiplexed and presented to a medical doctor or processed in an automatic monitoring system. Instructions from the doctor are then transmitted back to the mobile station through the communication link.

This technique has inherent characteristic problems which are different from those of conventional telemedicine or general data transmission using mobile communication. In this chapter, these problems are pointed out and technical solutions are presented to show the practical feasibility of the proposed technique.

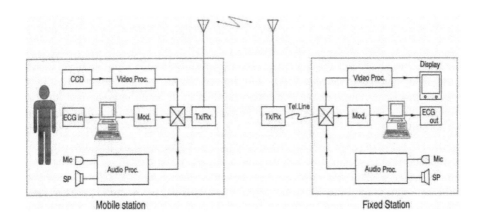

Figure 1.   Principle of telemedicine using mobile communication.

2.1. Channel Capacity

In general mobile communication such as the emergency radio of the fire department, the capacity of the transmission link is generally limited. It is typically 10 kbps - 100 kbps. This range of capacity is far below those required for the transmission of medically significant information, such as a color moving image. The transmission of a color video signal requires a transmission capacity of in the order of 1-10 Mbps with data compression for the moving image of proper quality. Thus, we must first examine whether any useful medical data-transmission can be performed with this limited capacity. For practical telemedicine using this limited transmission capacity, the following parameters were chosen and data compression techniques were applied.

[1] In telemedicine, visual information is of vital importance (Kohli, 1989). In some cases, high-quality colored images are required to check the colors of skin, lips, nails and blood, as well as the condition of the patient's body. In other cases, moving images are required to check the movement of a patient. In the former cases, still pictures are transmitted. By making the interval between each picture short (e.g., less than 1 min), slow movements of the patient can be understood. In the latter cases, if a picture with low image quality is sufficient, a monochromatic moving picture is transmitted. Various techniques have been developed to compress image data. The technique considered in this study is based on vector quantification of differential luminance and chrominance signals (Yamaguchi, 1985). With this technique, 10:1 compression is possible, and color still-pictures can be transmitted every 20 sec in 8 kbps.

[2] Audio signals are used primarily for communication between a doctor and a patient or crewmembers. Intelligibility is important, but an extremely high quality of sound is not required. Thus, in this study the audio signal is band-limited from 3 kHz to 1.5 kHz, and an adaptive delta modulation technique (Jayant, 1970; Jayant, 1974) is applied with a sampling frequency of 10 kHz. This allows a compression ratio of 4.8:1 and a bit rate of 10 kbps.

[3] For the biological signals to be transmitted, ECG and blood pressure were chosen as fundamental parameters to indicate vital signs. To transmit 3-channel ECG signals, a channel capacity of 4.8 kbps is required. Many techniques have been developed for ECG data compression (Jalaleddine et al., 1990), and it is not difficult to reduce data to 8:1. Systolic and diastolic blood pressures are measured every minute. Since the dynamic range and sampling rate are small, only a small bit rate and no compression are required.

The total capacity required for transmitting all of this data is about 19 kbps. Therefore, even if more capacity is required for other signals for multiplexing and redundancy, e.g., synchronization bits, parity bits etc., all this data can be transmitted in the practical capacity of a mobile communication link.

2.2. Reliability

In telemedicine, a transmission error can be fatal. Therefore, reliability is much more important in telemedicine than it is in other commonly used data transmissions. A high level of reliability can be obtained by introducing as much redundancy as possible in the signal transmission; e.g., by making the handshake transmission process complex and by establishing strict error-checking or error-control. However, such measures are difficult with the aforementioned small transmission capacity. If we use a satellite link, a significant propagation delay (about 0.5 sec) becomes another problem.

In this study, the following error-control techniques were applied. An automatic repeat request (ARQ) technique was employed for the ECG. Although ARQ is highly reliable in data transmission, retransmission of data produces a delay when an error is found. This problem can be solved by the technique described in the next section. There is no problem of retransmission if we use the method of forward error correction (FEC).

We employed FEC for blood pressure, and error-control was conducted according to the majority decision principle.

The reliability of each parameter was analyzed theoretically. First, an error interval (EI), or a mean time between errors, is defined as

$$EI = L_{data} / (E_{data} R_{data}) \tag{1},$$

where E_{data} is the probability that the data is in error, L_{data} is the data length [bit], and R_{data} is the data rate [bit/sec]. E_{data} depends on the coding method of the data.

In our system, EI's are theoretically calculated for ECG and blood pressure. The results are shown in Fig.2. The BER is assumed to be 10^{-3} in the worst possible case, which seldom happens. The BER of the communication link used in this study was about 10^{-5} in the usual situation. Thus, as can be seen in the figure, sufficient reliability is attained by the error control.

EI's are calculated for the image and audio signals in a similar way. An error in these data is produced once every 5-50 seconds. Therefore, an error appears in an image in this interval. The error is only in a small part of the image and can therefore be ignored. In the case of audio data, errors occur over a short period of time and do not affect the intelligibility of the conversation. The above data are presented to the human visual and auditory sensing systems, and a considerable error can be tolerated from the viewpoint of recognition. Thus, sufficient reliability is guaranteed in this system.

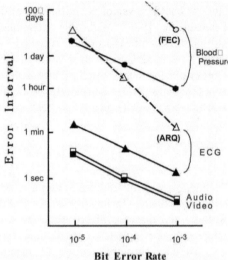

Figure 2. Theoretical analysis of reliability and improvements by error-control techniques.

2.3. Real-Time Operation

In telemedicine, real-time communication between moving and fixed stations is important. Recently the operation speeds of microprocessors and communication control units have increased greatly. This has made real-time data transmission easier, even with complex error-control procedures. However, if we use a satellite communication link, the propagation delay between the transmitter and receiver becomes a problem. If ARQ is used as the error-control principle, a delay is inevitable due to the time required for repeat-request signal transmission and data retransmission.

This problem can be solved by installing input and output buffer memories in both moving and fixed stations. The transmission time has to be made sufficiently shorter than the time corresponding to the buffer capacity. Since the speed of data transmission is faster than the speed of inputting and outputting of the original signal, there is no interruption in the output data. Detailed description of this technique can be found elsewhere (Shimizu, 1999).

As an example, transmission delay is evaluated in the case of a satellite communication link or a 24-kbps MSK (minimum shift keying) link. If the bit error rate and the round trip delay of the link are 10^{-3} and 0.6 sec, respectively, the mean delay is 94 bits or 0.073 sec. These are sufficiently small to cope with by the technique mentioned above. Therefore, the real-time operation is not impeded by the propagation delay.

2.4. Electromagnetic Interference (EMI)

The electromagnetic environment in a moving vehicle may be adversely affected by ignition noises, radio communication and radar equipment. Therefore, medical devices and instruments should be designed to be immune to such an unfavorable environment. At the same time, they must not interfere with the navigation equipment of the vehicle, particularly an airplane or a spaceship.

As to electric lines and cables, techniques have been established to suppress both conduction and emission noises. Thus, equipment used in emergency care does not generally have the problem of EMI. However, telemetry devices using radio waves are of primary concern. Although telemetry devices are useful for measuring and monitoring physiological parameters of the subject in telemedicine, they are vulnerable to the EMI and can also be the cause of the EMI.

The telemetry technique using indirect light transmission (Kudo et al., 1987) is one solution to the above problem. This technique is useful in an enclosed environment, such as the cabin of an airplane, ship or space-lab. Ambulatory measurement using indirect light transmission has been shown to be feasible, and its effectiveness was verified by the acquisition of physiological data from astronauts (Shimizu, 1995b).

2.5 Multiplexing Technique

To realize multiple data transmission through a single communication channel, we have to design a signal multiplexing technique. An analog multiplexing technique was newly developed to transmit an ECG and a vocal signal both in a real-time operation through a single vocal communication channel (Matsuda et al., 1993). For a digital multiplexing, the throughput of data transmission was analyzed for various bit-error rates, and the data format with an optimal efficiency was designed (Murakami et al., 1989). Table 1 shows the designed frame structure.

Table 1. Designed Frame Structure and Transmission Rates

Data	Rate (kbit/s)	Length (bit/frame)
Video	11.36	568
Audio	10.08	504
ECG	1.28	64
B. P.	0.02	1
Others	1.26	63
Total	24.00	1200

3. TELEMEDICINE IN AN AMBULANCE

3.1. Fundamental Principle

Figure 1 illustrates the principle of this technique (Shimizu et al., 1995a). The appearance of the patient is recorded by a CCD camera at the accident scene or in the ambulance. Color still-pictures are transmitted, and each scene is renewed every 20-80 seconds. Vital signs, such as an ECG and blood pressure, are sampled, digitized and multiplexed in a personal computer. They are fed into the transmitter through a modem, and the signals are then transmitted to a central fire station using an emergency radio link or a mobile telephone network.

At the fire station, the communication channel is connected to a telephone line, and the signals are transferred to an appropriate hospital. In the hospital, a medical doctor can view the color images and monitor the patient's vital signs in real time. The doctor then sends the necessary information to the paramedics in a mobile station through a vocal communication channel (Shimizu et al., 1995a; Shimizu, 1999).

The emergency radio link of a fire station and a mobile telephone network are used as wireless transmission media. In the former case, the signals from an ambulance are transmitted to a relay station on the top of a mountain in the suburbs of Sapporo City using a radio frequency (150 MHz band FM, 10 W). The signal is then transmitted from the relay station to the central operating room of the fire station through a microwave network of 12 GHz. In the latter case, signals are transmitted by radio waves (900 MHz band) from the ambulance to the nearest base station of the mobile telephone system.

3.2. Multiplexed Transmission of ECG

To examine the usefulness of this system, experiments were conducted on transmission of an ECG from a moving ambulance. The ECG was obtained from a subject using the portable electrocardiograph commercially available.

3.2.1. Analog Multiplexing

First, the analog multiplexing technique was examined. In the central city and suburban areas, the results were satisfactory except for when the ambulance passed through a tunnel or an underpass. Fig.3 shows an example of these results. When the communication link was not good, noises appeared in the re-

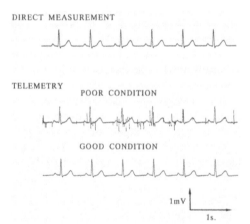

Figure 3. Results of analog ECG transmission from a moving car.

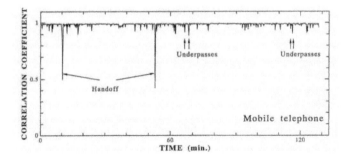

Figure 4. Temporal changes in the quality of received signals in analog ECG transmission.

ceived ECG. However, the major waveform of an ECG can still be recognized in a noisy signal. Fig.4 shows a typical example of temporal changes in the received signal quality. The ordinate is the correlation coefficient between the input ECG waveform of a transmitter and the output waveform of a receiver. Generally, the ECG waveform is sufficient for clinical diagnosis if the correlation coefficient is more than 0.8-0.9. As can be seen in Fig.4, the results using a mobile telephone line were satisfactory. The temporal degradation of signal quality at 10 and 50 minutes after the start of transmission seemed to be due to the hand-off phenomenon of a mobile telephone line. In the case of an emergency radio link, the degradation of signal transmission was relatively large (Shimizu et al., 1995a; Shimizu, 1999).

Through the analysis of the electric field strength of the received radio wave, the decrease in the reception level was found to be the major cause of this signal degradation (Shimizu et al., 1995a; Shimizu, 1999). This means that even in our experimental system, a reasonable quality of signal transmission can be expected as long as the reception level of the radio wave is maintained appropriately.

3.2.2. Digital Multiplexing

An experiment on digital ECG transmission was conducted, as well. Fig.5 shows theresults of three channel multiplexed ECG transmission in the intelligent mode. When the car was running in a usual condition, the quality of the ECG transmission was high. When the car passed through a tunnel, the communication link was temporarily cut and the ECG signal was lost for a short time, but real-time transmission was resumed when the car came out of the tunnel. In the experiment, the delay between the input signal of the transmitter and the output signal of the receiver was less than 5 seconds, which is acceptable for the purpose of emergency care.

To test the practical usefulness of this technique, a 3-channel ECG transmission was conducted from a moving ambulance to the emergency room of a hospital. This transmission

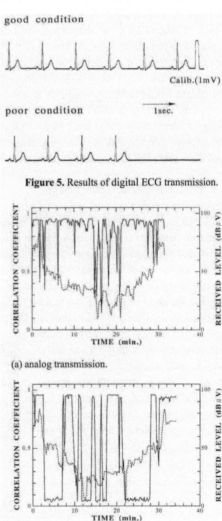

good condition

Calib.(1mV)

poor condition   1sec.

Figure 5. Results of digital ECG transmission.

(a) analog transmission.

(b) digital transmission

Figure 6. Comparison of analog and digital ECG transmissions using temporal changes of correlation coefficient (solid line) and received level of radio wave (dashed line).

was evaluated as "sufficiently useful" by a medical doctor specializing in emergency medicine.

3.3. Comparison of Analog and Digital Transmissions

To study the merits of analog and digital ECG transmissions, their characteristics were compared in transmission experiments from a moving car. Fig.6 shows temporal variation of transmission fidelity with changes in the strength of the received radio wave. The routes taken by the car were almost the same in the analog and digital ECG transmission experiments. The routes were confined to the central area of Sapporo city. The results of the experiments showed the characteristics of analog and digital ECG transmissions. In analog ECG transmission, the fidelity of data transmission varied with changes in the received radio wave. While in digital ECG transmission, the fidelity remained high until the strength of the radio wave dropped below a certain level.

Generally, if the correlation coefficient is more than 0.8, we can identify the ECG waveform from background noises and recognize the change in a heart rate by visual observation. Therefore, when the conditions of the communication link are not stable, such as in an emergency radio link, analog ECG transmission seems to be better. However, if the communication link is relatively stable, such as in a mobile telephone network, digital ECG transmission in an intelligent mode is more effective.

3.4. Transmission of Color Images

Although the need for ECG transmission from an ambulance has been widely recognized, there have been relatively few reports on the use of image transmission to show the condition of an emergency patient until very recently. The reason for this is the technical difficulties of practical image transmission in emergency cases. We have realized the significance of visual information for diagnosis by a medical doctor, and we have applied the image transmission technique to mobile communication (Shimizu et al., 1990; Shimizu et al., 1991; Murakami et al., 1994; Shimizu et al., 1995a).

Fig.7 shows the color image transmitted from a moving car under stable conditions, i.e., traveling at a speed of about 50 km/h on an asphalt-surface road. As can be seen from the figure, the transmitted image was of sufficient quality for understanding the patient's condition.

For comparison to extreme conditions, a color image was transmitted from the car traveling on a gravel road in a mountainous area. The vibrations of the car were so strong that the experimenters could not remain standing. The image was

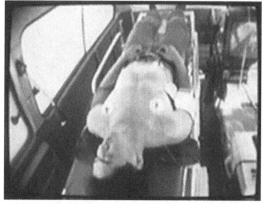

Figure 7. A color image transmitted from a moving car.

disturbed to some extent by noises, which appeared as white horizontal lines and as a change in the colors in a horizontal band. However, despite the noise, the quality of the images was still sufficient for understanding the patient's condition.

To examine the practical usefulness of this technique, transmission experiments were conducted under various conditions that were likely to be encountered in emergency cases. In the city area, the ECG and images were transmitted from various points; e.g., from a street among many tall buildings, from a narrow road between two tall buildings, under the overhead lines of streetcars, under overhead train lines, and from an underground parking lot. In the suburbs, signals were transmitted from residential areas, from highways, and from hilly areas. No problems for practical use were found in most of these locations. The only problems found were some long cuts in the communication link when the ECG and images were transmitted from underground parking lots or from behind a high mountain. The experiments were repeated using a mobile telephone line and an emergency radio link. In the emergency radio link, the communication link was relatively unstable and noise was greater than that in the mobile telephone line. However, except for this difference, the results obtained using these two communication links were almost the same.

Finally, the simultaneous transmission of ECG's and color images from a moving ambulance to the emergency room of a hospital was attempted. A medical doctor specializing in emergency medicine evaluated the quality of received signals, and he found it satisfactory for emergency practices.

4. TELEMEDICINE WITH SATELLITE COMMUNICATION

4.1. Satellite Link

This part of the system was developed in the cooperation of the Communications Research Laboratory of the Ministry of Posts and Telecommunications, the Electronic Navigation Research Institute of the Ministry of Transport, and the National Space Development Agency of Japan. Mobile stations have been developed for installation in ships, aircraft and automobiles. The ship station is small enough to be installed in a ship of about 30 tons. The weight of above-deck equipment with a 40cm short backfire antenna is about 45 kg. Countermeasures were taken against the fading effect caused by sea surface reflection.

The aircraft station was designed to be installed on B-747 cargo planes. The antenna system was designed to be mounted inside the roof of the B-747. The frequency shift due to the Doppler effect was compensated for.

The ETS-V satellite is a three-axis stabilized geostationary satellite (Murakami et al., 1994; Shimizu, 1999). On-board equipment for communication experiments consists of L-band and C-band antennas and transponders (Murakami et al., 1994; Shimizu, 1999). The fixed station, which has RF equipment and antennas for C-band and L-band links, was set up in a laboratory on the ground. The major specifications of the mobile stations and the satellite link can be found elsewhere (Murakami et al., 1994; Shimizu, 1999).

4.2. Medical Data Transmission System

Figure 8 shows the outline of the medical data transmission system. The data flow from the mobile station to the fixed station is as follows. In the vehicle shown in Fig.8(a), the audio, ECG (3 channels) and blood pressure signals are acquired at specified intervals from the patient and are encoded in a personal computer. The image signals are encoded by the image compression system, and the coded signals are sent to a personal computer. The image signals and other signals are multiplexed in time and fed to the modulator of the satellite link. The multiplexed data is sent to the satellite in a serial mode and transferred to the fixed ground station.

In the ground station shown in Fig.8(b), the signals are obtained from the demodulator of the satellite link and are demultiplexed in a personal computer. The image signals are sent to the image decompression system, and the picture is decoded and displayed. Other signals are directly decoded in the personal computer to produce the outputs. If an error in an ECG signal is detected in the decoding process, an ARQ signal is generated. The error correction of blood pressure signals is executed in the personal computer.

The data flow from the fixed station to the mobile station is as follows. In the fixed station (Fig.8(b)), the voice of the doctor is encoded in a personal computer. The encoded signals are multiplexed with the ARQ signals from the ECG error detector and are sent to the modulator of the satellite link. The signals are sent to the satellite in a serial mode and transferred to the mobile station.

In the mobile station (Fig.8(a)), the signals from the demodulator are separated and decoded. The audio signals are directly decoded to an audio output. ECG ARQ signals are sent to the encoder in the personal computer. The stored corresponding data is then transferred to the transmitter buffer. Automatic retransmission is achieved by transmitting these data again to the fixed station through the satellite link.

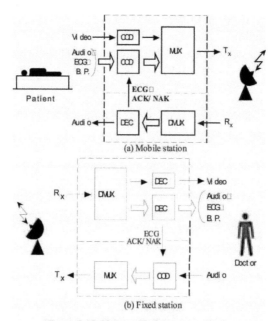

Figure 8. Multiple medical data transmission system

4.3. Transmission from a Navigating Ship

After the basic performance of the proposed system was verified by experiments in a fixed station, a transmission experiment from a navigating ship was conducted. The mobile station was installed in the training ship (1,400 tons) of our university. The fixed station was set up in the ground station of the Communications Research Laboratory.

In the experiment, a variable noise was added immediately before the demodulator of the satellite link to change the state of the communication link. The color image, audio signal, ECG, and blood pressure were transmitted separately, and the quality of the data after decoding was examined. The lower limit of the carrier-to-noise ratio (C/N_0) that would guarantee satisfactory quality of transmission for telemedicine was investigated for each type of information. Fig.9 shows some examples of the received images. When C/N_0 was 55.1 dBHz (Fig.9(b)), the shape of the object was clear and color reproduction was good in more than half of the picture area. As C/N_0 decreased further, errors in the image increased rapidly. Thus, the threshold value for image quality was determined to be 55.1 dBHz.

Table 2 summarizes the results of the experiments. The reliability of blood-pressure transmission was improved by applying the error correction technique, and satisfactory transmission was achieved up to 49.9 dBHz. This agrees with the results of the theoretical analysis (Fig.2).

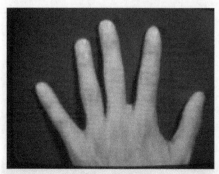

(a) C/N_0 = 65.5 dBHz

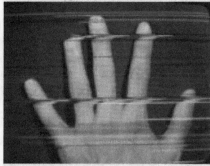

(b) C/N_0 = 55.1 dBHz

(c) C/N_0 = 53.1 dBHz

Figure 9. Color images transmitted from a navigating ship to a fixed station.

Table 2. C/N_0 Thresholds in Data Transmission

Data	Non-error Transmission (dB Hz)	Acceptable Transmission (dB Hz)
Video	57.1	55.1
Audio	56.9	55.0
ECG	55.0	N.A.
B. P.	49.9	N.A.

4.4. Transmission from an Aircraft in Flight

A transmission experiment was also conducted from an aircraft in flight. The mobile station was installed in a jet cargo plane (Boeing 747), and the fixed station was set up in a ground station. The flight route was along the great circle route from Narita, Japan (New Tokyo International Airport) to Anchorage, U.S.A. The transmission experiment was conducted in the same manner as that for the navigating ship. Fig.10 shows examples of transmitted images. Again, when the C/N_0 of the satellite link was more than 55 dBHz, satisfactory quality was obtained in the separate transmission of image, sound, ECG, and blood pressure.

A simultaneous transmission of image, sound, ECG, and blood pressure was attempted. With C/N_0 of more than 55 dBHz, good multiplexing and demultiplexing without any crosstalk were confirmed. In this experiment, change in the C/N_0 of the satellite link was less than several dBHz during an approximately 20-hour round trip flight from Narita, Japan to Anchorage, USA. The communication link was not blocked as it was in the case of the ship, and the feasibility of telemedicine in a flying aircraft was verified.

Detailed descriptions of the ship and the aircraft experiments can be found elsewhere (Ikegami et al., 1990; Ohmori et al., 1990; Sakasai et al., 1990; Taira et al., 1990; Shimizu, 1999).

(a) C/N_0 = 58.9 dBHz

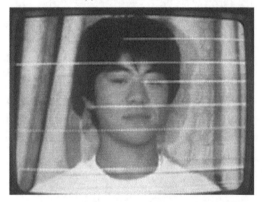

(b) C/N_0 = 54.0 dBHz

Figure 10. Color images transmitted from a flying aircraft.

5. EMERGENCY DETECTION USING MOBILE COMMUNICATION

5.1. Vital-sign monitoring

The above mentioned systems are useful when the emergency problem is detected properly. However, the appropriate detection of the emergency case is not always easy, particularly when the patient is out of a house or out of a hospital. To solve this problem we have developed a vital-sign monitoring system and a location system using mobile communication techniques (Shimizu et al., 2000a; Shimizu et al., 2003).

Figure 11 illustrates the principle of the proposed technique. It consists of two steps of telemetry. In the first step, the vital-sign or a pulse wave is detected by a photoelectric sensor at the root of a finger using a ring-type sensor. The pulse wave signal is transmitted from the ring to the mobile phone terminal by a near-field radio wave. The terminal is wearable and always carried by the subject both inside and outside the house. In the second step of the telemetry, the vital-sign signal is transmitted from the wearable terminal to a personal computer by a mobile phone network. The computer is placed in the house of the subject. It processes the received signal and analyzes the pulse wave. If any emergency is detected, an alarming process and an automatic reporting process are activated by the computer

Since the telephone link was originally designed for verbal communication, special modification has to be made to use it for biotelemetry. For example, if we use the network for continuous vital-sign monitoring, we have to occupy the public telephone network for a long time. Thus, we modified the system so that it can be used in different mode in a house. In this technique, PHS (personal handy-phone system, or a common cellular phone system in Japan) is used for outdoor data transmission. For indoor data transmission, the same portable PHS terminal serves as a transmitter in a cordless-phone mode. In this way, we can transmit a biological signal without occupying a public network as long as the

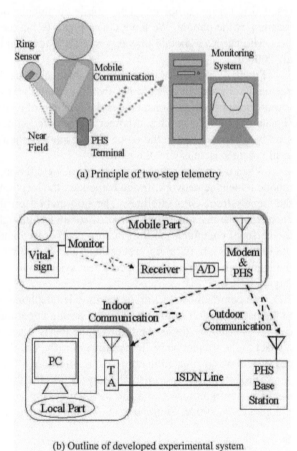

(a) Principle of two-step telemetry

(b) Outline of developed experimental system

Figure 11. Vital sign monitoring using mobile communication.

user is inside the house. This is preferable from an economical viewpoint, as well. The operation mode of the terminal is automatically changed between indoor and outdoor modes according to the position of the patient. Using this principle, we can monitor a biological signal both inside and outside of the house using a common terminal. In addition, the area of coverage extends as wide as the network distribution.

Another problem is the momentary disconnection of the telephone link due to the hand-over between base stations or due to the interference in a house. This is not a serious problem for the verbal communication, but can cause significant error in the vital-sign monitoring. Thus, the period required to re-establish the communication link after the momentary disconnection has to be analyzed, and an appropriate countermeasure has to be taken (Shimizu et al., 2000a; Shimizu et al., 2003).

5.2. Location system

In emergency cases, often the position of the patient could not be identified and it resulted in serious problems. The mobile communication technique can be used for the location of the patient. We have developed a technique to obtain the position of the patient who carries a mobile phone with a GPS function (Shimizu et al., 1998; Shimizu et al., 2000b).

Figure 12 illustrates the outline of this technique. The location system consists of a personal computer at a home (or institution) and a responder carried by a patient. The responder has two functions. One is the function of a GPS receiver to receive the location signals from GPS satellites. Another is the function of a terminal of a mobile phone to send back the location to the home. The responder should be small and light to be carried easily or unconsciously by the patient.

When a computer detects an emergency condition, it calls up the responder using a mobile telephone network. In the responder, the location of the patient is calculated using the signals from GPS satellites. The information on the location is automatically sent back from the responder to a personal computer through a mobile telephone network. In the personal computer, the information is processed and displayed on a CRT display. At the same time, an alarming process and an automatic reporting process are activated by the computer. We have developed a test system and conducted experiments to examine the feasibility.

The communication link of the mobile telephone is known to be fairly invulnerable to the bad condition of the weather and terrain conditions. However, the GPS radio wave can be affected by these conditions particularly when the responder is carried in close contact to the subject body. To examine the effect of these conditions, the performance of our system was evaluated in adverse conditions.

Through these experimental analyses, the feasibility and the practical usefulness of the vital-sign monitoring and the location systems were verified (Shimizu et al., 1998; Shimizu et al., 2000a; Shimizu et al., 2000b; Shimizu et al., 2003).

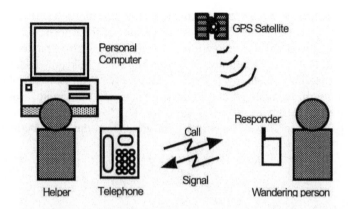

Figure 12. Principle of location by mobile communication.

6. CONCLUSIONS

With a view to providing emergency care in any place, a telemedicine technique using mobile communication was proposed. The practicality of this technique was investigated through technical considerations required to realize mobile telemedicine. Some problems with this technique were identified, and measures to resolve the problems were devised. Then, theoretical analysis verified the feasibility of the proposed technique. Different multiplexing techniques for the multiple medical data transmission by mobile communication were investigated. An experimental system that can simultaneously transmit color images, an audio signal, 3-channel ECG's, and blood pressure from a moving vehicle to a fixed station was developed. Experiments on the transmission of multichannel medical data from a moving ambulance, a navigating ship, and a flying aircraft were conducted. The results of these experiments verified the practical feasibility of the proposed technique. The mobile communication technique can be applied to the detection of emergency conditions, as well. The vital-sign of a patient can be monitored and his/her position can be located in and out of the house continuously.

In the practical application of this technique, there may be some legal problems: for example, whether medical treatment through a communication medium would be legally acceptable or not, and whether the transmission of medical data violates the protection of personal privacy. However, considering the emergency nature of this technique and the significance of the results, both problems seem to be either legally or technically solved.

This application of mobile communication to telemedicine is not confined as merely a proposal of one of the new techniques. It can also bring about methodological change in the concept of conventional telemedicine by changing it from static to dynamic and by enlarging its scope from a local area to a global or cosmic area. It may also have an impact on conventional emergency medicine in that it will open up a new field of application that applies to moving vehicles.

7. ACKNOWLEDGMENTS

The author wishes to thank the staff of the Communications Research Laboratory, Ministry of Posts and Telecommunications, and the staff of the Bureau of Telecommunication for their cooperation in the experiments. The author is also grateful to Professor Goro Matsumoto, Hokkaido University; Professor Tomohisa Mikami, Hokkaido University; Professor Nozomu Hoshimiya, Tohoku University; Professor Takeshi Hatsuda, Hokkaido Institute of Technology; and Professor Katsuyuki Yamamoto, Hokkaido University for their valuable advice.

A part of this research was supported by a grant from the Ministry of Education, Science and Culture, Japan.

8. REFERENCES

Grim, P., Feldman, T., Martin, M., Donovan, R., V. Nevins, and Childers, R. W., 1987, Cellular telephone transmission of 12-lead electrocardiograms from ambulance to hospital, Am. J. Cardiol. 60: 715-720.

Ikegami, T., Wakana, H., Kawamata, F., Ide, T., and Matsumoto, Y., 1990, Experiments on maritime satellite communications using ETS-V, Rev. Commun. Res. Lab. 36: 47-56.

Jalaleddine, S. M. S., Hutchens, C. G., Strattan R. D., and Coberly W. A., 1990, ECG data compression techniques - a unified approach -, IEEE Trans. Biomed. Eng. BME-37: 329-343.

Jayant, N. S., 1970, Adaptive delta modulation with a one-bit memory, Bell Syst. Tech. J. 49: 321-342.

Jayant, N. S., 1974, Digital coding of speech waveforms, Proc. IEEE 62: 611-632.

Kohli, J., 1989, Medical imaging applications of emerging broadband networks, IEEE Commun. Mag. 8-16.

Kudo, N., Shimizu, K., and Matsumoto, G., 1987, Optical biotelemetry using indirect light transmission, Biotelemetry, 9: 91-98.

Matsuda, S., Shimizu, K., and Yamamoto, K., 1993, Multiplexing technique of audio and medical signals for emergency radio, Jap. J. Med. Elec. Bio. Eng. 31: 74-78.

Moller, J., 1988, Health care via satellite, Computer Graphics World, 69-70.

Murakami, H., Shimizu, K., Yamamoto, K., Mikami, T., Hoshimiya, N., and Kondo, K., 1989, Medical data transmission using mobile satellite communication, Trans. IEICE, J72-B-II: 262-268.

Murakami, H., Shimizu, K., Yamamoto, K., Mikami, T., Hoshimiya, N., and Kondo, K., 1994, Telemedicine using mobile satellite communication, IEEE Trans. Biomed. Eng. 41: 488-497.

Ohmori, S., Hase, Y., Wakana, H., Taira, S., and Ide T., 1990, Experiments on aeronautical satellite communications using ETS-V, Rev. Commun. Res. Lab. 36: 27-37.

Sakasai, M., Arakaki, Y., Morikawa, E., and Matsumoto, Y., 1990, Experimental results of TDM/TDMA system via ETS-V, Rev. Commun. Res. Lab. 36: 57-62.

Shimizu, K., Murakami, H., Yamamoto, K., Mikami, T., Hoshimiya, N., Kadowaki, N., and Kondo, K., 1990, Multiple medical data transmission from mobile station using ETS-V, Rev. Comm. Res. Lab. 36: 105-111.

Shimizu, K., Murakami, H., Yamamoto, K., Mikami, T., Hoshimiya, N., and Kondo, K., 1991, Medical telemetry from moving vehicles using a communication satellite, Biotelemetry, 11: 242-245.

Shimizu, K., Matsuda, S., Saito, I., Yamamoto K., and Hatsuda, T., 1995a, Application of biotelemetry technique for advanced emergency radio system, IEICE Trans. Commun. E78-B: 818-825.

Shimizu, K., 1995b, Optical biotelemetry, Jap. J. Physical Fitness and Sports Medicine, 44: 317-324.

Shimizu, K., Kawamura, K., and Yamamoto, K., 1998, Development of location system for wandering person, Biotelemetry, 14: 203-208.

Shimizu, K., 1999, Telemedicine by mobile communication, IEEE Eng. Med. Biol., 18: 32-44.

Shimizu, K., and Suzuki, T., 2000a, Development of indoor-outdoor common biotelemetry technique, Biotelemetry, 15: 563-569.

Shimizu, K., Kawamura, K., and Yamamoto, K., 2000b, Practical considerations for a system to locate moving persons, Biotelemetry, 15: 639-645.

Shimizu, K., Suzuki, T., and Nishida, T., 2003, Development of vital-sign telemonitoring system using mobile phone communication, Biotelemetry, in press.

Taira, S., Ohmori, S., and Tanaka, M., 1990, High gain airborne antenna for satellite communications, Rev. Commun. Res. Lab. 36: 39-46.

Uhley, H. N., 1970, Electrocardiographic telemetry from ambulances, Am. Heart J. 80: 838-842.

Yamaguchi, H., 1985, Vector quantization of differential luminance and chrominance signals, IEEE Trans. Commun. COM-33:457-464.

FUTURE CHALLENGES AND RECOMMENDATIONS

Efthyvoulos Kyriacou[*] and Sotiris Pavlopoulos

Continuous and rapid Changes in Information technology and telecommunications lead to a new and improved type of infrastructure and services in the field of wireless telemedicine systems for emergency health care support.

Infrastructure can be divided into three main parts: Communications, Computers and Biosignal acquisition:

A) Communication for wireless emergency health care systems, until today, was performed mainly using mobile telecommunication systems such as the GSM which is a standard used almost everywhere in the world. The use of these networks was able to provide easy access to telemedicine services without having to set up a private network for communication and spending huge amounts of money. Despite these advantages GSM has limited bandwidth (up to 9.6 Kbps or up to 57.6Kbps for HSDC-High Speed Circuit Switched Data) (GSM web site) which adds a limit to the information transmitted. This fact sometimes results to the loss of information which could have proven useful for the health care professionals. During the last years, the introduction of new mobile telecommunication systems, like the GPRS system which provides much higher bandwidth (theoretically up to 171.2 Kbps) enabled the transmition of much more information which can be proved useful for the healthcare providers and crucial for patient's treatment. In addition future coming 3G mobile networks like the UMTS (UMTS forum) will be able to provide bandwidth up to 2Mbps something that will enable the transmition of more information like continuous 12 leads of ECG when monitoring cardiac patients from a moving Ambulance vehicle. Furthermore the introduction of new services like video telephony through wireless networks will be an addition that can help with communications between a health care provider (Nurse, Paramedics) and an expert doctor. Another important factor is the installation of wireless networks in Cities something that will be able to improve wireless health care systems; this is something that is being discussed and looks like it will be a reality during the next years. The use of such networks will be very important because health care providers will have immediate and high speed telemedicine access from anywhere in the area of a city.

---

[*] Efthyvoulos Kyriacou Department of Computer Science, University of Cyprus, 75 Kallipoleos str, P.O.Box 20537, 1678, Nicosia, Cyprus, e-mail:ekyriac@ucy.ac.cy.

The use of locating systems such as the GPS (Global Positioning System) and the GIS (Geographical Information Systems) will also be able to improve health care services. For example when a moving Ambulance vehicle is trying to reach a patient using the fastest route, or when an Ambulance vehicle carrying a patient is trying to get to the base hospital.

Other networks used for wireless health care telemedicine systems are the satellite networks. Even thought these networks can provide high bandwidth every where in the world, their use is limited because of the high installation and operation costs. They are used mainly to support areas where no other network can support, like isolated mountain areas, navigating ships or airplanes.

B) Changes in commercial computer systems are rapid and continuous. New systems are presented every day. Modern portable computer systems have smaller size and weight but provide almost the same computational capabilities as non portable computer systems. The use of these devices in wireless telemedicine application is something that was presented some years ago but capabilities were limited due to the size or the computational capabilities of the systems. Nowadays the introduction of portable devices like PDAs, Small Size laptops, Pen-Tablet Pc's is something that enables wireless telemedicine systems designers to create faster, better and smaller systems. Such efforts have already appeared and will continue to appear during the next years.

C) Biosignals acquisition is another technological field which affects wireless telemedicine systems. The collection of biosignals (Axisa, F. et al., 2004; Bolanos, M. et al., 2004; Wealthy project), such as ECG, until now was performed using expensive devices which could only be handled and supported by medical personnel. Nowadays the collection of biosignals, such as ECG, can be performed by very small devices. These are not always devices on their own but they might connect to a PC in order to display the signals or to a mobile phone in order to send the signals. They might be wearable, have the shape and weight of a necklace etc. These devices will enable the use of wireless telemedicine systems almost anywhere and at less cost. Such devices can be used for home care purposes much easier than the standard medical devices.

Healthcare services (The ass. of tel. serv. Providers) are the other major issue which will rapidly change during the next years because of the introduction of wireless telemedicine systems. Several issues have been addressed from the very beginning of telemedicine systems use. Legal, liability, ethical issues as well as the structure of services will have to change in order to use these systems. Starting from legal issues the introduction of new services will have to be covered by several laws National, EU laws in the case of European countries or Federal laws for the United States. These laws will have to cover all issues about the responsibilities during an emergency incident or the responsibilities when handling a home care incident.

Liability of systems will have to be covered by standards which will describe everything that a system should follow. Even though there has been significant effort in creating a standard for collecting and exchanging biosignals (ACR-NEMA, 1999; IEEE 1073.1.3.4, 1999) (like DICOM for images) no standard is widely used by the commercial biosignal monitors manufacturers. Such issues will have to be resolved in the future in order to cover liability and interoperability of medical devices.

On the other hand several ethical issues will have to be covered when using these systems; such issues concern the exchange of medical information through public networks thus having potential problems with security. This will be solved by securing the exchanged information and covering the identity of patients in the best possible way.

REFERENCES

ACR-NEMA, DICOM Supplement 30: Waveform Interchange, *Revision 1.1 Document for Letter Ballot: NEMA Standard* Pub., Feb. 1999.

Axisa, F., Gehin, C., Delhomme, G., Collet, C., Robin, O. , Dittmar, A., Wrist Ambulatory Monitoring System and Smart Glove for Real Time Emotional, Sensorial and Physiological Analysis, *Proceedings of the 26th Annual International Conference of the IEEE EMBS*, San Francisco, CA, USA • September 1-5, 2004, pp 2161-2164.

Bolaños, M., Nazeran, H., Gonzalez, I., Parra, R., and Martinez, C., A PDA-based Electrocardiogram/Blood Pressure Telemonitor for Telemedicine, *Proceedings of the 26th Annual International Conference of the IEEE EMBS*, San Francisco, CA, USA • September 1-5, 2004, pp 2169-2172.

GSM web site; http://www.gsmworld.com/

IEEE 1073.1.3.4, "Medical Device Data Language (MDDL) virtual medical device, specialized—Pulse oximeter," in *Standard for Medical Device Communications*, Apr. 1999.

The association of telehealth service providers; http://www.atsp.org/

UMTS forum; http://www.umts-forum.org

Wealthy Project: Wearable Health Care System, IST 2001 – 3778, Commission of the European Communities; http://www.wealthyist.com .

V. ECHOGRAPHY SYSTEMS AND SERVICES

Pierre Vieyres

Section Editor

SECTION OVERVIEW

Pierre Vieyres[*]

By minimizing or even eliminating the effect of distance, telemedicine comes to establish new standards for the delivery of health services worldwide. One of the major application areas is the provision of skilled medical care to patients who are in some way isolated from the specialized care they need. This is specially the case for patients living in isolated areas (rural regions, islands) who often have substandard access to specialty health care. Access to these services is only achieved with the use of applications that allow remote expert consultation providing a faster response in emergency and crisis situations. In the last decade, several telemedicine concepts and scenarios have been investigated and developed to fulfill this healthcare need for the benefit of isolated populations.

Ultrasound echography is a non invasive method that is very well adapted for the diagnosis of anomalies and is currently used in routine and in emergency cases at the hospital. This specialty health care method is an expert-dependant technique, difficult to access in areas with reduced medical facilities. As ultrasound imaging is getting further involved in the emergency medical or surgical decision making, there is a need to have access to this method in most of the isolated areas lacking of ultrasound specialists.

This section of four chapters introduces new operational concepts dedicated to remote echography and ultrasound image processing required to provide the distant expert with the best image quality, over the communication link, to make his/her diagnosis. Chapters 34 and 35 describe two major European projects, TeleInViVo and OTELO, and their specificity to remote tele-echography concept. To bring ultrasound healthcare to populations living in areas with reduced medical facilities, tele-echography implies the use of long distance communication links, either terrestrial or satellite, with bandwidth ranging from 64kb/s to 2Mb/s. These emerging telemedicine techniques require appropriate compression methods to send, from the distant patient side to the specialist, ultrasound images with good resolution and with relevant frame rate. These compression methods and graphical user interfaces needed by the specialists in the tele-

[*] Pierre Vieyres, Laboratoire Vision et Robotique, Université d'Orléans, 63 Ave de Lattre de Tassigny, 18020 Bourges Cedex – France. Pierre.Vieyres@bourges.univ-orleans.fr

echography concept are presented in chapter 36 and 37. The chapters are organized as follows :

- ∈ Chapter 34 starts with a short introduction to tele-echography, recent developments, current status and future perspectives. This is followed by a case study, at which the results of the TeleInViVo project is presented: A transportable telemedicine workstation has been designed, developed and evaluated for use in isolated areas such as islands and rural areas. The EU-TeleInViVo is a custom-made device integrating in one solid case a portable PC with telecommunication capabilities and a light, portable 3D ultrasound station. The system developed has low price, low weight, is transportable and non-radiating. The integrated workstation uses advanced software techniques to acquire 3-dimensional ultrasound data of a patient. The device has been successfully tested in different socio-economic conditions and adjusted accordingly to meet needs of developing countries and countries in transition. It currently comes in two versions, one fully portable, self-containing device, and a workstation version. A dedicated software package has been developed for the needs of the TeleInViVo project, based on the InViVo ScanNT software package developed at Fraunhofer IGD. During a teleconsultation both medical specialists, connected over a telecommunications network, are able to share the same interface and perform tasks visible in real-time, even on low bandwidth channels, to both ends on the same data set. The system supports various ways of communication, such as Internet, ISDN, common phone lines, GSM, etc. The technical principles of the system are described and data from the pilot trials at different sites are presented.

- ∈ The OTELO project is introduced in Chapter 35. OTELO system is a fully integrated end-to-end mobile tele-echography system for population groups that are either temporarily or permanently not served locally by medical experts. It is a dedicated and portable ultrasound probe holder robotic system with 6 degrees of freedom, associated with existing and new mobile communications technologies, and state of the art image compression techniques. Its goal is to reproduce the expert's hand movements during an ultrasound examination performed on a patient being located at a distance from the medical expert. The light robotic system, located at the patient site, is positioned on the patient, by a paramedical assistant, according to the medical expert information using a videoconferencing link. At the expert site, the ultrasound specialist controls the remote robot holding an ultrasound probe thanks to a dedicated input device and according to the received patient ultrasound images. The robotic system offers, in real time, good image quality back to the expert site where medical expert can make a preventive diagnosis. Clinical results performed in different clinical sites and with various communication links are presented.

- ∈ Mobile interconnectivity and telemedicine are important issues for achieving effectiveness in health care, since medical information can be transmitted faster and physicians can make diagnoses and treatment decisions faster. Chapter 36 presents the architecture and the design of the user interface environment for the remote monitoring and VR controlling of a mobile ultrasound examination. This environment is developed in the context of the OTELO IST project presented in chapter 35 of this section. The proposed user interface of the mobile tele-echography system consists of the following two parts:

- A novel 3D virtual environment for controlling and monitoring the remote mobile station; At the expert site, the proposed user interface employs the PHANToM^{TM} device in order to control the remote site robot and assist the expert performing the echography examination process. In order to perform the mobile tele-echography examination, the medical expert must be able to visualize the patient and the slave robot positioning during the medical examination, The available bandwidth of the wireless channel is first detected and evaluated in order to provide this important feature in the proposed user interface.

- The second part of the graphical user interface is the component for viewing, handling and processing medical image/video files. The system is organized according to the physician's needs. An image/video viewer is used for the visualization of the run-time medical image/video and several functions needed for the medical image processing are provided to be used by the physicians. Within this user interface, an image/video database is designed and linked to the user interface for the storage and retrieval of the medical image/videos. Finally, the set of the communication parameters determined by the user interface (image frame rate, image quality and size, etc) are presented.

•∈ The successful use of medical video transmission systems is often judged by their ability to deliver high quality video content over the available communications channels. The quality of medical video content is evaluated in terms of the perceived quality of objects of diagnostic interest. To achieve such quality, the video images are segmented into Video Object Planes (VOPs) that allow separate bandwidth allocation to different portions of the MPEG-4 based video. This feature, that is unique to object-based, allows us to allocate more bandwidth to objects that are of strong diagnostic interest, while objects of little diagnostic value are allocated significantly less bandwidth. The basic problem then becomes one of segmenting objects of diagnostic interest, and specifying the appropriate amount of required bandwidth. The system is being evaluated using the GSM and GPRS wireless communication channels. Chapter 37 provides a summary of the use of digital video in echocardiography and of object-based video compression. A new model is introduced for diagnostically-lossless compression of M-mode and a novel AM-FM representation for M-mode ultrasound is discussed.

All the new techniques presented here are only a sample of what is currently under a research and development phase. However, they represent today's trend in tele-echography, preferred by many for its low-cost and non-invasive characteristics, for bringing advanced ultrasound techniques in real time to isolated populations for a better healthcare.

MOBILE TELE-ECHOGRAPHY SYSTEMS – TELEINVIVO: A CASE STUDY

Georgios Kontaxakis[*], Georgios Sakas, and Stefan Walter

1. INTRODUCTION

Telemedicine, defined as practice of medicine at a distance, has been performed ever since humans learned to communicate: a messenger carrying over a medical advice from a distant expert to the bed of a patient has been always practicing telemedicine. By definition, therefore, telemedicine does not imply any technological breakthrough. It is however in the present days, with the breathtaking advancement in medicine, information technology, telecommunications, and biomedical sciences, that the term telemedicine emerged as one designating a new health care service, firmly related to the cutting edge of scientific and technological progression in these fields.

By minimizing or even eliminating the effect of distance, telemedicine comes to establish new basements for the delivery of health services worldwide. One of the major application areas is the provision of skilled medical care to patients who are in some way isolated from the specialized care they need. This is specially the case of isolated areas (rural regions, remote islands) and of crisis situation areas (ecological disaster, military conflict, etc.). Residents of these areas often have substandard access to specialty health care, primarily because specialist physicians are more likely to be located in areas of concentrated population. Access to these services is only achieved with the use of applications that allow remote expert consultation providing a faster response in emergency and crisis situations.

Tele-echography is a telemedicine modality preferred by many for its low-cost and non-invasive characteristics of capturing dynamic images of human body organs in real time. TeleInViVo is an international and intercontinental tele-echography project, which proved in practice that mobile tele-echography could dramatically improve the medical and health care provided to every person worldwide.

---

[*] G. Kontaxakis Universidad Politécnica de Madrid, ETSI Telecomunicación, Dpto. Ingeniería Electrónica, Spain E-28040, E-mail: gkont@die.upm.es.

2. BACKGROUND

The Fraunhofer Institute for Computer Graphics (IGD) has developed InViVo, a family of applications for the visualization, processing and analysis of volume data, mainly for medical imaging applications. The principal application area is 3-dimensional (3D) ultrasound imaging. By adding computer-supported cooperative work (CSCW, Kopsacheilis et al, 1997) and telecommunication modules to the existing InViVo application, IGD and its sister in USA Fraunhofer Centre for Research in Computer Graphics (CRCG, Providence, RI, USA) created the concept of TeleInViVo, a telemedicine platform allowing cooperative processing on the same data set by remote users (Coleman et al, 1996). A prototype portable version of this system, called MUSTPAC^{TM} (Littlefield et al, 1998) included TeleInViVo with a portable ultrasound scanner able to acquire 3D volume data. This device was field-tested in Bosnia in 1996.

Fraunhofer CRCG later developed further TeleInViVo applications for (non-medical) data and image acquisition, processing, and analysis (Jarvis et al, 1999). At the same time, Fraunhofer IGD leaded a Telematics Application Program (TAP #HC4021) project called TeleInViVo, supported by the European Commission under the 4th Framework Program for Research and Development. This project focused on the design, development, and large-scale trial of a transportable telemedicine workstation, integrating in one solid case a lightweight, portable 3D ultrasound station and a portable computer (PC) running the basic InViVo application with telecommunication capabilities. This chapter describes the work and results obtained in the framework of this project and in the following, the term TeleInViVo interchangeably refers to the project itself, to the proper software application or to the workstation in its two versions (fixed and portable).

This project was awarded the 2001 European IST Grand Prize Award, organized by the European Council of Applied Sciences and Engineering (Euro-CASE) and sponsored by the Information Society Technologies Programme of the European Commission.

3. TECHNICAL APPROACH

The use of advanced technologies allows the system to collect 3-dimensional ultrasound data of the patient, i.e. to perform an "echo-tomography". The doctor in the field scans the patient. With the built-in flexible telecommunication channel (from a simple phone line and Internet to ISDN, GSM or satellite) the acquired 3D-dataset are transferred to a remote expert, who can be virtually everywhere in the world. After data transmission both doctors are linked online over the telecommunication channel performing "virtual echography" on the 3D-data replicated in each site and viewing identical images on their screens in real-time and practically without lag, even through narrow-band telecommunication channels.

The solution implemented partly lies on the principle of "scanning the data and not the patient". The main idea is to establish a kind of "information equivalence" between the patient and the representation of that patient in digital form, so that there is little to no loss in examining the patient through the data representation, compared with what would be lost during a direct in-person examination. Given that it is faster and cheaper to ship information around than to ship people (patients and doctors), volumetric representations become compelling as a substitute for the live patient.

A significant advantage for the radiologist is that the volume representation can be "re-sliced" from any direction one might desire. This avoids the need to acquire a whole new set of images (requiring the patient to make another trip to the hospital), if for some reason the original set leaves 3D relationships ambiguous. Similar tele-echography applications have been recently presented (Heer et al, 2001), but did not include the mobile feature, unique for the TeleInViVo portable workstation, neither reported extensive clinical trials, apart from controlled feasibility experiments.

3.1. 3D Ultrasound

Figures 1 and 2 describe the technical principle for the freehand acquisition of 3D ultrasound images. The video exit of a conventional ultrasound device is linked to a frame grabber card on a PC. During acquisition all images displayed on the ultrasound device in the B mode are captured by the frame grabber card (30 Hz, NTSC) and saved in on the PC's RAM. At the same time, the local position and the orientation of the ultrasound transducer are recorded by a tracking system. This system consists of a transmitter generating a magnetic field and a tracking receiver defining position and orientation in relation to the tracking transmitter. The signals of the tracking transmitter and the tracking receiver are led to the PC by the evaluating unit (pcBird, Ascension Technologies, USA).

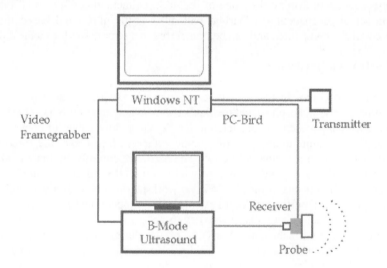

Figure 1. When ultrasound frame images are acquired, the spatial coordinates and orientation of the ultrasound transducer are transferred to the PC with the help of a magnetic tracking system.

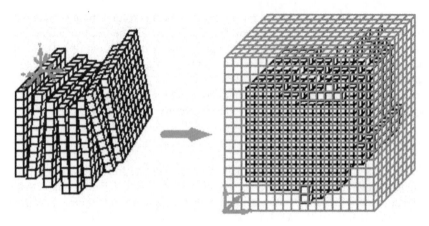

Figure 2. Schematic of the resampling scheme for a sequence of ultrasound frames to a regular volume grid.

The tracking receiver is permanently mounted onto the ultrasound probe. Both parts are adjusted to another by precise calibration. The receiver has the size and weight of a sugar cube and, therefore, does not hinder in any way the ultrasound probe movement respectively the operations of the doctor. The tracking transmitter has to be positioned at a max. distance of 60 cm from the patient's acquisition spot. The digitized sequences of the images are, together with position and orientation data, converted to a volume data set and stereoscopically displayed by one of the multiple methods of 3D visualization. This 3D data set of the patient - a "virtual patient" - is the ultrasound based information equivalent of the real patient and can be transferred to a remote site for medical diagnosis.

3.2. TeleInViVo Hardware

TeleInViVo comes in two versions, one portable device for field use and a fixed workstation for use in hospitals and medical centres, where portability is not critical.

The ultrasound system is connected to the PC via a video cable for connection of the ultrasound video output to the input of a frame grabber, which makes the video frames of the ultrasound device available to the PC. The hardware controls the functionality of the ultrasound board via a control line, which uses the RS232 serial interface of each component. Both versions of the TeleInViVo workstation consist of a personal computer PC, equipped with a pcBird tracking system, ISDN, modem, and frame grabber cards, a foot pedal and an identification hardware key (hasp).

Figure 3 shows a schematic overview of the design and the components of the TeleInViVo fixed workstation. The TeleInViVo fixed workstation is based on an ultrasound device already in operation in the trial site's radiology or ultrasound department. It extends the existing device first by the capability to acquire 3D ultrasound data sets and second by the necessary software and hardware for teleconsultations. TeleInViVo fixed workstations are located in well-equipped clinics where the necessity for a portable device is rather small.

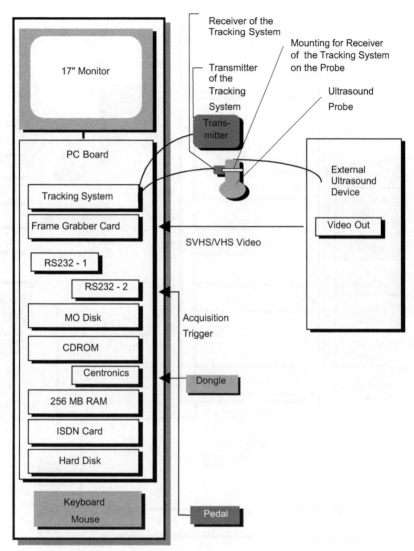

Figure 3. System overview of the TeleInViVo fixed workstation

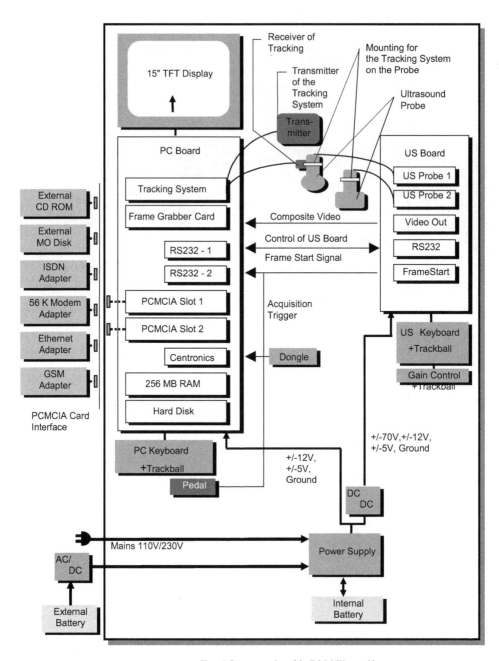

Figre 4. System overview of the TeleInViVo portable

Figure 5. The TeleInViVo portable device (left) and the fixed workstation (right).

The portable TeleInViVo workstation consists of two main components: a PC board (DSC, Langen, Germany) and an ultrasound device, (Pie Medical, Maastricht, Netherlands), both integrated in a solid aluminum case (GfM, Weiterstadt, Germany) and in an electronic board at the size of a typical PC main board. The connection of the ultrasound system to the PC follows the same principle than the one used for the fixed workstation for the acquisition of 3D ultrasound volume images. With a battery integrated in the power supply of the system it can be used without external fuse as a portable emergency system. Figure 4 shows a schematic overview of the design and components of the portable workstation. Both devices can be seen in Figure 5.

Having such volumetric data sets enables the physicist to make a diagnosis without the presence of the patient. This data set can be transferred over a communication line integrated in the PC (ISDN, modem, network) to a distant expert who runs another TeleInViVo application and who, after transmission, has all the information - the same 3D ultrasound volume data - available at his/her own application. Both can then collaboratively interact on the data and produce a common diagnosis.

3.3. TeleInViVo Software

The software part of the TeleInViVo workstation is based on the InViVo ScanNT® (Interactive Visualizer for Volume Data) software package (Sakas et al, 1995), which has been developed over several years at Fraunhofer IGD. It was mainly designed for fast and interactive visualization of volume data and various extensions have been developed covering several other aspects of medical data visualization. Together with the 3D ultrasound freehand scan extensions it has become a fully CE certified medical software.

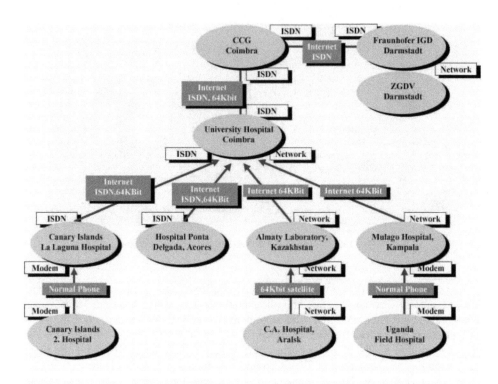

Figure 6. Schematic overview of the TeleInViVo network at all trial sites

The main work in software development for the TeleInViVo workstation is the extension of the existing 3D-ultrasound software to a collaborative application. This was achieved by the introduction of several communication layers in the software structure of the application. The topmost layer is the layer of the InViVo ScanNT® application itself. The next layer below is an adaptation layer that adapts the InViVo software to a communication layer, which hides further details of the communication from the adaptation layer and the application layer. InViVo has a modular structure that can be easily extended by "plugging in" additional modules. The adaptation layer is implemented by an InViVo module, which acts as an interface between the application and the communication layer. The communication layer is based on the HOUCOM software package (Schiffner et al, 1999), developed by Fraunhofer IGD and ZGDV.

Important is the issue of the transmission of large data sets over the (usually low) bandwidth available. A wavelet-based image compression algorithm (Villasenor et al, 1995; Li et al, 1999) has been implemented for the size reduction of the image data sets to be transmitted. The compression is lossy, but very efficient in cases where bandwidth is a crucial issue. Otherwise, a loss-less compression can be used for the data transmission. Log-files are implemented to monitor the process of the data transmission and the actions during a session mostly for statistical purposes.

The system supports various communication ways, such as Internet, ISDN, common phone lines, GSM, etc. The innovative idea of the system lies on the fact after the transfer of the 3D ultrasound data and during the on-line teleconsultation, only control signals (e.g. position of mouse, activation of buttons etc.) are transmitted over the network. Point-to-point connection is a truly distributed system, where an instance of the same application is running autonomously on each of the hosts. Each instance is then synchronized via the CSCW module, according to an interface sharing principle (Becker et al, 1998).

Only the actions introduced by one user are transferred to the remote location, where the second workstation locally calculates the corresponding image. Therefore, no bulky image data have to be transferred over the network, only a few bytes of control signals. Nevertheless the users see exactly the same image on their screens. The delay between two locations depends only on the latency of the intermediate network, which can be as low as that of a usual telephone line or even a GSM mobile phone.

4. SYSTEM EVALUATION

The TeleInViVo system has been evaluated at pilot trial sites during the years 1999-2000. Medical trials have been coordinated by the radiologists of the University Hospital of Coimbra (Hospitais da Universidade de Coimbra, HUC), Portugal. Portable TeleInViVo systems have been installed at the Hospital of Ponta Delgada (HPD) in the Azores Islands, Portugal, and CATAI (Centro de Alta Tecnologia en Analisis de Imagen) in La Laguna of the Tenerife Island in Canary Islands, Spain. Figure 6 shows a schematic overview of the TeleInViVo network at all trial sites with the installed workstations.

Over all locations more than 600 examinations have been performed, 334 of them using a standardized protocol allowing comparisons of results of different centres. Due to the quite high number of total examinations over locations distributed in three continents, we regard the trial as to be representative beyond cultural, ethnical and lingual barriers.

One of the major aspects of this project was the extension of European technology to third countries under the responsibility of UNESCO. Uganda and Kazakhstan were the selected trial countries, representing highly diverse environments in terms of technical capacity, telecommunications, and local human resources. In this context, TeleInViVo was constructed in the knowledge that it could enhance medical treatment in the developing world. Therefore, the project incorporated a large technical capacity building component, related to connectivity in remote areas, and cooperation with high-tech European research and development projects.

In Uganda two portable devices have been setup in Kampala (Mulago Hospital) and Nakaseke (Regional Health Centre and ITU Telecentre). In Kazakhstan a fixed workstation has been installed in Almaty and a portable device has operated in Aralsk, near the Aral Sea, an area of a great ecological disaster. Medical coordinator for these trials was again HUC.

The TeleInViVo project sets objectives to really enhance connectivity in areas which it was lacking previously, and so the project should provide lessons for the future. In Aralsk there was no sign of information technology during the first mission of the TeleInViVo project manager. However, trials were successfully registered using a 64Kbit satellite channel. In Nakaseke, the problems were greater, and the project encouraged linkage to the ITU telecentre wireless link-up. This kind of rural connectivity in Africa should be treated as a success in itself, considering the difficulties that preceded it.

For all trials, the medical staff involved in the acquisition process evaluated, qualitatively, the acquisition easiness graded in a scale ranging from 0 (difficult) to 2 (easy to perform), and the patient constraints using also a grade scale (from 0-difficult to 2-good). The images were always read and discussed by two radiologists either in the on-line sessions or in the off-line sessions. The transmission time of the 3D data and the time of analysis and clinical discussion were registered.

According to the reading protocols the radiologists involved in the trials performed a qualitative evaluation regarding diagnostic quality of the volumes using a gradual scale (from 0=bad to 2=good quality) and registered the degree of expected anatomic coverage.

Diagnostic performance was evaluated retrieving the number of diagnostically conclusive examinations and the number of correct diagnosis. The confidence to produce a remote diagnosis was also evaluate by assigning a diagnostic level of confidence for diagnosis ranging from 0=low to 3=very high. The accuracy of remote diagnosis was obtained through the confrontation with the previous diagnosis, already known by the radiologists who made the acquisitions.

4.1. Trials with Ponta Delgada

The study population consisted of 48 patients, referred to the Imaging Department from 23 November 1999 to 18 October 2000 with a break during the second half of 2000.

Of the 3D data acquired, 126 acquisitions were sent to HUC during 28 teleconsultation sessions. Fifty-five acquisitions have been sent off-line (the data were sent before the start of the teleconsultation). In all the volumes the transmission was via ISDN, except of 2 cases that have been sent via Internet connection.

All the 3D data was sent as 4MB volumes, being 102 from the upper abdomen and 24 from pelvis. There were pathologic findings in 111 acquisitions, with predominance of the hepatic/gallbladder problems, followed by kidneys lesions and pelvic pathology. Several acquisitions showed more than one alteration/lesion. Ten volumes were considered normal and 17 were discarded, because of their poor quality. The average time of transmission via ISDN was very short (1´39"). The average time for clinical discussion was 3´50´´ with a large range between 1 and 26 minutes.

The easiness to perform an acquisition was considered difficult in the majority of the cases (71%) and easy to perform only in 9%. The patient constraints were classified as medium in 38% and good in 24% of the acquisitions. The diagnostic quality of the 2D images obtained from the volumes was considered good and fair in 87% of the acquisitions. There were only 13% of bad acquisitions. The confidence to produce a diagnosis was very high or high in 87% (110) of the acquisitions. Only in eight volumes the confidence was assigned 0 (none). The diagnostic performance was very high (88%). There was no recourse to the 3D volume displayed in the upper right corner of the screen to produce a diagnostic.

When the radiologists compared the proposed diagnostic of the reading site with the previous diagnosis of the patients, there were 115 true positive, 9 true negative, 13 false positive, and 11 false negative diagnoses, which results in 90% of sensitivity, 44% of specificity, and 82% of accuracy.

4.2. Trials with CATAI

Medical staff of Canary Islands involved in the trials with HUC works in the Department of Obstetric and Gynaecology at the local University Hospital. The staff there is experienced ultrasonographists and has interest in 3D US. Usually they scanned the patient in their Department and sent the data to CATAI by ISDN. There, the 3D volumes were stored to be transmitted to HUC for discussion.

The study population consisted of 55 patients, all females, referred to the Department of Obstetrics and Gynaecology between 13 July 1999 and 31 May 2000. The gynaecologists collaborating with the project acquired the data and in co-operation with the CATAI staff, transmitted 89 of those acquisitions, which have been discussed in 39 teleconsultation sessions, 43 of them on-line and the others 46 off-line. Seven volumes were sent via ISDN, the rest 82 were transmitted by Internet. From the 3D volumes, 77 were received at 4 MB and 12 in 16 MB mode.

In the data received, 62 acquisitions were obstetric, 21 gynaecologic, 5 from upper abdomen and 1 breast. From the transmitted acquisitions, pathology was identified in 30 of them, whereas the rest 49 were normal. The average time of transmission using Internet was 5′38" for volumes at 4MB varying in a range from 1 to 15 minutes. The average time of transmission using ISDN was 4′10" varying in a small range of 4-5 min.

The average time for clinical discussion that is, the total time spent for analysing all available data, chatting, producing a diagnosis, and reporting was about 16 min and 55 sec, ranging from 1 to 60 minutes. The diagnostic quality of the 2D images derived from the 3D volumes was considered fair and good in 85% of the cases discussed, with only 15% of cases being considered not satisfactory. Only in 11 cases the confidence to produce a remote diagnosis was considered null. On the contrary, it was considered high or very high in 74% of the received volumes.

When the diagnosis proposed by the reading site was compared with the clinical, pathological, follow-up or the previous diagnostic of the patients, there were 27 true positive, 46 true negative, 2 false positive, and 4 false negative diagnoses, corresponding to 96% specificity, 92% sensitivity, and an accuracy of 93%.

4.3. Trials with Kazakhstan

Trials HUC-Kazakhstan (Almaty Diagnostic Centre) took place from 24 May 2000 to 30 November 2000. The study population consisted of 128 patients. Of the 3D data acquired, 182 acquisitions were sent to HUC during 26 teleconsultation sessions. All the trials were on-line and the transmission was by Internet.

All 3D datasets were sent at 4MB volumes, being 144 from the upper abdomen, 29 from pelvis, 2 obstetric, and 7 from the retroperitoneum. There were pathologic findings in 161 acquisitions, with predominance of the hepatic/gallbladder problems, followed by kidneys lesions and pelvic pathology. Some of the acquisitions showed more than one alteration/lesion. Twenty-one volumes were considered normal and 179 acquisitions (98%) had sufficient quality to be discussed.

The average time of transmission via Internet was 8′40″. However, there was a big difference between the average transmission time of the volumes received with 75% of quality (5′09′) and the average transmission time of the volumes received with 100% of quality (10'37"). The average time for clinical discussion was 6'48" with a large range between 1 and 25 minutes.

The easiness to perform an acquisition was considered medium in all the cases where this information was available (12% of the volumes acquired). The patient constraints of the same volumes were classified as medium in all of them. The diagnostic quality of the 2D images obtained from the volumes was considered good or fair in 98% of the acquisitions. There were only 2% (three) of bad acquisitions.

The confidence to produce a diagnosis was very high in 81,6% of the acquisitions and high in 14,9%. Only in five volumes the confidence was assigned 1 and there was no volume with 0 (none). The difference in the confidence to produce a diagnosis between the volumes received with 75% and 100% of quality was not significant.

When HUC doctors compared the proposed diagnostic of the reading site with the previous diagnoses of the patients, there were 182 true positive, 19 true negative, 11 false positive, and 5 false negative diagnoses, with 96% of sensitivity and 90% of accuracy.

4.4. Trials with Uganda

Trials HUC-Uganda (Mulago Hospital) took place from 6 April 2000 to 30 November 2000.

The study population consisted of 105 patients, referred to the Department of Radiology in Mulago Hospital. The radiologists collaborating with the project acquired the data and in co-operation with computer specialists sent to HUC 186 acquisitions. There were 29 teleconsultation sessions, all of them on-line. All image volumes were transmitted via Internet.

All 3D volumes were received at 4 MB. In the data received, 82 acquisitions were from the upper abdomen, 57 gynaecologic, 30 obstetric, 8 retroperitoneum, 2 from the heart, and 2 from the abdominal wall. From the 186 acquisitions received, only 14 were normal. In the upper abdomen the most frequent pathologies were of the liver, kidneys, and spleen. The majority of the lesions in the liver were focal, namely metastasis. There was a high percentage of uterus and ovarian pathology in the pelvic acquisitions. The most frequent lesions in the ovaries were cysts and in the uterus were myomas. In the obstetric volumes there were ectopic pregnancies, aborts, and malformations.

The average time of transmission using Internet was 7'56" for volumes at 4MB varying in a range from 1 to 18 minutes. However, while the volumes received with 100% of quality had an average time of transmission of 9'45" , the acquisitions received with 75% of quality had an average time of transmission of only 3'33".

The average time for clinical discussion, that is, the total time spent for analyzing all available data, chatting, producing a diagnosis, and reporting was 13'17", ranging from 1 min up to 120 min. The easiness to perform an acquisition, according to the grade scale of the table below, was considered medium or easy in 95,4% of the cases. Indeed, there were only three cases with difficult acquisition. The patient constraints were classified as medium in 34,5% and good in 59,8% of the acquisitions.

The confidence to produce a diagnosis was very high or very high in 84% of the acquisitions (151). There were 21 volumes with 0 confidence. The diagnostic performance was good, with 94,4% of the imaging data considered diagnostically conclusive. When the diagnoses proposed by the reading site were compared with the clinical, pathological, follow-up or the previous diagnostic of the patients, there were 155 true positive, 18 true negative, 9 false positive, and 20 false negative diagnoses, corresponding to 66% specificity, 77% sensitivity, and an accuracy of 80%.

5. DISCUSSION

The system developed has low price, low weight, is transportable, non-radiating, and supports a very large range of applications varying from gynaecology over pathology to abdominal scans. It is a medical teleconsultation workstation, able to provide health care service where this is not possible by the usual means (ecological disaster areas, remote rural areas or isolated islands). Since advanced telecommunication means are usually not available or are very expensive in such locations, the system has been designed to work with over simple low bandwidth telephone lines.

Medical care can be provided through this system almost anywhere at anytime. The physicians and health care organizations are therefore able to guarantee high quality care, without the patient having to travel and being carried over long distances. Ultrasound supports a very large range of applications varying from gynaecology to cardiological examinations and is currently the only economically and practically affordable imaging modality. On the other hand, recent developments in ultrasound technology allowed the introduction and use of compact, portable and, more recently, handheld ultrasound scanners, which can be easily upgraded to acquire 3D ultrasound volume data.

The recent advances in Information Technology allow today the transmission of images, video, voice, and data from remote locations to a centre of expertise. The TeleInViVo system has been designed to provide high-tech diagnostic applications based on telecommunications technology scalable from simple telephone lines and Internet connections to ISDN, GSM and satellite systems. It makes also optimal use of today's powerful but still low cost and robust PC systems and the impressive advances in ultrasound technology with the dramatic reduction of the size of modern ultrasound scanners and the simultaneous spectacular improvement of their capabilities.

Apart from being a telemedicine device, the TeleInViVo brings to these remote areas the revolutionary advantages of 3D ultrasound. This is a new emerging technique, which is maximally contributing to the recent uprising of this imaging modality, suppressed during the past years by other (but expensive, not readily available, and sometimes invasive) imaging methods such as CT, MRI or nuclear medicine. Hundreds teleconsultation sessions have been performed until now between HUC and the rest of the trial sites, as well as within these sites themselves, proving that the system is reliable, accurate and allows detailed and correct diagnoses to be carried out from a distance.

The results presented in the previous section compare very well with the results of a previous much more limited study, within USA (Rosen et al, 1999), published at the same time the TeleInViVo trials were taking place. That study evaluated the diagnostic accuracy of radiologists (speaking the same language and working at the same country) interpreting static (and not 3D) ultrasound images both at the on-site outpatient unit and a remote academic medical centre. This study reported an accuracy of 92% and 94% for the on-site and remote radiologists, respectively. TeleInViVo trials reached an accuracy of 90% for intercontinental trials (Portugal – Kazakhstan) and up to 93% for trials between Portugal and Spain (Canary Islands).

Despite this multitude of technologies and concepts integrated in one single compact device, the users who are participating in the initial implementation phase have found the system quite easy to learn and use. The TeleInViVo is an existing and operating device since the early 1999. Its implementation today extends on three continents (Europe, West Africa, and Central Asia) and spans over six time zones. Its operation has been tested in well-developed areas in Europe and major cities elsewhere (e.g., Almaty in Kazakhstan), as well as in remote locations such as islands (Canary and Azores) and poor regions under development (Uganda, Aralsk in Kazakhstan).

Tangible benefits from the project's implementation (with direct monetary value) are:

- Savings from reduced travel costs of specialists engaged in teleconsultation sessions;
- Savings from reduced travel costs of patients;
- Savings on hospital accommodation of patients that can be diagnosed remotely;
- Savings on hospital processing costs of patients that can be diagnosed remotely;
- Savings due to provision of health care in remote clinics or mobile health units versus expansion of urban or regional hospitals (i.e. the difference in the construction and running costs of facilities).

Intangible benefits (with a definite perceived value but with actual value not directly able to be determined) include:

- Better opportunity for consultations and second opinions resulting in avoidance of delays or costly mistakes;
- Reduced waiting time and transfer delays which can in some cases prevent serious complications;
- Reduced loss of income for patients who need not travel;
- Reduced expenses for family members who otherwise accompany the patient
- Improved effectiveness of specialists: broader reach with more patients being seen due to reduced travel;
- Improved overall health-care management, both internally and externally;
- Increased collegial support to medical personnel working in remote and isolated areas, resulting in increased job satisfaction;
- Improved teaching and learning possibilities and opportunities.

Looking further in the indirect benefits from a large-scale implementation of the TeleInViVo project one can easily list:

- Increased revenues to equipment providers, hospitals, telecommunication services providers and the like;
- Enabling specialist and technical personnel to increase their knowledge and qualifications;
- Facilitating decentralization of care and distribution of competence;
- Promoting maximization of scarce central resources (e.g. specialists, diagnostics apparatus and computers).

TeleInViVo is therefore a great example of how recent advances in medical imaging, especially mobile tele-echography, with the help of advanced telecommunications, can be applied to provide better health care and diagnosis, improve patient handling, reduce health service costs and save lives.

6. ACKNOWLEDGMENTS

Although this chapter has only three authors (or better editors), we want to emphasise that none of these results could become possible without the invaluable contributions of the following persons: L. Beolchi, L. Teixeira, O. Ferrer-Roca, J. Ponte, R. Cluzel, J. Skillen, T. Besimensky, L. Almeida, B. Cardoso, C. Martins, D. Ziyashewa, H. Dlamini, J. Willems, A. Ramakers, J. Stärk, N. Sewankambo, I. Matovu, M. Mayanja, F. Kakaire, Ch. Mutyaba, A. Zhanazarova, D. Ostashko, S. Karpov, J. Grushin and W. Rösch. We express our deep appreciation for their altruism, enthousiasm and pathos during the project.

7. REFERENCES

Becker, E., Hoyss, R., and Grunert, Th., 1998, A system for cooperative work in the medical domain, *Proc. 24th IEEE Euromicro Conf.*, **2**, 1061 -1067.

Coleman, J., Goettsch, A., et al., 1996, TeleInViVo: Towards collaborative volume visualization environments, *Computers and Graphics*, **20**(6), 801.

Heer, I.M., Strauss, A., Müller-Egloff, S., and Hasbargen, U., 2001, Telemedicine in ultrasound: New solutions, *Ultrasound Med. Biol.*, **27**(9), 1239-1243.

Jarvis, S., Barton, R., and Coleman, J., 1999, TeleInViVo: A volume visualization tool with applications in multiple fields, *MTS/IEEE OCEANS '99 - Riding the Crest into the 21st Century* (Conf. Record), **1**, 474 - 480.

Kopsacheilis, E.V., Kamilatos, I., Strintzis, M.G., and Makris, L., 1997, Design of CSCW applications for medical teleconsultation and remote diagnosis support, *Med. Inform.*, **22**(2), 121-132.

Li, X., Hu, G., and Gao, S., 1999, Design and implementation of a novel compression method in a tele-ultrasound system, *IEEE Trans. Inf. Tech. Biomed.*, **3**(3), 205-213.

Littlefield, R.J., Macedonia, Ch. R., and Coleman, J.D., 1998, MUSTPAC™ 3-D ultrasound telemedicine/ telepresence system, *IEEE Ultrasonics Symp. Proceedings*, 1669-1675.

Rosen, M.P., Levine, D., et al., 1999, Diagnostic accuracy with US: Remote radiologists' versus on-site radiologists' interpretations, *Radiology*, **210**(3): 733-736.

Sakas, G., Walter, S., Hiltmann, W., and Wischnik, A., 1995, Fetal visualization using 3D ultrasonic data, *Comp. Ass. Radiol.*, Springer Verlag, Berlin, 241-247.

Schiffner, N., and Ruehl, Ch., 1999, HOUCOM framework for collaborative environments, *SPIE Intl. Symp. on Voice, Video & Data Comm.* (Conf. Record), **3845**, 413-425.

Villasenor, J., Belzer, B., and Liao, J., 1995, Wavelet filter evaluation for image compression, *IEEE Trans. Imag. Proc.*, **2**, 1053-1060.

A TELE-OPERATED ROBOTIC SYSTEM FOR MOBILE TELE-ECHOGRAPHY: THE OTELO PROJECT

Pierre Vieyres[*], Gérard Poisson, Fabien Courrèges, Natalie Smith-Guerin, Cyril Novales, and Philippe Arbeille

1. INTRODUCTION

In routine examination, various pathological situations like "abnormal heart rate, pericardic collection, cholecystis, renal lithiasis, normal and ectopic pregnancies, ovarian cyst, acute appendicitis, phlebitis…" may occur in subjects without any serious medical history; on the other hand after a trauma, the medical doctor may look for an abdominal organ lesion with blood collection. Ultrasound echography and Doppler are non invasive methods that are very well adapted for the diagnosis of such anomalies and are currently used in routine and in emergency cases at the hospital.

As ultrasound imaging is getting further involved in the emergency medical or surgical decision making, there is a need to have access to this method in most of the inhabited isolated places. Isolated places defined as areas with reduced medical facilities could be small secondary hospitals, a few kilometres away from the university hospital, a dispensary in Africa or in the Amazonian region, in polar areas, or in a moving vehicle such as an emergency vehicle, a boat, a plane or even the International Space Station (ISS).

However, the main drawback of the ultrasound technique is that the quality of the examination highly depends on the operator's skills. The technique is operator-dependant: a few degrees modification of the ultrasound probe orientation on the patient's skin could lead to miss or loose the investigated organ. Furthermore, many small medical centres and isolated sites do not have the appropriate well trained sonographer on hand to perform the first echography from which the evaluation of the emergency could be done. Also, the cost of providing an operator 24 hours a day is seldom justified at many centres. In these cases, in order to offer to a patient a quick and reliable diagnosis as early as possible, his transfer, from the isolated site (secondary

---

[*] Laboratoire Vision et Robotique, , 63 Ave de Lattre de Tassigny, 18020 Bourges Cedex – France. E-mail: Pierre.Vieyres@bourges.univ-orleans.fr.

hospital, island) to the expert centre, may be required even though it might be problematic and expensive. It has to be noted that in approximately 50% of the cases, patients who are transferred to the hospital centre at night, to receive an ultrasound examination, are sent back home a few hours later. Therefore, the only way to avoid these constraints and to guarantee a reliable ultrasound examination is to control it via tele-echography scenarios.

Such a response is proposed by OTELO, a multidisciplinary and crossed expertise European Project (OTELO-2001). It is dedicated to the development of an advanced tele-echography robotic system which will bring to population groups, that are not served locally either temporarily or permanently by medical ultrasound experts, preventive care support using the latest ultrasound investigation techniques. The objective of the OTELO project is to enable a medical ultrasound expert, located at an expert site (e.g. the closest university hospital), to perform an echography on a distant patient located at an isolated site and to make a reliable echographic diagnosis. The OTELO tele-echography concept supposes that there is only a non ultrasound specialist (e.g. a paramedic assistant) at the isolated site and it includes a standard communication link (terrestrial or satellite), between the two sites, offering a reasonable bandwidth to enable the transfer of ultrasound images, ambient images and robotic controls data. Thus, the tele-operated robotic arm system, if present in these institutions with a lack of ultrasound experts, will enable the ultrasound specialist, from the expert centre and via appropriate communication links, the control of any ultrasound investigation at any time (24h a day) and anywhere. In this chapter, we will focus on the robotic aspect of the OTELO tele-echography chain. For ultrasound image processing associated with telemedicine applications and 3G wireless connectivity results, see also chapters 34, 36, and 1 respectively.

2.   ROBOTIC TELE-ECHOGRAPHY STATE OF THE ART

Two main approaches have been studied to demonstrate the tele-echography concept:

   a-   the tele-medicine approach where ultrasound images are collected at the patient site by an expert and, after an image processing step, are sent to the expert centre where eventually collaborative work can be performed. Since the mid-nineties several projects around the world have involved imaging data transmission for telemedical applications. For some of them, videoconference was used between two experts, one of whom was performing the echography examination. Both experts could simultaneously discuss the ultrasound image, and the expert, distant from the patient, could suggest a different probe orientation to his peer for better observation and analysis of the area of interest (Wootton, 1997; LOGINAT project, 1998). In other projects, dedicated workstations were developed to acquire ultrasound images data, and then compress and analyse them at a distant site (Lee, 1994). In 2000, the major European project TeleInVivo developed a similar approach: the echography was performed by a clinical expert standing next to the patient; and then ultrasound data were sent via satellite to a data base station for processing (Kontaxakis, 2000). Chapter 34 of this book can be referred to for this approach and for more exhaustive references.

   b-   the tele-robotic approach where the images are collected via a dedicated robotic system controlled at a distance by an ultrasound expert. In this case, only a paramedic assistant is present next to the patient to comfort him, to assist in the set up of the robotic mechanism or to hold the probe holder structure. A non exhaustive list of these approaches are presented here. In 1998, developed under the MIDSTEP project, a telescanning demonstrator based on a tele-operated industrial robot holding an ultrasound probe, was used to take a biopsy with the help of the ultrasound images (De Cunha, 1998). In 1999, a seven degree-of-freedom spherical industrial robot (the Hippocrate project), equipped with a force sensor, was controlled to maintain an ultrasound probe in contact with a patient's carotid. During a necessary learning phase, the robot was guided along the carotid artery by the medical expert. During the second phase the robot was able to follow the same path at a chosen speed and ultrasound topography was recorded to evaluate the arterial wall (diameter, plaque, compliance): the project is currently under clinical evaluation (Degoulange, 1998). In 1996, a similar concept had been introduced by Salcudean to provide a better user interface for ultrasound technicians. A safe counterbalanced robot was designed for carotid examination, using visual servoing for motion in the plane of ultrasound (Salcudean, 1999). During the year 1998, a first world premiere robotic tele-echography experiment was successfully performed between France and Nepal, via Inmarsat link, for the purpose of the SYRTECH project (Gourdon, 1999a, 1999b). This project was followed by the development of a dedicated 4 degree-of-freedom tele-echography robot (the TERESA project) with the support of the European Space Agency (ESA) (Courreges, 2001). More recently, a mechanical structure driven by McKibben actuators has been developed to drive a probe holder system for obstetrical applications under the TER French national project (Vilchis, 2002). A parallel robotic structure is also being developed by Masuda, to control the ultrasound probe over a large body area (Masuda, 2002). These two latter structures, even though still in the research phase, are very promising for the robotic tele-echography concept.

Most of these approaches, except the last four research projects mentioned above, require the presence of an ultrasound specialist nearby the patient, or are merely tele-medicine approaches where data are transmitted via communication networks as described in point a). Also, some of these robotic structures use industrial voluminous robots, some of them being driven by still, non-proofed actuators for long duration standard or emergency medical applications. The OTELO project, in continuation of the SYRTECH and TERESA projects, presents an alternative to the robotic structures, described above, as it is a dedicated probe holder system not requiring the presence of an ultrasound specialist at the location where the ultrasound examination is being performed. The robotic aspect of the OTELO tele-echography system is presented here and has been studied as a "master-slave" system. However, as this system is dedicated to a specific medical application, we will use the term "expert-patient" system instead. The OTELO system, more thoroughly described in section 4, includes: a probe holder robot and an echograph device at the patient site; a fictive probe at the expert site; a videoconferencing system connecting the two sites.

3.   FROM USERS REQUIREMENTS TO THE OTELO ROBOTIC CONCEPT

3.1. Study of Ultrasound Probe Motion

Before designing a dedicated mechanical structure of the probe holder robot able to remotely extend the medical expert's hand movements, a study has been realised in order to characterise the ultrasound probe positioning and motions during a standard echography examination. The analysis of ultrasound examinations helped to quantify the positions and velocities of the probe used by the medical specialist during these examinations. Figure 1 shows the required probe positioning and the various positions that the patient has to be set in for various ultrasound investigations: cardiac, abdominal and renal. From the movements of the real ultrasound probe, analysed during an echography examination near the patient, we defined standard movements that would have to be followed by the remote probe during the tele-operated echography. Characteristics of these movements are prime aspects taken into account in the design of the robot, for the remote probe positioning and for its handling during a given investigation.

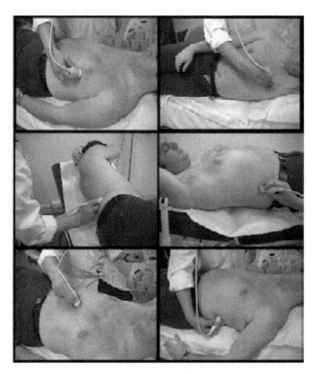

Figure 1. Probe and patient positioning for cardiac, renal and abdominal investigations (from top left to bottom right)

Some specifications of the mechanical structure are directly defined by the motions characteristics performed by the sonographer during a standard ultrasound examination.

During these investigations, interesting points have been reported:
- During the organ investigation, the expert maintains a continuous contact between the probe and the skin, even when the probe is applied on the ribs and the irregular abdominal skin.
- The maximum force exerted by the real ultrasound probe on the patient's skin varies from 5N to 20N for safety and comfort and for having the assurance of a reliable diagnosis.
- The probe is positioned on the area of interest where only rotations are performed around a contact point with the skin. Figure 2 shows possible rotations of the ultrasound probe. Once the probe is positioned on the area of interest, then a continuous contact is kept between the tip of the probe and the skin.
- The contact point is chosen with accuracy and based on anatomical references.
- To reach the best angle of incidence, the probe axis is moved inside a conical space and rotated around its axis. The apex of this cone is the contact point (see Figure 2).
- During most of the investigations, the inclination of the probe axis, from the normal direction of the skin, is most of the time smaller than 25°. Although, in some cases and for some specific organs investigations, such as in bladder investigation, the probe axis can be up to 60° from the normal direction of the skin.

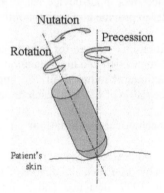

Figure 2. Three identified rotations of the ultrasound probe during a standard echography examination

Other specifications of the robot are due to its specific use that are listed below:
- Distant site, distant medical expert, communication link and transmission delay between the two sites.
- The patient will eventually be assisted by a non ultrasonographer specialist.
- The paramedic has to be able to position the robotic system on any anatomical area requested by the expert.
- The system has to be designed to be easily manipulated by any paramedical personnel.
- The ultrasound probe holder system has to maintain continuous contact of the probe with the patient's skin during the examination to minimise ultrasound signal attenuation, therefore offering the best image quality to the expert.

There are several conditions arising from the use of a robotic system in medical applications:
- •The safety in the functioning of the robotic system.
- •The quality of the communication link which should provide the necessary information for optimal control of the remote device by the expert.
- •A good image quality to make a reliable pre-diagnosis.

Based on the information gathered among echography medical experts, and related to the ultrasound techniques and gestures, the tele-echography system has to satisfy the following specifications:
- •The patient station has to be easy to transport next to the patient as well as easy to handle during the set up phase and the examination phase.
- •The robot, that will be positioned on the patient's skin, has to be light weight not to be uncomfortable for the patient and powerful enough to follow in almost real time the control data corresponding to the expert's hand movements that are being sent from the expert station.
- •At the expert station site, the medical specialist has to control the remote system with a specific input device that he can adapt quickly with, that is, a system resembling a standard ultrasound probe; this input device has to record in real time the expert's hand movements in order to provide him with the full control of the distant ultrasound probe orientation.
- •Finally, the communication link, chosen for the tele-echography system, should offer a reliable point to point connection and high quality of service between the expert and patient sites using a maximum bandwidth of 256 kbps, that could be two terrestrial ISDN basic rate interface (BRI) of 128 kbps each or new wireless system.

3.2. Robot probe holder mechanical specifications

According to the study of ultrasound probe motions and of the constraints due to the tele-operated chain, a portable compact six degree-of-freedom (6 DOF) probe holder robot has been patented in 1999, designed and being manufactured in 2002 in the framework of the European OTELO project (see Figure 3). It is a first prototype of a serie of three robots, all dedicated to tele-echography. For this first OTELO prototype, the objective is to validate the concept of a 6 DOF portable robot. During year 2003, two other prototypes will be designed that will integrate new functionalities and geometric modifications for the mechanical structure, based on the first phase results in the medical environment.

In response to users specifications, the chosen kinematics structure is a light weight and portable 6 DOF mechanical system that enables to manipulate the ultrasound probe without displacement of the contact point on the skin (see Figure 3). The mechanical characteristics are:
- •Two translations to move the probe end tip on the region of interest, to allow the expert to adjust the position of the contact point above the organ to be investigated within a range of 5 cm in the two x and y directions. The x-y-translation device is located at the top of the robotics structure; this solution minimises the inertia effects on the mobile parts of the robot and limits the probe oscillations around the desired

position.
- • Three rotations of concurrent axes to give the probe the desired orientation, to allow the probe to rotate inside a conical space, of 60° vertex angle, keeping a point of contact with the patient's skin.
- • One z translation to modify the contact force between the probe end tip with the patient's skin. It allows the probe to be continuously in contact with the skin and to monitor the applied force via a specific force sensor. The range of this displacement is 3 cm.

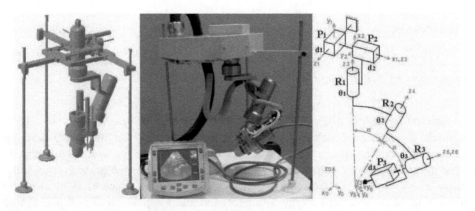

Figure 3. From left to right: CAD representation of OTELO first prototype - OTELO first prototype (resting on temporary supporting "legs") - Kinematics structure of the OTELO first prototype

Direct drive motors were chosen for the three rotations and step by step motors for the translations. Motors capabilities (power, response time, accuracy) permit the articulations to move the ultrasound probe as quickly as a medical expert's hand can do it. The mass of the power transfer units are limited thanks to the use of small DC electrical actuators. Furthermore, the control laws of the articulation positions, based on the implementation of Inverse Geometric Models (IGM) of the robot, were found to be very efficient for the OTELO first prototype. Singular configurations of this non-redundant first prototype have been identified and taken into account in the control laws.

This IGM model is, as its name indicates, established by inverting the Direct Geometric Model (DGM). This latter model gives the probe posture, position/orientation X, as a function of the joint positions q called generalised coordinates:

$$X = f(q) \hspace{3cm} \text{(DGM)}$$

where $q = [\theta_1, \theta_2, \theta_3, l_1, l_2, l_3]^T$

and $X = [X_1, ..., X_6]^T$

θ_1, θ_2, θ_3 represent the three coordinates of the rotational articulations and l_1, l_2 and l_3, the three coordinates along the translations, corresponding to the x, y and z displacements of the ultrasound probe. $[X_1, ..., X_6]$ characterises the six desired position and orientation coordinates of the probe in the Euler frame. The relation f between the probe posture X and the joints positions is expressed by the transformation matrix T_6^0 (q).

For $U_0 = \begin{bmatrix} s_x & n_x & a_x & p_x \\ s_y & n_y & a_y & p_y \\ s_z & n_z & a_z & p_z \\ 0 & 0 & 0 & 1 \end{bmatrix}$, Uo a standard form of the transformation matrix, we

can determine, by identification between T_6^0 (q) and Uo, the coordinates of $X = [P_x \ P_y \ P_z \ s_x \ s_y \ s_z \ n_x \ n_y \ n_z \ a_x \ a_y \ a_z]^T$ which are composed by the three director cosines s, n, a elements. By inverting the DGM relation we obtain:

$$q = f^{-1}(X) \qquad\qquad \text{(IGM)}$$

The mathematical resolution of this system, described below, gives the joints positions for a desired posture (X) of the probe. Firstly, the three rotations are determined. From the chosen mechanical structure of the first prototype, we note that θ_2 could have two different values for the same desired posture of the probe (we have the choice of two aspects for the robot configuration). From the choice of θ_2 obtained by minimising the displacement energy during the path following, the joints positions θ_1 and θ_3 are obtained. These angles being determined, we can easily determine l_1, l_2 and l_3.

In order to better control the robot near the singular configurations and limit their effects, e.g. greater displacements of the rotating robot articulations, we use a local control law, dedicated to the specific kinematics structure of the probe holder robot.

The OTELO first robot prototype has been manufactured. A 3.5-5 MHz ultrasound probe of Pie Medical Tringa 50s has been chosen to be integrated on this first prototype for its compact size and its general applications (see Figure 4). However, OTELO mechanical structure is designed to accept several currently manufactured ultrasound probes for various applications based on the users requirements and experts advice from various specialities.

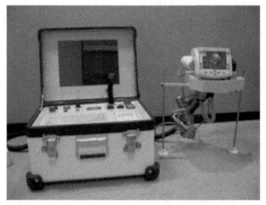

Figure 4. OTELO first prototype for the patient station. The portable PIE Medical ultrasound device TRINGA50S with a 3.5-5 MHz ultrasound probe are shown with the portable power and control unit

3.3. The expert station input device

The control system used at the expert station is defined as the input device. At the expert station, the medical specialist has to be able to control the distant robot to perform the echography act with the feeling of performing his task as if he were next to the patient. The input device, designed for the specificity of the remote ultrasound scan application, was developed based mainly on the criteria presented by Gosselin (2002); the input device has to present the best performance criteria so that the expert-patient system is "transparent" in order for the expert to have the feeling of performing his task directly in the remote environment, that is, doing the echography on the patient. It is a compact device enabling easy integration with the expert station. Also, to maintain the standard workspace the medical expert is used to, the input device, also called the passive fictive probe, is designed to present the same mechanical geometry as a standard real ultrasound probe (see Figure 5). The system is intrinsically safe and ergonomic as it resembles a real ultrasound probe. It is an off-the-shelf component minimising cost and simplifying maintenance. The passive stiffness induced by a standard spring set inside the fictive probe gives the input device the ability to generate a force resolution close to the one exerted by the real probe on the patient's anatomical chosen area of interest.

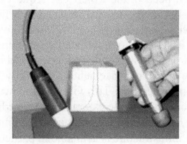

Figure 5. Input device: passive fictive probe designed for the OTELO expert station shown with a 3.5–5MHz ultrasound probe attached to the robotic probe holder system that is located at the patient station

The passive fictive probe is associated with a 6 degree-of-freedom localisation sensor that enables the following of the expert's hand movements within a desired workspace. Due to the OTELO robotic probe holder structure, the reachable workspace is about 5cm x 5cm, which is sufficient according to users requirements to search for a given organ. Furthermore, this input device provides the expert with z displacement along its axis, allowing the expert to exert a controlled force on the patient's skin up to a maximum value of 20N along the probe axis. This displacement is of great importance to maintain a continuous contact of the remote real ultrasound probe with the patient's skin and to enable the force control, by the expert, throughout the echography examination.

3.4. The OTELO tele-echography system

Three main parts can be identified for the overall OTELO telemedicine chain (see Figure 6):

   a- The expert station where the ultrasound expert receives, on a control monitor and in almost real time, the ultrasound images of the patient's organ. Based on these images, he can modify the orientation of the remote real ultrasound probe by changing the position of the fictive probe. The changes can be a slight rotation along the fictive probe axis, an inclination from the vertical axis, a rotation around the vertical axis or an x-y displacement within the chosen workspace. A graphic user interface (see M-Health book chapter 36), allows the specialist to record, zoom and freeze the received ultrasound image to make a first diagnosis. The expert can communicate with the remote patient via videoconferencing.

b- The patient station is equipped with the robotic probe holder system, an ultrasound device and a videoconference link. The 6 degree-of-freedom mechanical structure can hold several types of currently manufactured ultrasound probes to be found in various secondary hospitals or dispensaries. It receives the positioning data from the expert fictive probe localisation sensor, and sends back to the expert station the force exerted by the real probe on the patient's skin. During the tele-operated diagnosis, the robot is maintained over the patient by a paramedic. At the beginning of the examination, the paramedic positions the probe holder robot on the patient on a chosen anatomical point, according to the expert information. This point is initially fixed by the expert at about 3 cm around the appropriate area. When needed, the expert can request a different positioning of the overall mechanical structure to analyse or search for a different organ. The paramedic assistant, who continuously hold the probe holder robot during the tele-examination, may be asked to move the whole structure according to the expert information and based on a set anatomical chart defined by the users requirements clinical scenario. The robot displacement is carefully followed by the distant expert by the videoconference link. This videoconference link also offers a closer relationship and comfort between the expert, the paramedic and the patient.

c- The communication link between the two stations can be of two kinds:

• terrestrial when ISDN infrastructure is available in particular for homecare market
• satellite (mobile or fixed), when coverage is possible, in order to reach isolated populations such as islands or Amazonian regions.

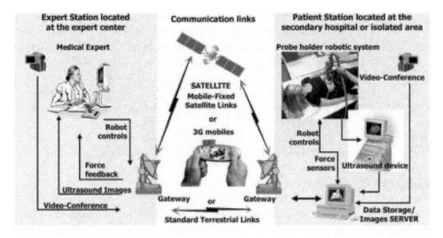

Figure 6. The overall OTELO tele-echography chain

3.5. First results on a group of patients

Tests have been carried out in a clinical environment at the UMPS Unit (University of Tours, France) for the overall tele-echography chain evaluation, after agreement by the Health State Committee (CCPPRB n°2002/07) (Arbeille, 2003). The expert was sitting at his control desk at the expert station and the patient was situated in a far away room within the same hospital. The two stations were connected by two internal ISDN lines, giving a total bandwidth of 128 kbps. The positioning of the robotic system by the paramedic on the patient was supervised by the expert from the ambient images monitored via videoconferencing. The expert might ask the operator to adjust the setting of the echograph device, especially gain and depth, as no remote control of the ultrasound device was then available.

Echographic examinations were performed on 20 patients to obtain longitudinal and transverse views of four groups of organs. This corresponds to the minimal request of organ visualisation for diagnosis in the most frequent emergency situations: (a) digestive symptoms (liver, gallbladder, pancreas and kidney); (b) after physical trauma such as car accident (aorta, right and left kidney, liver, spleen, bladder); (c) pelvic and urinary symptoms (bladder, kidneys, prostate or uterus and ovaries); (d) cardiac symptoms (cardiac four chamber view). These echography examinations were followed by a second conventional echography performed by a sonographer of the hospital specialising in the organ to be investigated. These ultrasound views obtained by the remote robotics device have been compared to longitudinal and transverse views of the same organs in standard ultrasound examination conditions. The heart and spleen were not visualised in two and four out of the 20 cases, respectively. The mean duration of the robotized tele-echography was approximately 50% longer (27+/-7 min for three to four organs) than in direct echography of the patient.

Post trauma have been determined on aorta, kidneys, liver, spleen and bladder in 80 % of the cases by the robotic device compared to standard examinations. The four cardiac chambers views have been obtained in 90 % of the cases by the robotic device compared to classical examination. Digestive symptoms on liver, gall bladder, pancreas and kidney have been detected both by the remote and the classical examinations in all cases (20/20). Urinary/pelvis symptoms on bladder, kidney, prostate, uterus and ovaries have been seen with both remote and classical examinations in all cases (10/10). These first results show the feasibility of that device and the possibility to obtain good views from the remote site. It allows the expert to make a comparable diagnosis than the one made with a standard echography act.

4. CONCLUSION

The OTELO tele-echography concept is a light portable system allowing an ultrasound expert to perform an ultrasound analysis on a patient in a site remotely located from the expert centre. This European project, conducted by nine partners, includes several disciplines such as mechanical design of the probe holder, robot control laws, graphic user interface, image coding and communication links.

The first dedicated prototype is a 6 DOF serial structure with three rotations for the orientation of the probe, two translations for its precise positioning on the region of interest and one translation along the probe axis for maintaining a continuous contact force on the patient's skin. This robot moves the real ultrasound probe on the patient's skin, in almost real time, in accordance to the movements performed by the distant expert with the fictive probe.

Patient ultrasound images, compressed before the transmission with JPEG-LS and JPEG-2000 techniques (Delgorge, 2002; Triantafyllidis, 2002), are acquired during the examination and transmitted in almost real time to the expert. Studies have been successfully conducted using existing terrestrial IDSN links; they will be extended to fixed satellite and new mobile 2.5/3G communication means (Voskarides, 2002) and comparison will shortly be performed in term of clinical performance, cost, bandwidth, availability and Quality of Service.

Echography is a widespread technique, well known to the public. OTELO tele-echography will have applications in many heath sectors, for example: provide better access to healthcare in remote areas, allow the maintenance of local infrastructures, such as small maternity units supported by large expert centres. It offers an alternative for medical centres, with lack of experts, to bring quality healthcare to isolated patients. The benefit of the industrial prototype would be to lower the cost of ultrasound scan exams by performing examinations in local infrastructures, to reduce the patient transportation costs to the expert centre by avoiding it when it is not required, or the transportation of the clinical expert to the local infrastructure. From a professional point of view, this tele-operated ultrasound probe-holder could also be used as a demonstrator of the probe holding techniques to train future clinical ultrasound specialists.

5.   ACKNOWLEDGMENTS

All authors would like to thank all OTELO partners for their valuable contribution to the achievement of the OTELO European project (ITS-2001-32516) presented here : Sinters, Elsacom, Kell, Ebit-Sanità, Hospital Clinic Barcelona, Tours University, Kingston University, CERTH, Orléans university.

6.   REFERENCE

Arbeille, Ph., Poisson, G., Ayoub, J., Vieyres, P., et al., 2003, Echographic examination in isolated sites controlled from an expert centre using a 2D echograph guided by a robotic arm, *Journal of Ultrasound in Medicine and Biology, Vol29, n°7, pp993-1000.*

Courreges, F., Smith, N., Poisson, G., Vieyres, P., Gourdon, A., Szpieg, M., Merigeaux, O., 2001, Real-time exhibition of a simulated space tele-echography using an ultra-light robot, *6th International Symposium on Artificial Intelligence, Robotics and Automation in Space,* Montreal Canada.

Degoulange, E., Boudet, S., Gariepy, J., Pierrot, F., Urbain, L., Megnier, J., Dombre E., and Caron, P., 1998, Hippocrate: an intrinsically safe robot for medical application, *Proc IEEE/RSJ Intl. conf. on Intelligent robots and systems,* Victoria B.C., Canada, pp. 959-964.

Delgorge, C., Rosenberger, C., Vieyres P., and Poisson, G., 2002, JPEG2000, an adapted compression method for ultrasound images ? a comparative study, *Proc of the 6th World Multiconference on Systemics, Cybernetics and Informatics,* Florida, vol IX, pp. 536-539.

De Cunha, D., Gravez, P., et al., 1998, The MIDSTEP system for ultrasound guided remote telesurgery, *Proc. of International Conference of the IEEE Engineering in Medicine and Biology Society,* Hong Kong, pp. 1266-1269.

Gourdon, A., Poignet, Ph., Poisson, G., Vieyres, P., and Marché, P., 1999, A new robotic mechanism for medical application, *IEEE/ASME International Conference on Advanced Intelligent,* Atlanta, pp33-38.

Gourdon, A., Vieyres, P., Poignet, Ph., Szpieg, M., and Arbeille, Ph., 1999, A tele-scanning robotic system using satellite, *European Medical & Biological Engineering Conference*, Vienna, Vol II, pp1402-1403.

G. Kontaxakis, S. Walter and G. Sakas, "EU-TeleInViVo: An integrated portable telemedicine workstation featuring acquisition, processing and transmission over low-bandwidth lines of 3D ultrasound volume images", 2000; *Proceedings IEEE EMBS International Conference on Information Technology Applications in Biomedicine (ITAB 2000)*, Editor-in-Chief Swamy Laxminarayan, Arlington, VA, USA, pp. 158-163.

Gosselin, F., Riwan A., 2002, Virtuose 3D, nouvelle interface haptique à retour d'effort, Techniques de l'Ingénieur, Paris; http://www.techniques-ingenieur.fr/affichage/dispIntro.asp?nGcmid=IN1.

Lee, W., Kim, Y., Gove, R. J., Read, C. J., 1994, MediaStation 5000: integrating video and audio, *IEEE Multimedia*, Vol. 1, No. 2, pp. 50-61.

Masuda, K., Tateichi, N., Susuki, Y., Kimura, E., Wie, Y., and Ishihara, K., 2002, Experiment of a wireless tele-echography system by controlling echographic diagnosis robot, *Proc. of International Conference of Medical Image Computing and Computer-Assisted Intervention, MICCAI 2002,* Springer-Verlag, LNCS 2488, pp.130-137.

Salcudean, S.E., Zhu, W.H., Abolmaesumi, P., Bachmann, S., and Lawrence, P.D., 1999, A robot system for medical ultrasound, *Proc. of International Symposium of Robotics Research*, USA, pp. 152-159.

Triantafyllidis, G.A., Tzovaras, D. and Strintzis, M.G., 2002, Blocking artifact detection and reduction in compressed data, *IEEE Trans. on Circuits and Systems for Video Technology*, Vol.12, No.10, pp. 877-890.

Vilchis, A., J. Troccaz, Cinquin, Ph., et al., 2002, Experiments with the TER Tele-echography robot, *Proc. of International Conference of Medical Image Computing and Computer-Assisted Intervention, MICCAI 2002*, Springer-Verlag, LNCS 2488, pp.138-146.

Voskarides, S., Pattichis, C.S., Istepanian, R., Kyriacou, E., Pattichis, M.S., Schizas, C.N., Mobile health systems: a brief overview, *Proc. of SPIE AeroSense [4740-25], 2002.*

Wootton, R., Dornan, J., Fisk, NM., Harper, A., Barry-Kinsella, C., Kyle, P., Smith, P., Yates, R., 1997, The effect of transmission bandwidth on diagnostic accuracy in remote fetal ultrasound scanning, *Journal of Telemedicine and Telecare*, Vol. 3, pp. 209-214.

French national patent n° 9903736, 1999, Robot à trois degrés de liberté et à un point fixe, Laboratoire Vision et Robotique - University of Orléans.

LOGINAT Project, 1998; http://www.healthnet.lu/Situations_internationales.pdf

OTELO project, 2001; http://www.bourges.univ-orleans.fr/otelo/site.html

USER INTERFACE ENVIRONMENT AND IMAGE COMMUNICATION IN MOBILE TELE-ECHOGRAPHY

George A Triantafyllidis[*], Nicolaos Thomos, George Nikolakis, Dimitrios Tzovaras, George Litos, and Michael G. Strintzis

1. INTRODUCTION

Mobile interconnectivity and telemedicine are important issues for achieving effectiveness in health care, since medical information can be transmitted faster and physicians can make diagnoses and treatment decisions faster. In this context, the OTELO project (OTELO, 2001) aims to develop a fully integrated end-to-end mobile tele-echography system using an ultra light, remote-controlled robot, for population groups that are not served locally by medical experts. An expert located in the expert center will do the echographic diagnosis. There will be only a "non-sonographer" person in the isolated site and the wireless transmission system will be the only link between the two sites. At the master station site, the clinical expert's role is to control and tele-operate the distant robot by holding a fictive probe.

This paper focuses on two specific and strongly linked components of the OTELO tele-echography system which are very important for the overall efficiency of the system. Specifically, this paper proposes the architecture for the ultrasound image communication as well as the user interface for the visualization of the ultrasound images and the remote monitoring and VR controlling of the mobile ultrasound examination

The scheme of the ultrasound image communication is based on the following scenario: Due to bandwidth restrictions a lossy ultrasound video is real-time received in the expert's site using the proposed scalable wavelet-based video compression algorithm. Moreover, whenever the expert feels that he should have a still image of a better quality (for better inspection and storage in the patient's record), a still image is then losslessly compressed using JPEG-LS (Weinberger et al., 2000) and received by the expert.

[*] George A Triantafyllidis Informatics and Telematics Institute (ITI-CERTH), 1st Km Thermi-Panorama Road, Thermi-Thessaloniki GR-57001, Greece - gatrian@iti.gr
* This material is based upon work supported by the IST program of the EU in the project IST-2001-32516 OTELO

The proposed ultrasound video coder offers Fine Granular Scalability (FGS) (Van der Schaar , 2000). Fine granularity means that the enhancement layer can be decoded at any length. Practically, FGS guarantees a minimum acceptable video quality without drift, over heterogeneous (different capacities) networks (i.e. ISDN, ATM, Satellite Link, and GSM) while offering a visually lossless quality when there is available bitrate. Telemedicine can greatly benefit from advanced tools for scalable video coding. Consider our scenario of a physician receiving an ultrasound image sequence which must be examined from a distant location. The sequence may be transmitted via communication channels with different bandwidths, such as a terrestrial cellular network, a satellite link or even a fiber-optic. In a real-time system which captures and encodes an echography, it is practically infeasible (due to delay and processing power constraints) to encode it in different rates to serve all possible communication links. Due to FGS, the present image sequence coder is able to offer scalability and bandwidth adaptivity.

Regarding the proposed user interface in the expert site, it consists of the following two main parts:

- ❑ An environment for the display of the ultrasound image/video and the environmental video, providing the appropriate tools for allowing the expert to perform the remote diagnosis.
- ❑ A novel virtual reality user environment for controlling and monitoring the remote mobile station: a virtual reality 3D environment is designed to provide the expert with visual and control feedback (if needed) during the mobile examination.

Recently, there have been a lot of efforts (Guerraz et al, 2002; Abolmaesumi et al., 2002; Preusche et al., 2000; D'Aulignac et al., 1999) to employ and integrate virtual environments in medical applications. In our case, a 3D virtual environment is developed to provide the expert with a realistic simulation of the remote ultrasound examination. The force feedback control and the detailed visualization of the remote examination are the two basic aspects of the proposed virtual environment.

The paper is organized as follows: In section 2, the ultrasound image communication is described providing a description of the compression schemes used for the dynamic and still ultrasound images. Section 3 presents the overall user interface. A detailed analysis will be given regarding the design and the implementation of the user interface in order to adapt the system to the physician's needs for the ultrasound examination and to reduce complexity in human - machine interaction. In subsection 3.1, the necessary functions for the post-processing of the received ultrasound image are described, while in subsection 3.2, we present the virtual reality 3-D environment and the force feedback scheme. In section 4, we describe the specific scenarios for the ultrasound image and robotic control communication depending on the available bandwidth and present experimental results. Finally, the conclusions are drawn in section 5.

2. ULTRASOUND IMAGE COMMUNICATION

As mentioned in the introduction, we employ for the ultrasound image and video compression the algorithms of JPEG-LS and a novel scalable wavelet-based video coder. In the following, the basics of JPEG-LS are presented and the proposed scalable video coder is described.

2.1 Lossless image compression with JPEG-LS

Lossless image coding schemes are designed particularly for medical image applications where information loss is less tolerant. The most popular lossless image coders are the JPEG-LS (Weinberger et al., 2000) and the JPEG2000-LS (Christopoulos et al., 2000). A description of JPEG-LS coder is presented in Fig. 1. It is widely known for its simplicity, low complexity and advanced performance. JPEG2000-LS is a lossless version of the recently finalized JPEG 2000 standard and is based on a reversible wavelet transform (Calderbank et al., 1999). Although its performance is almost equivalent to JPEG-LS, in practice JPEG2000-LS is used only in cases of compound images due to its significant higher complexity in comparison to JPEG-LS.

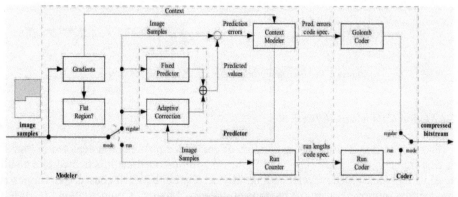

Figure1. JPEG-LS: Block diagram

The algorithm at the core of JPEG-LS is the LOCO-I (Weinberger et al., 1996) (LOw Complexity LOssless COmpression for images). In a LOCO-I the coefficients are processed in a predefined scanning order (*i.e.* raster scan). Thus the procedure is invertible at the decoder side. The two distinct components of the LOCO-I are modeling and coding. In the modeling part, inferences for the value of a coefficient are made from the previously scanned samples by assigning a conditional distribution to it. This probability assignment consists of the following: a prediction step in which the value of a coefficient is guessed from the previously processed data, the determination of a context and a probabilistic model for the prediction residual.

2.2 Lossy video compression with the proposed scalable wavelet-based coder

A novel ultrasound image sequence coder is proposed which is based on the 2D wavelet transform. The proposed coder is endowed with the property of Fine Granular Scalability (FGS) (Van der Schaar, 2000), as depicted in Fig. 2. The bitstream comprises of two layers: the base and the enhancement layer. Due to the FGS property, the enhancement layer can be truncated and decoded at any length.

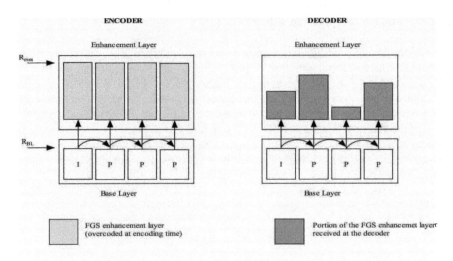

Figure 2. FGS scalability structure at the encoder and decoder.

2.2.1 Motion Compensated Prediction

The proposed system for the generation of embedded bitstreams is depicted in Fig. 3. The first frame in each GOP (Group of Pictures) is intra-coded i.e. it is treated like a still image) using block-based wavelet coding. The correlation between consecutive frames is subsequently removed using Overlapped Block Motion Compensation (OBMC) (Watanabe et al, 1991). The reference frames used to calculate motion vectors are the original frames in order to ensure that the highest possible precision is achieved in the estimation of the motion vectors. Motion vectors are losslessly coded (Boulgouris et al., 2001).

Using the previously estimated half-pixel accurate motion vectors, the procedure for the generation of the stream for the inter frames continues as follows: initially, the first inter-frame (noted as P) is compensated. A version of the intra-frame (first frame in a GOP, noted as I), reconstructed using only the base layer of the coded bitstream, is used as reference for the compensation process. The prediction error is derived by subtracting the compensated prediction from the original inter-frame. The prediction error is wavelet transformed and coded into layers. A version of the error frame is reconstructed using only the base layer of the coded bitstream and then added to the compensated frame. The resulting inter-frame (instead of the original), reconstructed using only the base layer, will serve as the reference frame for the compensation of the next inter-frame. The same procedure is iterated until all frames in a GOP are treated.

Using the methodology described above, the proposed wavelet image sequence coder generates a layered bitstream. The bitstream is composed of a base layer, which is used for the derivation of the reference frame in the motion compensation process, and a complementary enhancement layer that is used to improve the quality and can be truncated at any point. This method has been adopted to ensure that the motion compensation process performed at the encoder can be identically replicated at the decoder even if only the base layer is received. Otherwise, in case the decoder (Fig. 4) is unable to use the same reference frames, errors would accumulate in the decoded video sequence causing a distortion usually termed as drift (Arnold et al., 2000). With the proposed methodology, the possibility of facing drift at the decoder is eliminated and thus, a reconstructed sequence of high quality is obtained even if only the base layer is received. In our coder the minimum possible quality, which corresponds to the base layer, is equal to a half ISDN line (64 Kbps). The wavelet coefficients are coded using a simple bitplane encoder which employs the context models in EBCOT (Taubman, 2000).

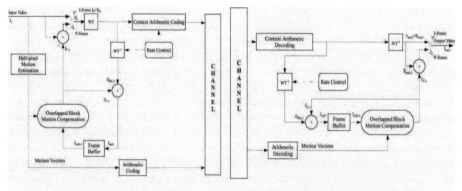

Figure 3. Block diagram of the coder.   **Figure 4**. Block diagram of the decoder.

2.2.2 Block-based wavelet coding of motion compensation residuals

The intra-frame and the motion compensated residuals are decomposed using a wavelet transform based on the 9-7 bi-orthogonal filter bank (Antonini et al., 1992). The transmission of information is done in a bitplane-wise manner starting from the Most Significant Bit (MSB) to the Least Significant Bit (LSB). Within each bitplane, subbands are encoded in a predefined scanning order from the lowest to the highest resolution.

For each subband, first the coefficients whose most significant bit lie in the bitplane currently coded are identified by comparing them to a threshold $T=2^n$ where n is the index of the bitplane that is being coded. If a coefficient becomes significant then its sign is coded. This process is called significance identification. Similarly, the refinement layer is defined as the one containing the n-th bitplane of coefficients found significant in previous passes. In order to achieve greater coding efficiency, the significant identification pass is divided in two layers: the layer containing predicted significant coefficients and the layer containing predicted insignificant coefficients. The transmission order of the layers is predicted significant, refinement and predicted insignificant.

The number of symbols that have to be coded is reduced if a single bit is coded for each subband indicating if all coefficients in a subband are insignificant. In case a coefficient is found to be significant, the subband is split into smaller code-blocks. The same process is applied to the code-blocks. The smallest permitted code-block size is equal to the lowest frequency subband size. The symbol stream generated as described above is coded using adaptive arithmetic codes. The context modeling strategy in

EBCOT (Taubman, 2000) is followed for the coding of all layers. The sign of significant coefficients is entropy coded using a single arithmetic model. The final bitstream is a succession of layers which very often are independent to each other.

2.2.3. Advantages of the proposed coder for medical applications

The FGS guarantees efficiency for the compression and transmission of ultrasound image sequences over heterogeneous (different capacities) networks (i.e. ISDN, ATM, Satellite Link, and GSM) while offering a visually lossless quality when there is a high bitrate channel. Provided that the base layer is received, our scheme does not suffer from temporal quality degradation due to drift, since only the base layer is needed for the exact replication of the motion compensated prediction that took place during encoding. Furthermore, any additional portion of the enhancement layer can be decoded at arbitrary length and not at some specific rates, as in other video coding schemes. Additionally, wavelet block-based schemes are inherently more robust to bit errors that may occur in the bitstreams, in comparison to zero-tree like methods. This is due to the fact that the block-based approach allows the independent coding of blocks of wavelet coefficients and therefore facilitates the localization of corrupted streams. Also, the produced bitstream is valid for transmission over packet erasures channels and error-prone wireless networks, since it can be easily endowed with error resilient tools. Finally, as shown in Table 1, the proposed video coder achieves higher performance (in terms of PSNR) in comparison to H.263.

Table 1. PSNR comparison (in dB) of H.263 coder and the proposed video coder

Bit Rate	15 fps		25 fps	
	Proposed	*H.263*	*Proposed*	*H.263*
32 Kbps	40.11	39.19	38.34	37.42
64 Kbps	42.05	41.49	40.27	39.86
128 Kbps	44.15	43.73	42.86	42.16
256 Kbps	46.38	45.52	45.01	44.23
512 Kbps	48.43	47.34	47.12	45.67

3. USER INTERFACE

As discussed in the introduction, the proposed user interface contains the part of two video inputs: one for the ultrasound video and the other for the environmental video. Also, it contains the 3D user environment simulating the examination and providing details and more accurate visualization of the examination (see Fig. 5).

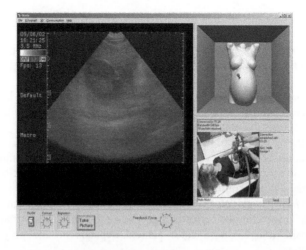

Figure 5. Screenshot of the user interface

Whenever, the expert requires a high quality still image of the ultrasound image (for better inspection and/or storage to the patient's record), he is able to press the "Take picture" button and receives the corresponding - losslessly compressed - ultrasound image (see Fig. 6).

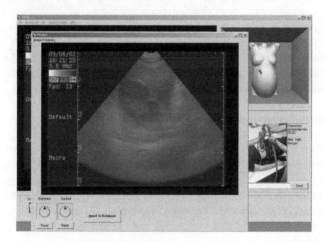

Figure 6. Screenshot of the user interface

3.1 Ultrasound image processing functions

The first part of the graphical user interface is the component for viewing, handling and processing medical image/video files. The system has been designed according to the physician's needs. An image/video viewer is used for the display of the run-time ultrasound image/video and several functions (needed for the ultrasound image processing) are provided to be used by the experts. Specifically, these functions (Collyda et al., 2002) are:

- ❑ *Zooming / Creating of region of interest*: To make an image appear larger by re-displaying the image at different resolutions. Higher resolutions will make the image appear larger, easier to read.
- ❑ *Brightness and Contrast*: Brightness and contrast both regulate the light/dark balance of the image. Contrast makes the colors in an image sharper and more defined. Brightness controls the black/white value of the image.
- ❑ *Resize*: Enlarging the size of the image.
- ❑ *Histogram equalization*: Histogram equalization can be used to improve the visual appearance of an image. It is a contrast enhancement technique to obtain a new enhanced image with a uniform histogram.
- ❑ *Pseudocoloring*: Pseudocoloring is a technique to artificially assign colors to a gray scale image. This technique performs a gray level to color transformations. The result is a composite image whose color content depends on the gray level to color transformations.
- ❑ *Rotation*: In digital picture manipulators, rotation is the turn of an image on one of its axes.
- ❑ *Invert colors*: Invert produces a negative of the image. It works for both color and grayscale images.
- ❑ *Darken and Lighten*: Darken increases the black value for all the colors in an image. Lighten does the exact opposite of Darken. It increases the white value for all the colors in an image.
- ❑ *Image enhancement processing algorithms*: a) *Noise Reduction* is used in video and images to remove fine-grain interference (snow). b) *Sharpen* makes slightly blurry pictures more detailed. If an image has been captured slightly out of focus, the use of sharpen make it clearer. c) *Edge enhancement and detection* uses the difference in color between the background color and the foreground color.

3.2 Virtual reality 3D environment

In this section the virtual reality environment for controlling and monitoring a remote ultrasound examination is described. At the master station site, the expert's role is to control and tele-operate the distant mobile robot by holding a force feedback enabled fictive probe. The PHANToM™ fictive probe (PHANToM, 2001) can provide sufficient data for the control of the remote site robot. Moreover, the use of this device offers the capability to use force feedback to assist the expert performing the echography examination process.

The medical expert must be able to visualize the patient and the slave robot positioning during the medical examination. In order to provide this critical feature in the proposed user interface, the available bandwidth of the communication channel is first detected and evaluated:

- ❑ In the case of a broadband communication connection providing sufficient data rate for a real-time environmental video transmission, the user interface provides the expert with the feature of viewing and monitoring directly the remote robot and the patient, using a direct video transmission link between the two sites.
- ❑ In the other case of a weak communication connection, where the channel does not allow a transmission of a stable real-time environmental video (added to the

ultrasound video), the expert must have an alternative visual feedback. In order to achieve this, the expert can use and adapt a 3D object geometry in order to represent the part of the body which is examined. This accurate virtual model of the examined area can be also used to determine the force feedback of the PHANToM™ device since the communication link is weak and there is no force information sent by the slave robot.

In the proposed virtual reality 3D environment a menu provides the user/expert with the ability to select between a variety of actions and functionalities. Most of these functionalities can be selected from a toolbar on the top of the window (in the standalone application which we examine now). Specifically, the user can view a representation of the workspace. In the workspace area there is a cone, which represents the probe in the virtual space (the PHANToM™ position), a small cube that represents the position of the remote site robot probe (the ultrasound probe position), and the 3D representation of the body part that is examined. The user in the expert site can simply import appropriate 3D geometries from VRML files (see Fig. 7). The models automatically fit in the workspace. The user can also set the size of the model in the virtual world and rotate it to any desired viewing angle (see Fig. 8).

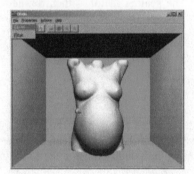

Figure 7. A VRML model is imported in the scene

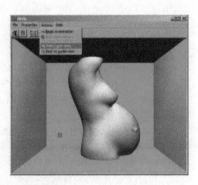

Figure 8. Change in the orientation of the model

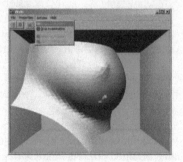

Figure 9. Select/zoom in the examination area.

The expert can select the examination working area using the PHANToM™ Stylus and can start/stop the examination process (see Fig. 9). When the expert selects the examination area an automatic scaling occurs so that the active PHANToM™ workspace size can access all the workspace of the slave robot, using a uniform scale factor. When the expert starts the examination, the force feedback is applied to the expert via

PHANToM™ depending on the force measured by the slave robot located at the remote station. The 3D model (especially when this is an accurate model) can be also used as an alternative source for providing this force feedback input to the PHANToM™ device, in cases of weak communication links. For safety reasons, if the PHANToM™ probe penetrates the 3D graphical model over a threshold value, an alarm sound is activated to warn the expert for possible communication or other failure.

The use of a master probe and slave robot brings up the problem that they have different kinematics structure and different workspaces. The problem of kinematics can be solved by using a generic interface (e.g. Cartesian control (Ortmaier et al., 2000)). Such an interface is provided with the PHANToM™ library (called GHOST®). Using indexing or scaling can solve the problem of different workspaces. Indexing means that the movement is not scaled, but we have a variable offset (index) of the master's position within the slave's workspace. In this case, the different dimensions of the master probe and slave robot do not influence force feedback. The main problem of indexing is that it is not very comfortable to move the slave robot from one manipulation area to another. If scaling is applied only to the positions and the forces felt by the human remain the same as measured at the remote site (Preusche et al., 2000), then the appearing stiffness of the environment changes due to the scaling factor s. The scaling factor is then defined by equation (1):

$$x_{slave} = s \cdot x_{PH} \tag{1}$$

If scaling is performed only to the position, then:

$$K_u = \frac{f_{Env}}{\Delta x_{slave} \cdot \frac{1}{s}} = s \cdot K_{Env} \tag{2}$$

where K_u is the stiffness felt at the operator side, f_{Env} the force measured, Δx_{slave} the displacement of the slave robot and K_{Env} is the stiffness of the environment. In order to avoid changes in stiffness the force feedback vector can be calculated using the formula,

$$f_{PH} = K_p(x_{PH} - x_{slave}/s) + K_f f_{Env}/s \tag{3}$$

where f_{PH} is the force vector sent to the haptic device, K_p represents a virtual coupling between the master and the slave system, x_{PH} is the position of the haptic device probe, x_{slave} is the position of the remote robot probe, s is the scaling factor, K_f is a control parameter (normally equal to one) which can be tuned down if communication delay affects the stability of the system and f_{Env} is the measured force vector in the remote site. The first part of the sum is hysterisis force. The hysterisis force provided to the expert does not exist in the real world. The reason of its existence is to avoid having different positions between the probe in the virtual and the real world.

Another important issue needed in order to make the prototype software credible to use for performing echography examination, is to make the virtual and the real workspaces identical. In order to achieve this we need to ensure: a) The local and remote site workspaces are identically oriented. b) The model of the body part has the same shape as the examined body part. c) The workspaces are synchronized so that the working area of the remote robot is the working area of the master probe. In the virtual reality environment the expert must be able to use a virtual 3D model that corresponds to the actual part of the patient's body. This can be achieved using a large database of several models or using parametric models for all parts of the human body. Another solution, used in our case, is to use a "sculpturing" 3D scanning, method. In this case the operator in the remote site moves the probe of the remote robot, manually, on the area that is to be examined. The position and the force applied on the body measured by the robot are used to create a point mesh. That point mesh is used for creating a 3D representation of the body part. In order to create a solid part from the point mesh Delaunay triangulation is used. There are several triangulation methods, such as Divide and Conquer (Amato et al., 1996), Sweepline (Edelsbrunner et al., 1989) and incremental insertion (Devillers et al., 1998), which enable the creation of 3D solid geometry using a point mesh.

4. COMMUNICATION SCENARIOS AND EXPERIMENTAL RESULTS

In order to control the haptic device, it is important to define the requisite input data rate so that the control and the force feedback provided to the expert can be realistic and at least not seductive for the examination process (see Table 2). In order to assure an efficient, comfortable and realistic force sensing simulation, the force update rate must be higher than 50Hz (20 msec transmission delay) (Burdea et al., 1996).

Table 2. Interface requirements.

Transmission	Interface requirements		Application values	
	Transmission delay	Bandwidth	Transmission delay	Bandwidth
Control	10 msec	100Hz	10 msec.	100 Hz
Force Feedback Sensing	20 msec	50-100 Hz	10 msec.	100 Hz

The priority order in our system is firstly the ultrasound video, then the master's probe position data (transmitted from master to slave), and finally the force feedback and the robot position feedback data (transmitted from slave to master). The system is useable when it provides at least the real-time video and the probe position data communication. Specifically, when the delays in video and position data communication are not significant (less than 1 sec per minute) then the system is considered acceptable and usable. In cases that the communication of the force/position feedback data brings significant delay to the system, we prefer not to send this feedback and have all the available bandwidth for the ultrasound video and the probe position data communication. We use then the 3D model of the virtual reality environment to estimate the force/position feedback. In order to choose the appropriate communication model (video + position, or video + position + feedback) the bit rate and the quality of the communication channel are measured. For the simulations a packet erasure channel is used. These channels erase independently packets with a probability p (packet error rate - PER). In case of an erased packet the position of the erasure is known to the receiver side. However, if the packet is not erased, it is received with probability one. Communication simulations for PER of 10% and 1% and variable bit-rates have been tested. The delay time per minute of active communication is calculated in each case. To perform the tests we allocate a bit-rate for the video transmission, a bit-rate for the position transmission (master to slave) and the rest of the bandwidth for the force feedback and the position feedback transmission (slave to master). Fig. 10 presents the delay times for PER=10% and Fig. 11 presents the delay times for PER=1%. The x-axis represents the bandwidth assigned to the position data transmitted from the master site to

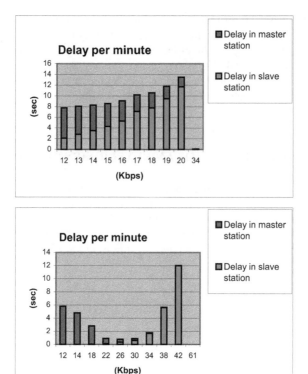

Figure 10. Delay times for 64Kbps and 96Kbps communication channels using 30Kbps and 35Kbps respectively for video transmission (PER=10%). The x-axis represents the bandwidth assigned to the position data send from the master site to the slave robot.

Table 3. Working platform selection depending on the bit rate for PER 10%.

Channel bit rate Kbps (PER=0.1)	Video (bit rate) Kbps	Position (bit rate) Kbps	Force/position Feedback (bit rate) Kbps	Average Delay per minute (sec)
56	30	26	-	0.46
64	35	29	-	0.32
96	35	30	31	0.87
128	40	40	48	0.06
>128	40	40+	48+	~0

the slave robot and y-axis represents the delay time. The remaining bandwidth (total – (video + position)) is assigned to the force/position feedback data transmission. Large values of "delay in slave station" mean that the feedback information is not sufficient enough to provide realistic feedback to the expert. In that case the communication model of video and position is selected and the force/position feedback is estimated locally from the 3D model. When large values of "delay in master station" occur, then the system cannot provide real-time data for the positioning of the robot, so the system becomes unusable. The proposed bandwidth selections depending on the quality and the bandwidth of the communication channel are presented in Tables 3 and 4. When the channel bandwidth is less than 64Kbps, it is preferable to use one-way communication from the master to slave robot so that the delay times remain under 1 second per minute.

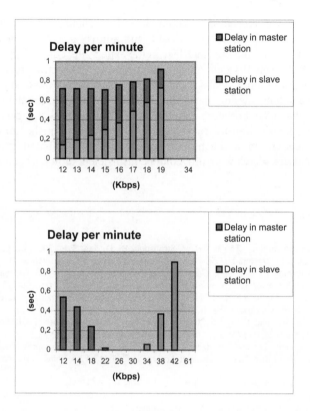

Figure 11. Delay times for 64Kbps and 96Kbps communication channels using 30Kbps and 35Kbps respectively for video transmission (PER=1%). The x-axis represents the bandwidth assigned to the position data send from the master site to the slave robot.

Table 4. Working platform selection depending on the bit rate for PER 1%.

Channel bit rate Kbps (PER=0.01)	Video (bit rate) Kbps	Position (bit rate) Kbps	Force/position feedback (bit rate) Kbps	Average Delay per minute (sec)
56	30	26	-	0.0
64	35	29	-	0.0
64	30	12	22	0.73
96	35	30	31	0
128	40	40	48	0
>128	40	40+	48+	0

5. CONCLUSIONS

In this paper, we presented a communication scheme for the transmission of ultrasound images and video, using JPEG-LS and a novel scalable wavelet-based video coder. Also, we described the user interface in the expert site, containing not only the part of the display of ultrasound video/images, but also a new 3D virtual reality environment for better visualization, force feedback control and inspection of the examination. Finally, we tested different transmission scenarios based on the available bandwidth in several packet erasure channels. For each given bandwidth and packet erasure channel, we concluded to a specific transmission scenario which supports the remote ultrasound examination procedure in a much more effective way compared to other possible transmission scenarios which were proved to be insufficient or not useable.

6. REFERENCES

Abolmaesumi P., Salcudean S. E., Zhu W.-H., M. Sirouspour R., DiMaio S. P., 2002, Image-Guided Control of a Robot for Medical Ultrasound, *IEEE Transanctions on Robotics and Automation*, 18(1), pp.11-23.

Amato N. M. and Ramos E. A.. 1996, On computing Voronoi diagram by divide-prune-and conquer, *in Proceeding of 12th Annual ACM Symposium on Computational Geometry*, pp. 166-175.

Antonini M., Barlaud M., Mathieu P., and Daubechies I., 1992, Image Coding Using Wavelet Transform, *IEEE Transanctions on Image, Processing*, 1(2), pp. 205-210.

Arnold J. F., Frater M. R., and Wang Y., 2000, Efficient Drift-Free Signal-to-Noise Ratio Scalability, *IEEE Transanctions on Circuits and Systems for Video Technology*, 10(1), pp. 70-82.

Boulgouris N.V., Tzovaras D., and Strintzis M.G., 2001, Lossless Image Compression Based on Optimal Prediction, Adaptive Lifting and Conditional Arithmetic Coding, *IEEE Transactions on Image Processing*, 10(1), pp. 1-14.

Burdea G. C., Force and touch feedback for virtual reality, (Eds:Wiley, 1996).

Calderbank R., Daubechies I., Sweldens W., and Yeo B. L., 1998, Wavelet transforms that map integers to integers, *Applied and Computational Harmonic Analysis*, 5, pp. 332-369.

Christopoulos C., Skodras A. and Ebrahimi T., 2000, The JPEG2000 Still Image Coding System: an Overview, *Applied and Computational Harmonic Analysis*, 46(4), pp. 1103-1127.

Collyda C., Kyratzi S., Triantafyllidis G.A., Boulgouris N.V. and Strintzis M.G., 2002, The Graphical User Interface for the Otelo Tele-Echography System, *in Proc 6th World Multiconferece on Systemics, Cybernetics and Informatics*, Orlando, USA.

D'Aulignac D., Balaniuk R., 1999, Providing reliable force-feedback for a virtual, echographic exam of the human thigh, *in Proc. of the PHANToM™ Users Group Workshop*, Boston, USA.

Devillers O., 1998, Improved Incremental Randomized Delaunay Triangulation, *in Proc. of the 14th Annual Symposium on Computation Geometry ACM*, 106-115, Minneapolis, USA.

Edelsbrunner H. and Guibas L. J.., 1989, Topologically sweeping an arrangement, *Journal of Computer and System Sciences*, 38, pp. 165-194.

Guerraz A., Hennion B., Thorel P., Pellissier F., Vienne A., Belghit I., 2002, A Haptic Command Station for Remote Ultrasound Examinations*, in Proceedings of the 10th Symposium on haptic Interfaces for Virtual Environments & Teleoperator Systems*.

Ortmaier T. and Hirzinger G., 2000, Cartesian Control in the dlr minimaly invasive surgery telepresence scenario, *in Proceedings of IEEE/RJS International Conference on Intelligent Robots and Systems*.

OTELO project, 2001, IST-2001-32516, http://www.bourges.univ-orleans.fr/Otelo/.

PHANToM™ device, 2001, http://www.sensable.com.

Preusche C., Hirzinger G., Scaling Issues for Teleoperation, 2000, *in Proceeding of the 5th PHANToM™ Users Groups Workshop*.

Taubman D., High Performance Scalable Image Compression with EBCOT, 2000, *IEEE Transactions on Image Processing*, 9(7), pp. 1158-1170.

Van der Schaar M., All fgs temporal-snr-spatial scalability, 2000, *in Contribution to 54th MPEG Meeting*, m6490, La Baule, France.

Watanabe H. and Singhal S., 2001, Windowed motion compensation, *in Proceeding of SPIE*, 1605, pp. 582-589.

Weinberger M., Seroussi G., G. Sapiro, 2000, The LOCO-I Lossless Image Compression Algorithm: Principles and Standardization into JPEG-LS, *IEEE Transaction on Image Processing*, 9, pp. 1309-1324.

Weinberger M., Seroussi G., Sapiro G., 1996, LOCO-I: A Low Complexity, Context-Based, Lossless Image Compression Algorithm, *in Proceeding of IEEE Data Compression Conference*, Snowbird, Utah, USA.

OBJECT-BASED ULTRASOUND VIDEO PROCESSING FOR WIRELESS TRANSMISSION IN CARDIOLOGY

Paul Rodriguez V[*], Marios S. Pattichis, Constantinos S. Pattichis, Rony Abdallah, and Mary Beth Goens

1. INTRODUCTION

The transmission of medical video over wireless channels requires the use of scalable Codecs, with error-resilient and error-concealment functionalities. A fundamental challenge in developing new methods for effective scalable coding of medical video is the careful and meaningful assessment of the coding performance. In this Chapter, we introduce a new, diagnostic approach to help guide the video compression process. A new segmentation method, based on multidimensional Amplitude-Modulation Frequency-Modulation (AM-FM) is introduced for segmenting motion-mode ultrasound video into diagnostically meaningful objects. Rate-distortion curves are provided to provide a measure of video quality-scalability within the framework of object-based coding.

There has recently been a number of studies assessing the effect of video compression on diagnostic accuracy (Ehler et al., 2000; Garcia et al. 2001; Karson et al., 1996; Main et al., 2000; Soble et al., 1998; Thomas et al., 2002). These clinical studies advocated the use of digital video technologies over video tapes. However, they did not present any new video compression methods, or even show the use of existing video compression technology that is appropriate for wireless transmission.

In the image processing literature, effective video compression techniques are described as perceptually-lossless. This implies that no visual information is lost in the compression process. In medical image quality assessment, an effective test is to perform a blind test, asking the readers to select the better image between two unknown images (the compressed and uncompressed). If the readers select either video image an equal

[*] Paul Rodriguez V. Department of ECE, Room 229-A, ECE Building, University of New Mexico, Albuquerque, NM 87131-1356, USA. prodrig@eece.unm.edu

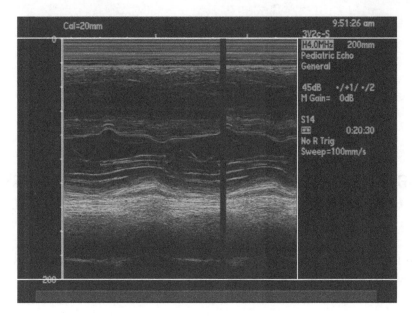

Figure 1. An Motion-mode (M-mode) video image. The thin white lines define boundaries between objects of interest. In the upper rectangular region, patient information has been deleted.

number of times, or if the compressed images are selected more often than the uncompressed images, then the compression scheme is deemed effective.

The compression assessment problem is even more difficult in wireless video compression. The requirement to achieve very low bitrates leads to visual artifacts that are often unacceptable. Also, if high compression ratios are required, the requirement to achieve perceptually-lossless compression maybe unattainable.

Clearly though, we may be able to accept visual artifacts if they have no measurable impact on clinical practice. We are thus led to introduce a new method for assessing video quality that we will call diagnostically-lossless. We assume that human readers are used for obtaining diagnosis. Then, perceptually-lossless videos are also diagnostically-lossless. Thus, we expect that diagnostically-lossless video compression can lead to improved video compression rates since the diagnostically-lossless video compression ratio will be at least as high as the one achieved in perceptually-lossless compression.

The task of defining diagnostically-lossless video compression is very challenging. To recognize the challenges, consider the case where a sonographer obtains videos at a remote site, and then transmits them wirelessly for diagnostic assessment. The basic idea is to adopt a protocol that will allow effective diagnostic assessment after the video has been compressed. Yet, at the time of compression, there is no way of assessing all possible diagnostic outcomes.

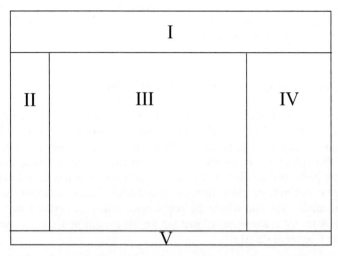

Figure 2. M-mode video image partition into text objects and the main video image.

As we shall describe in detail, the development of diagnostic compression methods requires the use of effective segmentation methods. We introduce a new, multidimensional AM-FM approach that can simultaneously track all the cardiac walls of diagnostic interest.

In Section 2, we provide a summary of the use of digital video in echocardiography. This is followed by a summary of object-based video compression in Section 3. In Section 4, we introduce a new model for diagnostically-lossless compression of M-mode. A novel AM-FM representation for M-mode ultrasound is discussed in Section 5. We use the AM-FM model to derive a segmentation system for video objects of interest. The results are given in Section 6, and concluding remarks are summarized in Section 7.

2. DIGITAL VIDEO TRANSMISSION IN CARDIOLOGY

A summary of video compression technology is beyond the scope of this Section, and can be found in an earlier Chapter of this book by Pattichis et al. (2005). In this Section, we summarize clinical studies that studied the use of digital video in echocardiography. None of these studies looked into the possibility of developing a video compression software system based on the concept of diagnostically lossless compression. Instead, they provided a summary of testing existing technologies.

The use of digital video in echocardiography has been studied at-least since 1996, as discussed by Karson et al (1996). In 1996, Karson et al established that a 20:1 JPEG compression for still images did not degrade diagnostic accuracy. Soble et al. and Garcia et al. (Soble et al., 1998; Garcia et al., 2001) argued that MPEG-1 video delivered the same diagnostic accuracy as the super VHS format. Real-time transmission of MPEG-2 over ATM at 2 Mbps was studied by Main et al. (2000). A multi-site study was reported by Ehler et al. (2000). More recently, Thomas et al. (2002) argued for the wide adoption of digital echocardiography based on the DICOM standard.

3. A SUMMARY OF OBJECT-BASED VIDEO COMPRESSION

Object-based coding has been recently adopted by the MPEG-4 video compression standard (ISO/IEC., 1999, IS 14496-1; ISO/IEC., 1999, IS 14496-2, ISO/IEC., 1999, IS 14496-X). It is widely accepted that effective video compression at very low bitrates can only be achieved through object-based coding. This approach allows us independent rate control on different parts of the video. Yet progress on this subject has been hindered by the lack of effective methods of video image segmentation. Lau et al report an object-based video segmentation method where the segmentation was carried out manually by Lau et al., (2000). A recent Wavelet-based method for ultrasound image compression has been introduced by Chiu et al (2001).

The basic ideas behind object-based video coding can be demonstrated using the examples in Figures 1 and 2. In Figure 1, a single video frame of an M-mode video is presented. Around the primary video image, there are four text-only regions labeled I, II, IV, and V in Figure 2, and outlined with extended white lines in Figure 1. We think of these video regions as video objects that will be independently encoded. In this case, it is clear that the difference between the text objects and the video objects will suggest that different Codecs should be used for encoding these rather distinct objects. For example, the text objects are very slow to change, while the video object changes substantially through time. Thus, it appears that more bandwidth must be allocated for the video images, as opposed to the text objects.

On the other hand, it important to note that the lack of change in the text objects of the video will result in zero motion vectors, which in term will lead to less bandwidth allocated to these objects. Yet, if a special text encoder was used for text objects as opposed to the standard DCT encoder, we would expect significant improvement in text object compression efficiency. Thus, a key principle for achieving compression efficiency is to select an appropriate Codec for each object.

In our research, we develop an object-based segmentation system based on a diagnostically lossless compression system. The basic idea is to segment out the part of the video image that is not of diagnostic significance. Then, by discarding the part of the video that is irrelevant to diagnosis, we can reduce the required bandwidth in the ratio of the discarded video to the entire video. The scheme is explained in detail in Section 4.

4. DIAGNOSTICALLY-LOSSLESS MOTION-MODE (M-MODE) VIDEO COMPRESSION

We begin this Section with an introduction to the basic anatomical features of Motion-mode (M-mode) ultrasound video. We then proceed to define what we mean by diagnostically-lossless compression, as applied to compression of M-mode video.

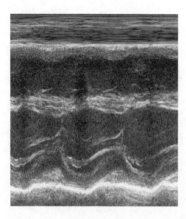

Figure 3. A basic Motion mode (M-mode) video image used in echocardiography.

The basic segmentation structure of a Motion-mode (M-mode) ultrasound video image used in echocardiography is shown in Figure 3. The basic principle in M-mode video is that we image the reflectivity of a single line of pixels through time. In Figure 3, as we move from the top to the bottom, we are moving through a spatial coordinate, from the point nearest the ultrasound probe to the point that is furthest away from the probe. From left to right we have the time evolution of the spatial line through time.

Video images such as the one depicted in Figure 1 are typically used for assessing heart function. To recognize the video segments that are of diagnostic interest, each part of the image has been annotated. First, we have the near field that is of little diagnostic value. This is followed by the right ventricular wall, and the interventricular septum that separates the left from the right ventricle. The left ventricular chamber includes the mitral valve apparatus. The left ventricular chamber ends with the left ventricular posterior wall. This is followed by the epicardium. For assessing heart function, we are interested in clear images of the lower wall of the interventricular septum and the endocardial surface of the left ventricular posterior wall (LVPW).

4.1 Diagnostically-lossless Compression of M-mode Video: A Case Study

To help define regions of diagnostic interest, an M-mode video image is segmented into rectangular regions as shown in Figure 4. These regions represent video objects that we have selected for diagnostically-lossless compression. In the following discussion, we discuss the diagnostic relevance of each region.

We note that the near-field and the image portion beyond the epicardium is of no diagnostic interest. Thus, regions 1 and 3 are of no diagnostic value.

The use of M-mode video in clinical diagnosis requires that we are able to clearly measure distances between wall boundaries. Thus, we require that the video compression algorithm does not degrade the definition of the wall boundaries. This is difficult to define in the sense that wall-boundaries themselves tend to be noisy, and in such cases,

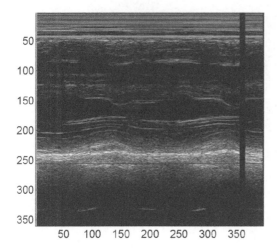

Figure 4. Object-based segmentation of M-mode video. This video image was segmented using a frequency-modulation (FM) model.

sonographers are expected to extrapolate so that they get a well-defined, continuous line throughout the image.

For assessing the left ventricular function, we consider the shortening fraction, as defined by Snider at al. (page 196, 1997):

$$SF = \frac{LVDD - LVSD}{LVDD} \times 100,$$

where:

LVDD is the left-ventricular end-diastolic dimension

LVSD is the left-ventricular end-systolic dimension

SF is the shortening fraction (in percent).

The shortening fraction has a normal mean value of 36% with an effective range between 28% and 44%, independent of age and heart rate.

Thus, for evaluating the shortening fraction, it is not necessary to encode the walls during the entire cardiac cycle, but only during the time-instances when the posterior wall of the interventricular septum and anterior wall of the left ventricular wall come together (systole), or are furthest apart (diastole). Even though it is possible to achieve greater compression ratios by taking advantage of this observation, we will instead encode the entire wall region.

This avoids the problem of having to develop a detector for systole and diastole, and also the problem of having to develop an encoder for curvilinear boundaries (shape-adaptive coding). Developing solutions for these problems is definitely possible. However, the complexity of a computer-assisted video compression system that incorporates these observations would increase substantially. This is due to the time that the sonographers would have to spend on verifying the detection of systole and diastole in every cycle, as opposed to simply verifying that the two object lines include the walls of interest.

5. M-MODE ULTRASOUND VIDEO OBJECT SEGMENTATION USING AM-FM DEMODULATION

The development of general image representations for object-based coding presents many challenges. There is a need for a spatio-temporal model that can describe (i) sharp edges that occur between objects, (ii) 3D object deformations or due to imaging of different slices through time, (iii) imaging artifacts such as ultrasound shadowing, and (iv) the presense of significant additive or multiplicative noise (specle noise in ultrasound). Edges are usually associated with very sharp transitions in image density, that traditional edge detectors try to capture by detecting large image gradients. For ultrasound images, there is a large number of such gradients associated with different reflecting boundaries. Furthermore, the boundaries turn to be broken, resulting in discontinuous edges that are difficult to link or track. Continuous-object deformations suggest that satisfactory edges can be achieved by developing methods that can assume continuity while removing the high levels of noise. We begin with a discussion of well-known methods for noise reduction, and then continue with multi-resoultion and Wavelet based methods. We then present a novel Amplitude-Modulation Frequency-Modulation method that can be used to model edge continuity while providing for an effective methods for dealing with image noise.

Traditional bandpass filtering methods are usually employed for reducing image noise by rejecting out-of-band components, which are assumed to be rich in noise, while containing less signal power. In practice, one starts by identifying the frequency bands where most of the signal power is present. The rest of the frequency bands are expected to be mostly due to noise and are thus rejected through bandpass filtering. The primary weakness of the approach is the assumption that signal components are made up of sinusoids that have infinite spatial extend, since the same bandpass is used throughout the image.

Wavelet theory introduced multiresolution methods where images were modeled in terms of objects of finite extend, occurring at different resolutions. A basic assumption of Wavelet based filters is constant-Q filtering, where the ratio of the central frequency radius to the filter bandwidth. As a result, bandpass filters for higher frequency components occupy larger bandwidths, and correspondingly, they can be implemented using a smaller number of filter coefficients. It is important to note that these filters, with a reduced number of coefficients, also exhibit very high spatial resolution. The situation is reversed for low-frequency components. The bandpass filters have smaller bandwidths and exhibit very low spatial resolution. Thus, in order to describe the edges of larger objects, there is a need to combine different resolutions. This complicates the models since it is no longer possible to work at a fixed scale.

The introduction of multidimensional Amplitude-Modulation Frequency-Modulation (AM-FM) methods attempts to overcome these limitations. Continuous-space and continuous-scale variations are modeled through the product of a spatially varying positive amplitude function and an FM signal. Throughout an object, the AM-FM model

Step 1. Compute "Analytic Image" using 2-d FFT
Step 2. Apply bandpass channel filters
Step 3. Compute AM-FM parameters over each channel
Step 4. At each pixel, reconstruct AM-FM image using the channel that
produces the maximum amplitude estimate.

Figure 5. An overview of an AM-FM demodulation system.

acts like the fundamental harmonic from Fourier Analysis. We assume slowly varying Amplitude Modulation (AM) and rely on Frequency Modulation to capture most intensity variation. The model is especially effective on images with a large number of ridges, such as Motion-mode ultrasound described in this Chapter. The fundamental advantage of the AM-FM representation is that it allows for a single model to describe the combined output from multiple bandpass filters. Both ridge orientation changes and variations in ridge spacing are captured through the Frequency Modulation model, while image intensity variations are captured in the amplitude.

5.1 An AM-FM Model for M-mode Ultrasound

In this subsection, we introduce a video object segmentation system that is particularly well suited for M-mode video segmentation. To develop an AM-FM model for M-mode ultrasound, we return to Figure 1. In M-mode ultrasound video, the video is formed by plotting the beam reflection intensity along a line through time. The resulting video objects are made up of cardiac wall boundaries that evolve through time.
There are significant challenges with modeling the walls, including broken boundaries, continuous-space deformations of the walls and speckle noise. To recognize a suitable AM-FM series for the M-mode ultrasound video image, we consider the row-column coordinate system $y-t$ where y is used for indexing the rows, while the time coordinate t is used for indexing the columns. Along the y-coordinate, the ultrasound video image is "cutting through" cardiac wall boundaries, showing a single line of the 3-D object points. Along the time-coordinate t, it is assumed that the motion of the same 3-D wall points are observed through time. To track the wall motion, we consider the use of a curvilinear coordinate system that models wall-motion through a frequency modulation process, while variation in the actual wall material is described through an amplitude modulation process. We model M-mode video in terms of an AM-FM series expansion given by Rodriguez and Pattichis (2002), Pattichis (1998), and Havlicek (1996):

$$f(y,t) = \sum_n C_n a(y, t) \cos(n\phi(y,t)),$$

where ϕ denotes the phase function, $\cos(n\phi(y,t))$ denotes the FM harmonics, $a(.)$ denotes the amplitude function, and C_n represents the coefficients of the AM-FM harmonics. We also define the instantaneous frequency as $\nabla\phi$, and its magnitude by the vector magnitude $\|\nabla \phi\|$. In what follows, we set $x = t$ to represent images in the familiar $x - y$ coordinate system.

5.1 Real-time AM-FM Demodulation

The basic AM-FM demodulation algorithm is summarized in Figure 5. Here, we define the AM-FM demodulation problem, as one of having to estimate the amplitude $a(x, y)$, the phase $\phi(x, y)$, and the instantaneous-frequency vector, as defined by Havlicek (1996):

$$\nabla \phi(x, y) = [\partial \phi / \partial x, \partial \phi / \partial y]^T ,$$

from the original image.

To estimate the phase, we apply a collection of bandpass channel filters that cover the two-dimensional frequency plane at different orientations and magnitudes. The AM-FM parameters are then estimated over each channel. If we let the impulse responses of the bandpass channel filters be denoted by $g_1, g_2, ..., g_R$, and let f denote the input image, the filtered output images are given by $h_1, h_2, ..., h_R$ satisfying $h_i = f * g_i$ where $*$ denotes two-dimensional convolution. We then obtain estimates for the instantaneous frequency and the phase using (see Havlicek (1996)):

$$\nabla \phi_i(x, y) \cong \text{real}\left\{ \frac{\nabla h_i(x, y)}{j h_i(x, y)} \right\}, \text{ and } \phi_i(x, y) \cong \arctan\left\{ \frac{\text{imaginary}\{h_i(x, y)\}}{\text{real}\{h_i(x, y)\}} \right\}.$$

Let $G_1, G_2, ..., G_R$ denote the frequency responses of the channel filters. Using the instantaneous frequency estimate $\nabla \phi(x, y)$ and the frequency response of the channel G_i, we estimate the amplitude over each channel using

$$a_i(x, y) \cong \left| \frac{h_i(x, y)}{G_i(\nabla \phi(x, y))} \right|.$$

In the Dominant Component Analysis (DCA) algorithm, from the estimates for each channel filter, we select the estimates from the channel with the maximum amplitude estimate

$$c_{\max}(x, y) = \underset{i}{\text{argmax}} \; a_i(x, y).$$

Hence, the algorithm adaptively selects the channel filter with the maximum response. This approach allows the model to quickly adapt to singularities in the image. The performance of the AM-FM demodulation algorithm in one and two-dimensions has been studied by Havlicek (1996). A discrete-space algorithm is also described by Havlicek (1996). Next, we present a number of modifications to the basic algorithm that allow us to achieve real-time performance. In addition, we discuss how to achieve robustness.

5.2 Real-time AM-FM Demodulation Implementation

In Step 1 of Figure 5, the 1D analytic signals are computed for each pixel-column in the image. We write:

$$\mathbf{X}_{A, i} = \mathbf{X}_i + \mathbf{H}\mathbf{X}_i, \qquad i = 0, 1, 2, \ldots, M-1.$$

where $\mathbf{X}_i$ denotes the i-th column in the image, M is the number of columns, and H denotes the Hilbert transform.

The continuous-time analytic signal has a Fourier Transform that is equal to the Discrete Fourier Transform of the original signal for non-negative frequencies, and is zero for negative frequencies. This observation leads to a straight-forward implementation using 1D FFTs.

For the bandpass filter implementation, we use a separable design. The separable implementation is efficient because the M-mode ultrasound video itself could be thought of as separable. This observation has been addressed in detail in Rodriguez and Pattichis (2002). In essence though, this is due to the fact that a product of vertical and horizontal modulations effectively capture the changes in the M-mode video.

Following AM-FM demodulation, a sum of 20 FM harmonics is taken as an approximation to the walls of the video image (see Rodriguez and Pattichis (2002)). The sum of the pixels along each row of the harmonics image is used for detecting the near-field and the epicardium (see Figure 6). The sums from all the rows are filtered by a 19-point median filter, followed by simple thresholding. This detection method was found to be very robust for segmentation (see Figure 6).

6. RESULTS

The rate distortion curves are summarized in Figure 9. We show the rate-distortion curves for: (i) method 1: single-object MPEG-4 video compression, (ii) method 2: MPEG-4 compression using four text objects and a video object, and (iii) method 3: MPEG-4 compression using FM-segmentation to derive a Region of Interest within the video object plus the four text objects. The video objects were encoded using the MPEG4IP software that is freely available (see MPEG4IP, 2003). In all Figures, the letters 1, 2, and 3 are used to identify the method that was used. Also, the dotted-line that overlaps with the rate-distortion curve of method 3 represents the result of multiplying the bitrate of method 2 by the ratio of the number of pixels in the FM-segmented image, divided by the number of pixels in the original image. It represents the theoretical performance improvement by using method 3.

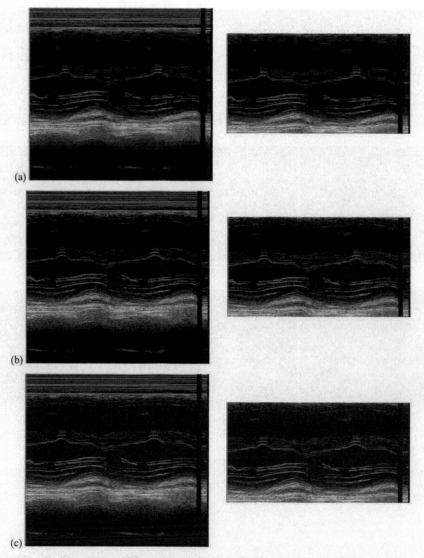

Figure 6. M-mode video objects at different bitrates. In (a), both objects are shown at 50kbps, in (b) they are at 500kbps, and in (c) both objects are at 1000kbps.

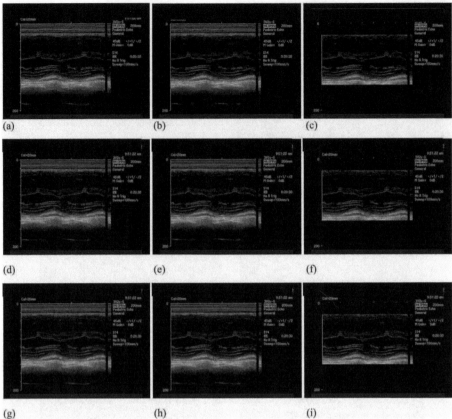

Figure 7. M-mode video reconstructed with and without objects at different bitrates. In (a), (d), and (g) the entire video is compressed at 50Kbps, 500Kbps, and 1000Kbps. In (b), (c), (e), (f), (h), and (i), all the text objects are compressed at 10:1 using still image compression of the first frame. In (b), (e), and (h), we show the results of compressing the main video object at 50Kbps, 500Kbps, and 1000Kbps. Similarly, in (c), (f), and (i), we show the results of compressing the extracted region of diagnostic interest at 50Kbps, 500Kbps, and 1000Kbps.

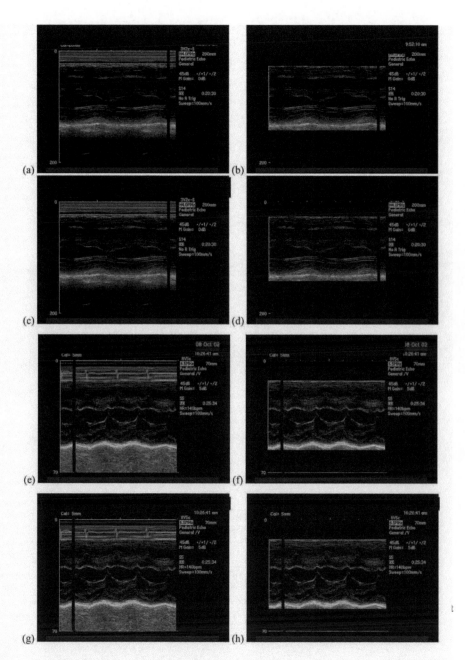

Figure 8. M-mode video object-based reconstruction. In (a), (b), (e) and (f), the effective bitrate is about 200Kbps. In (c), (d), (g), and (h), the effective bitrate is about 500Kbps.

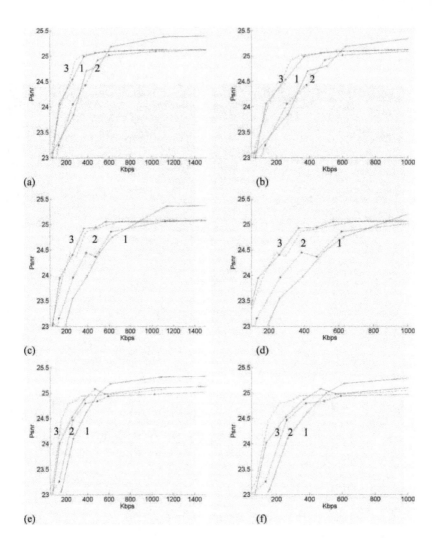

Figure 9. Rate-distortion curves using methods 1, 2, and 3. Method 1 is MPEG-4 without objects, method 2 is MPEG-4 with four text objects and one video object, while method 3 is MPEG-4 with four text objects and a segmented video object. The "leftmost" dotted curves show the result of multiplying the rate-distortion curve of method 2 by the ratio of the number of pixels in the segmented region divided by the entire video image region. It demonstrates that the FM segmentation method 3 achieves a near optimal improvement over method 2 that is proportional to the reduction in the number of video pixels used in method 3. Figures 9(a) and 9(b) correspond to the images in Figures 6 and 7. Figures 9(e) and 9(f) correspond to Figures 8(a)-8(d), while Figures 9(c) and 9(d) correspond to Figures 8(e)-8(h).

For quantitative measurement of the compressed video quality, we compute the Psnr at a variety of bitrates. The Psnr is defined by

$$\text{Psnr} = 10\log_{10}\left(\frac{255^2}{\text{mse}}\right)$$

and measured in decibels (dB). In the formula for the Psnr, mse stands for the mean square error, defined as the average squared difference between the uncompressed video and the original (uncompressed) video intensity, while 255 represents the peak image intensity. For our study, Psnr values between 23dB and 25dB were found to be of interest (see Figure 9). A Psnr of 25dB represents a mean drop in image intensity of 14.34 (5.6% of 255), while a Psnr of 23dB represents a mean drop in image intensity of about 18.05, (7.1% of 255). Pnsr is not generally accepted as a good measure of perceptual image quality. Instead of using a Psnr measure over the entire image, we only use the Pnsr over the video region of diagnostic interest.

As defined earlier, the Region of Interest (ROI) was estimated using FM-segmentation for extracting the M-mode video region that lies between the near field and the epicardium. Examples of the computation of the ROI are given in Figures 6, 7, and 8. In Figure 6, the rightmost images represent the ROIs. Furthermore, in the examples shown, all text objects were compressed using still-image compression of the first video frame, sacrificing the clock updates (which can be readily restored).

To allow a critical evaluation of the rate-distortion curves, we also used perceptual evaluation of the compressed video. The purpose of the experiments was to determine the minimum bitrate below which the compressed video quality was unacceptable. The Cardiologist was allowed to review each video individually or compare videos against each other in order to determine whether the effect of compression was impacting the clinical use of the video. Two perceptual tests were carried out. In the first test, the reader was asked to assess the quality the displayed video. In the second test, the reader was asked to pause the video and to examine whether any distortion in the wall definition was likely to impact the diagnostic accuracy of the compressed video. The second test is in line with the standard practice of pausing the video playback before measurements on the video image are made. For both tests, three videos were tested at a variety of bitrates.

For both tests, at high bitrates, 1000Kbps and above, no significant distortion was observed. To compare with other studies, we note that 500Kbps corresponds to a compression ratio of 50, 500Kbps corresponds to a compression ratio of about 120, and 200Kbps corresponds to a compression ratio of about 280. The fact that no significant distortion is observed at high bitrates is in agreement with the rate-distortion curves of Figure 9, where the Psnr appears to be flattened beyond 1000Kbps. As discussed earlier, at such low compression ratios, there is less than a 5% drop (from the peak value of 255) in image intensity between the compressed and the uncompressed videos. This very simple rule appears to be a good predictor of the diagnostically lossless region.

For the first perceptual test, significant artifacts were observed in the neighborhood of 400Kps-500Kbps. This is in agreement with the sharp drop in the rate-distortion curves shown in Figure 9. For the second perceptual test, the minimum acceptable bitrate range varied from 100Kbps to 300Kbps.

Thus, the cardiologist was able to detect artifacts in the continuous video display at a higher bitrate than in the still video image display. The continuous video display test is associated with a perceptually lossless requirement, while the still image video display test is associated with a diagnostically lossless requirement, where the basic question is whether there is sufficient information for a diagnosis. As expected, the perceptually lossless requirement was found to be stricter than the diagnostically lossless requirement.

As seen in Figures 9(b), 9(d), and 9(f), over the diagnostically-relevant range, method 3 provided the same Psnr performance as both methods 1 and 2, but at a significant reduction of the bitrate (approximatedly 40% in all cases). The reduction in the bitrate is directly proportional to the reduction in the number of pixels used in the video object images of method 3 (see the rightmost images in Figures 6 and 7). At higher bitrates though, method 1 dominated. However, for wireless communications applications, there is obvious preference for the maximum acceptable compression achieved by method 3.

7. CONCLUDING REMARKS

The Chapter has focused on the use of diagnostically-guided video compression. Based on clinical practice, video objects of interest were defined, and novel AM-FM segmentation methods were developed for extracting the objects. For each object, we examined quality (or SNR) scalability through rate-distortion curves.

It is important to note the strong differences between the video objects in both coding requirements and their relevance to diagnosis. They range from text objects where only the clock text is changing, to near-field ultrasound video that is of no diagnostic value, to the main M-mode ultrasound video that includes the cardiac walls. It is also important to note that the near-field video, and the region beyond the epicardium would require high bandwidth for quality transmission. Yet, they are of no diagnostic interest.

The different video objects require dramatically different types of Codecs. For the text objects, a text-based encoder that will first convert the text video objects into ascii followed by an alphabet based compression scheme such as LZW will produce optimal encoding. Furthermore, once the text has been converted to ascii, critical video content information associated with the acquisition can be extracted and used for digital libraries or other applications. For the m-mode video images, little bandwidth needs to be allocated to the near field or to the regions beyond the epicardium. In this case, most of the bandwidth will be allocated for tracking the cardiac wall motion. The coding gain is directly proportional to the number of pixels of the non-diagnostic objects divided by the total number of pixels.

The difficult problem of developing quantitative measures that will be in agreement with perceptual experiments requires further investigation. However, over the region of diagnostic interest, compressed videos with mean intensity drop of less than 5% from the peak intensity value were found to be indistinguishable from the original (uncompressed) videos.

This Chapter has focused on M-mode ultrasound. Further research is needed for establishing diagnostic measures and video objects of diagnostic interest for 2D-mode and the Doppler modes. It is also important to consider automated systems that can simultaneously, optimally compress video from combinations of different modes.

8. REFERENCES

Chiu, E., Vaisey, J., and Atkins, M. S., 2001, Wavelet-based space-frequency compression of ultrasound images, *IEEE Transactions on Information Technology in Biomedicine.*, **5**:300-310.

Ehler, D., Vacek, J. L., Gowda, M., and Powers, K. B., 2000, Transition to an all-digital echocardiograpgy laboratory: a large, multi-site private cardiology practice experience, " *Journal of the American Society of Echocardiography*, **13**:1109-1116.

Garcia, M. J., Thomas, J. D., Greenberg, N., Sandelski, J., Herrera, C., Mudd, C., Wicks, J., Spencer, K., Neummann, A., Sankpal, B., and Soble, J., 2001, Comparison of mpeg-1 digital videotape with digitized svhs videotape for quantitative echocardiographic measurements, *Journal of the American Society of Echocardiography*, **14**:114-121.

ISO/IEC., 1999, IS 14496-X:Information technology-coding of audio-visual objects (MPEG-4).

ISO/IEC., 1999, IS 14496-1:Information technology-coding of audio-visual objects-part 1: systems (MPEG-4 systems).

ISO/IEC., 1999, IS 14496-2:Information technology-coding of audio-visual objects-part 2: visual (MPEG-4 video).

Havlicek, J. P., 1996, AM-FM Image Models. Ph.D. diss., The University of Texas at Austin.

Karson, T. H., Zepp, R. C., Chandra, S., Morehead, A., and Thomas, J. D., 1996, Digital storage of echocardiograms offers superior image quality to analog storage, even with 20:1 digital compression: results of the digital echo record access study, *Journal of the American Society of Echocardiography*, **9**:769-778.

Lau, C., Cabral, J. E., R. Jr, A. H., and Kim, Y., 2000, MPEG-4 Coding of Ultrasound Sequences, *Proceedings of SPIE Medical Imaging 2000*, **3976**:573-579.

Main, M. L., Foltz, D., Firstenberg, M. S., Bobinsky, E., Bailey, D., Frantz, B., Pleva, D., Baldizzi, M., Meyers, D. P., Jones, K., Spence, M. C., Freeman, K., Morehead, A., and Thomas, J. D., 2000, Real-time transmission of full-motion echocardiography over a high-speed data network: impact of data rate and network quality of service, *Journal of the American Society of Echocardiography*, **13**:764-770.

MPEG4IP, 2003; (2003) http://mpeg4ip.sourceforge.net/.

Pattichis, M. S., 1998, AM-FM Transforms with Applications, Ph.D. diss., The University of Texas at Austin.

Pattichis, M. S., Cai, S., Pattichis, C. S., and Abdallah, R., 2005, An overview of digital video processing algorithms, chapter in M-Health: Emerging Mobile Health Systems, R.H. Istepanian, S. Laxminarayan, and C.S. Pattichis, eds., Kluwer Academic/Plenum, New York.

Rodriguez, P. V. and Pattichis, M.S., 2002, Real-time AM-FM Analysis of Ultrasound Video, invited in the 45th IEEE Midwest Symposium on Circuits and Systems 2002, Tulsa, Oklahoma, August 2002.

Soble, J. S., Yurow, G., Brar, R., Stamos, T., Neumann, A., Garcia, M., Stoddard, M. F., Cherian, P. K., Bhamb, B., and Thomas, J. D., 1998, Comparison of MPEG digital video with super VHS tape for diagnostic echocardiographic readings, *Journal of the American Society of Echocardiography*, **11**:819-825.

Snider, A. R., Serwer, G. A., and Ritter, S. B., 1997, *Echocardiography in Pediatric Heart Disease*. Second edition, St. Louis, Missouri, Mosby-Year.

Thomas, J. D., Greenberg, N. L., and Garcia, M. J., Digital Echocardiography, 2002: Now is the time, *Journal of the American Society of Echocardiography*, **15**:831-838.

FUTURE CHALLENGES AND RECOMMENDATIONS

Pierre Vieyres[*], Gérard Poisson, George Triantafyllidis, Marios S. Pattichis, and Georgios Kontaxakis

A technological revolution in health care is underway, triggered by advances in information technology (IT) including modern telecommunications, information processing capability and miniaturised health diagnostic technology. This powerful combination permits advances in high-bandwidth applications for clinic and patient-based telemedicine techniques and low-bandwidth applications for the mass market, which will allow the individual citizen to take greater responsibility for his or her own health. Substantial new markets are opening up, for example, in:

- personal sensor technology, for integration into fixed and mobile consumer products (e.g. home computers, personal digital assistants),
- communications infrastructure, used in health maintenance and disease prevention,
- centralised diagnostic services, virtually connecting the patient and the service provider reducing the need for travel
- universal medical databases, available to the service provider and patient alike.

The above briefly and clearly illustrate the importance in this technological revolution, often called "telemedicine", in science, technology, market development, health and patient care. The recent advances in IT allow today the transmission of images, video, voice and data from remote locations to a site of expertise. Today the world experiences the birth of a new, revolutionary discipline in medicine, or maybe more accurately the vertical upgrade of all medical disciplines at a higher technological level.

This chapter presents future challenges and recommendations, describing briefly the forthcoming technologies, road blocks, and envisioning the advances of the area covered in the section on echography systems and services. Emerging medical image and video processing technologies are firstly approached. Their development is not only foreseen as a mere isolated research theme but has to be included in a wider spectrum of telemedical

---

[*] Pierre Vieyres , Laboratoire Vision et Robotique, Université d'Orléans, 63 Ave de Lattre de Tassigny, 18020 Bourges Cedex – France. Pierre.Vieyres@bourges.univ-orleans.fr.

activities. Therefore, this is followed by challenges and future benefits of the expanding field of remotely operated robotic systems applied to the medical world and more specifically to tele-echography systems.

From the structural point of view of a telemedicine chain, challenges in images compression and dedicated tele-operated robotics will be involved both at the "expert" and the "patient" stations. The expert station receives and analyses image data, and controls remote systems (i.e. webcam, robot) in the distant environment. At the patient station, medical and electronic data are retrieved using the available tool (i.e. ultrasound devices, webcam, robot) and are sent to the expert center using an appropriate communication link. The research perspectives on echography systems and services will have to take into account this available communication bandwidth to transfer good quality images and reliable control data between the two stations.

Currently, video compression bandwidth is primarily allocated to track changes between video frames. Clearly, dynamic images as opposed to still images will require significant bandwidth for effective transmission. Yet, for medical image compression, bandwidth allocation should be distributed according to diagnostic significance, not just for tracking changes. Emerging wireless video object technology allows us to segment video images into "objects" of interest. Yet, defining objects of diagnostic significance remains an open area of research. In addition, it is important to note that pre-processing of noisy video images, such as ultrasound images, can result in smooth video images that can again yield significant reductions in bandwidth.

Again, diagnostic measures should be developed to help assess the effectiveness of any pre-processing method. To compute and track video objects of diagnostic significance, there is a need for automated segmentation methods. In medical imaging, the segmentation problem is further complicated by partial-volume artifacts. These artifacts may be effectively addressed through video-object (2D + time) segmentation. However, the validation of video-object segmentation methods can be very expensive since ground truth may require the work of multiple graders over several frames. For the validation task, the challenge is to develop semi-automated segmentation methods that allow the graders to grade several images over a relatively short period of time.

Further research can be conducted to develop more efficient coding techniques, which effectively remove the inherent dependencies of the ultrasound images, to enhance the resulting image quality. Although an extended survey on different scenarios for the ultrasound image communication according to the available bandwidth has been carried out, the measure used for the evaluation of image quality seems to be inappropriate for the diagnostic usability of the whole system. Future subjective evaluation by medical experts is necessary to provide a full assessment of the proposed communication and visualization system.

Also, telemedicine applications such as robotised tele-echography implies the development of advanced user interface, i.e. the technological mediation that has to be transparent so the system user will feel completely immersed in the remote environment and be able to interact with it as though actually there. Several research groups are currently investigating transparent telepresence for this technological mediation. Medicine relies on a strong human interaction, i.e. medical specialists with patients, and remote telemedicine applications have to take this specification into account for the development of the user interface. This will help in maintaining a close relationship between patients and doctors, in enhancing applications of new concepts and making them accepted by the general public and the medical specialists. More specifically, regarding the visualization system of robotised tele-echography, the definition of an accurate 3D dedicated human model is an open research field. Techniques involving the initialization (i.e. "sculpturing") of a generic 3D model will be investigated. They should allow the specialist, immersed in a virtual environment with a virtual patient (representing the remote one), to improve the accuracy of his/her medical gestures on the distant real patient.

Although most tele-echography applications and services rely on the mere transmission of data, i.e. ultrasound images, using appropriate bandwidths, robots are becoming an essential tool to support and to extend the medical gestures on remote locations where there is a lack of medical experts. Even though the medical robotic concept is well accepted, the full achievements of most remote medical applications lies many decades in the future; however some of them, such as robotised tele-echography could soon be available. Their implantation in medical facilities will be eased thanks to the non invasive specifications of ultrasound techniques and its acceptance by the general public. Today, the utilization of robots in telemedicine applications is in a preliminary phase. Through several projects and within the framework of many medical pilot phases, benefits of telemedicine technical applications for isolated places, either by geographical considerations, medical issues or in the case of catastrophic situations, have been clearly demonstrated. Efficient communication, i.e. links with good quality of service (QoS), for digital medical signals and ultrasound images in particular, is the principal vector of successful distant medical diagnosis.

Dedicated robots used in the existing telemedicine loop opens new application horizons that are currently inaccessible by the simple transmission of image data. Robotized teleoperated echography constitutes a characteristic example of this application extension in the telemedicine field : ultrasound echography technique, due to its expert-dependent specification, defines the robot as a prime element in the teleoperating chain. However, the success of the robot integration will be effective only if it presents an advanced performance level. The dedicated medical robot, moreso than the industrial robot, thus awaits many improvements in hardware products (miniaturized components, lighter and more resistant composite materials, and the reliability of mechatronics elements). Accurate controls of the distant robotized gestures will be necessarily related to the performance levels of the chosen robot controls. The development of new control methods, dedicated to communication lines with delay, and predictive control techniques, constitute without any doubt, future research paths for the improvement of the remote medical gesture path following.

Moreover, even though from a technical point of view, many robotic telemedicine applications are about to be marketable, psychological aspects of the medical actors, i.e. doctors and patients, still remain to be addressed accordingly. Within this approach, it is necessary to evaluate what the real needs are for the patient, his benefits and the overall cost benefits for the community. At this point in time, as soon as scientists and medical staff will overcome together technological road blocks and psychological barriers, the robot should become an accepted and effective link in the medical field for better healthcare management overall. Robotic telemedicine applications require input from a large spectrum of technologies and disciplines such as mechatronics, telecommunications, image processing, medicine and ergonomics. Achievement of these goals will be a strong benefit to numerous applications ranging from places with reduced medical facilities and isolated places to remote hazardous environments.

Significant progress for telemedicine and applied robotics to telemedicine with the emerging technologies of today are possible in the next decade. However, the achievements of theses goals will be possible thanks to interdiscipline and transverse activities that will have to be accepted and also recognized by the whole scientific community as a real applied research objective.

All above recommendations regarding technology and future research should be also placed in the proper socio-economic framework. Echography systems traditionally represent the low-cost and easy-to-use basic infrastructure for medical health centres, which make them appropriate and within reach practically throughout the world. Further developments in this field therefore should maintain the profile of ubiquitous presence, without departing from one of the main objectives of telemedicine, which has always been the provision of quality health care to underserved, remote region or areas with minimum economic and financial resources.

For countries with limited medical expertise and resources, telecommunications services have the potential to provide a solution to some of these problems. Tele-echography has the potential to improve both the quality of and access to health care regardless of geography. Health-care professionals can be more efficient. Tele-echography offers solutions for emergency medical assistance, long-distance consultation, administration and logistics, supervision and quality assurance, and education and training for health-care professionals and providers.

Population growth and the emergence of new health problems are increasing the demand for health services and for more expensive treatments. However, increasing demand and rising health-care costs have not been matched by the funding of the health sector in most developing countries. Health infrastructure — buildings and equipment, medical staff, drugs and vehicles — is central to good health care and requires high investments. Health services must also be integrated, cost-effective and available to the people who need them. The adoption of sound policies and strategic plans that guarantee the provision of high quality, sustained and integrated health services to the population are challenges faced by most governments of developing countries today. To meet this challenge, governments and private health-care providers must make use of existing resources and the benefits of modern technology.

ACKNOWLEDGEMENT

The section editor would like to address a special thanks to all authors who, thanks to their respective expertise and enthusiasm, contributed fruitfully to the successful writing of the "Echography Systems and Services" section.

VI. REMOTE AND HOME MONITORING

Chris D. Nugent

Section Editor

SECTION OVERVIEW

Chris D. Nugent[*] and Dewar D. Finlay

INTRODUCTION

In this section the role of wireless and mobile communication technologies supporting healthcare management is exemplified. All cases illustrate how the convergence of modern technology with healthcare practices can have the resulting intended effect of improved patient care. Three of the chapters describe the role of mobile communication technologies in situations where conventional healthcare practice is not possible namely the management of disaster situations and remote and extreme environments. The remaining two chapters focus on specific medical modalities and the role that mobile communication technology can offer in improving the healthcare and patient management process.

It is evident that natural or non-natural disasters can have devastating effects at a social level. Disaster management protocols are designed to address the havoc of the situation and facilitate regaining control as quickly as is possible. Many rescue personnel are involved in such situations and co-ordination of their activities both at the site, centrally and nationally are key to the overall success of the operation. As identified by Widya et al., one of the major challenges in disaster management is addressing the management of the process through improved usage of Information and Communication Technology (ICT). Widya et al. have presented adaptable models with the ability to 'virtually' separate the roles of those involved in disaster situations and assign them to places of need without physically moving them and hence offering immediate support. Doarn et al. in their chapter also recognise the importance of ICT in disaster management and stress the necessity of prior preparation as the key to management of a real disaster. Coupled with the design of the communications infrastructure, Doarn et al. present the need and detail of effective planning and the education and training of such plans.

---

[*]Chris D Nugent, University of Ulster, Jordanstown, Northern Ireland, BT37 0QB. Email: cd.nugent@ulster.ac.uk.

∈

Effective usage of telemedicine in disaster situations is presented in the form of a number of case studies.

Remote situations can be characterised by separation distances in excess of hundreds or even thousands of kilometers. In Nicogossian *et al.*'s chapter the need to effectively manage healthcare delivery in remote, extreme and hostile environments such as Mt. Everest and the Antarctic is presented. In this study it is shown how communication technologies can provide effective healthcare management and successful support where otherwise no support, or significantly delayed support, would be evident.

On a more specific application level the chapters by del Pozo *et al.* and Nugent *et al.* show how the effective integration of mobile communication technologies can be used in chronic disease management and medication management respectively. In these it is shown how 'off the shelf' mobile technologies can be adapted through an underpinning, adaptable and interoperable Internet based communications infrastructure to provide management of healthcare and assessment by medical professionals. Key to these approaches is the mobile aspect to both the patient and the medical professional solutions and their entry to the ICT based healthcare model through their convenience and choice of technological solution.

In all of the following chapters, the role of ICT and mobile communications has been identified as the potential key to improvement upon current processes. It is shown how that as the technology has evolved over the last century, so to has its role in healthcare and with this, improved processes and remote support which can be offered.

CIVILIAN TELEMEDICINE IN REMOTE AND EXTREME ENVIRONMENTS

Arnauld E. Nicogossian[*], Desmond J. Lugg, and Charles R. Doarn

1. INTRODUCTION

Remote and extreme environments are characterized as inhospitable places on Earth or in space where humans require a variety of resources to support their needs. For millennia, humans have ventured into places on earth that have challenged their very existence. Today these environments include Mt. Everest, Antarctic research stations, the jungles of the Amazon, or a space-based platform in low earth orbit. These environments present difficult problems when it comes to providing medical support. The advent of modern telecommunications has provided the foundation for individuals in these environments to be linked to an area of greater concentration of expertise that may not be resident with them. The integration of telecommunications and information systems in the practice of medicine referred to as telemedicine, telehealth, or e-health has been ongoing for nearly a century. The use of the radio was integral to support for research expeditions on winter stays on the Antarctic continent from 1911. The use of telecommunications during the early days of space flight provided the necessary information to ascertain the health status of the astronauts. As the 20th Century came to a close, telecommunications and the ability to be connected has greatly enhanced the ability to address medical care in remote and extreme environments. Many of the technologies used in supporting such medical response have been evaluated and validated in test beds, demonstrations, or real-life experiences. The following will highlights the many challenges of conducting telemedicine in such environments.

2. BACKGROUND

There are many places around the world that are considered extreme, remote, and hostile. Geography, seasonal variations, environment, and accessibility are the key

---

[*]Arnauld E Nicogossian, Distinguished Research Professor, School of Public Policy, George Mason University 440 University Drive, MS 3C6, Finley Building, V: (703) 993-8217, F: (703)993-2284, E-mail: anicogos@gmu.edu

characteristics which best describe such places. For instance, the continent of Antarctica occupies a huge surface of the Earth. A continent almost totally covered by snow and ice, Antarctica is the one of the most isolated places known to human kind. Although our ability to travel to and from established scientific stations on the continent is improving, access to it still remains a challenge. Nothing more illustrates this than the inability to rescue individuals in need of medical attention at the time a rescue is needed. Often there is not a feasible way to get to the continent because of weather. Arrival by ships in the winter can prove difficult due to frozen seas. In addition, transportation by air is a challenge. Air transportation from Christ Church, New Zealand to Antarctica is nearly eight hours and is often not available because of weather.

During the age of discovery, (the 15th to 18th centuries) Europeans circumnavigated the globe. Principally this was done with the idea of commerce conquest first and knowledge second; however, it characterized human kind's hunger for adventure, riches and exploration. (Roberts 1993) The same can be said about Livingston's exploration of the interior of Africa in the mid 19th century and humanities quest to reach the stars. In the 20th Century, nearly all the landmasses and oceans of the world had been explored and charted. With the coming of faster and reliable communications and mechanisms for traveling, people began to explore remote and isolated places on the Earth and in space. These environments provide sizeable challenges when as serious need to medical care arises.

In the late 1700's, Captain James Cook's voyage aboard the Endeavor or even Ferdinand Magellan's circumnavigation of the globe in 1524 required a surgeon be on board the ship to serve as the doctor. In addition, the ships often had to stop along the way to replenish stores with supplies, including medicine when available. Today, such bold adventures take individuals to new depths and new heights. This of course includes individuals who wish to ascend Mt. Everest, explore the jungles, or sail the ocean single handedly. Just as it was five centuries ago, one has to be prepared for living in extreme and remote conditions.

With the advent of telecommunications, principally the radio, isolated populations were no longer isolated. In fact the concepts of telemedicine began in earnest from Australian Research Stations in the Antarctic in the 1911.

During the past 100 years, different means of telecommunications have been integrated into the delivery of healthcare. This approach has become known as telemedicine. It has proven to be a unique tool in supporting healthcare needs in such areas. The barriers of distance, geography and time are no longer insurmountable as they were perhaps only 10 years ago.

As the validation of the capabilities and successes of the integration of telecommunications and information systems into healthcare continues, it has become apparent that the most remote and extreme environment may in fact be the patient's home.

3. TELECOMMUNICATIONS NEEDS

With the introduction of telecommunications in the mid to late 18th century and its initial distribution to limited areas, more populations that once were isolated were no longer out of reach. Although telecommunications distribution up through the 20th century continued to provide increased access, principally because of the Internet, wireless communications and a constellation of ever-capable satellites, there were still large areas where telecommunications were limited and remain so in the early 21st century. This has been attributed to many factors including technology availability, technology capability, financial – a means to support the acquisition, construction, sustainability, and in many cases the sheer differences in geography.

The 21st century technological breakthroughs will produce intelligent interfaces and make these once insurmountable barriers transparent. The introduction and declining cost of wireless telecommunications is providing capabilities for many areas where traditional terrestrial infrastructures are prohibitive. For example, digital wireless service is much more available in developing countries in Africa than legacy phone lines strung on poles for hundreds of kilometers in places like the U.S. and Europe. Travel through any city in Russia, Eastern Europe or Africa reveals far more cell phones usage than say in New York, London, or Madrid. In addition, telecommunication services that are available in some places are so severely degraded, that it is often not usable. Figure 1 illustrates Internet and telephone wiring at a hospital in Prishtina, Kosova.

Telecommunications technologies are a critical tool in providing access to information throughout the world. This is very apparent in the U.S., where there are large areas of the country that might be hard to reach by road or rail, yet they can gain access to the Internet through an 'Information Superhighway'. Edwin Parker wrote in a report to the TVA Center for Rural Studies, University of Kentucky in May 1996 that increased access would have tremendous impact on commerce and economies of rural areas. (www.vtac.org/Tutorials/rural2.htm) The same can be said about healthcare in these rural areas. Table 1 illustrates the kinds of telecommunications systems available today.

In rural regions of Alaska in the 1970's, the National Aeronautics and Space Administration (NASA) provided simple telecommunications support to remote villages. This system was used in part to support health. (Doarn 1998, Nicogossian 2001) Such efforts have been made in several instances throughout the world in the past 100 years. Telecommunications is the "glue" that binds the various technological elements together to form integrated systems that perform required functions including healthcare. If telecommunications can be used to support health delivery requirements then effective planning, effective training and implementation will be required.

Figure 1. Typical wiring infrastructure at a hospital in Kosova.

Table 1. Types of data communications available in the late 20th Century.

Modality	Primary Use	Issues
Copper	The most common medium for existing telecommunications	Legacy system, limited bandwidth, subject to signal degradation as distance increases. Supports Digital Subscriber Line (DSL). Shielded Coax supports cable modem
Fiber	Communication backbones, long-haul circuitry	Large bandwidth capacity, less subject to interference, expensive
Terrestrial Broadband Wireless	Microwave, laser	Primarily used for short to medium range point-to-point connections such as between cellular towers, buildings, etc. requiring line-of-sight
Cellular		
• ∈AMPS/CDPD	Analog cellular with data capabilities overlay	Original cellular services being supplanted by CDMA, TDMA and others
• ∈PCS/CDMA	Most prevalent in the U.S. with less than 100 million subscribers	Newer technology than GSM and also incompatible
• ∈GSM/TDMA	Most prevalent in Europe with ~450 million subscribers	Embraced by European community in early 1990's and incompatible with CDMA
Wireless Ethernet	Wireless LANs	Broadband capability requiring access points to connect to a fixed network
Bluetooth, Home RF	Short range networks, personal area networks	Limited range but provides peer-to-peer ad hoc communications
Geosynchronous Satellite	Broadband or mobile (maritime) communications	Requires special receivers and antennas depending on the type of service, data requirements and carrier
Low Earth Orbit Satellite	Voice, Paging	Utility and usage anywhere on the globe. External antenna required – Low data rates
High Frequency/ Short Wave Radio	Amateur voice communications	Low range. Can support packet data at very low rates.

4. EFFECTIVE PLANNING

In the 21st Century one would give no thought at all to traveling to some distant location without having some kind of telecommunications system available. From the simplest form of a telephone at the destination, a cell phone in you personal items or a more complex system such that it is fully independent of what ever services may or may not be available. For example, during a cross-country trip in most regions of the world by train or car one would have access to phone service. However, if one was going to the Outback of Australia, to the Himalayan Highlands or the remote jungle of Central Africa, telecommunications may not be available once you arrived at the destination. In addition, there are different kinds of technology as well as different standards that may make the phone you have inoperable. To illustrate, the cell phone network used in the U.S. is different from that used in Europe.

In planning a mission to a remote and extreme environment, one must have an effective, well thought out plan for telecommunications support. If healthcare is a requirement either for the population in the region or those traveling to a remote region, there must be a telecommunications system to support the needs. There are many different kinds of telecommunication systems and technologies that can be brought to bear. The choices are not as simple has buying technology off the shelf and implementing it.

Prior to the implementation of any communication system to support health, a needs assessment must be conducted. This can be as simple as reviewing the level of technology is available and what the cost is. It must also include a review of what other infrastructure is available to support the telecommunications capabilities, such as electrical power.

There are several approaches to providing telecommunications support in remote regions (see table 1). To illustrate, consider the use of a portable satellite phone in a remote village in Kenya. In the summer of 2002, NASA's Commercial Space Center for Medical Informatics and Technology Applications Consortium (MITAC) at Virginia Commonwealth University in Richmond, Virginia, U.S.A. (a center of excellence in telemedicine) provided a telemedicine capability to physicians and students who were supporting humanitarian assistance in the area of infectious disease in remote regions of Kenya. This team was linked back to Richmond via a Thrane and Thrane International Maritime Satellite (INMARSAT) satellite phone. This unit required battery power for operation. Through effective planning, a power system was designed and deployed using a 3-foot by 4-foot deployable solar panel. This capability provided an opportunity to use the Sun's energy to charge the car battery, which in turn was used to charge the satellite phone and other devices.

Such planning leads to a state of readiness. Existing, reliable and robust telecommunications systems can provide a strong foundation for response. During NASA's Spacebridge to Armenia, a telecommunication system was deployed to remote, devastated areas of Armenia to support health recovery. This operational system was easily and rapidly deployed to central Russia in response to a natural disaster. (Houtchens 1991) This was accomplished through effective planning and state of readiness.

5. EXTREME ENVIRONMENTS

To illustrate how telemedicine and telehealth have been deployed and used in remote and extreme environment, the following activities are presented. In each example, telecommunications and effective planning played and continue to play a key role in supporting telemedicine in remote and extreme environments.

5.1 Mt. Everest

Figure 2 illustrates the remoteness and austere conditions at Mt. Everest. This image is of the Khumba Ice Fall. Everest Base Camp is at 17,500 feet and serves as the gathering point for researchers and climbers who want to ascend the world's tallest mountain. The Himalayan Mountain Range is geographically isolated. The trek to Base Camp takes several days. In addition, once you are there or have gone even further toward the summit, it becomes more difficult to return rapidly. This is especially important if one is in need of medical care.

In 1998 and 1999, researchers at NASA's MITAC located at Yale University developed a program to evaluate telemedicine and wireless physiological monitoring of climbers on Everest. In order to support telemedicine needs, the team established a robust telecommunications network. This network combined long-distance and short-range wireless communications technologies to provide seamless data transfer abilities regardless of location and infrastructure. A highly portable and rugged battery-powered 900mHz radio network for local area connectivity was used to permit technicians at Base Camp to receive and transmit data from explorers in real-time over several miles of inhospitable and undeveloped terrain.

Figure 2. Kumba ice fall on Mt Everest.

The data from Base Camp was regularly forwarded with real-time and store-and-forward protocols over INMARSAT satellite communications systems. The INMARSAT network allowed 64 kilobits per second (Kbps) of Internet Protocol (IP) and Integrated Switched Digital Network (ISDN) connections from any number of commercially available field terminals. In this work, a Thrane & Thrane M4 INMARSAT satellite phone was used. At Base Camp, MITAC employed a custom-built computer system to connect two 64Kbps channels, which provided 128Kbps ISDN to support video-teleconferencing. This increased bandwidth made possible better video and audio transmission in real-time, which was used on a daily basis as the Everest physicians presented medical rounds to monitoring physicians at Yale University.

In addition, this system was utilized to provide real-time tracking of physiological parameters of climbers as well as real-time interaction to support medical intervention. On several occasion such a system was used to support clinical decision-making. In one instance, a climber was suffering from severe eye pain. A MITAC researcher, who had been trained in ophthalmology, was able to transmit images of the eye to consulting ophthalmologists at Yale for diagnosis and recommended treatment. This resulted in the climber being sent down in elevation so that the environmental conditions of climbing higher in elevation would not do further injury to the climber. (Angood 2001, Satava 2001, Harnett 2001)

Clearly the integration of a robust telecommunications system in such an environment as Everest provided a strong, reliable telemedicine link to an established medical clinic for guidance and management of health issues.

5.2 The Jungle

Medical care relies on the management and movement of information to provide a high level of patient care. In less developed regions of the world many of the modern communication systems that are routinely used to convey information are nonexistent or unreliable. In many areas, even the plain old telephone (POTS) landline may not be available. In remote jungle areas, such as central Ecuador, these very challenges are further exacerbated.

For many decades the only form of communication between remote jungle villages was high frequency (HF) short wave radio. This capability has been the main stay of voice and text messaging to remote areas and the high seas since early in the twentieth century. The transmission of text information has improved greatly with the integration of computer technology and modern digital modulation techniques.

NASA's MITAC and others have explored, evaluated and validated various telecommunications tools for use in telemedicine to remote villages of the Andean Mountains and the Amazon Rainforest. A remote hospital located on the edge of the Amazon rainforest needed a simple but reliable telecommunications link back to the regional capital city. The medical team in the jungle wanted to send preoperative, postoperative reports, and medical records to surgical teams that regularly visit the remote rainforest hospital. A standard POTS telephone is located in the hospital, however it is not uncommon to find it out of service for extended periods of time. However, the hospital had an HF radio transceiver for voice communications throughout the region. The HF radio was used to transfer patient medical information to the regional medical clinic and the visiting surgical teams. (Praba-Egge 2003)

The existing 100-watt HF radio transceiver, a radio modem and simple wire antenna were used to communicate from the rural hospital to the regional medical center. Data was transmitted using a modern radio frequency modem system on the same voice frequency that already was in use in the 31-meter band. A 12Kb file consisting of ten medical records was successfully sent error free in approximately 15 minutes. The data was received by an almost identical radio system located in the regional capital's medical clinic. The records were then transferred via the Internet to the Fundacion Cinterandes in Cuenca, Ecuador for review.

The communications equipment utilized had a relatively low cost (< $1,000 USD). It consisted of a commercial ICOM HF radio transceiver capable of 100 watts of output power, a HAL DSP 38 radio data modem and a B&W broadband dipole wire antenna. The radio equipment and antennas were installed in approximately 4 hours at each location. Both systems functioned as planned with the first communications attempt. The system continues to be in operation after two years of use.

Similarly communications using a satellite phone provided an opportunity for preoperative planning. American surgeons, visiting Ecuador, collaborate with the Fundacion Cinterandes in conducting surgical clinics in remote regions of Ecuador. This is often accomplished using a mobile facility (Figure 3). A telemedicine capability was introduced into this facility to permit interaction in real-time between medical personnel in the truck and medical expertise located elsewhere. This system has permitted the transfer of medical images, live anesthesia monitoring, and interaction between surgeons on three different continents. All accomplished using low bandwidth, wireless telecommunications. (Doarn 2002, Otake 1998, Rosser 1999, Rosser 2000)

Figure 3. Mobile surgical facility used in Ecuador.

5.3 Antarctica

Sir Douglas Mawson's Australasian Antarctic Expedition (AAE) from 1911-14 pioneered the use of wireless telecommunications in Antarctica. Although less than fully reliable as illustrated by Mawson's summary of his second wintering year in 1913, the communications gave Antarctic physicians the potential for practicing telemedicine and consulting with colleagues outside of Antarctica. (Mawson 1914) In 1912, while Mawson was establishing wireless stations in Antarctica and on the sub-Antarctic island of Macquarie, and radio contact between Antarctica and Australia, a proposal to establish a Flying Doctor Service was being discussed in Australia. Action on this was delayed as it was realized that a doctor in an airplane would be of little use to those isolated in the vast hinterland of Australia unless they could communicate with the doctor. It was not until 1927 that the advent of Traeger's 'Pedal Radio' allowed the Royal Flying Doctor Service (RFDS) to materialize into the great system that endures today.

The establishment of the Australian National Antarctic Research Expeditions (ANARE) in 1947 with arguably the most remote groups on Earth, who are totally physically isolated during the Antarctic winter with no means of evacuation for eight to twelve months, and limitations of a sole medical practitioner, meant that communication with health professionals in Australia was extremely desirable to aid in diagnosis and distant management. The evolution of telemedicine on ANARE and its use, together with the philosophy, which has underlined such practice, has been described by Sullivan and Lugg and Lugg. (Sullivan 1995, Lugg 1999)

Initially HF radio was available to the doctor as either Morse Code telegrams or more rarely by voice. This system was frequently subject to the blackouts of polar cap absorption (PCA), which could last up to a week. Despite the slowness and difficulties of the system, many Antarctic medical practitioners were assisted by specialists in Australia. The medical archives of the Australian Antarctic division are full of rather poignant reminders of this era. Pardoe's case of a ruptured intracranial aneurysm at Mawson in 1961 highlights the value of the primitive telemedicine. (Pardoe 1965) With no previous neurosurgical experience, using some instruments manufactured on the isolated outpost, and telegrams of instruction from a neurosurgeon in Australia, Pardoe carried out a life saving craniotomy.

A further innovation in 1967 was the use of facsimile transmission as an aid to diagnosis and treatment. (Lugg 1974) The equipment, initially installed for sending scientific data, was similar to that used for the commercial transmission of pictograms by newspapers. Clinical photographs of injuries and other conditions, microscopic specimens, serial X-ray photographs of fractures, EKGs and temperature charts were sent from Antarctica. It was a laborious process involving the preparation of positive pictures, which were loaded onto a drum. This in turn was spun at high speed and transmitted over a two-way radio circuit, which needed synchronization by operators at both ends. Negatives were received in Australia, where they were then converted back to positive images. The system was the mainstay of telemedicine until its demise with the introduction of satellites in the 1980s. Then a Two Way Image Transfer System (TWITS), developed out of necessity, allowed pictures either from video or still digital cameras to be sent. Benefits of the new system were better resolution, and for the first time, color.

Radiotelephone (RADPHONE), a system designed for maritime use, allowed ANARE access to the world telephone networks from the 1970s. Only available at certain times it was subject to blackouts and had poor voice quality. In emergencies special links had to be established. Taylor and Gormly have described a number of cases that used the system to advantage. (Taylor 1997) The introduction of the INMARSAT to ANARE allowed a 24-hour/day voice link not subject to PCA. Although this was of tremendous value, it was very expensive. A newer system of leased circuits on an International Telecommunications Satellite Organization (INTELSAT) satellite, called ANARESAT gave less expensive, uninterrupted telephone and data circuits, as well as email, facsimile, and later access to the World Wide Web.

Personal computers coupled with commercially available digital cameras and scanners, added to imagery systems enhancing color, contrast, sharpness and size and compressed formats such as Joint Photographers Experimenters Group (JPEG) and Graphics Interface Format (GIF), provide doctors with a simple, user friendly, low technology telemedicine facility. This is in addition to voice discussions with consultants outside Antarctica. Good quality medical pictures of trauma, dermatology, dentistry, pathology, including microscopic views of blood cells, bacteria, malaria parasites and forensic material), X-rays, EKGs and other clinical records have been transmitted.

Associated with the pictures is a Health Register of all events occurring on ANARE; diagnosis, treatment including procedures performed and pharmaceuticals prescribed are coded and available to Head Office medical staff. (Sullivan 1991) With a database of over 3,000 scientists and support staff (over 2,000 person years) valuable and real time epidemiological research can be performed. Innovative research, including space-analogue studies in psychology and behavior along with education and continuing training are now available to the isolated medical practitioners. (Wood 2000)

Television was transmitted from ANARE stations in 1988 but because of current bandwidth restrictions, video is not available. The lack of video is not seen as a problem and as bandwidth becomes available then new systems will be introduced. The time-honored approach on ANARE has been for medical developments to be introduced as communications are able to support them; diagnostic ultrasound is one potential area for ANARE and current portable hardware and training of medical generalists in using the equipment are under constant review.

ANARE telemedicine has many unusual factors such as the highly screened population, sole medical practitioner, isolated practice, and fact costs of system have been small add-ons to the overall communication costs. From an outcome measured strictly in health terms it has been highly successful and has improved healthcare for the expeditioners. Reviewing 9,000 consultations on ANARE in a 10-year period from an annual population as described in Table 2, 400 were referred to Australia through the telemedicine system. There were no evacuations in winter and <10 in the summer months when evacuation by ship or aircraft of another nation were able to be accomplished As evacuations are costly in both logistic terms and loss of scientific program, it is important to decrease the need for evacuations; telemedicine has assisted the doctors to do this on ANARE. Currently a summer air service between Australia and Antarctica is being planned.

Table 2. Annual number of ANARE staff at four Antarctic stations

WINTER	80
ADDITIONAL SUMMER	350
TOTAL	430[a]

a Equates to 196 person years

With 18 nations operating in Antarctica there are many other telemedicine systems operating. The United States has a policy of evacuation of severe injury or illness and has had a number of highly publicized cases needing evacuation and which required pre-evacuation assessment by telemedicine. (Hyer 1999, Nielsen 2001, Ogle 2002) In a recent injury, telesurgery was used. (USA Today July 5, 2002) In 2002, a contract was signed between the University of Texas Medical Branch at Galveston (UTMB) and the logistical support contractor for the National Science Foundation in Antarctica to provide specialty medical services via telemedicine. A project conducted under the program of the Commission of the European Communities entitled ARGONAUTA provides telemedicine to Italian, German, Argentinean, and Chilean stations in Antarctica, while France and United Kingdom have active programs of their own. (Siderfin 1995) Antarctic groups as a whole have been most innovative in telemedicine over many decades and a case study has been made of these telemedicine operations with potential application to other extreme environments. (Sobel 2000)

5.4 Space Exploration

Since its inception, the space exploration program has relied heavily on telecommunications to gather medical and engineering information from spacecraft and their crewmembers to ascertain the progress of the mission and ensure successful outcomes. The information obtained can be placed into three broad categories: human health, system function and status of the environmental situation within and outside the habitable confines. (Doarn 1998, Nicogossian 2001) All parameters are displayed in an Earth-based facility called the Mission Control Center. Some of the parameters are also displayed to the space crews for navigational purposes. All parameters are monitored very closely for deviations and corrections either automatic or manual are entered periodically to maintain the reliability and safety of the mission. Both the American Space Agency NASA and the Aviation and Space Agency, for Russian Federation (and former Soviet Union), developed similar capabilities and approaches in telemedicine.

NASA's early space missions, including the exploration of the Moon, continuously monitored electrocardiography, respiration rate, and body temperature during all phases of flight and all activities including those external to the spacecraft. Additional physiological parameters were also monitored, including the space suit temperatures, computed metabolic rates, blood pressure, food and water intake, urine production and any medical symptoms that were periodically verbally reported by the crewmember(s).

With the advent of the Space Shuttle Program and the Soviet/Russian Space Stations Salyut and MIR, this practice of continuous biomedical monitoring was abandoned in favor of periodic in-depth medical evaluations. This practice resulted from a better understanding of the human physiological response to space flight, availability of countermeasures and therapeutic means and medical training of space crews. Yet because of the importance of a safe habitat to crew health and function, environmental parameters such as total cabin, partial oxygen, partial carbon dioxide, carbon monoxide, water and humidity pressures, radiation levels and detection of pyrolisis products were performed on a continuous or near real-time basis. Figure 5 illustrates the model used to support space mission design. Telemedicine is an integral part of this design.

In the early space missions a modest number of approximately 100 parameters were monitored. Today as the International Space Station Alpha (ISS), in low Earth orbit, nearly 10,000 parameters are available for monitoring. Automation and informatics to understand and analyze cluster of events requiring intervention are being implemented. To this end protocols and procedures used in telemedicine are being standardized among the partners involved in ISS operation. Additional Mission Controls have been built in Europe, Canada and Japan, requiring the addition of telemedicine links for consultations and medical monitoring. End effectors such as computerized medical records, automated fluid analysis and imaging systems have been added to the physician's toolbox for better practice of telemedicine in space.

ISS is probably the most remote and isolated human outpost in the world today and presents many internationally-driven medical challenges, such as time zone differences, language and cultural barriers, which are systematically overcome and improved upon by physicians around the world dedicated to the practice of aerospace medicine. NASA accounts for almost 18 years of cumulative human space flight, mostly short duration, and over 850 individuals in space, with no mission termination because of illness, which could not be handled in space. Soviet Union/Russia accounts with about 44 years of cumulative mostly long duration missions 4 months or longer, with over 90 individuals in space, and three medical evacuations from space stations. Most of the astronauts and cosmonauts enjoy healthy aging with excellent quality of life despite the exposures to the hostile environments of space. This record speaks quite well for the proper use of telemedicine and preventive health programs in the space activities.

Human Space exploration has brought many dividends back to Earth improving our quality of life. We can safely say that telemedicine progress has been one of its many contributions.

6. SUMMARY

Integration of telemedicine into remote and extreme environments offers unique tools and services to those segments of the population. The examples cited here, are just a small sampling of the growing body of work being done around the world. They illustrate the diversity of application. Telemedicine has been validated in these environs and it has been an effective tool in the delivery of healthcare and medical education. Better management of medical issues can be achieved through a range of telecommunications capabilities.

7. REFERENCES

Angood, P. B., Satava, R., Doarn, C., Merrell, R. and the E3 Group. 2001, Telemedicine at the top of the world; the 1998-1999 Everest Extreme Expedition, *Telemed J E-Health* **6(3)**:315-325.

Cold Science. Telemedicine operation performed at South Pole. USA Today, July 5, 2002.

Doarn, C. R., Nicogossian, A. R. and Merrell, R. C., 1998, Application of telemedicine in the United States Space Program, *Telemed. J.* **4(1)**:19-30.

Doarn, C. R., Fitzgerald, S., Rodas, E., Harnett, B., Praba-Egge, A. and Merrell, R. C. 2002, Telemedicine to integrate intermittent surgical services in to primary care, *Telemed. J. E-Health* **8(1)**:131-137.

Harnett, B. M., Satava, R., Angood, P., Merriam, N. R., Doarn, C .R., and Merrell, R. C. 2001, The benefits of integrating Internet technology with standard communications for telemedicine in extreme environments, *Aviat Space Environ Med* **72(12)**:1132-1137.

Houtchens, B. A., Clemmer, T. P., Holloway, H. C., et. al. 1993, Telemedicine and international disaster response: medical consultation to Armenia and Russia via a telemedicine spacebridge, *Prehosp Disaster Med* **8(1)**:57-66.

Hyer, R. N. 1999, Telemedicine experiences at an Antarctic station. *J Telemed and Telelecare*, **5(1)**:87-9.

Lugg D. J. 1999, Telemedicine: the cornerstone of Australian Antarctic medical Practice. In: Walker J, et al, eds Proceedings of the 7th National health Informatics Conference, Hobart, 29-31 August, 1999. Melbourne, Health Informatics Society of Australia, pp. 13-17

Lugg D. J. 1974, Facsimile transmission from Antarctica as an aid to diagnosis and treatment. *Med J Aust,* **1**:472-474.

Mawson, D. 1914, Australasian Antarctic Expedition 1911-1914. *Geographical Journal*, 280

Nicogossian, A. E., Pober, D. F., and Roy, S. A. 2001, Evolution of telemedicine in the space program and Earth applications, *Telemed J E-Health* **7(1)**:1-15.

Nielsen J. and Vollers, M. 2001, Icebound. New York, NY, Hyperion.

Ogle J. W. and Dunckel,G. N. 2002, Defibrillation and thrombolysis following a myocardial infarction in Antarctica, *Aviat Space Environ Med,* **73(7)**:694-698.

Otake, L. R., Thomson, J. G., Pershing, J. A., and Merrell R. C. 1998, Telemedicine: Low bandwidth applications for intermittent health services in remote areas, Letter to *JAMA,* **280(15)**:1305-1306.

Pardoe. R. 1965, A ruptured intracranial aneurysm in Antarctica, *Med J Aust,* **1**:344-350.

Praba-Egge, A., Hummel, R. Stewart, N., Doarn, C. R., and Merrell, R. C. 2003, Remote telemedicine services by high frequency radio link. *J Clin Eng* **28(1)**:55-61.

Roberts J. M. 1993, *A Short History of the World.* Oxford University Press, New York, NY.

Rosser, J. C., Bell, R. L., Harnett, B., Rodas, E., Murayama, M., and Merrell, R. 1999, Use of mobile low-bandwidth telemedical techniques for extreme telemedicine applications, *J Am Coll Surg* **189(4)**:397-404.

Rosser, J. C., Prosst, R. L., Rodas, E. B., Rosser, L. E., Murayama, M., and Brem, H. 2000, Evaluation of the effectiveness of portable lo w-bandwidth telemedical applications for postoperative follow-up: initial results, *J Am Coll Surg,* **191(2)**:196-202.

Satava, R. M., Angood, P. B., Harnett, B., Macedonia, C. and Merrell, R. C., 2001, The physiological cipher at altitude: telemedicine and real-time monitoring of climbers on Mount Everest. *Telemed J E Health* **6(3)**:303-313.

Sullivan, P., Gormly, P. J., Lugg, D. J., and Watts, D. J., 1991, The Australian National Antarctic Research Expeditions Health Register: Three Years of Operation. In Postl B, et al eds. *Circumpolar Health 90.* Winnipeg: University of Manitoba Press: pp. 502-504.

Sullivan, P. and Lugg, D. J. 1995, Telemedicine between Australia and Antarctica; 1911-1995. *SAE Technical Paper Series 951616,* pp. 1-6.

Siderfin, C. D. 1995, Low-technology telemedicine in Antarctica, *J Telemed and Telecare* **1(1)**:54-60.

Sobel, A. L. 2000, A case study of Antarctic telemedicine operations: Potential applications to space medicine, *Aviat Space Environ Med* **71(1)**:95.

Taylor D. and P. J. Gormly, P. J. 1997, Emergency medicine in Antarctica, *J Emerg Med,* **9:**237-245.

Vermont Telecommunications Application Center web site www.vtac.org/Tutorials/rural2.htm. Visited February 15, 2003.

Wood, J. A., Hysong, S. J., Lugg, D. L., and Harm, D. L. 2000, Is it really so bad? A comparison of positive and negative experiences in Antarctic winter stations, *Environment and Behavior* **32:** 85-111.

TELEMATIC REQUIREMENTS FOR EMERGENCY AND DISASTER RESPONSE DERIVED FROM ENTERPRISE MODELS

Ing Widya[*], Pieter Vierhout, Val M. Jones, Richard Bults, Aart van Halteren, Jan Peuscher, and Dimitri Konstantas

1. INTRODUCTION

One of the prime objectives of disaster response management is to gain control of the disaster situation as rapidly as possible. Observations have shown that coordination and communication between response teams during disasters is essential for gaining control of the situation, but is often inadequate in past disasters (Auf der Heide, 1989) and often suffers from up-scaling difficulties. Examples were the September 11th attack in New York in 2001 (BBC News, 2002) and the explosion on the premises of a professional firework assembly site, which completely destroyed or severely damaged some 480 houses and some 100 small-medium sized enterprises in the town of Enschede in the Netherlands in 2000 (Oosting, 2001).

The different teams involved in disaster response not only have their own specialties, responsibility and (hierarchical) scope of authority, they frequently differ in their ways of working. The Enschede disaster response, for example, involved several fire brigades (also from neighboring areas), site controlling police teams, crisis coordination teams and medical emergency services teams, which moreover received assistance from across the border in Germany. This scenario illustrates the fact that a disaster response process has to be viewed as a complex and multifaceted distributed system involving many teams and requiring sophisticated coordination and communication facilities.

Disaster response needs to be dynamic and adaptive. It has to cope with a range of events, from near disasters to very large-scale actual disasters involving a chain of rapidly escalating accidents. It also has to cope with dynamic (rescue) processes. Previous disasters have shown, on one hand, that many off-duty rescue services staff

[*]Ing. Widya, Centre for Telematics and Information Technology, Faculty of Electrical Engineering, Mathematics and Computer Science, University of Twente, P.O. Box 217, 7500 AE Enschede, the Netherlands. E-mail: widya@ewi.utwente.nl

spontaneously volunteers at the site of the accident. The inability to coordinate the many off-duty volunteers was explicitly identified as a problem in a report on the New York September 11th disaster. To make best use of these extra resources, the volunteers need to be identified, coordinated and deployed effectively within team structures. On the other hand, communication infrastructure may be severely damaged or congested at the scene of the disaster. However, due to the advancement of Information and Communication Technology (ICT) several communication alternatives may be available, e.g. GSM, UMTS, terrestrial and satellite links, and including ad-hoc or rapidly deployable networks (Midkiffs and Bostian, 2002). These (rapidly deployable) communication infrastructures and alternatives are valuable in disaster response (Doarn et al., 2004).

A major challenge in medical emergency services and disaster response is how to improve the management process using ICT. For example, how to select dynamically the computing and network resource alternatives (which possibly are scarcely available at a scene of a disaster) that match with the coordination and communication needs of the medical emergency or disaster response teams.

This chapter explores the potential of mobile telematic services to support medical emergency and disaster response, in particular it addresses the requirements of the communication support for medical emergency or disaster response teams.

Requirements of telematic services for disaster response have been investigated for example in Meissner (2002). In this chapter, we derive the requirements using Enterprise Models. We apply a primarily top-down approach using the notion of viewpoints introduced in the ODP (Open Distributed Processing) framework (Blair and Stefani, 1998). We start with Enterprise models for their emphasis on justification purposes. Particularly, we describe the models using generic organizational notions like tasks, roles and teams which are responsible for the tasks, and agents who act in certain roles to perform the tasks. These Enterprise models are inspired by the educational models describe in Widya (2002) and the model for healthcare processes described in Luzi (1997). This healthcare model however defines a guideline-oriented conceptual model of healthcare units for workflow formulations of healthcare processes and healthcare process reengineering. Instead of addressing the details of tasks and inter task relations relevant for workflow and process reengineering, we focus on the (refined) modeling of teams and agents. This focus meets our objective to derive the requirements of telematic services that match the communication needs in disaster response. However, readers interested in disaster management information systems and tools may consult FEMA (2001).

Furthermore, the Enterprise models have been specified in a way to capture the invariants of healthcare and disaster response processes and for this reason it can be viewed as a generic meta-model in the context of object-oriented technology (Holz et al., 2001). We therefore believe that our model can be specialized for different disaster specialties. We validate our Enterprise models by applying them to derive the requirements for emergency services of the tele trauma team trial of the European project MobiHealth (Jones et al., 2004).

The Enterprise models proposed also separate roles responsible for the tasks from the agents assigned to those roles. This separation of functional entities from the physical entities has the advantage that roles can be virtually (and immediately) moved to the scene of the disaster without the need to transport all agents to the scene. This is a simple form of an augmented reality environment (Azuma, 1997; Milgram and Kishino, 1994). It brings on-site reality at the scene of the disaster into the scope of command and control of off-site teams or team members. In emergency services, trauma specialists (i.e. agents) at care centers may retain their established way of working as much as possible if augmented reality brings the scene of accidents to their scope of monitoring and control. This property is essential for the acceptance of ICT deployment as added value services, for example if compared to the deployment of significant different ways of working.

The remainder of the chapter is organized as follows. The next section discusses the Enterprise models. For validation of the models several issues of healthcare and disaster scenarios are presented in Section 3. Section 4 discusses the requirements for the telematic support which can be derived from the models and which conforms to the scenarios. Finally, Section 5 provides the conclusion of this chapter.

2. ENTERPRISE MODEL

This section describes the Enterprise models for disaster response and healthcare processes (in particular, emergency services). We use the Unified Modeling Language (UML) (Booch et al., 2001) to express these models. First, we analyze the considered processes from an organizational enterprise perspective to achieve a generic model which can cope with a broad range of emergency or disaster events. Then, we refine the model further to reveal event specific modeling details like the dependencies between modeled entities which instantiations are distributed in the augmented reality environment, for example Body Area Networks (BANs) of patients' vital sign sensors in emergency events (Jones et al., 2004) or BANs of firefighters in disaster response (Meissner et al., 2002). In the context of the ODP framework, this means that the refined models possess properties of a Computational viewpoint model. Further elaboration of these dependencies will in turn reveal the requirements of the communication support for emergency services or disaster response (cf. requirements for ODP Engineering viewpoint model).

2.1. Task Structure

In this chapter, a healthcare process is viewed as a workflow of objective-driven tasks aimed to improve or to restore a patient's condition and, conversely, a task is an embodiment of healthcare activities (Luzi et al., 1997). Similarly, disaster response is viewed as a workflow of objective-driven tasks aimed to control disasters.

In a typical healthcare scenario, we can distinguish between diagnostic tasks and treatment tasks (Hobsley, 1979). Moreover, different diagnostic approaches can be applied in healthcare. The selected approach strongly influences the diagnostic task and the breakdown of this task. In the differential diagnosis approach, a diagnosis is labelled in terms of the pathological disease. The list of diseases matching the symptoms and signs in a previous diagnostic stage has to be narrowed-down, for example using additional tests. This may eventually lead to the cause. On the other hand, in the working diagnostic approach, a diagnosis is labelled in terms of symptoms and signs. Refinement is then sought in order to determine the subsequent activities. A sequence or a partial order of diagnostic activities refines the diagnosis towards the etiological or pathological cause. Preference for a certain approach depends amongst others on the medical specialty.

If the diagnostic hypothesis is confirmed, a treatment task may be initiated. This task may also be decomposed into several subtasks, e.g. treatment planning, execution and evaluation. These (sub-) tasks are often chronologically intertwined with the diagnostic tasks, since the validity of a diagnosis may need to be monitored continuously as this validity affects the treatment and moreover patient illness may develop in other directions.

Similar structure of tasks can be identified for disaster management. Disaster response has to be prepared in a planning phase (disaster preparedness in FEMA (2001)). Intertwined tasks across different disaster management phases have been identified by Auf der Heide (1989), since disasters are different and traditional division of activities and resources of routine emergency management are often unsuitable.

2.1.1 Hierarchical Task Structure:

An Enterprise model which can cope with hierarchically structured tasks is needed for the modelling of diagnostic processes which apply the differential approach and also for the modeling of inter- and intra-organizational disaster response processes. Moreover, if the model also can cope with partially ordered tasks, it will also be suitable for diagnostic working approaches. The need to cope with both kinds of structuring of tasks has been identified in other domains, e.g. in task-oriented educational processes (Widya et al., 2002) and enterprise models of telecommunication network architectures (Yates at al., 1997). Accordingly, we reuse those task structures to model healthcare tasks as well as scalable disaster response tasks (Figure 1). For clarity of the figures, methods and attributes of classes are not shown in the UML class diagrams.

Figure 1 shows that a task is either a leaf-task or a true compositional aggregation of tasks. The aggregated tasks are constrained by the association class compositional-constraint, which for example represents the constraints on the permitted partial order of these tasks, on the diagnostic narrowing of the differential approach or on the decomposition of the medical objective of the encapsulating task.

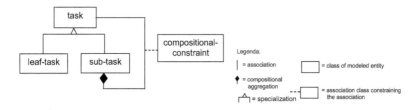

Figure 1. Healthcare and disaster response task structure

The model can cope with arbitrarily deeply nested tasks, including flatly structured tasks. The hierarchical depth of a task structure not only depends on the complexity of the task modelled, it also depends on the task context such as the urgency of the task. The following citation illustrates the need for a flat, monolithically structured task, which contains a large set of activities in a complex medical emergency scenario (Argyle, 1996):

> "In dealing with the trauma victim, the physician must treat as he or she gathers information. The approach cannot be routine "take a history, do an exam, order some tests, make a diagnosis, then treat the patient." Therapeutic interventions must be made "on the fly," before the full evaluation can be completed. .... For example, the combination of low blood pressure, unilaterally decreased breath sounds, and respiratory distress triggers a

response from the physician. A chest tube is placed immediately, rather than waiting until an x-ray can "prove" the diagnosis."

The flexibility of the model in respect of the hierarchy and the intertwining of tasks makes the model suitable for disaster management tasks. These tasks have to cope dynamically with a range of scenarios, scalable from near accidents to large disasters involving chains of accidents, for which urgency may also be essential.

2.1.2 Refined Healthcare Task Structure:

In the context of healthcare, a leaf-task embodies the work-item activities associated with the care of a patient's health, for example observation, interview, preparation, treatment or hypothesis validation activities. An activity of a task may require resources such as oxygen saturation, ECGs and blood pressure data (modelled by class resource and its components in the refined model of tasks shown in Figure 2 and Figure 4, respectively). It may also require (logistical) facilities such as an operating theatre or an intensive care unit (class facility). From the perspective of a medical specialty, a patient is a subject which receives care of the specialty at a reference point (i.e. a particular time and location). A component of leaf-task is therefore care-receiver, which has to be assigned to a patient by the association class patients'-schedule, which contains the reference point.

The model also reveals the modelling strength of UML compositional aggregations, for instance if the treatment task is an operation, the instantiation of the leaf task components guide to plan a facility (i.e. an operation theatre). On the other hand, if the association class patients' schedule is not properly implemented by corresponding data flow diagrams or medical protocols, patients scheduling will be sensitive to human assignment mistakes like authentication mismatch between a task and a scheduled patient. Therefore, the class patients' schedule needs to be dependent (in UML sense) on scheduling policies which refer to medical protocols (e.g. provide patients with identification labels, check the labels before treatment and check patients' name) or authentication protocols (this dependence is not shown in the figure).

In a firefighting scenario, resources of on-site firefighters' tasks may represent resident register information, building or site maps and facility may represent tools such as voice conference or e-messaging applications for command and control guidance from an off-site command and control center (cf. Meissner et al., 2002). Disaster response leaf-tasks do not posses a care-receiver component, but instead a deliverable component (not shown in the figure) which represents results of a task in an information system sense. For example, firefighting reports which may further be used as resources of consecutive tasks (i.e. assigned via an association class to a consecutive leaf-task). On the other hand, resources of off-site leaf-tasks in a firefighting scenario may represent live video clips of the site of the fire, processed read-outs of on-site sensors measuring gas compound emissions or casualty statistics. Without being present at the scene of the accident, the off-site assisting or controlling team members, which depends on remotely located resources, may therefore extend their scope of assistance and control onto the scene of the accident in the augmented reality environment.

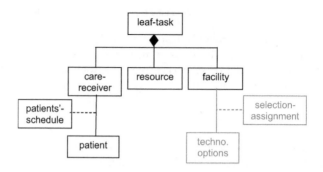

Figure 2. Leaf task structure and patient-task assignment

2.2. Task – Team Structure

As with educational tasks (Widya et al., 2002), disaster management or healthcare tasks are organized and performed by agents acting in certain roles and in these roles these agents are responsible for the tasks. In healthcare, the class role represents a medical role (e.g. an expert in anaesthesia) and eventually has to be assigned to an agent on duty (e.g. an anaesthetist on duty). In the assigned role, this agent has the responsibility to perform a certain medical task with the authorization rights associated to the role (represented by authorization in Figure 3). In disaster response, these assignments and authorization may need to be specified or updated dynamically during disaster response, due to the different conditions of disasters and pre-planned activities of tasks in disaster preparedness process may be unsuitable (Auf der Heide, 1989). One may view this issue as an overlapping of phases, the disaster preparedness and the disaster response.

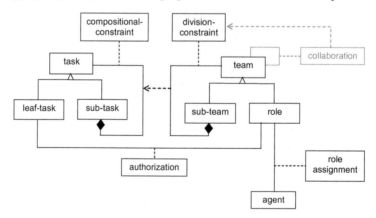

Figure 3. Task – Team structure for healthcare processes and disaster response

Since complementary roles can split responsibility for a task, especially for complex and aggregated tasks, we introduce the class team either as a role or a compositionally aggregated team. The aggregation is constrained in a multi-way manner by the association class division-constraint. In healthcare for example, division-constraint may impose how a medical assistant role relates to the role being assisted and specify e.g.

whether an assistant anaesthetist at a point of care has the right to place a chest tube in a trauma case when the off-site anaesthetist is at the hospital. In disaster management, division-constraint may impose the use of a common frequency band and radio technology to enable communication between teams of different disciplines (e.g. fire brigade, police, medical teams). Add-on guidelines or policies for coordination and communication may also be incorporated in this association class, for example the preferred set of communication tools and technology.

In large-scale accidents furthermore, several teams may perform the same task regime concurrently, for example in disasters where a number of rescue teams are deployed and where each team follow the same rescue regime and policies (the black dashed arrow in Figure 3 models this property).

2.3. Team-Team Structure

The association class collaboration (piggybacked in grey in Figure 3), which is a refined part of task-team structures, specifies how teams are related to one another. Amongst others it represents the coordination and communication aspects between teams as is imposed by the association class division-constraint (dashed arrow in the figure). For example, the provisioning of a BAN incorporating communication devices tuned on a specific frequency band or wireless technology as imposed or prioritized by division-constraint. The grey-colored UML construct also models the breakdown of the (multiparty) association class division-constraint into the simpler association class collaboration.

Due to the nested structure of teams, collaboration also represents the intra-team coordination and communication, e.g. referring the rescue protocols that should be used between an on-site agent at a point of care and an off-site agent at the control center, e.g. as illustrated earlier by the healthcare case involving the on-site assistant anaesthetist and the off-site anaesthetist. The class facility of a leaf-task (Figure 2) represents the agreed communication tool, including its configuration setting.

2.4. Task Resource and BANs

The resources of a task represent for example ECGs in emergency services or chemical emission graphs in firefighting. These resources typically depend on the context of the task, for example, the objective of the task (e.g. validate cardiac arrest or validate rescuer safety) and indirectly the role responsible for the task. Since roles are conceptual modelling entities and the agents assigned to these roles may be located remotely from the roles, assistant agents co-located with the roles have to be appointed to perform assisting activities. Accordingly, we distinguish between (functional) activities related to the selection of task resources (i.e. the selection of the vital signs (e.g. ECGs) and their quality) and the (physical) activities related to the acquisition of the required data (i.e. the selection, attachment and the calibration of the (physical) vital sign sensing devices, e.g. the selection of a high sensitive 3 leads ECG set in emergency services or the selection of chemical emission sensors in firefighting). In the healthcare context, the class media is task resource oriented and the class medium is device oriented (Figure 4).

At task resource level, the association class media-BAN-attachments models the assignments of a relevant and coherent set of medical data (i.e. aggregation of medium entities, incl. their quality) to the resource component of a task (Figure 4). This assignment has to be in line with the authorization rights of the role (dependence to class authorization represented by the dashed arrow in the figure). The class media-BAN-attachments also specify the relations between the media, for example, the priorities and the synchronization between the vital signs in emergency services or chemical emission measurement data in firefighting.

Analogously, device-BAN-attachments represent the correspondence of devices (e.g. vital sign sensors) to the (medical) data represented by the class medium. Especially relevant in healthcare, permissions to attach these devices also depend on the authorization rights of the role.

In a firefighting scenario, a two levels BAN assignments as proposed here can be used to equip a firefighter with emission sensors to enable an off-site member of the fire brigade to monitor fire exhaust for safety of the rescuers.

Both association classes are isomorphically connected to guarantee the correspondence between the devices and the vital signs (emission readings, resp.). Although a one-to-one association may model this isomorphism, we use two strongly connected homomorphisms to capture the dynamics of these attachments, the ripple-up of device attachments to the task resource level (e.g. in case that the request for a device has been communicated out of band via an audio channel to the agent that attached the devices) and the ripple-down of medical media selections and attachments in case assignment has been initiated at task level (via a media-BAN attachments attribute), respectively.

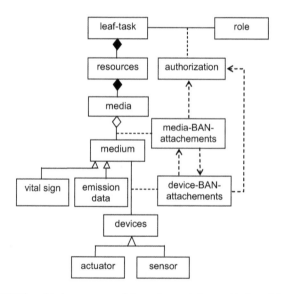

Figure 4. BAN based task resources and the assignment classes at task- and device-levels

The previously discussed separation of task-level and acquisition-level issues also provides a richer set of implementation options in a distributed environment of asymmetrical resource means with respect to wireless communication and computational processing resources. It enables the selection of (device/acquisition-level) communication alternatives available during disaster response that matches as closely as possible with the (task-level) facility specification, e.g. "use (and when needed hand-off to) an high fidelity voice conference facility" (grey colored construct in Figure 2 and cf. Hesselman (2003)). This separation is moreover common in healthcare standardization, which distinguishes between physical devices and virtual devices (Norgall, 2000).

3. TELEMATIC NEEDS IN EMERGENCY AND RESCUE SERVICES

To validate the benefits of the Enterprise models in the derivation of the requirements for the telematic system, we examine some of the issues in distributed management of disasters, in particular using the experience from the firework disaster. We also analyze the trauma team scenario valid in the Netherlands and used in the project MobiHealth (Jones et al., 2004). This project conducts healthcare trials to asses use of next generation wireless networks and mobile BANs for in- and out-door healthcare.

3.1. Trauma Team Scenario

If a road accident happens, the Ambulance Coordination & Dispatch Centre receives a request for ambulance assistance and contacts a regional ambulance service which serves that particular area. This service dispatches an ambulance to the scene of the accident with an "ambulance team" of paramedics on board. After a first assessment of the situation, the paramedics may request assistance from a mobile trauma team at a hospital and when considered necessary members of this team (e.g. an anaesthetist or a traumatologist) may travel to the scene in a second ambulance. In this scenario, we identify three sets of actors: the patient, the paramedics, and the members of the trauma team (at the HealthCare Centre and in the second ambulance).

Bi-directional audio-visual conversation facilities will be needed between the trauma team members and the paramedic or patient at the point of care. Moreover, upstream data transfer facilities will also be needed to convey vital signs of patients to the HealthCare Centre. On the other hand, low bandwidth downstream messaging services are required for command and control of the multimedia devices of paramedic headset, for example to move or zoom a paramedic's headset camera towards the patient or the scene of the accident to relieve the paramedics from burdening activities.

3.2. Disaster Scenario

The management of the firework disaster in Enschede in the Netherlands mentioned earlier did not only involve teams of different specialties (fire brigades, police, emergency services and coordination teams), it also operated within a layered organisational structure involving local, regional and national teams (Oosting, 2001). However, the regional and national teams had a facilitating role in the disaster response. In this response, the national police team deployed mobile communication units on the scene of the accident and local teams benefited from their experience with large-scale coordination and control.

The disaster initially started with harmless looking fires at the premise of the professional firework assembly site, but it evolved dramatically with two consecutive demolishing explosions. Before the explosions, management of the disaster operated adequately in terms of coordination and communication. The Alarm & Dispatch Centre responded immediately by dispatching firefighting units, informed the Police Dispatch Room and the Ambulance Coordination & Dispatch Centre. The on-site fire brigade officer and the ambulance paramedic correctly up-scaled the disaster management by requests for assistance, however, were unaware of the danger due to a lack of crucial premise-risk information.

Slow response of the Alarm & Dispatch Centre to drastically up-scale the disaster response within half an hour after the explosions hampered efficient control, amongst others because afterwards the center became a bottleneck overloaded by public requests for help and also requests from off-duty rescue volunteers asking for information and instructions. Moreover, outcall traffic of the center was jammed amongst the burst public traffic in the public communication infrastructure. For example, resulting in a late on-site deployment of mobile communication units, while the national emergency network that interconnects important instances did not span over the site of the accident, which required a wireless infrastructure connected to the fixed emergency network.

In hospitals, first aid and medical trauma teams were quickly operational (e.g. 6 operating rooms were operational within half an hour). This due to the many off-duty staff that were alarmed by radio and television news bulletins, telephone calls from colleagues and also by the visibility of the disaster itself, the explosions and the large wreath of smoke. However, these teams passively wait for arriving patients and were unaware of what could be expected and how long to be on stand by after the burst of incoming patients. Even the hospital disaster management plan was instantiated on own initiative by the hospital without having received the request to do so by the Ambulance Coordination & Dispatch Centre. Relevant feedback information for these medical teams came scarcely from news agencies, typically not under the scope of control of the crises coordination team until several hours after the explosion. The very seriously injured patients who arrived by personal cars or the patients who arrived by ambulance were treated adequately in accordance with the common working practices. However, the more serious cases amongst the many other patients who came individually using all available doors to enter chaotically the hospital were invisibly drowned within the large group of minimally traumatized patients.

Experience from management of this firework disaster reveals the following communication oriented needs:

- facilities for controlled deployment of off-duty rescue volunteers in a distributed environment;
- facilities for distributed inter and intra team coordination and communication, off- and on-site, e.g. to acquire for an overview of the situation to improve efficient control and to enable up-scaling and also down-scaling of the disaster response;
- facilities to improve time-to-control of disaster situations in the envisioned augmented reality environment;

4. TELEMATIC SUPPORT REQUIREMENTS

This section describes the derivation of the requirements for the telematic system to support distributed disaster response. It mainly addresses the requirements that originate from the Enterprise models discussed earlier, in particular, the UML classes which associate entities that are possibly remotely located, e.g. teams and agents, devices and task resources. Healthcare processes in the setting of the trauma team scenario will be used as a vehicle to better illustrate the derivation and to extract concrete scenario specific requirements. Results are however valid for several other rescue processes as indicated in a later section.

On the other hand, feasibility of the requirements depends on the capability constraints of the underlying ICT infrastructure, such as the constraints of a (public) wireless infrastructure and the limitation of BAN resources. This section therefore starts with a brief description of the infrastructure.

4.1. Application Environment

Figure 5 shows the Computational or Engineering level components of the healthcare system instantiated in a setting for the trauma team scenario. The MBUs (Mobile Base Units) are the gateways between patients' BAN vital sign sensing devices or the audio-video headset devices of the paramedic BAN and a secure networked infrastructure deployed on top of the public wired and wireless network infrastructure, such as the Internet and a providers' GPRS or UMTS infrastructure. The Surrogate Server (Dokovsky et al., 2003) intermediates as a high-level secure gateway between the HealthCare Centre node attached to a wired network and the BANs. Besides controlling secure interconnections this server among others bridges the performance hicks of the wireless communication services, as surrogate objects representing BANs run on this server.

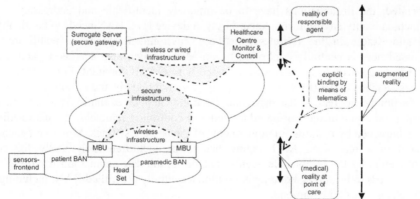

Figure 5. Mobile and wireless environment of the trauma case

In this setting, we observe the disjoint realities at the HealthCare Centre and at the point of care. Secure mobile and wireless telematic services bind the two realities onto an augmented reality, enabling availability of vital signs measured at the point of care at the HealthCare Centre.

The requirements for this distributed system derived from the (refined) Enterprise model, in particular the association classes media-BAN-attachments, device-BAN-attachments and (team to team) collaboration can be categorized amongst others into the following kinds:

- ∈ *Control-plane requirements*: i.e. the requirements associated to the set-up of the bindings of the agent-to-agent communication and the remote monitoring of vital signs (or other types of monitored data in other rescue scenarios) that bridge the spatial gap between location of the role and the associated agent. This explicit binding involves the following issues:
  - o how to find the peer entity in the (wireless) distributed environment: e.g. naming & addressing and discovery (or registration) issues;
  - o how to align the intentions: e.g. (negotiation) issues related to the readiness or willingness to set-up a binding;
  - o how to align the capability of the bindings: e.g. (negotiation) issues related to the type and the quality of service of the binding that matches to the requirements of the applications;
- ∈ *Usage-plane requirements*: i.e. the requirements associated to the use of the binding in respect of the coordination and communication activities in the application domain.

4.2. Control-plane Requirements

Some of the requirements for the BAN that are associated to the set-up of the augmented reality environment by an explicit binding of the remote point of care reality towards the scope of control of the off-site agent are the following:

4.2.1. Naming & Addressing:

For a UML association class to specify which instantiations of class entities are associated, the instantiations have to be uniquely identifiable and addressable in the augmented reality environment. Accordingly, a device level BAN needs to be identifiable and addressable in the (wireless) network environment. Since a healthcare BAN discussed here is centrally controlled by an MBU, which also acts as a gateway, this MBU in turn needs to be addressable in the wireless environment. Furthermore, if the devices attached to the BAN need to be addressed individually and remotely, they have to have a unique address in the particular environment. For example, if a camera head-mounted on a paramedic can be addressed and controlled remotely, the paramedic will not be burdened by requests for adjustments of the camera settings. Moreover, each ECG channel of a multi-lead ECG setting has to be identified individually for correct interpretation by the responsible agents. Therefore, the data acquisition front-end of the devices needs to be addressed uniquely within a centrally controlled BAN, including each of the channels of this front-end.

4.2.2. Plug-and-Play Ripple-up:

If the MBU as a gateway is powered off, the BAN and its components are unknown in the networked environment. The MBU start-up should therefore contain a push mechanism (e.g. embodied by an announce or a registration protocol) to enable its discovery in the augmented reality environment.

Quality of Service mechanisms and multi-protocol systems may be further included to enable capability alignment within the network and the selection of available infrastructure technology (e.g. telephony, radiotelephony, GSM, UMTS) in case some of the systems ceased operation after a disaster.

4.2.3 Adaptable Communication:

In outdoor cases, communication resources are typically scarce. Amount, quality and coherence of the data to be exchanged have to match with the limitations of the communication channel capability. Set up of tailorable and adaptable telematic services is typically required in these environments to better cope with bandwidth limitations and variations in communication errors like data loss and channel dropouts. Buffers, data prioritising, synchronization mechanisms and data acknowledgements may be needed in these environments, including hand-off mechanisms to communication channels matching the task objective if alternative technologies are available at the scene (see also Figure 2).

4.3. Usage-plane Requirements

Although retrieval services in the form of a tightly coupled query-response mechanism satisfy remote monitoring applications, BAN intermittent availability and the way task resources are acquired (i.e. indirectly via collaboration of off-site and on-site agents), we prefer a less tightly coupled mechanism to retrieve BAN data.

4.3.1. Upstream Push Mechanism:

Dependence of media-BAN-attachments to device-BAN-attachments requires a push mechanism, such as a plug-and-play or a call-back mechanism, which enables the visualization (remote monitoring) of vital signs, for which a medical agent has attached corresponding sensors at a remote point of care. However, the agents may also communicate some vital sign data via an audio or video channel, e.g. patients' facial colour description as a trend vital sign for brain oxygenation (cyanosis).

Additionally, a (block-based) streaming protocol may be needed to upstream sets of coherent continuous-time vital signs or multimedia data. This also generates the need for several preservation mechanisms like (time and event) synchronization, data priority and data accuracy when using layered compression.

4.3.2 Downstream Messaging Mechanism:

Dependence of device-BAN-attachments to media-BAN-attachments also requires a messaging method that ripples down the choice for a specific task resource to the MBU to alert a medical agent or a patient at the remote point of care to attached a corresponding sensor or to forward a configuration setting to a sensor or actuator device. Remote camera control described earlier also emerges from this dependence.

4.3.3. Conversational Communication:

The requirements for the BAN in respect of the team collaboration or patient to team communication depend on the healthcare context. A GSM based channel often satisfies the requirements for communication between a paramedic and a trauma specialist in the trauma case. On the other hand, conversation between a patient and a specialist may need better fidelity. In chaotic or stressful situations, high fidelity audio channels help patients to calm down or be reassured, video channels may improve these conversations further.

4.4. Refined Trauma Team Scenario Requirements

The scenario described in the previous section also shows the interactivity of the communication between the remotely located agents including the transfer of the clinical patient data to the healthcare centre. This means that the data transfers are often dependent on one another and transfer delays therefore should not exceed the boundaries given by human factor studies. This constrains the audio-video communication quality in the wireless environment, it also constrains the quality of the vital signs which can be transferred. For example, an accurate 12 leads over-sampled ECG set already requires a 64 kbits/s UMTS channel (approximately, 8 time-slots of GPRS Coding Schema 1) for real-time communication.

Moreover, several types of vital signs for emergency services are interdependent, together they form a unit of interpretation due to the typical indirect way of measuring patient condition. For example, oxygenation of the brain is usually estimated from oxygen saturation (O2sat) measured at a fingertip surface by beaming for blood colour, respiration rate including a CO2 (pCO2) measurement, temperature, blood pressure and heart rate or ECG measurements. However, as a first indication (trend sign) of oxygenation the patient's facial colour is observed for signs of cyanosis. In the estimation of brain oxygenation, validity of the measured O2sat depends on the patient's temperature. Moreover, heart rate as a trend sign of ECG has a higher priority than ECG. So the latter may be dropped or deferred in case of communication dips. The previously mentioned vital signs are typical for emergency services (Leisch and Orphanoudakis, 1999; Gagliano and Xiao, 1997).

Some other vital signs can be communicated by the paramedics to the trauma team at the HealthCare Centre via audio or video channels, for example pupil reaction, facial colour and also the (Glasgow Coma) trauma score. Other important information for a remotely located trauma team is still pictures or video of the scene of the accident to give an indication of patient's condition from the damage of the vehicle and the patient's positions in the vehicle. Today, polaroid pictures are taken and are handed over on the patient's arrival at the hospital A&E department. Wireless transfer of these types of information from the scene will enable teams and facilities to be prepared before patient's arrival, and thus improve time to treatment of the patient on arrival at the hospital.

4.5. Disaster Scenario Requirements

Requirements for telematic support to facilitate the issues identified in the disaster scenario analysed are:

- $\in$ Volunteering off-duty but on-site rescuers have to be identified immediately (i.e. assigned as agents) and deployed in teams to make best use of their availability. Uniquely addressable BANs that include telematic services for example with electronic registration services facilitate the deployment of these rescuers;
- $\in$ Distributed intra and inter team coordination and communication is crucial in disaster control. Telematic services, mobile BANs of sensors and communicating devices and network infrastructure (incl. ad-hoc, wireless and possibly supported by mobile base stations) may facilitate this aspect. The telematic support requirements related to the two levels of BAN

assignment discussed for the trauma team case is also suitable for distributed control in the case that an on-site rescuer carries devices controlled and monitored by off-site team members;

- •∈ Modelling separation of roles and agents enables on-site roles to be assigned to off-site agents. For example, an augmented reality environment enables the leader of the crises coordination team to visually receive an impression of the accident including statistic data while the person (e.g. the mayor of Enschede) is delayed in the traffic jam which often occurs in early stages of disasters.

- •∈ Dynamic improvement or fine-tuning of coordination and communication guidelines in up-scaling disasters may benefit from comprehensive enterprise models. Association classes and dependencies enable rippling through of new instructions towards roles or agents in the team hierarchy.

These telematic support requirements are similar to the requirements in the trauma team scenario case.

5. CONCLUSIONS

To achieve full control of disaster situations as rapidly as possible is a prime objective of disaster response. This chapter discusses the telematic service requirements for a mobile and wireless environment of BANs incorporating mobile devices (e.g. medical and multimedia devices) to support and to improve distributed disaster response in respect of time to control for example. These requirements are derived using a primarily top-down approach in-line with the ODP viewpoints framework.

The developed and proposed Enterprise models capture the organisational invariants of healthcare processes in UML compositional constructs in terms of roles, agents and tasks. This chapter emphasizes the modelling of roles and agents instead of tasks and task relations because neither process reengineering nor developing an information system for disaster management or healthcare is in the scope of the work. The modelling separation of roles and agents reveals the opportunities to apply telematic services because roles could be dispatched virtually to a point of care (or rescue) while some of the associated agents remained off-site at the command and control or healthcare center, yielding a simple form of an augmented reality environment. Further, the hierarchical structure of tasks and teams, flexibly constrained by association classes in the model, fits in scalable scenarios like disaster management and moreover, it retains, as much as possible, current ways of working of specialized teams.

The elaborated requirements of the trauma team scenario and the examined requirements for the more general disaster scenarios show the feasibility and added value of the mobile and wireless environment to improve rescue and emergency services. However, the derived requirements and the capabilities of the underlying mobile and wireless infrastructure seem to have mismatching characteristics. Provided (wireless) communication channels are often asymmetrical configured with a higher downstream bandwidth, but the addressed applications often require high bandwidth upstream channels. Moreover, the applications typically fetch data sets from the mobile devices, which have limited resources and are intermittently online. Internet protocols, which apply the client-server paradigm with typically a thin client and heavy server, could not be used in a straightforward manner in this inverted producer-consumer problem. A challenge for future work is amongst others the further development of a generic BAN which can be specialized for different specialties and the Internet protocol stack for the support of scalable multimedia services in networks comprising different technologies, including mechanisms to automatically discover alternative computing or network resources and select an alternative that matches with the task objective.

6. REFERENCES

Argyle, B., 1996, "New York Emergency Room RN: General Approach to Trauma", http://www.nyerrn.com/er/t/g.htm;

Auf der Heide, E., 1989, *Disaster Response: Principles of Preparation and Coordination*, CV Mosby St. Louis, on-line book at http://coe-dmha.org;

Azuma, R.T., 1997, "A Survey of Augmented Reality", in *Presence: Teleoperators and Virtual Environments*, 6(4), pp. 355 – 385;

BBC News: world edition, 2002, "Failures in NY attack response", August 3rd, http://news.bbc.co.uk/2/hi/Americas/2170907.stm;

Blair, G., and Stefani, J-B., 1998, *Open Distributed Processing and Multimedia*, Addison-Wesley;

Booch, G., Rumbaugh, J., and Jacobson, I., 2001, *The Unified Modeling Language User Guide*, Addison-Wesley;

Doarn, C.R., Nicogossian, A.E., and Merrell, R.C., "Telematic Support for Disaster Events", *M-Health Emerging Mobile Health Systems*, Istepanian, R.H., Laxminarayan, S., and Pattichis, C.S. (eds.), Kluwer Academic/Plenum Publ.;

Dokovsky, N., Halteren, van A., and Widya, I., 2003, "BANip: enabling remote healthcare monitoring with Body Area Networks", *FIDJI2003 International Workshop on Scientific Engineering of Distributed Java Applications*, Nov. 27-28, Luxembourg;

FEMA, 2001, FEMA Information Technology Architecture 2.0, http://www.fema.gov/library/ita.shtm;

Gagliano, D.M. and Xiao, Y., 1997, "Mobile Telemedicine Testbed" in *Proc. 1997 American Medical Informatics association (AMIA) Annual Fall Symposium*, pp.383- 387, National Library of Medicine Project N0-1-LM-6-3541;

Hesselman, C., Eertink, H., Widya, I. and Huizer, E., 2003, "A Mobility-aware Broadcasting Infrastructure for a Wireless Internet with Hotspots", *The 1st ACM Int. Workshop on Wireless Mobile Applications and Services on WLAN Hotspots MWASH'03*, P. Kermani (ed.), Sep. 19th, San Diego CA, p.103-112;

Hobsley, M., 1979, *Pathways in Surgical Management*, Edward Arnolds Pub. Ltd, London;

Holz, E., Kath, O., and Born, M., 2001, "Manufacturing Software Components from Object-Oriented Design Models", 5th IEEE Internat. Enterprise Distributed Object Computing Conference EDOC 2001, Sept 1st, Seatle, USA;

Jones, V. et al., 2004, "MobiHealth: Mobile Health Services based on Body Area Networks", *M-Health Emerging Mobile Health Systems*, Istepanian, R.H., Laxminarayan, S., and Pattichis, C.S. (eds.), Kluwer Academic/Plenum Publ.;

Leisch, E. and Orphanoudakis, S. 1999, "HECTOR Solutions" in *1st International Conference on Health Emergency Telematics*, Sevilla, Spain, March 15–17, 17 pp.;

Luzi, D., Ricci, F.L., and Serbanati, L.D., 1997, "Extending the Standard Architecture for Healthcare Units: The Guideline Server", in *New Technologies in Hospital Information Systems*, J. Dudeck et al. (Eds.), IOS Press, pp. 95 – 101;

Meissner, A., Luckenbach, T., Risse, T., Kirste, T., and Kirchner, H., 2002, "Design Challenges for an Integrated Disaster Management Communication and Information System", *The 1st IEEE Workshop on Disaster Recovery Networks DIREN2002*, June 24th, New York City NY;

Midkiff, S.F., and Bostian, C.W., 2002, "Rapidly-Deployable Broadband Wireless Networks for Disaster and Emergency Response", *The 1st IEEE Workshop on Disaster Recovery Networks DIREN2002*, June 24th, New York City NY;

Milgram, P. and Kishino, F., 1994, "A Taxonomy of Mixed Reality Visual Displays", *IEICE Transactions on Information Systems*, vol. E77-D, no. 12;

Norgall, T., 2000, "VITAL – towards a Family of Standards for Open Medical Device Communication" in *XIII International Conference Computing in Clinical Laboratories*, Milan, 21-23 September;

Oosting, 2001, "*De Vuurwerkramp: Eindrapport*", Commissie Onderzoek Vuurwerkramp, Enschede/den Haag, Feb. 28th, ISBN: 90-71082-67-9, (*in Dutch, incl. an English version of the summary* "*Final Consideration*"), http://www.minbzk.nl/contents/pages/00001947/eindrapport_oosting_201.pdf;

Widya, I., Volman, C., Pokraev, S., De Diana, I. and Michiels, E., 2002, "Enterprise Modelling for an Educational Information Infrastructure", in *Enterprise Information Systems III*, J. Filipe et al. (eds.), Kluwer Academic Publ., pp. 231 – 239;

Yates, M., Takita, W., Demoudem, L., Jansson, and R., Mulder, H., 1997, "TINA Business Model and Reference Points – version 4.0", May 22nd, http://www.tina.com;

TELEMATIC SUPPORT FOR DISASTER SITUATIONS

Charles R. Doarn*, Arnauld E. Nicogossian, and Ronald C. Merrell

1. INTRODUCTION

Disasters, whether as a result of a human catalyst or by natural forces, can devastate a community's ability to sustain life as normal or be responsive to societal needs. Each year, nature's fury plays havoc on many nations. Moreover, humankind has figured out horrendous ways of destroying life. This is evident to every reader and does not require in-depth discussion.

The common theme in all disasters is that infrastructures of all kinds are destroyed or in some way impaired such that they cannot be used to support basic needs like medical care. Of course in many parts of the world a disaster can be far worst than in others. A hurricane in South Florida can result in billions of dollars in damage and yet the loss of life is low. Conversely a hurricane in Central America can result in many more deaths. This mostly is attributed to the existence of more comprehensive early response systems in the United States (US). Irregardless of the disaster's cause, the resulting damage to the basic infrastructures for communications, food and water supplies, and medical capabilities are disrupted, overtaxed or simply no longer available. With regard to medical response in disasters, delayed response can exacerbate the situation, resulting in epidemiological problems. (CDC web site 2003) Additionally, those remaining after the disaster may experience posttraumatic stress disorder.(Laor et al., 2002; Livanou et al., 2002)

Effective telecommunications and planning can mitigate some of the predictable outcomes; however, the response must continue until fully independent services are restored. (Garshnek et al., 1999) The medical response should be an effective telecommunications system for information management and direct medical intervention. Such a system can be comprised of pre-positioned or rapidly deployable mobile assets

---

*Charles R Doarn, Executive Director, Center for Surgical Innovation, Research Associate Professor, Department of Surgery, University of Cincinnati, SRU 1466, Cincinnati, OH 45267-0558 V: (513) 558-6148, F: (513) 558-3788, E-mail: charles.doarn@uc.edu

with robust backend systems integrated with overall disaster relief. This simple integration of telecommunications with medical response, effectively known as telemedicine or telehealth, can provide the foundation for rapidly restoration of vital medical services in disaster situations. This integration has been effectively used in many disasters over the past 20 years. The lessons learned from these events have led to enhanced protocols, processes, and mobile solutions. These solutions have been successfully applied during peacetime to provide and enhance medical health care delivery to remote or underserved populations and regions.

2. BACKGROUND

Prior to wide distribution of telecommunications and international travel, disasters often were events that occurred somewhere else, usually devastating the area. News of such an event often would pass into obscurity. In a world that is evermore linked by telecommunications and mass media, images of events are shared around the world in an instant. This near instant communications permits the global community to gain knowledge of events and develop a response to them. Although natural disasters are not preventable, preparedness could be more effective. Data from the World Health Organization (WHO) as illustrated Figure 1 indicates the number of people affected, killed or displaced by natural disasters since 1988. (Red Cross web site, 1999) Although dated, this graph represents those areas of the world that are affected by disasters.

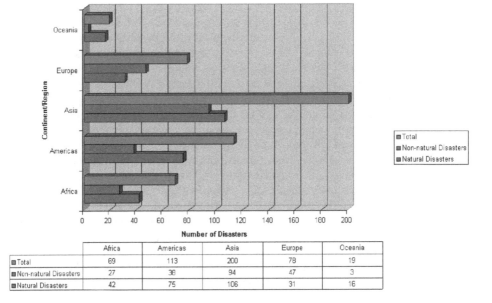

	Africa	Americas	Asia	Europe	Oceania
▣ Total	69	113	200	78	19
▣ Non-natural Disasters	27	36	94	47	3
▣ Natural Disasters	42	75	106	31	16

Figure 1. Annual average number of reported disasters by region and type 1988 – 1997.

In almost all cases, the first action in a disaster is to assess the situation and establish communications. During the past several decades, telemedicine has been shown to be effective in disaster planning and response. The use of telecommunications assets of the National Aeronautics and Space Administration (NASA) in the aftermath of earthquakes in Mexico City in 1985 and Armenia in 1988 are testaments to the possibilities of how telecommunications can be effectively used for planning and disaster response. (Nicogossian et al., 2001)

There are several organizations that respond to disasters. The level of response is often predicated on where the disaster happens as well as what kind of disaster it is. Humanitarian assistance is provided by organizations such as the Red Cross, Red Crescent, or the United Nations in the aftermath of natural disasters. (Red Cross web site, 2003; Red Crescent, 2003; WHO, 2003) These organizations often respond worldwide. (Youn, 1995) Various elements of the US Department of Defense (DOD), North Atlantic Treaty Organization (NATO), and others often provide resources in many places around the world in the immediate aftermath of a disaster.

The severity of the disaster can also impact the scope and breadth of response. In the US, the aftermath of the events of September 11, 2001, resources from many different Departments, Agencies at all levels of government were marshaled to respond. Events like the earthquake in Armenia or an oil spill of the coast of Africa often result in international response efforts. (Maheu et al., 2001) Certainly outbreaks of disease, especially those promulgated by acts of terrorism will require international coordination in effective response to minimizing the spread of disease. The same is true for chemical or poison surveillance with regard to health. (Butera et al., 1997)

In each and every case, the health and safety of the victims and responders is paramount. The following presents key issues that must be addressed in integrating telemedicine and telematic support in disasters.

3. EFFECTIVE PLANNING

The overall goal must be to provide a rapid response to a disaster area to support medical communications and strengthen the ability to respond to medical and public health needs through telemedicine. (Garshnek et al., 1999) To accomplish this objective, effective planning, including education and training, must be undertaken. (Llewellyn, 1995)

Effective planning provides unique opportunities for preparedness and response. Access to information is vitally important regardless of the kind or nature of it. Basic information can provide responders an idea of what is needed in the way of relief supplies, extent of damage or in the case of medical care – real time interaction and management of medical events. (Scott, 1991) Access to information is a significant step in planning.

Another extremely important aspect of effective planning is access to information and knowledge. Several organizations including the Centers for Disease Control, Red Cross, WHO, etc. have established web sites which abound with a plethora of information about disaster preparedness and education. Individuals all over the world gain access to this information using telemedicine. (Chandrasekhar et al., 2001)

Effective planning is comprised of two activities: (1) Assessment and Education and (2) Training. To be most effective, these steps must be taken before a disaster ever happens. Preplanning and pre-positioning of assets is vital to rapid and effective response. Assessment can be further broken down to needs and technology.

Interoperability of existing systems can also be of tremendous benefit in responding in an appropriate manner. Telemedicine or telematic tools that can be brought to bear should be in place prior to events. Such systems must be mobile and rapidly deployable. (Cabrera et al., 2001) Continuous training or simulation of events is critical to readiness.

3.1 Assessment

Assessment must begin prior to disaster occurrence. This critical step involves many different facets. These include: 1) an understanding of the kinds, nature and severity of natural disasters that might occur in the region; 2) what kind of infrastructure exists for addressing mass casualties and medical care capabilities, provision of potable water, communications, etc.; 3) what legal or political issues are potential barriers for rapid response; 4) what kind of technology is readily available and sustainable; and 5) systems integration – bringing it all together in a cohesive system.

Each of these requires extensive review of previous events and how the response was effective or ineffective. Certainly one organization's or one group's response can be a good metric on evaluating responsiveness. Most governments, regardless of level, have data on existing infrastructures. Most even have systems or processes that tap into these infrastructures for emergency response. However, in some cases this infrastructure can be so severely damaged or in some cases destroyed that there no longer was any infrastructure to support the population's basic needs. This was the case in the Armenian earthquake in December 1988. There was no system with which to respond with. It was apparent that systems had to be made available to support the medical and other needs.

There have been several instances when the authors of this chapter were approached by one or more Agency or Department of the US Government concerning a telemedicine solution in the aftermath of a natural disaster. Such events have occurred outside the borders of the U.S. In several cases, the response from the foreign government was simply that the event would be handled within their own country. The rational was one of politics. In the Armenian response, the details and plans took several months to work out because of political and legal issues. Certainly, these issues must be resolved in rapid fashion to ensure an effective response. An assessment of this landscape is critical to development and implementation of appropriate responses.

Often the solution to a particular scenario is based on the use and integration of technology. However, a technical solution can be pervasive at the site of disaster, yet to complication to actually operate. A review and knowledge of the technology that is already in place or available can provide a solid foundation for planners on the level and sustainability of the response. To illustrate, one can use a natural disaster in Central America. In this event all roads and power have been severely impacted. A technology solution involving a dependence on power would be adversely impacted because that infrastructure would not be able to support the solution.

Once planners have established a review and assessment of all needs, capabilities, and other barriers, then they must develop a systems approach to respond affectively. Such a system would include a plethora of steps such as pre-positioning of assets and preparedness training.

3.2 Education and Training

Clearly societies or population centers around the world have dealt with disaster preparedness in a variety of ways and levels. Many countries have well-funded agencies of responding to disaster, which have very good education and training programs. In the US, such organizations include the Federal Emergency Management Agency (FEMA), The Center for Disease Control, and the recently created Department of Homeland Security. The Red Cross, Red Crescent, and the WHO have similar capabilities. Of course the military in many countries also plays a role in responding to disasters and has training program that are brought to bear.

Disasters can happen virtually at any time or place. If the population is not prepared or unclear as to how to respond, the problems are further exacerbated. It may take time to marshal the resources to respond accordingly. Therefore, it is important to have a robust education and training capability. This will help those responding have a greater ability to respond. Training can be focused on specific subject matter, such as restoration of potable water, epidemiology, or health services. Many of these educational services can be found on the web. (CPHD web site, 2003) Some are tailored to meet specific needs.

It is important to be aware of culture and language. This is of extreme importance as it relates to health services provided by someone through telemedicine who is not in the region or from the region. (Garshnek et al., 1997)

A sound education and training program must be supported by continuous review and simulation. Simulation of events is a highly effective approach for training and readiness. Airline pilots continually simulate aircraft operations. Simulations provide a tremendously valuable tool for refinement of skills and processes. (Satish et al., 2002; Balch et al., 2001) It provides an opportunity for a comprehensive review of what can be done and how easily it is done. Moreover, it improves the capability of the response team. Simulations, conducted on a large scale, provide a unique opportunity to review the entire response system. Simulations provide teams or individuals an assessment with which to gauge their capability and ability to perform a serious of tasks.

Just as important to education, training, and simulation, is access to knowledge. The Internet provides a great portal to ever growing body of information. The Internet can be reached from anywhere in the world using wireless technologies or hard-wired facilities. This permits rapid access to information on public health, dedicated training sites for disaster response, and can provide valuable information on the region. It also becomes a very valuable tool in supporting medical education and disaster simulations. (Akada et al., 2000)

4. EFFECTIVE TELECOMMUNICATIONS

The ability to communicate in real-time is of high value with regard to disaster response. Telecommunications provides access to information that is reliable and timely. Often the first thing that is to be reestablished in a disaster is the telecommunications. The level of telecommunications can vary depending on several factors including need and region. Telecommunications can be supported by satellite communications such as C-Band, Ku-Band, Ka-Band, as well as cellular and broadband. These capabilities can be made available by a variety of providers and costs. In many instances, the military may respond but this is limited to 2-5 days.

Telecommunications in disaster can be overwhelmed. The events of September 11, 2001, in New York City, rapidly caused the telecommunications to be interrupted. This was due in part to demand exceeding capacity and physical damage or destruction of the infrastructure. In such events, the system responding to a disaster may have depended upon something that was thought to be there no matter what. This of course was not the case and what capabilities remained were rapidly overwhelmed.

In order to respond effectively and appropriately, telecommunications must be present for planners to assess damage and need. This same telecommunications system can be of value in supporting telemedicine. NASA has provided telecommunications support for disaster response in several areas over the past 3 decades. This telecommunications capability has provided unique capabilities to responders in assessing the severity of the situation and effectively planning a response. This has included logistics and supply of needed resources. (Doarn et al., 1998; Garshnek et al., 1991)

The Japanese have used satellite communications as an effective tool in disasters and emergencies. (Otsu et al., 1986)

5. INFORMATICS

The concepts of telemedicine have been around for hundreds of years. It has only been recently that the term telemedicine, telehealth or telematics has been part of the spoken word. These concepts are part of a larger discipline – medical informatics. Medical informatics is the management and study of medical information through integration of information systems, linked by telecommunications.

This provides a valuable foundation for training and education. Data and knowledge collected from previous telemedicine or telematic responses to disaster provide a solid base to develop models to evaluate future scenarios in training and simulation. Such work conducted by US Air Force demonstrates how informatics can be used in response to war, terrorism, or disaster. (Teich et al., 2002)

The ability to evaluate and mine large databases for trend analysis provides ability to look at issues that were not readily available in the past. Rapid access to such databases can be of tremendous value in planning and responding effectively. Access to information in NASA's Spacebridge to Armenia provided physician responders on the ground in Armenia the necessary information to respond to the medical needs of the individuals affected by the earthquake.

6. WHAT HAS WORKED

Over the past several decades, telemedicine has been integrated into disaster response quite effectively. Responding to each disaster, regardless of cause can be different and challenging. Today, telemedicine plays a key role in response. This is principally as result of the lessons learned from those events that have already passed.

Such efforts have included international responses to earthquakes or hurricanes and to war or other human-made disasters.

The Germans have initiated the integration of telemedicine in the emergency medical service system (EMSS) as it relates to mass casualty incidents. Using telemedicine to support medical response can provide better field management prior to arrival at definitive care. (Plischke et al., 1999) In fact, many emergency medical services around the world utilize telecommunications and telemedicine to support medical response, medical management, and/or prescreening of surgical intervention. (Doarn et al., 2002; Orlov et al., 2000)

These systems work because they have been effective in responding to a need. They were planned and implemented according to well structured protocols and were sustainable.

6.1 Spacebridge to Armenia

In 1988, an earthquake struck the Soviet Republic of Armenia, killing tens of thousands and destroying key infrastructures. The US Government through the Department of State and NASA developed a plan with Soviet authorities, including the Union of Soviet Socialist Republics (USSR) space program to develop telecommunications capability or a 'spacebridge' to link the affected region of Yerevan, Armenia with medical facilities in the US and in the USSR. Although the two were collaborating in space-based activities, it took several months to develop and implement a plan for providing a response. During this time, agreements had to be reached; protocols had to be developed; assessments had to be conducted of capabilities on all sides, American, Armenian, and Russian; and logistics had to be made. Once this was in place it was very easy to respond to a second disaster, a gas explosion in Ufa, Russia. The system was adapted for rapid response.

The 'spacebridge' demonstrated on a grand, international scale how telecommunications could be used effectively to support medical response in a disaster situation. During the conduct of this 'spacebridge', 209 clinical cases were discussed between medical personnel from the US, Russia, and Armenia. (Houtchens et al., 1993) Of these 209 cases, the diagnosis or treatment plan was altered by nearly 30%. The program was considered highly successful and very useful tool in providing response in the aftermath of a natural disaster. However, the implementation was not without challenge.

Clearly, far more effective and rapid response systems and agreements should be in place. The lessons learned from this event were shared in the First International Conference on Telemedicine and Disaster Medicine at the Uniformed Services University of the Health Sciences (USUHS) in Bethesda, Maryland, in 1991. This symposium brought together leading experts of the day to discuss telemedicine and disasters. A final report was prepared. (NASA Pub NP-207, 1993) Out of this event came a plan to continue evolving telemedicine as an effective tool for international telemedicine. Two additional spacebridges were conducted in Russia, one was satellite based and the second was Internet-based. (Doarn et al., 2003; Angood et al., 1999, Acuar et al., 1999) These projects were very successful and they provided the foundation for a second international meeting on Telemedicine at USUSH. The proceedings of this meeting were published as a monograph in Volume 19, Issues 1, 2, & 3 of the Journal of Medical Systems in 1995. (J of Med Sys special edition, 1995)

6.2 Hurricane Floyd

Hurricanes are a potent danger to those populations that live along the Atlantic coast of North and Central America. Hurricanes Andrew, Mitch, Camille, and Floyd have wrecked havoc along the eastern seaboard as well as hundreds of meters inland. In the U.S., there are systems that address the evacuation and response to potential disaster. This often minimizes impact on death but not on property damage. The converse of this often is the opposite in areas of the Caribbean. Here the loss of life and the ability to rapidly respond varies greatly from that of the US.

Hurricanes result in massive flooding and subsequent isolation of populations. In 1999, Hurricane Floyd ravaged North Carolina, resulting in destruction of roads, bridges, and infrastructure. The Telemedicine Center at East Carolina University (ECU) School of Medicine developed and implemented a rapid telemedicine response system to support the needs of the affected population. Using HAM radio communications, remote sites were linked to the ECU Telemedicine Center for planning, clinical consultations, and training. Such a system alleviated the strain on resources, permitting them to be used for other activities. (Balch, 1999; ECU Telemedicine web site, 2003)

6.3 Operation Strong Angel

In the summer of 2000, the US Department of Defense including the U.S. Navy, the United Nations (UN) Office of Humanitarian Assistance, the UN Commissioner for Refugees, UNICEF and military participation from Australia, Canada, the UK and many Pacific Rim countries gathered in Hawaii to conduct a series of exercises called 'Operation Strong Angel'. (Operation Strong Angel web site, 2003)

The premise of the exercise was to simulate a humanitarian disaster, where huge numbers of refugees had been displaced. These individuals required medical support and logistics for basic supplies. Overall the vision of this initiative was simply to 'further the collaborative efforts between multi-national military forces and civilian humanitarian organizations to assist populations in crisis." A number of additional participants focused on the telemedicine aspects of the broader Strong Angel project. The development of a distributed medical intelligence capability lead by the ECU Telemedicine Center was supported by NASA's Johnson Space Center, the US Navy, the Medical Informatics and Technology Applications Consortium (MITAC) – a NASA Research Partnership Center at Virginia Commonwealth University, and several other international entities.

The distributed medical intelligence system served as a medical communications network that supported the transfer of voice, video and data information from remote vessels participating in the exercise to and from medical facilities on the US mainland. This allowed medical personnel on these remote vessels to receive real-time voice, video and data medical consultation from care providers at hospitals on the US mainland as they treated disaster victims. This concept provided a telemedicine link from a refugee camp to a site where medical expertise and information could be obtained.

This exercise demonstrated that a well-developed and rapidly deployable system to support telemedicine can be highly effective in supporting disaster relief or humanitarian response in the case of a refugee camp.

6.4 Response in Kosova

The world has seen plenty of war and the affect it has on populations and infrastructure. In many instances, large groups of people are force out of their homes become refugees in another country or region. The Balkans war in 1998-1999 and its impact on peoples of Albanian descent further exacerbated an already terrible situation. The MITAC and several individuals championed the concept of establishing telemedicine as an effective tool in responding to the refugees spread out over the region of Kosova. It was envisioned that telemedicine and the technology and systems that support it could be of great value in rebuilding key medical infrastructure for both care and education.

British military physicians used simple tools – digital cameras and e-mail to transmit medical cases to the United Kingdom and the US for comment and consultation. This began a dialogue and a momentum with MITAC that eventually lead to the creation of the Kosova Foundation for Medical Development and the International Virtual E-Hospital. From 2001-2002, the concept was conveyed at several international meetings, including the G-8 Subproject 4 final meeting in Berlin in November 2001. The outcome was the creation of a Telemedicine Center of Kosova at the Medical School of Prishtina. This center was financial supported by the European Union. The Center was officially opened in October 2002 during the First Balkans Telemedicine Symposium. (Int'l Virtual E-Hospital web site, 2003)

This demonstrated the willingness of many countries to work together to be prepared to respond to disaster. This disaster in the Balkans was not necessarily measured in morbidity or mortality but in terms of infrastructure for teaching medical students. Prior to the liberation of Kosova, ethnic Albania medical students and other medical professionals were trained in the mountains, in homes, and in all places not affiliated with traditional academic centers. Telemedicine and distance learning were embraced as unique tools to support this human tragedy.

Similar issues are on the forefront in Afghanistan, Iraq, and other hot spots around the world, where humanitarian assistance can be benefited by application of telemedicine.

7. FUTURE

Telemedicine has been demonstrated, evaluated, and validated as highly effective tool in responding to disasters. The concepts of telemedicine continue to evolve and with continuous improvements in telecommunications technologies and computer systems, training, preparedness, and ability to respond rapidly it will be an integral part of disaster response.

It has been suggested under the G-8 leadership in health as well as other international experts and groups that an international systems for responding to emergencies should be established that would be multilingual and available 24/7. This concept is currently in place in parts of the world but it is not responsive to all needs of all situations. Nevertheless, there must be documentation and agreements in place prior to a response thereby minimizing the time to respond and broadening the reach and scope. With the development of such an effort, there must be common standards and common protocols. A lot of work has been done in the past regarding this, including the lessons learned from the Spacebridge to Armenia, the international conferences, the G-8 Subproject 4 –

Interoperability of Telemedicine Systems, and other activities. (Nerlich et al., 2002; Lacroix et al., 2002)

One of the most chilling thoughts that affects the entire human race is the concept that a nation state or terrorist group could utilize chemical, biological, or nuclear material to cause a disaster of untold proportions. Many countries are taking notice and have begun to develop disaster preparedness plans for response to such events. In the US, the new Department of Homeland Security is marshaling resources to prepare for such an event. Telemedicine is an integral part of this effort.

It is of tremendous value to have the ability to interact between different centers of excellence for planning and executing response as well as supporting education and training and sharing of data and information.

Telemedicine is a tool for the practice of medicine. It is a valuable tool and remains a key tool for disaster planning and response.

8. REFERENCES

Angood, P. B., Doarn, C. R., Holaday, L., Nicogossian, A. E., and Merrell, R. C., 1999, The Spacebridge to Russia Project: Internet-based telemedicine, *Telemed J.* **4**(4):305-311.

Aucar, J. A., Doarn, C. R., Sargsyan, A. E., Samuleson, B. A., Odonnell, M. J., and DeBakey, M. E., 1999, Case report: Use of the Internet for international post operative follow-up, *Telemed J.* **4**(4):371-374.

Balch, D. C. and West, V. L., 2001, Telemedicine used in simulated disaster response, *Stud. Health Technol Inform* 81:41-45.

Balch, D., 1999, Telemedicine helps with Hurricane Floyd disaster relief, *Telemed Today.* **7**(5):5-6.

Butera, R., Locatelli, C., Gandini, C., Minuco, G., Mazzoleni, M.C., Giordano, A., Zanuti, M., Varango, C., Petrolini, V., Candura, S. M., and Manzo, L., 1997, Telematics equipment for poison control surveillance. Its application in the health management of relevant chemical incidents, *G. Ital. Med. Lav. Ergon.* **19**(2):42-59.

Cabrera, M. F., Arredondo, M. T., Rodriquez, A., Quiroga. J., 2001, *Proceedings of the AMIA Symposium.* S:86-89.

Chandrasekhar C. P. and Ghosh, J., 2001, Information and communication technologies and health in low income countries: the potential and the constraints, *Bulletin of the World Health Organization* **79**(9):850-855.

Doarn, C. R., Nicogossian, A. E., and Merrell R.C., 1998, Application of telemedicine in the United States Space Program. *Telemed J.* **4**(1):19-30.

Doarn, C. R., Fitzgerald, S., Rodas, E., Harnett, B., and Merrell, R. C., 2002, Telemedicine to integrate intermittent surgical services in to primary care, *Telemed J E Health.* **8**(1):131-137.

Doarn, C. R., Lavrentyev, V., Orlov, O. I., Grigoriev, A., Nicogossian, A. E., Ferguson, E. W., and Merrell, R. C., 2003, Evolution of telemedicine in Russia: Influences from the space program. A ten-year summary, *Telemed J E Health, 9(1):103-109.*

Garshnek V. and Burke F. M., 1999, Telecommunications systems in support of disaster medicine: applications of basic information pathways, *Ann Emerg Med.* **34**(2):213-218.

Garshnek V. and Burke F. M., 1999, Applications of telemedicine and telecommunications to disaster medicine: Historical and future prospectives, *JAMIA.* **6**(1):26-37.

Garshnek, V., Logan, J. S., and Hassell, L. H., 1997, The telemedicine frontier: going the extra mile. *Space Policy* **13**(1):1- 11.

Garshnek, V., 1991, Applications of space communications technology to critical human needs: rescue, disaster relief, and remote medical assistance, *Space Commun.* **8**(3-4):311-317.

Houtchens, B. A., Clemmer, T. P., Holloway, H. C., Kiselev, A. A., Logan, J. S., Merrell, R. C., Nicogossian, A. E., Nikogossian, H. A., Rayman, R. B., Sarkisian, A .E., and Siegel, J. H., 1993, Telemedicine and international disaster response: medical consultation to Armenia and Russia via a Telemedicine Spacebridge, *Prehospital Disaster Med,* **8**(1):57-66.

Lacroix, A., Lareng, L.. Padeken, D., Nerlich, M., Bracale, M., Ogushi, Y., Okada, Y., Orlov, O. McGee, J., Wootton, R., Sanders, J. H., Doarn, C.R., Prerost, S., and McDonald I., 2002, International concerted action on collaboration in telemedicine. Recommendations of the G-8 Global Healthcare Applications Subproject 4, *Telemed. J. E. Health.* **8**(2):149-57.

Laor, N., Wolmer, L,, Kora, M., Yucel, D., Spirman, S. and Yazgany, Y., 2002, Post traumatic, dissociative and grief symptoms in Turkish children exposed to the 1999 earthquake, *J. Nerv. Men.t Dis.* **190**(12):824-832

Livanou, M. Basoglu, M., Salcioglu, T., and Kalendar, D., 2002, Traumatic stress response in treatment-seeking earthquake survivors in Turkey, *J. Nerv. Ment. Dis.* **190**(12):816-823.

Llewellyn, G. H., 1995, The role of telemedicine in disaster medicine, *J of Med Sys,* **19**(1):29-34.

Maheu, M. M., Whitten, P., and Allen, A., 2001, *E-Health, Telehealth, and Telemedicine*, Jossey-Bass, A Wiley Company, San Francisco, CA.

Multiple authors. Special Issue: Second NASA/USUHS International Conference on Telemedicine. *J. Med. Syst.*, **19**(1-3):3-304 (1995).

Nerlich M. L., Balas E. A., Schall T., Stieglitz S. P., Filzmaier R., Asbach P., Dierks C., Lacroix A., Watanabe M., Sanders J. H., Doarn C. R., and Merrell R .C. 2002, Teleconsultation practice guidelines: Recommendations of experts in telecare. A guideline from G8 GHAP Subproject 4 Group, international concerted action on collaboration in telemedicine, teleconsultation guideline consensus process, *Telemed. J. E. Health.* **8**(4):411-418

Nicogossian, A. E., Pober, D. F., and Roy, S. A., 2001, Evolution of telemedicine in the space program and earth applications, *Telemed, J. E. Health.* **7**(1):1-15.

Okada, Y., Haruki, Y., and Ogushi, Y., 2000, Disaster drills and continuous medical education using satellite-based Internet, *Methods Inf Med.* **39**(4-5):343-347.

Orlov, O. I., Drozdov, D. V., Doarn, C. R., and Merrell, R. C., 2000, Wireless ECG monitoring by telephone, *Telemed J E Health.* **7**(1):33-38.

Otsu, Y., Choh, Y., Yamazaki, I., Kosaka, K., Iguchi, M., and Nakajima I., 1986, Experiments on the quick-relief medical communications via the Japan's domestic communication satellite CS-2 for the case of disasters and emergencies, *Acta Astronaut.* **13**(6-7):459-466.

Plischke, M., Wolf, K. H., Lison, T., and Pretschner D. P., 1999, Telemedical support of prehospital emergency care in mass casualty incidents, *Eur J Med Res.* **9**(4):394-398.

Satish U. and Streufert, S., 2002, Value of a cognitive simulation in medicine: towards optimizing decision making performance of health care personnel, *Qual. Saf. Health Care.* **11**(2):119-120.

Scott, J. C., 1993, Communications infrastructure requirements for telemedicine/telehealth in the context of planning for and responding to natural disasters: Considering the need for shared regional networks, *International Telemedicine/Disaster Medicine Conference, December 1991. Proceedings* edited by H. C. Holloway, A. E. Nicogossian, D. F. Stewart, *NASA Publication NP-207.*

Teich, J. M., Wagner, M. M., Mackenzie, C. F., and Schaefer K. O., 2002, The informatics response in disaster, terrorism, and war, *J Am Med Inform Ass.* **9**(2):202-203.

Young, F. R., 1995, Telemedicine's role in domestic catastrophic disaster preparedness and response, *J Med Sys,* **19**(1),175-187.

American Red Cross (Washington DC, February 12, 2003); www.redcross.org/services/disaster.

Center for Public Health and Disaster (Los Angeles, January 27, 2003, www.ph.ucla.edu/cphdr/projects.html.

Centers for Disease Control/National Center for Global Health (Atlanta, February 12, 2003); www.cdc.gov/nceh/emergency/disasterepedimiology/description/surveillance.htm.

International Telemedicine/Disaster Medicine Conference Proceedings, December 1991. Editors HC Holloway, AE Nicogossian, and DF Stewart, NASA Publication NP-207, (1993).

International Virtual E-Hospital (Prishtina, Kosova, January 2003); www.IVEhospital.org.

International Federation of Red Cross and Red Crescent Societies. *World Disasters Report* (Continental Printing, Geneva, 1999)

Operation Strong Angel (San Diego, February 11, 2003); http://www.strongangel.org/.

Red Crescent (February 12, 2003); www.ifrc.org/disasterpreparedness.

The East Carolina University Telemedicine Center (Greenville, February 11, 2003); http://www.telemed.med.ecu.edu/.

World Health Organization (New York, February 12, 2003); www.who.int/en/

REMOTE MONITORING FOR HEALTHCARE AND FOR SAFETY IN EXTREME ENVIRONMENTS

Val Jones[*], Nadav Shashar, Oded Ben Shaphrut, Kevin Lavigne, Rienk Rienks, Richard Bults, Dimitri Konstantas, Pieter Vierhout, Jan Peuscher, Aart van Halteren, Rainer Herzog, and Ing Widya

1. INTRODUCTION

In this chapter we examine the potential use of remote health monitoring using Body Area Networks (BANs) to support individuals who are working or pursuing recreational activities in extreme environments.

Monitoring can be defined as (periodic or continuous) measurement and observation of dynamically changing physical attributes of a system. The observing agent may be a human or may be another system. A monitoring system may be coupled with a control system, where output of sensors is used to effect changes to the system via a feedback control loop. To give a simple example, a central heating control system incorporates monitors (temperature sensors) whose readings are used as feedback to control the operation of the thermostat, which effects changes in the process such as switching the pump and/or heating element off and on. Sensors simply 'listen', whilst actuators cause some physical effect such as activating a pump.

In health monitoring the monitored systems are human physiological systems and the attributes measured are variable parameters of those systems. An example would be monitoring of the functioning of the cardiovascular system by measuring ECG, oxygen saturation and blood pressure. In *remote monitoring* the sensor data is transmitted to a distant location for some kind of processing.

A combination of technological developments have made remote monitoring of physiological signals a feasible option in healthcare. These advances include: development of low-cost lightweight miniature biosensors; new non-invasive and minimally invasive measurement techniques; small but powerful mobile devices

---

[*] Val Jones, Centre for Telematics and Information Technology/Faculty of Electrical Engineering, Mathematics and Computer Science, University of Twente, PO Box 217, 7500 AE, Enschede, The Netherlands, v.m.jones@ewi.utwente.nl.

incorporating both computation and communication functionality (e.g. programmable mobile phones, PDAs with telecommunications extensions, computer-cell phone hybrids); and advances in wireless communication technologies (both short and long range).

Benefits associated with remote patient monitoring include enhanced freedom for the patient, reduced demand for hospital beds and the ability to control situations remotely location in order to prevent, detect or manage accidents and emergencies.

Many patients require monitoring for short or long periods of time. The reasons are manifold; some examples are: information gathering for the purpose of patient assessment (e.g. recording and analysis of the cardiac rhythm by means of the electrocardiogram in order to confirm diagnosis of arrhythmias); the immediate detection of medical emergencies signaled by changes in vital signs such as blood pressure or blood oxygen saturation or other biosignals, allowing rapid response when alarm conditions are raised; and routine measurements needed for management of chronic conditions (e.g. testing blood glucose in patients with diabetes mellitus). Nowadays the possibility of uploading sensor data over a telephone or Internet link means that patients who formerly had to be hospitalized for periods of monitoring can now be monitored at home. This alternative frees up hospital beds and staff (and thus reduces costs) and at the same time gives patients more freedom and better quality of life during periods of monitoring. Wireless communications bring the prospect of further freedom since in principle patients can now move freely anywhere in the world during monitoring, thanks to the development of wearable sensors and assuming global wireless coverage to enable transmission of biosignals.

In different situations different kinds of monitoring are appropriate. Some clinical conditions may call for signals to be measured without interruption and a continuous readout generated. An example would be measurement of blood pressure in a patient undergoing surgery, where BP is continuously measured by means of a catheter inserted into an artery. In other cases signals need to be sampled with a certain periodicity. Sampling rates can vary depending on the variable measured and the purpose of the monitoring. A typical sampling rate for oxygen saturation would be 5 times per second; for EMG sampling rates would be at least 2000 samples per second. This is also referred to as *continuous monitoring*. In other cases measurement may need to be done only on a daily or weekly basis.

The sampled biosignals may require real time processing. The processing may be done locally, as with a simple biofeedback device which gives real time auditory feedback and can be used to train patients to manage stress. Another example is the implanted cardiac pacemaker, which continuously measures heart rhythm and generates an electrical pulse in order to stimulate the cardiac muscle in case of too slow a heart rhythm. Another example is an automatic internal cardioverter/defibrillator (ICD), which acts if the heart rhythm exceeds a certain rate by delivering an electrical shock in order to restore the normal heart rhythm. These are examples of local processing of biosignals. If however the application is running remotely, at the hospital or health call center for example, then there may be a need for synchronous communications (generating an *always on* requirement for the communications). If post-hoc processing is sufficient then asynchronous communications will suffice and wireless communications will only be needed if there are other factors present (e.g. too much data is captured to be stored locally on the patient-worn equipment, or an alarm function is needed). An example of this is the continuous recording of heart rhythm over a period of several weeks for the diagnosis of infrequently occurring arrhythmias in patients with recurrent collapses using a recording device with limited data storage capacity. When a collapse occurs, data transmission through telephone lines from the device to the hospital enables evaluation of the cardiac rhythm during the collapse. Typically the *always on* feature is needed where a patient is at risk from sudden serious crises which would require rapid intervention. An example is the use of telemetry applied to in-patients with potentially lethal arrhythmias in the coronary care unit or on the ward. In the latter case, patients are free to walk around but their ECG is monitored continuously and instantaneously transmitted to a central control monitor. The monitor is observed around the clock by nurses trained to interpret the data and to act immediately in the event of a life threatening arrhythmia.

2. BODY AREA NETWORKS FOR REMOTE MONITORING OF PATIENTS

A wide range of patient monitoring equipment is used in healthcare. Some patient monitoring applications are described in another chapter of this book entitled *MobiHealth: mobile health services based on body area networks.*

Wearable monitoring devices are also marketed for use in various work or recreational settings to monitor a range of parameters. Use of heart rate monitors during fitness training is commonplace. Some devices incorporate motion sensors or environmental sensors. An example is the use of drop sensors and gas sensors for workers in hazardous environments such as chemical plants. An example of environmental sensors in recreation is the dive computer used by many scuba divers which displays parameters such as water temperature, current depth, maximum depth, dive time, pressure, ascent rate and time-to-fly.

In the remainder of this section we introduce the concept of the Body Area Networks (BANs), and discuss the MobiHealth (MobiHealth, 2002) approach to remote monitoring.

2.1. Body Area Networks

We define a BAN as "a collection of (inter) communicating devices which are worn on the body, providing an integrated set of personalised services to the user" (WWRF, 2001). According to our definition a Body Area Network (BAN) incorporates a set of devices which perform some specific functions and which also perform communication. Communication amongst the elements of a BAN is called *intra-BAN communication*. If the BAN communicates externally, that is with other networks (which may themselves be BANs), this communication is referred to as *extra-BAN communication*.

2.1.1. Intra-BAN Communications

Intra-BAN communication may be transported over a wired or a wireless medium. Wired options include copper wires, optical fibres and various 'wearable computing' solutions. Wireless options include infrared light, microwave, radio and even skin conductivity.

2.1.2. Extra-BAN Communications

Extra-BAN communication may be based on wireless technologies. A range of wireless communication technologies is available including Bluetooth, WLAN, GSM, GPRS and UMTS. Bluetooth and WLAN (Wireless Local Area Network) are short and medium range communication technologies whilst GSM, GPRS and UMTS are wide area technologies, offering the possibility for the wearer of the BAN to roam freely outdoors within the coverage area.

2.1.3. Body Area Networks for Healthcare

The *Body Area Network* is a generic concept which can have many applications in various domains. When the devices of a BAN measure physiological signals or perform other actions for health-related purposes we can call this kind of specialization of Body Area Networks a *health BAN*. As a further specialisation, a BAN whose devices are biosignal sensors enables patient monitoring. If we combine health BANs with *extra-BAN communications* this enables remote monitoring of patients. In the following section we briefly describe the approach of the MobiHealth Project to the use of health BANs for remote monitoring of patients.

2.2. The MobiHealth Body Area Network

The vision of the MobiHealth project was to develop an open extensible BAN platform which allows integration of different health functions by means of a plug-and-play approach and thus to enable transmission of biosignals to a distant healthcare location. This enables remote management of chronic conditions and detection of health emergencies whilst giving the wearer the freedom to move anywhere in the world. Further details on the MobiHealth project and the MobiHealth BAN can be found in another chapter of this book entitled *MobiHealth: mobile health services based on body area networks*. An extension to the MobiHealth trauma application to scale up to disaster scenarios is described in another chapter in this section of the book entitled *Telematics Support for Disaster Situations*.

2.2.1. Architecture

The MobiHealth BAN consists of a central unit known as the MBU (Mobile Base Unit) and a set of devices, which may include sensors, actuators, and various multimedia devices. The MBU handles coordination, local computation and communication functions. The sensor data is processed by a sensor front-end before being transmitted to the MBU. A range of front-ends can be associated with an MBU, enabling customization of the BAN. The MBU currently used in the MobiHealth trials is based on the HP iPAQ platform. Future plans include porting to a cell phone platform such as the Sony Ericsson P800.

2.2.2. Devices

The generic BAN can be customized with different sensor sets for different applications. The main devices used in the MobiHealth trials were sensors, e.g. electrodes for measuring ECG, pulse oximeters for measuring oxygen saturation and calculating heart rate, sensors for measuring NIBP (non invasive blood pressure) and motion sensors.

2.2.3. Intra-BAN Communications

The MobiHealth approach allows for wired or wireless communications within the BAN, or a combination of the two. The BAN currently uses a combination of wired communication from sensors to sensor front end (later this could be replaced by smart sensors and wireless Zigbee technology) and Bluetooth.

2.2.4. Extra-BAN Communications

In the MobiHealth trials extra-BAN communication was via UMTS and GPRS. In many remote locations however the only alternative is satellite communications, since terrestrial links are not available everywhere.

2.3 Clinical Settings in the MobiHealth Trials

The overall goal of the MobiHealth project was to test the ability of 2.5 and 3G infrastructures to support value added services. Trials were conducted in four European countries in the areas of acute (trauma) care, chronic and high-risk patient monitoring and monitoring of patients in home-care settings. In this chapter we examine some possible applications of the MobiHealth concept for non-patients.

2.4 Applicability in a Wider Context

As well as serving patients, remote monitoring can also benefit non-patients going about their daily life activities. A special case of this is the use of monitoring to support work or sport in remote or extreme environments. In the following section we examine the role that monitoring can play in improving safety in extreme environments.

3. MONITORING FOR HEALTH AND SAFETY IN EXTREME ENVIRONMENTS

Extreme environments place unusual demands on equipment and human physiology, making even normal activities more complicated and difficult than usual. If people get into difficulties the evacuation process may be lengthy and complicated. Since activity in these environments is physically demanding, it is relatively simple to take preventative measures such as resting, if early warning of a problem can be given through *monitoring*. Monitoring *and transmission* of biosignals can also enable more sophisticated telemedical assistance. In this chapter we focus particularly on underwater and polar environments.

Extreme environments can be considered as environments in which one or more of the following characteristics apply: a) the environment imposes severe difficulties with respect to human existence, often requiring the aid of special equipment for survival and basic functioning; b) environmental conditions make equipment functioning and communication difficult; and c) evacuation of patients or casualties is made especially difficult because of environmental conditions. This definition includes the obviously extreme settings such as deep water diving, mountain climbing or crossing the ocean solo on a sailing boat. Our definition also covers extreme situations in daily communal life such as firefighters entering burning buildings (*extreme* by our definition due to the problems of evacuation, also extreme due to environmental conditions such as heat, poor vision and toxic gases), or even swimming in a swimming pool (extreme due to limits in the functioning of monitoring equipment and transmission of data underwater).

In the current chapter we focus on two of the more remote and severe environments (underwater and polar environments), since solutions fitting these cases can be adapted to less exotic situations.

3.1 The Underwater Environment – Monitoring for Scuba Dvers

Scuba divers face many health hazards including nitrogen narcosis, oxygen toxicity, decompression sickness and lung over-expansion injury. Monitoring of the diver's physiological status and transmission of biosignals to a remote medical facility can aid in prevention and management of diving accidents.

The underwater environment fits our definition of "extreme" on all three criteria: human existence is impossible without the supply of a breathing gas mixture (air, NitrOx, or other technical mixture), weights are required to remain submerged, a protective diving suit is almost always needed, and even movement and vision require special equipment (fins, mask). Operation of any electronic equipment is difficult, firstly due to the requirement for encasing it in a water-proof and pressure-proof housing, and secondly due to the conductivity of the water which inhibits the use of some types of sensors. Underwater wireless communication is much more difficult than through the air and methods such as cabled communications or acoustic transmission are usually preferable. Evacuation of patients injured in diving operations is rarely easy. Since the breathing gas supply is limited the duration of the dive is planned and therefore known in advance. As a result notice of a problem underwater rarely takes more than an hour after its occurrence. Unfortunately, this type of "notification" (that is, the diver fails to surface within their known air capacity time) always means a serious and life threatening problem. Most divers do not carry any means of communication with the surface and rely on the assistance of their dive buddy.

Once a problem is detected, evacuation has three independent stages - getting the victim safely to the surface; carrying the victim to shore/boat and out of the water; and evacuating the victim to a medical center or a hospital capable of treating diving related problems. Even the last stage of the evacuation is complicated by the fact that, although air evacuation may be possible, the rescue aircraft must fly at very low altitudes to avoid (further) decompression effects on the casualty.

3.1.1 The Underwater Environment and its Hazards for Human Physiology

Human beings are poorly adapted for underwater activities and cannot survive without the aid of life support systems. Even with life support many of the diver's senses and capabilities are grossly impaired and a diver's ability to perform underwater is only a fraction of his/her ability on dry land.

The most dramatic effects of the underwater environment on human divers occur due to immersion and the rapid changes of ambient pressure which occur with depth. These two factors can cause a variety of related problems ranging in severity from mild to life threatening. Included among them are exposure and hypothermia, reduction of vital capacity of the respiratory system, over expansion of the lungs and other aspects of barotrauma (pressure injuries) and the various manifestations of decompression sickness (DCS).

At depth or under elevated ambient pressure, some of the gases breathed by divers dissolve into body tissue, in accordance with Henry's Law. During the decompression phase tissue becomes over-saturated with inert gas and the process of elimination of gas from the tissue takes place. Over-saturation beyond a certain value (critical over-saturation) can result in bubbles of inert gas forming in tissue and blood. The value of critical over-saturation as well as the rate of gas intake and elimination varies according to difference in tissue, and environmental and physiological factors.

During descent micro bubble formation does not occur, since all nitrogen is dissolved in the blood. It is during and after ascent that the micro bubbles emerge in the tissues and are transported through the blood to the lungs. These micro bubbles may cause local problem in joints, bones, and other tissue or through embolism in for instance the lungs or brain. Micro bubbles can be detected by ultrasound or doppler.

DCS has many different manifestations and is best categorized according to the organ or tissue effected. Brain, inner ear, spine, peripheral nerves, respiratory system, gastro-intestinal tract, musculoskeletal structure and skin are among the sites where DCS may manifest, with symptoms as variable as back pain, headache, apathy and malaise, and haematological changes. In most cases, the onset of symptoms will take place within six hours of ascent but there are well-recorded cases where symptoms of DCS have appeared after more than 24 hours. On the other hand, mostly in severe cases, symptoms may appear during ascent.

Over-expansion of the lungs (Pulmonary Barotrauma of ascent.) A diver using any kind of breathing apparatus is breathing gas at a pressure equal to the ambient pressure. By inhaling and exhaling continuously the pressure in the diver's respiratory system is constantly equalized to the ambient pressure, with no ill effects. This is true even if the ambient pressure is changing due to changes of depth.   If however a diver fails to equalize the pressure in the lungs during ascent, because he holds his breath or due to various medical conditions which prevent adequate venting of excess gas through the bronchioles, the gas trapped in the lungs will expand. Expansion of gas in the respiratory system can result in stretching of alveoli at a first stage, which would be followed by alveoli rupture if pressure is not relieved.

There are a number of phenomena associated with ruptured alveoli, all of which are dangerous, and some of which are severely life threatening. They are: inadequate gas exchange through the blood gas barrier in the alveoli due to damaged lung tissue, mediastinal and subcutaneous emphysema, pneumothorax and arterial gas embolism. These are among the most common diving related medical traumas.

The supply of oxygen to the diver has to be secured in order to survive and to perform labour, moreover the removal of carbon dioxide has to be secured. At greater depths, the use of compressed air is not suitable because of the risk of nitrogen narcosis, so other gases have to be used to avoid this problem.

Other risks to the diver are ear barotrauma, hypothermia, exhaustion, dehydration and decompression sickness. Other hazards result from currents, poor visibility, water pollution and toxic marine animals. Simple rupture of an eardrum can lead to vertigo and disorientation.

Depending on the situation, a diver may get his supply of oxygen from a surface station, and speech communication with the surface can be possible using full face masks and the umbilical which delivers the gas mixture. In many cases however, divers are self-supporting, carrying a limited amount of breathing gases, and communication may only be possible to another diver visually or, in case of poor visibility, through a line connecting two divers.

It is sometimes difficult to assess the actual condition of the diver. Especially in cases of nitrogen intoxication, the self assessment of the diver is impaired by a lowered and blurred consciousness. This can cause major judgment errors with possibly fatal outcome.

In addition to specific diving illnesses, the physical demands of diving and exposure to conditions which are malign to the diver - pressure, temperature, limited air supply - can trigger other medical events during a dive. For example heart attacks occur during diving with higher frequency than on land (comparing a population matched on health status, levels of activities, etc.). These events are considerably more deadly when occurring underwater than on land. This is simply because of the complications involved in detecting the problem and in fast but safe evacuation of the patient to a proper medical facility. Statistics on accidents and deaths show that cardiovascular events (heart attacks) cause 20 to 30 percent of all deaths which occur during scuba diving (source: Divers Alert Network). In most cases, early detection and termination of the dive could have prevented the event.

So, early warning to a diver of the possibility of an imminent medical situation can give the diver or his buddy the opportunity to take preventive measures. Early notification to the surface (to a diving monitoring post, a boat, or a shore base) of a diving emergency and of its nature, can trigger proper evacuation procedures which can mean the difference between life and death to the casualty.

The physiological measurements which would be most useful in order to assess the condition of the diver include heart rate, respiration rate, body temperature, oxygen and carbon dioxide content in the respiratory air, oxygen saturation and pressure and carbon dioxide pressure in the blood. Also useful would be blood pressure, dehydration level and microbubble formation.

3.1.2 Sensors

The obvious environmental conditions affecting underwater sensors are the wet (and sometimes saline) environment, temperature, pressure and poor visibility. These problems can be solved and many underwater sensors are in regular use by marine scientists engaged in environmental monitoring and as basic functions on existing dive computers. Divers are accustomed to carrying a considerable amount of gear and the diving suit itself provides a fairly stable framework or scaffold for positioning of sensors. The air-providing regulator, which is by definition in the mouth of the diver, provides another possible platform for sensors.

Due to the short duration of most dives, equipment does not require a long-life power supply and it is possible to recharge batteries between dives. Oxygen saturation level of haemoglobin can be measured optically. Positioning would aid location of the divers and could be transmitted in the form of a beaconing signal.

At the time of writing a few top-of-the-range dive computers offer breathing rate as one display feature. One wrist worn cardiac monitor targeted at the sports and training market is advertised as being water resistant to depth of 20 metres, but apparently has not been tested for divers.

3.1.3 Communications

Wireless communications are technically more difficult underwater than on land. Some short range wireless solutions are commercially available, for example wireless transmission of tank pressure from a tank mounted sensor to a wrist-worn dive computer. However the underwater environment limits the possibilities for long range wireless data communications at reasonable bandwidths. Wireless optical transmission is rather impractical underwater due to limitations on visibility. Low frequency radio is used by the military but requires high power and is not practical for recreational diving. Acoustic (audio) communication is the underwater wireless transmission mode of choice for recreational diving, offering limited bandwidth but good range. For voice communication various 'buddy phone' products are commercially available based on accoustic transmission (for example using ultrasonic, upper single sideband transmission). These can operate down to at least 40 metres in seawater with a range of anything up to 500 metres depending on sea conditions, and can be used diver to diver or diver to surface. However the price may be beyond the reach of many recreational divers. Data communications are a different story especially where transmission of high volume data such as certain biosiognal data or audio-video is required. If wired communication is possible this offers by far the best alternative for extra-BAN communication. A wired communication link can for instance be fully integrated with a diver's support line. This is not a problem where divers use surface supply breathing mixture (as is the case in many commercial diving operations), where an additional wire can be easily added to the diver's existing umbilical but it is not an option for divers using SCUBA (Self Contained Underwater Breathing Apparatus) gear. In cases where use of a cable is impossible or impractical, communication will need to be based on acoustic modulations and coding and compression of information. Such an acoustic signal could be received by a manned installation, such as a dive boat or a barge, or by a receiver attached to a surface buoy, where the signal could be converted to electromagnetic form and transmitted from the buoy through the air. Acoustic signals using low power high frequency transmitters have a relatively short range (in theory up to hundreds of metres, but usually tens of metres); hence there needs to be a receiving station within this range.

One of the limitations of acoustic systems is the difficulty of deriving accurate location information based on acoustic signals. Though such systems do exist, they are mainly deployed during accurate geographical surveys and other such operations requiring precise location control. Therefore the emergency function of the BAN system, the activation of which implies a call for immediate help, will need to produce a beaconing type acoustic signal. Obviously a matching detector will need to be available to the surface/rescue team. If a buoy based system can be used for relaying from the surface, a GPS location could be added to its transmissions topside. Diver carried portable location equipment is now commercially available which can operate when the diver is at the surface. It is intended to give divers short range communications so they can find their way back to a dive boat, or so the support team onboard can locate a diver in need of help. Surface voice communication systems to link divers in the water and surface support teams are also available. For longer range topside communication offshore the global solution is satellite transmission.

Location information is difficult underwater (GPS is not available underwater) but underwater location systems are available using a triangulation system, which requires (at least three) underwater installations at the dive location, in other words the systems relies on a fixed underwater infrastructure. This solution is suitable for certain kinds of dive sites where professional or recreational diving occurs regularly within the same area.

Intra-BAN communication is preferably integrated into the diving suit and other diver worn equipment. Wires (copper, optical fibre) are options. The special environmental factors of wetness, salinity and pressure have to be taken into account.

3.2 Monitoring in Extreme Cold Environments

Of the three criteria for extreme environments (severe hardship of human survival, on equipment functioning and communication, or on evacuation), polar regions present examples of two of the criteria and the Antarctic continent of all three; human survival requires special gear for constant protection as well as for performing the most simple of tasks. Communications are fairly straightforward in populated regions of the Arctic circle and often can rely on satellite coverage, but in more remote areas of the Arctic circle and in the Antarctic heavier systems using iridium satellites provide the backbone of communication. Evacuation can be very difficult, always requiring special equipment ranging from helicopters to specialised snow vehicles. Even so, weather conditions frequently prevent evacuation operations and during wintertime many field locations are totally cut off.

The extremely harsh climate of the world's polar regions imposes a spartan existence upon visitors, natives and local people, the land and sea based workers, mariners, scientists and others who live and operate there. The northern hemisphere polar regions, north of the artic circle, are sparsely populated with remote villages and towns, in contrast human activity in Antarctica is generally restricted to research stations and remote camps. However, with the increasing tourist movement in the Antarctic, and in agreement with the Antarctic Treaty, travel within the continent is rapidly expanding beyond the well-planned and monitored scientific expeditions to include also many groups of tourists and adventurers. Travel and research in these regions generally occurs during the summer months, which are characterized by 24 hour sunlight and warmer temperatures. During the dark winter months northern polar towns and villages remain active whilst in Antarctica the continental population plummets and 'winter-over' individuals are restricted to research stations.

The record for the coldest temperature on the earth's surface - –89.6°C - was measured at Russia's Antarctic research station Vostok on July 21, 1983. The mean temperatures in the Antarctic interior range from – 40°C to –70°C during the coldest month, and from -15°C to -32°C during the warmest month. On the coast, temperatures are considerably warmer: -15°C to -32°C in the winter, and from $+5^{\circ}$C to -5°C in the summer, with the Antarctic Peninsula experiencing the highest temperatures year round.

Arctic regions are somewhat milder with average summer temperature in northern polar regions around 7.2°C, and annual rainfall less than 4 inches; but nonetheless winter conditions and winter storms can halt any outdoor human activity.

The difficulty of evacuation from polar regions means that intervention via telemedicine can be vital, hence the potential value of communication of biosignals. Though primary importance should be given to monitoring vital signs and emergency related signals, a biomonitoring system which could measure and transmit a wide range of parameters is a vital prerequisite to remote surgery. The need for such remote surgical intervention was demonstrated in the tele-surgery performed in July 2002 at the South Pole research station, executed by the South Pole's physician with expert assistance given via a telemedicine link-up with Massachusetts General Hospital.

In cases where evacuation is needed, accurate positioning is needed for an evacuation team to locate the casualty. This could be via a beaconing/pinger type signal, however transmission of accurate GPS coordinates would be more useful for the long-range approach of the evacuation team.

3.2.1 The polar environment and its hazards for human physiology

The polar environment presents a range of challenges to human physiology. Hypothermia, simply defined as having a lower than normal temperature, is perhaps surprisingly not the most common hazard to human physiology in polar regions. It is, however, the easiest hazard which can be monitored via a sensor kit, primarily via measurement of core body temperature. A mildly hypothermic individual will have a temperature around $32.7\text{-}35^{0}$C, while a profoundly hypothermic person will have a core body temperature of 32.2^{0}C or lower. When the body's core temperature approaches 32.2^{0}C, the chance for core body systems failing increases. As core temperature drops, the body's muscular system becomes stiff and weak. The heart, being a muscle, reacts in the same way, and the number of beats per minute will begin to fall until, in a state of deep hypothermia, a heart rate of 20 beats per minute or less is not uncommon. Heart rate is an ancillary measurement that can indicate hypothermia.

To combat hypothermia, and to prepare for emergencies, people working in polar regions tend to carry a large amount of equipment, from protective gear to special support instrumentation. Carrying such a weight, in addition to the physical demands of moving and working in harsh conditions, can rapidly bring exhaustion and in extreme cases dehydration. In severe cases hypothermia can develop. Unfortunately, exhaustion and dehydration affect human physiology in synergy with hypothermia, hence exacerbating the patient's condition.

Since every venture outside shelter is physically demanding, any pre-existing medical conditions, and especially heart, lung, circulation and muscular problems, have a higher chance of manifesting. We have already seen that this is true also in the underwater environment.

The emergency related measurements we would like to monitor are divided into two categories: a) General distress signals, which should be measured, when possible, everywhere; b) Specific signals and vital signs that may warn of health related problem common to the specific environment at hand (in the polar region mainly hypothermia).

Of the first (general distress signals) one would like to measure blood pressure, heart rate, pulse shape, respiration/ventilation rate, dehydration level and temperature. Pulse shape would be useful as a trigger for immediate alarms, however it is not clear how this can be measured. Current devices work on heart rate. Nor is it clear how dehydration level can be measured. Hypothermia related signals are core body temperature and temperature of skin areas which are usually protected such as the chest and lower abdomen. In addition GPS positioning should be available for transmission at all times.

A BAN that is capable of transmitting emergency medical information may also be used for other medical situations. Stations in the polar regions are often isolated and therefore expert medical advice and telemedicine intervention can be critical. For such cases, small specialised sensor sets can be added to measure specific information. For example, in pregnancy and delivery situations, monitoring of contraction rate, maternal and foetal pulse rates, and of blood flow in amnionic cord or placenta would add important information if remote consultation is needed. In addition the system should be able to transmit audio and video signals from mobile and stationary medical facilities/ and also from vehicles.

3.2.2 Sensors

One function of the BAN is to alert its wearer of the approach or development of a potential medical emergency so that precautionary measures can be taken. If the situation reaches a critical stage then an alert to a medical alert and evacuation center needs to be transmitted. In such a case the exact location of the patient needs to be transmitted and in addition any medical information that will assist in the evacuation, and enable provision of first level medical treatment. The sensor sets and BAN for extreme cold environments can be the same as for the ordinary BAN applications, with the added proviso that battery packs should be of arctic specification.

3.2.3 Communications

The poles provide unique communication challenges. In both polar regions handheld radios, HF radios and satellite phones are available. The limitation of a handheld radio is the number of repeaters in position while HF radios require large antennas. Cellular phone technology is available is some parts of the northern polar region but coverage is inconsistent. Satellite communication could be used in both regions. The best solution would seem to be a combination of technologies based on satellite coverage but switching opportunistically to terrestrial technologies where available (this kind of solution is variously referred to as "beyond 3G", "3G and beyond" and "4G").

4. CONCLUSIONS

The use of Body Area Networks together with various wireless communication technologies offers one approach for remote monitoring. Convergence of communication and computation functionality, specifically seen in programmable cell phones and in pocket computers with telephony functionality, gives a choice of implementation platforms for Body Area Networks with extra BAN communications carried over (terrestrial) wireless telephony or via satellite links. As sensors get smaller and cheaper and self-powering devices become a reality, telemedicine services based on these technologies will become more viable and more convenient to use.

As well as serving patients in ordinary circumstances, remote monitoring has a special role to play in helping to maintain health, and even save life, for people who happen to live, work or play in various extreme environments.

5. ACKNOWLEDGEMENTS

Our thanks go to the European Commission, who fund the MobiHealth project under the IST FP5 programme, and to the MobiHealth and xMotion partners. Thanks are also due for the support of: the University of Twente, (especially APS group, Department of Computer Science and CTIT); the Inter University Institute, Eilat, Israel; and Ben Gurion University, Beer Sheva, Israel, who supported Val Jones' sabbatical leave early in 2002 during which part of this research was conducted, thanks too to the many colleagues at UT, BGU and IUI who contributed knowledge, ideas and support during those three months.

6. REFERENCES

BlueTooth, 2003; http://www.bluetooth.org/

van Dam, K, S. Pitchers and M. Barnard, 'Body Area Networks: Towards a Wearable Future', Proc. WWRF kick off meeting, Munich, Germany, 6-7 March 2001; http://www.wireless-world-research.org/.Jones, V. M., Bults, R. A. G., Konstantas, D., Vierhout, P. A. M., 2001a, Healthcare PANs: Personal Area Networks for trauma care and home care, Proceedings Fourth International Symposium on Wireless Personal Multimedia Communications (WPMC), Sept. 9-12, 2001, Aalborg, Denmark, http://wpmc01.org/, ISBN 87-988568-0-4

Jones, V. M., Bults, R. A. G., Konstantas, D., Vierhout, P. A. M,. 2001b, Body Area Networks for Healthcare, Proceedings Wireless World Research Forum meeting, Stockholm, 17-18 September 2001; http://www.wireless-world-research.org/

Konstantas, D., Jones, V. M., Bults, R. A. G., Herzog, R., 2001, MobiHealth – innovative 2.5 / 3G mobile services and applications for healthcare, Thessaloniki.

Lymberis et al., Smart Mobile Applications for Health Professionals, 2003, IN M-Health: Emerging Mobile Health Systems, Robert H. Istepanian, Swamy Laxminarayan, Constantinos S. Pattichis, Editors, Kluwer Academic/Plenum Publishers.

MobiHealth, 2002; http://www.mobihealth.org

Schmidt, R., 2001, Patients emit an aura of data, Fraunhofer-Gesellschaft, www.fraunhofer.de/english/press/md/md2001/md11-2001_t1.html

Widya, I., Vierhout, P. A. M., Jones, V. M., Bults, R. A. G., van Halteren, A., Peuscher, J. and Konstantas, D., Telematics Support for Disaster Situations, 2003, IN M-Health: Emerging Mobile Health Systems, Robert H. Istepanian, Swamy Laxminarayan, Constantinos S. Pattichis, Editors, Kluwer Academic/Plenum Publishers.

Wireless World Research Forum, 2001, The Book of Visions 2001: Visions of the Wireless World, Version 1.0, December 2001; http://www.wireless-world-research.org/

xMotion , 2002; http://www.ist-xmotion.org/

ZigBee Alliance, 2003; http://www.zigbee.org/

Zimmerman, T.G., 1999, 'Wireless networked devices: A new paradigm for computing and communication', IBM Systems Journal, Vol 38, No 4.

CHRONIC PATIENT'S MANAGEMENT: THE COPD EXAMPLE

Francisco del Pozo[*], Paula de Toledo, Silvia Jiménez, M. Elena Hernando, and Enrique J. Gómez

1. INTRODUCTION

Care of chronic patients within any health care model around the world shows two main characteristics: first, it is responsible for a very significant part of the total health care costs and second, their intervention paradigms, based mostly on acute care by episodes, adapt with difficulties to the needs of these kind of patients. This situation opens a wide margin of potential improvements through new paradigms to cope efficiently with these needs, including, as a minimum:

1. Care should be provided at any time and any place, depicting the need for efficient, ubiquitous and secure extra-institutional care and
2. Efficient coordination tools should be provided to all health professionals dealing with each single patient, to allow the implementation of patient centred continuous shared-care spaces.

This change of the traditional care delivery structures is in progress over recent years, resulting in an increasing emphasis on: maintaining the health of citizens through proactive disease management, health education (i.e. disease awareness and lifestyle), preventive programmes and citizen self care programmes, with an increased involvement of community based care providers.

The Chronic Disease Management Model (CDMM) described in previous papers of the authors (Del Pozo 2002; De Toledo 2002), is our vision of that paradigm; well in agreement with the Continuum of Care Model recently worked out by the World Health Organization (WHO) (Pruitt 2002). A model that pays particular attention to the human resources and organizational implications, identified as the bottle-neck of the shift to the new paradigm of care; where the pivotal aspects will be the new types of interactions

[*] Francisco del Pozo. Grupo de Bioingeniería y Telemedicina, E.T.S. Ingenieros de Telecomunicación, Universidad Politécnica de Madrid, Ciudad Universitaria s/n 28040 Madrid, Spain

between actors in a technology enabling framework: 1) health care professionals working from different levels of care; and 2) between patients and professionals.

For the sake of clarity of the presentation that follows we concentrate on a specific group of chronic patients, chronic respiratory diseases, an important burden on healthcare systems worldwide (Pauwels 2001), which is expected to increase over the forthcoming two decades (Murray 1997), particularly due to chronic obstructive pulmonary disease (COPD). Winter outbreaks of COPD exacerbations mostly occurring in elderly people with concurrent chronic co-morbidities often generate dramatic increases of hospital emergency room admissions with subsequent dysfunctions in the healthcare system. It is estimated that hospitalizations of COPD exacerbations represent approximately 70% of the overall costs associated with the management of the disease (Strassels 2001).

A first feasibility analysis of home-based services to prevent conventional hospitalizations of COPD exacerbations was reported in by Gravil et al (1998). Three subsequent controlled trials (Skwarska 2000; Cotton 2000; Davies 2000), also conducted in the United Kingdom, have demonstrated both safety and cost reduction when these type of services are applied to selected COPD patients. There are also some experiences demonstrating the benefit derived from the use of home care multidisciplinary teams interventions (Stewart 1999).

In the following sections we present the technological platform developed to support the Chronic Disease Management Model (CDMM) for COPD patients, following the generic criteria of the new chronic care paradigm and taking advantage of the possibilities of the new information and communication technologies, notably wireless technologies.

The platform has been mostly developed within the European Union partially funded project CHRONIC (IST–1999–12158), with some contributions from HealthMate (IST–2000–26154). The system was developed to demonstrate the adequacy of the platform to the CDMM proposed and the cost-efficiency of the services built on top of it.

2. CHRONIC DISEASE MANAGEMENT PLATFORM

The system developed to support the Chronic Disease Management Model has been designed to configure an environment that includes two main functionalities:

1.   A virtual and ubiquitous cooperative working space, a continuum of care space, to coordinate at any time and place all professionals of the multidisciplinary care team (primary care personnel, specialists based at different hospitals, etc.). A space where professionals procure common targets centred on the patient, with predefined and agreed care plans, making use of tools to optimise the efficiency of all available resources and tasks.
2.   A multi-access system that grants access to patients and professionals to any service available adapted to his/her better convenience and needs, including telemonitoring services if required by the patient.

To support these functionalities, the system developed integrates, as shown in Figure 1, two main elements: A **Chronic Patient Management Centre (CPMC)** that provides all the services needed to manage the chronic patients and a set of **Remote Access Units** for patients and professionals to access the available services at the CPMC, wherever required: hospital, primary care centre, while visiting the patient, the home, etc.

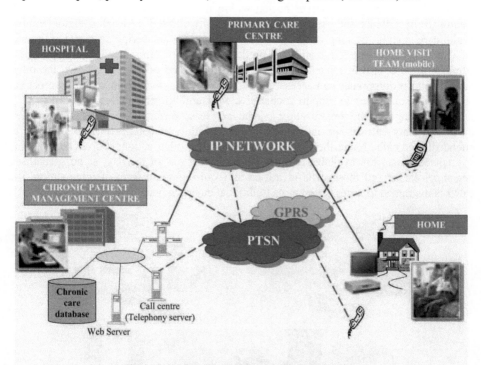

Figure 1. Chronic Disease Management Model platform

Connectivity among all agents involved is based on a multi-access approach to allow professionals and patients the use of the access device most appropriate to their specific needs and context of use. Figure 2 shows possible accesses to the services coordinated through the CPMC. It includes both, fixed lines based access devices: PC-based set top box, web TV or web phone; and wireless: GSM or GPRS terminals, PDAs (Personal Digital Assistants), specific purpose terminals, etc.

Whenever remote monitoring programs are included within the care plan, connectivity and protocols to allow data acquisition and pre-processing of data are provided. Both wired and wireless options, based on open and interoperable solutions, have been implemented to allow the interfacing of a wide variety of sensing devices.

2.1 The Chronic Patient Management Centre (CPMC)

The CPMC integrates several modules: a call centre, with human operators and an automated voice response unit, access to the chronic patient record, a module for the reception and analysis of the biomedical parameters remotely monitored, a cooperative and coordinated working space, access to educational material and a module supporting televisit. The structure is outlined in Figure 3.

The Call Centre is a 24 hours a day, 7 days a week service to provide immediate answer to any demand by patients, to manage, route and filter incoming calls. The service is supplied by a telephony server hosted within the CPMC. Calls received at the Call Centre, from patients and professionals, are initially filtered by a teleoperator without health care specific training or an automatic autonomous answering system. On the one hand, patients are requested to answer a predefined questionnaire, prepared to identify the administrative or health related nature of the demand and of its urgency. On the other hand, health professionals can access the Call Centre from their centres, during a patient home visit or anywhere to support cooperative work and information retrieval tasks.

Depending on the organisation of the services, different specialised personnel: nurses, primary care or specialised doctors are on duty at certain times of the day to attend filtered calls, for further advice to the patient or other actions. For COPD patients specialised nurses are available at working hours; the rest of the day, as no specialised personnel can attend those calls identified as requiring some form of intervention, the caller is instructed to access alternative traditional emergency centres.

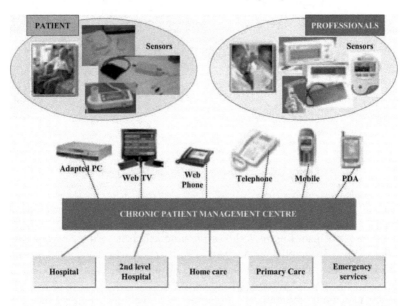

Figure 2. Multi-access possibilities

The Patient Management Module grants access to the chronic patient record through a web-based interface and according to pre-agreed access rules to each member of the multidisciplinary team involved in the care and follow-up. Data are classified in four different categories:

- ∈ Patient electronic record to monitor, control and evaluate each patient care plan: fragility report, treatment, working plans, discharge reports, etc.
- ∈ Information related to the patient patterns of use of the CPMC: calls handled, patient's answers to the questionnaires, state of the reported problems, etc.

- ∈ Telemonitoring data: biomedical data and signal records and their statistical descriptions provided by any automatic analysis system installed.
- ∈ Health professional usage of the system: visits schedule, pending answers to previous calls, etc.

All these sections can be retrieved, reviewed and updated from any point of care: primary or specialised level, as well as from any internet access terminal, provided we follow the appropriate predefined secure protocols.

Monitoring data gathered by the remote units are sent to and stored by a telemonitoring server hosted within the CPMC. This process generates a great amount of data, forcing professionals to increase the time dedicated to visualise and analyse data from his/her patients. It has been demonstrated that this fact represents one important barrier to the acceptance of the technology and its deployment, being in some cases the direct cause for rejecting the system[12]. To avoid this problem an automatic analysis of monitoring data (Knowledge Management Tool) has been developed to detect specific patient status, including alarm situations, whenever the selected signal features deviate from pre-established ranges of variability, alarms that are also sent to the right person on duty, using the messaging engine.

A key component in chronic patient care is patient education, which helps raising patients' responsibility and self-care abilities. It is focused, as well, on health care professionals, mainly primary care and home visit teams, who can also benefit from specialised information. Access to educational material is available on a video-on-demand basis through web-based interfaces, off-the-shelf tools or specifically designed user interfaces (to cope with the specific patients' needs).

Professionals can perform televisits with their patients. Televisits take place during videoconferences and allow monitoring, enabling supervision and correct sensor probe placing.

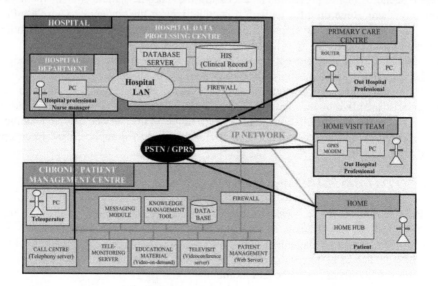

Figure 3. Chronic Patient Management Centre structure and access

Finally, the centre provides all the mechanisms necessary to cope with security requirements inherent to health systems as well as data interfaces to the health information systems of the institutions that the system is connected to. Data are encrypted for transmission, in any local or remote access, by means of standard protocols (https) and digital certificates that provide security across the shared working space. A security server, hosted within the CPMC, provides a Public Key Infrastructure (PKI) for issuing, generating and maintaining certificates and certificate revocation lists (CRLs), and additional protections, such as hardware tokens and Secure Shell protocol. In any case, final access to the system is granted using login and password that allow different access levels, each of them with different sets of functionalities and information accessible.

2.2 Remote access units

Two different remote access units have been developed. One of them, the **Patient Unit**, is used by the patient at home or elsewhere to access any remote service available, as well as to provide local connectivity to the biomedical monitoring devices that could be required to follow up his/her status. The other one, the **Mobile Home Visit Units** are used by Home Visit Teams (health professionals engaged in the multidisciplinary team in charge of the patient at home). These units support: the ubiquitous access to the chronic patient information databases, the use of specific applications required to monitor biomedical parameters, the tools for team coordination (scheduling, team awareness, responsibility transfer, etc.) and those other tasks for cooperative patient assessment, care plan, QA, etc.

A clear trend of the remote access units towards mobile terminals, whenever possible, is evident. In our opinion, this should be integrated with a multi-access concept. At the professional side the use of portable PC first and now PDAs characterises the present scene of health professionals accessing the Chronic Patient Management Centre from anywhere. Regarding the patient, it is equally clear the tendency to wireless Patient Units, motivated by the development of special purpose wireless terminals to handle the acquisition of patient data, to offer friendly user interfaces and to provide connectivity at the wide area level. The current research emphasis on wearable sensors and body area wireless networks is also pushing in this direction.

It has to be taken into account that, whenever wireless terminals are used to access the services specified by the Chronic Disease Management Model, an exception has to be made: video based consultations and virtual home visits services, which cannot be handled by present wireless technology. Obviously, 3G based applications are an option in the proximate future.

2.2.1   *PDA based Mobile Home Visit Units*

PDA terminals can be seen with present state of the technology as the best mobile option to be used ubiquitously, at any time and place, by the health professional within the Chronic Disease Management Model. PDAs are terminals that, together with any other Internet access system, either wired or wireless, can be used to accommodate to the technology available at any place and moment.

It is important to stress that the user interfaces for the PDA unit have been carefully designed to suit the environment of use where they will be working, minimising the number of user interactions needed to carry out any task , adapting the amount of information shown to the size of the screen and minimising the free text introduction sections.

Figure 4 presents the professional terminal showing the system prepared to acquire a patient spirometry during a visit to a patient at home, and the graphical results once the spirometry has been performed and validated. The nurse, during the visit, can also access to previous results to assess patient progress and care plan targets. If required he/she can send messages to the members of the team to trigger other actions such as a tele-consultation. Finally the professional can fill in patient reports and update his/her personal agenda.

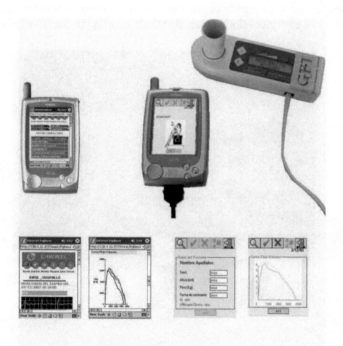

Figure 4. PDA-based Home Visit Unit

2.2.2 Patient unit

The Patient Unit (Jimenez 2002) (Figure 5) has been implemented based on a central personal device (called home hub), a specific PC with a VCR look to improve acceptability and a set of sensors. Sensors included are: pulse-oxymeter, three leads ECG monitor, spirometer, blood pressure unit and glucometer. The first two are wireless to allow recording regimes not requiring any physical restriction of the patient within their home. The home hub is connected to a standard TV set and to the appropriate telecommunications network and supports all patient interaction with the CPMC.

The application supporting the services has been optimised to provide a simple user interface based on simple self guiding and easily seen designs on the TV screen and a zapping remote control with only three buttons (up, down, enter).

A light version of the unit has also been developed substituting the central device (PC) by a PDA. Chronic care applications based on PDA patient unit can support most of the services experienced with the wired system, namely:

1. Web-based access to the Chronic Patient Management Centre:
   - ∈ To connect at anytime through the centre with his/her carers for consultation
   - ∈ To connect to the institutions, by text messaging or voice.
   - ∈ To get any information related to the personalised disease management program or his/her clinical health record.
   - ∈ To trigger the alarm services available.
2. Retrieval of general information, as well as education material or rehabilitation programs.
3. Interoperability to enable the integration of a variety of monitoring commercial devices or newly developed ones into the telemedicine shared-workspace.

Figure 5. Patient Unit

It has been shown, however, that wireless terminals, are not very useful for COPD patients, a group that, with few exceptions, is mostly home restricted, generally very fragile and with difficulties to deal with user interfaces as those of the PDA commercially available.

3. CONCLUSIONS

The Chronic Disease Management Platform for COPD patients has been installed at two different sites: Hospital Clinic in Barcelona (Spain) and Katholieke Universiteit

Hospital in Leuven (Belgium). Evaluation has been carried out on different aspects of the system as well as on its clinical use.

2.3 Phases of the evaluation

The main research has been conducted at Barcelona and has been structured in different phases. The first phase consisted of small pilots to evaluate the technical quality of the system, its usability and acceptability, as well as to characterise the use made by patients and professionals.

The results of this phase have been very positive showing that all the system elements (Call Centre, Web Patient Management Module, Home Visit Units and Patient Unit) have been scored very high in their technical quality, usability and acceptability.

Following this feasibility evaluation a first clinical trial was performed. This trial did not include the Patient Unit for organisational reasons (difficulties in providing the adequate technical maintenance of the system at the patients' homes and restrictions in the availability of telecommunications infrastructure at the home for a quick installation after discharge). All the other components of the platform were used in the clinical evaluation.

2.4 Clinical trial

To evaluate the impact on the health and quality of life of patients, a hospitalisation prevention clinical trial was prepared (Hernandez 2003). The aim of the trial was to validate the efficacy of a minimum intervention after an admission resulting from an exacerbation episode on COPD patients. The intervention consisted of an educational session lasting for one and a half hour at the time of discharge, the elaboration, in a single home visit 24 to 72 hours after discharge, of an individualised action plan and the accessibility to the system's Call Centre. This trial was performed at a third level hospital (Clinic Hospital, Barcelona, Spain), with the collaboration of professional teams from primary care centres with no specialisation in respiratory diseases. All professionals used the web Patient Management Module and the Home Visit Units to share patients' information and follow the care plan.

157 COPD patients were recruited for this trial, which lasted for one year. Patients were recruited during a hospital admission session caused by an acute episode of the illness. These patients were divided among the intervention group (67 patients) and the control group (90 patients).

The results (Table 1) showed a decrease in the number of readmissions and an outstanding increase of the number of patients that did not need to be hospitalised again during the year following a first admission. Furthermore, the patients of the intervention group stayed longer at home before the first readmission. A remarkable increase in the health related quality of life (HRQL) was also found.

Beneficial interventions of the system include: enable the sharing of information of a patient across the multidisciplinary care teams involved, closer follow-up based on home visits (helped by professional remote units and sensors); improvement of the communication between patient and care team, telephone interventions and finally patient empowerment by education.

Table 1. Hospitalisation prevention trial

	Minimum Intervention	Control	p
Patients that did not need a readmission (% patients)	51,7%	33,3%	0,04
Patients readmitted more than once	46,9%	65,2%	0,03
Patients readmitted more than twice	21,9%	29,2%	0,35
Number of readmissions	$0,90 \pm 1,28$	$1,33 \pm 1,73$	0,04
Number of emergency department visits	$0,36 \pm 0,98$	$0,54 \pm 1,12$	0,15
Mortality	20,3%	16,9%	0,67

2.5 Further evaluation

Future experiments with the system are in progress as an extension of the trial described above, enhancing: 1) the number of patients included, 2) the number of pathologies treated (it is our intention to provide a single technical infrastructure for the follow up of most prevalent chronic conditions), 3) the provision of Patient Units to the most fragile patients to improve follow-up and self management through telemonitoring and patient education.

To achieve a real and practical use of the Patient Unit it will be provided with an interface with 3G communications network, to make feasible its quick and simple installation at the patient's home after discharge. The use of the Patient Unit will benefit the patient with a closer follow-up based on telemonitoring and televisits, as well as with access to remote learning tools.

Although final conclusions are not yet available, the proposed ubiquitous care systems have been proved as an effective/efficient alternative to conventional care systems in selected patients. However, this goal can only be achieved with the support of the right organisational health care model, oriented to continuum of care, and the appropriate telemedicine tools, tailored to this model.

4. AKNOWLEDGEMENTS

This work has been partially funded by the EU Commission: projects Chronic (IST-1999-12158) and HealtMate (IST-2000-26154); the Spanish Scientific and Technology research Council (CICYT) project Middlecare (TIC-2002 -02129) and Telemedicine Network of Excellence (G03/117).

We express our warm gratitude to our clinical colleagues of the Hospital Clinic in Barcelona (Spain): Josep Roca, Albert Alonso and Carmen Hernández.

5. REFERENCES

Cotton MM, Bucknall CE, Dagg KD, Johnson MK, MacGregor G, Stewart C et al. Early discharge for patients with exacerbations of chronic obstructive pulmonary disease: a randomized controlled trial. Thorax 2000; 55(11):902-906.

Davies L, Wilkinson M, Bonner S, Calverley PM, Angus RM. "Hospital at home" versus hospital care in patients with exacerbations of chronic obstructive pulmonary disease: prospective randomized controlled trial. BMJ 2000; 321(7271):1265-1268.

De Toledo P , Jimenez S, and del Pozo F. A telemedicine system to support a new model for care of chronically ill patients. Journal of Telemedicine and Telecare. 2002. 8(Suppl 2) : S2:17-9

Del Pozo, F., De Toledo, P., Jiménez, S., Roca, J., Alonso, A.,. "Un sistema de telemedicina para un nuevo modelo de atención continuada y compartida de pacientes crónicos". Sociedad Española de Informática de la Salud, nº XXXIV Ed. Informática y Salud. 2002

Gravil JH, Al Rawas OA, Cotton MM, Flanigan U, Irwin A, Stevenson RD. Home treatment of exacerbations of chronic obstructive pulmonary disease by an acute respiratory assessment service. Lancet 1998; 351(9119):1853-1855.

Hernández C., Troosters T., Casas A., Alonso A., García J., Barberá JA., Antó J., Toledo P., Pozo F., Rodríguez-Roissin R., Decramer M., Roca J. Efectos de una intervención mínima en la prevención de hospitalizaciones de pacientes con EPOC. Arch Bronconeumol. Jun 2003; 39(supl 2):44.

Jiménez S., De Toledo P., Del Pozo F., Laborel B., La unidad de paciente de un sistema de atención domiciliaria para el tratamiento y seguimiento de enfermedades crónicas. Proceedings of the XX Congreso Anual de la Sociedad Española de Ingeniería Biomédica (CASEIB) 2002. pp 513-515

Murray CJ, Lopez AD. Global mortality, disability, and the contribution of risk factors: Global Burden of Disease Study. Lancet 1997; 349(9063):1436-1442.

Pauwels RA, Buist AS, Calverley PM, Jenkins CR, Hurd SS. Global strategy for the diagnosis, management, and prevention of chronic obstructive pulmonary disease. NHLBI/WHO Global Initiative for Chronic Obstructive Lung Disease (GOLD) Workshop summary. Am J Respir Crit Care Med 2001; 163(5):1256-1276

Report of the Workshop on Home Care Technologies for the 21st century. Jack Winters, William Herman. 1999. http://www.hctr.be.cua.edu/HCTWorkshop

Sheri Pruitt, Steve Annandale, JoAnne Epping-Jordan, Jesús M.Fernández Díaz, Mahmud Khan, Adnan Kisa, Joshua Klapow, Roberto Nuño Solinis, Srinath Reddy and Ed Wagner. Innovative Care for Chronic Conditions: Building Blocks for Action. Organización Mundial de la Salud (OMS). Informe número WHO/MNC/CCH/02.01. ISBN 92-4-159-017-3. 2002

Skwarska E, Cohen G, Skwarski KM, Lamb C, Bushell D, Parker S et al. Randomized controlled trial of supported discharge in patients with exacerbations of chronic obstructive pulmonary disease. Thorax 2000; 55(11):907-912.

Stewart S., Narley J., Horowitz J. Effects of multidisciplinary, home-based intervention on unplanned readmissions and survival among patients with chronic congestive heart failure: a randomised controlled study. Lancet 1999; 354:1077-1083.

Strassels SA, Smith DH, Sullivan SD, Mahajan PS. The costs of treating COPD in the United States. Chest 2001; 119(2):344-352.

A MOBILE TELEMEDICINE WORKSPACE FOR DIABETES MANAGEMENT

M. Elena Hernando[*], Enrique J. Gómez, Angel García-Olaya, Verónica Torralba, and Francisco del Pozo

1. INTRODUCTION

In Western societies, diabetes and its complications is still causing a tremendous amount of suffering in over 5% of the population and remains a major health problem being responsible of up to 8% of national health care expenditure (Waldhäusl, 2001). Diabetes is a chronic disease characterised by a sustained elevated blood glucose level, caused by a reduction in the action of insulin secretion where related metabolic disturbances generate severe, acute and long-term complications that are responsible for premature death and disability (de Leiva et al. , 1995).

To obtain the best care outcomes, diabetic people should receive medical care from a physician-coordinated team (American Diabetes Association, 2003). Such teams may include physicians, nurses, dieticians, pharmacists and mental health professionals with expertise and interest in diabetes. This collaborative and integrated shared care approach requires that individuals with diabetes assume an active role in their care.

Glycemic control is fundamental to the management of diabetes. The benefits of the intensive insulin therapy have been well established, demonstrating the reduction of long-term complications (American Diabetes Association - 2, 2003). Nowadays it is a reality that a well-treated diabetic patient can expect to have an almost normal life span although the achievement of the therapeutic goals requires a tight control of patients in their self-monitoring blood glucose levels and day-to-day insulin adjustment, requiring in many cases a better patient empowerment and education. Nevertheless, the fulfilment of the current guidelines in diabetes management (American Diabetes Association, 2003) implies a significant increase in the amount of patient data to be monitored, increasing physicians and nurses workload and raising immediate health care costs.

Due to its multifactorial and systemic character, diabetes mellitus has been considered a paradigm of chronic disorders which has led to an extensive application of

[*] M. Elena Hernando, Grupo de Bioingeniería y Telemedicina, Universidad Politécnica de Madrid, E.T.S.I. Telecomunicación. Ciudad Universitaria s.n., 28040 MADRID, Spain.

information technologies in diabetes care (Andreassen et al., 2002). These applications have been classified taking into account several approaches such as the system end-user, namely the patient or the physician supporting them with tools to support their daily tasks in diabetes care (Gómez et al., 2002). Nowadays telemedicine provides an integrated approach to those tools, which enhances co-operation between users, information and knowledge sharing.

Over the last years, both the search for increasing the possibilities offered by telemedicine in diabetes care and the need to enhance patient empowerment has led to a growing number of experiences covering several technological approaches and, scenarios. Earlier experiences aimed to facilitate the remote monitoring of patients from home by the transmission of computerised blood glucose profiles to the hospital (Marrero et al., 1989; Ahring et al., 1992; Billiard et al., 1991). Most of interactive telemedicine services were delivered using a distributed approach integrating "Patient Units" and "Medical Workstations" (del Pozo et al., 1991; Lehmann, 1996; Gómez et al., 1996; Bellazzi et al., 2002). The Patient Unit system (PU) was used by patients during their daily living and could be implemented on a personal computer or a palmtop computer. The Medical Workstation system (MW) was used by physicians and nurses at the hospital and was implemented as PC-based applications. Nowadays telemedicine is still using similar approaches to those defined more than ten years ago (Biermann et al., 2002), although technology evolution has enabled the development of advanced PU and MU systems based on more powerful, portable and easy to use terminals and applications, such as hand-held electronic diaries with touch-screen (Tsang et al., 2001), video telephones (Edmonds et al., 1998), or web-based prototype systems (Bellazzi et al., 2002; Nigrin and Kohane, 2000).

The rapid growth and development of information technologies during recent years in the areas of mobile computing, computer-telephony integration (CTI) and mobile Internet is shaping a new technological scenario of telemedicine and shared care systems. Today the seamless integration of available mobile and wireless technologies allows to build a new scenario for telemedicine and shared care in diabetes in which the traditional concepts of PU and MU are blurred, suffering a profound transformation evolving to a new multi-access, mobile, universal and ubiquitous workspace concept for diabetes care ((Hernando et al., 1999).

This chapter presents a further step in the field of telemedicine and shared care services for diabetes management. It starts defining the common requirements for telemedicine services and shared care system in diabetes management. These requirements have been used to build a mobile telemedicine shared care workspace based on a multi-access platform, implemented with a multi-agent architecture, that provides a universal access to a wide range of services both for patients and for professionals. This architecture has been developed under the research project named the "M2DM Project" that is partially funded by the European Commission. The technology involved in the telemedicine workspace was evaluated within a one-year clinical trial involving 160 diabetic patients carried out in four European hospitals in Spain, Italy and Germany. The chapter finishes drawing some conclusions about the current telemedicine and shared care experiences and the possibilities for the future of mobile technologies in diabetes care.

2. REQUIREMENTS FOR TELEMEDICINE SERVICES AND SHARED CARE IN DIABETES MANAGEMENT

The analysis of requirements have been divided into sections devoted first to telemedicine services and secondly to shared care systems.

2.1. Telemedicine Services in Diabetes Management

A telemedicine system for chronic care has to cater for the main requirements for physicians' decision making. It must provide enough information to enable assessment of the patient's condition and must present the relevant patient clinical data to define a therapeutic change. In the case of diabetes mellitus the responsibility of data collection is shared between patients, who record data during daily life (self-monitoring data) and the healthcare personnel, who capture data during hospital visits (hospital patient record). The telecare service makes available the patient's self-monitoring data whenever a doctor wants or needs to assess the patient's state. Therefore, telemedicine changes the conventional clinical protocol and implies to define when the patient status assessment must be done using a telemedicine protocol.

The services to be provided by a telecare system for diabetic people are:

1. *Telemonitoring*: It is the most basic service and allows physicians the remote control and monitoring of patient's data between two face-to-face hospital encounters. The actions involved in this service are the following: firstly, the patient annotates in an electronic logbook (a mobile device such as a glucometer or a PDA) the self-monitoring data (glycaemia values, insulin changes, diet changes, exercise data, illness and medication data and, in general, any data relevant to diabetes care). Afterwards, the user sends these data to the physician using any of the methods allowed by the system (glucometer with modem, telephone, PDA, etc.) and finally, the system notifies the physician that the patient has downloaded data and there is available for carrying out an overview of the data as well as generating alarms.

2. *Telecare*: This service allows the remote care of the patient supported through two complementary services: teleconsultation and supervised care.

    •∈ *Teleconsultation*. This service allows message exchange (text and voice) between patients and doctors. It is an asynchronous facility: patient sends a message to the doctor and doctor answers within a short time. The aim of the message can be asking for advice, explanation of some data sent, etc. As a result of the process, the physician can decide to modify the patient treatment, to set a new visit, to encourage him/her to continue the same treatment, etc.

    •∈ *Supervised care*: This service provides the patient with a "supervised autonomy" on diabetes self-management. Patients carry out day-to-day therapy changes (glycaemia levels and if it is necessary the patient modifies the insulin doses or the diet) and send this information to the physician. The physician reviews in "background" the monitoring data, the therapy changes decided by the patient and their effects. Finally, the physician validates the patient's decisions or proposes further adjustments of the treatment. The patient receives, if required, the corresponding medical advice.

3. *Intelligent patient data monitoring*. The telecare service can be complemented with Knowledge Management tools that analyse patient's data, when received at the server, and generate automatic intelligent alarms to focus doctors' attention on those points where non-desired deviations in the metabolic control are found.

2.2. Shared Care services in diabetes management

The success of shared care programmes in diabetes management leans on the viability to provide powerful clinical and monitoring data collection/access platforms and appropriate collaboration information systems to support care team members, including the patient (Kindberg et al., 2000). In this sense, the application of Computer Supported Collaborative Work (CSCW) current theories and tools to diabetes management can be a valuable aid to achieve the goal to create new information technologies and telecommunication workspaces to bear diabetes shared care (García-Olaya et al. 2002).

The services to be provided by a shared care workspace for diabetes management are:

- ∈ *Multi-channel messaging services*: to allow users to send and receive messages in any format despite of the original message's format (automatic conversion of e-mail to voice, voice to text, etc.).
- ∈ *Shared information dissemination and request*: to send data automatically to users interested on them using both pull and push methods (i.e. referral letters). This information can include EHR data as well as data related to user's activities, etc.
- ∈ *Group and activity awareness*: to integrate an event and activities notification service that allows users to know at anytime the activities other users are performing or will perform. This service requires to log data about users activities, to generate automatic messages to notify them and to send this information upon request. These notifications should be fully configurable: what event was triggered, by what user, affecting who, at what time, where, etc. The aim of this service is to aware every user of the most relevant events occurring at the lowest cost (measured in terms of time, effort, etc.). This service is associated to the automatic generation of intelligent alarms notified to users using any of the possible ways (e-mail, voice mail, SMS…) pre-selected by recipient users.
- ∈ *Shared agenda (scheduling meeting/visits)*: this facility let users to share an agenda complementing the above services. It can include a temporal reminder service.

3. TECHNOLOGICAL BUILDING BLOCKS OF THE MOBILE WORKSPACE

3.1. The multi-agent architecture

The mobile telemedicine workspace system defined here supports a telecare service that allows ubiquitous telemonitoring, teleconsultation and supervised care of diabetic people. At the other hand, the mobile telemedicine workspace is devoted to shared care and supports a common information space in which all care team members execute collaborative tasks by seamless sharing and exchanging clinical and organisational information and knowledge.

The complexity and flexibility required to build the mobile workspace is achieved with the definition of a middleware architecture that copes with several technical requirements: 1) it allows the independent integration of different access methods (mobile phone, PDA, web, etc.); 2) it facilitates the use of heterogeneous software and/or hardware solutions to allow each Health Care Organization to configure the number and the kind of services they want to offer to their own users; 3) it supports the development of the collaborative tools supporting the care team work in an efficient manner; 4) it creates a continuous record of the users actions to monitor all the collaboration processes involved in shared care to further improve the quality of care.

This proposed architecture is a multi-agent, open and distributed platform that yields on three components: the Agents, the multi-access Organizer and the database.

The agents collaborate to guarantee homogeneous access by the users to the system services. They allow users to access the information in a transparent way whatever access terminal or combination of terminals they choose. The agents are classified into "Communication Servers" that are in charge of communications with the different user terminals integrated within the architecture, and "Application Servers" that perform data analysis and data processing. The number of agents present in a specific system is not limited a priori and it only depends on the kind of terminals that will be used at each site. The rules for interoperability between different agents are independent of the development platform and are coordinated by the "Multi-access Organizer".

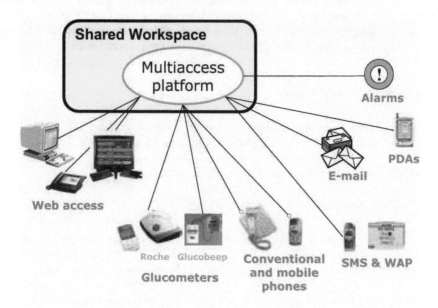

Figure 1: The mobile multi-access telemedicine workspace

The Multi-access Organizer is a message-oriented middleware module that is in charge of coordinating the interoperability between all the agents integrated into the architecture. The communication between the Organizer and the agents is done using "event messages" encapsulated into TCP/IP frames. The Organizer manages the message distribution, the receipt confirmation, and the error handling processes. Additionally the Organizer includes: Event Registry Services that enable the establishment of an event registry and event monitoring services and Smart Routing Services that ensure the message gets delivered to the appropriate recipients in the correct sequence.

The cooperation process between the Organizer and the agents is triggered by the later ones that activate events when they detect some pre-programmed situations, as for example: one user logs into the system and interacts with his/her data; one physician updates any item in the patient's care process; one agent concludes a specific automatic data analysis. The Organizer behaviour is accomplished in three different steps: 1) The Organizer receives a message from each architecture agent whenever an event happens; 2) The Organizer records the event and checks in the database which is the next action to be done and which is the agent that should perform it (this can be dynamically configured so the user is able to change the Organizer's behaviour); 3) The Organizer sends a message to the final agents indicating the author of the event.

The number and nature of the events is not fixed and can be defined at any system implementation according to their preferences or specific care protocols. During the exploitation of the system, the multi-agent architecture allows the inclusion of new events without affecting previous service performance. Likewise, the action the Organizer triggers when a certain event is received is fully configurable.

The database is composed of a common set of tables including users information (personal, contact, roles, groups and notification preferences), patient's medical data (therapy, monitoring and basic visit information), messaging data (user's mail boxes and messages logs) and system management data to rule the service (service configuration and event logs). This generic database structure is complemented with the Electronic Health Record that is hospital-dependent.

3.2. Workspace user applications

Users can interact with the common workspace using several terminals such as Web, PDAs and mobile phones. The user applications running on these terminals have a set of common functions both for patients and the health care professionals that enable the provision of the diabetes management services described above.

3.2.1 Electronic patient logbook for data collection

The electronic patient logbook shows monitoring data collected by patients such as blood glucose readings associated to their related events. Physicians can consult patient's data to assess patient's state and to support their decisions when a therapy adjustment is required.

Patients can enter information about modifications in the prescribed insulin doses, modifications in diet, illness, etc. The effort needed to enter self-monitoring blood glucose readings is minimised allowing the automatic data download from commercial glucometers. Data transmission can be done in three different ways: 1) through the use of a smart modem developed by the Roche company where patients can connect their glucometer and data are seamless transmitted to the hospital after pressing a single button; 2) through the connection of the blood glucose meter to the serial port of a user terminal (PC or PDA) that synchronises afterwards with the central server; or 3) through the use of a universal device (Glucobeep™) that is able to obtain monitoring data from several glucose meters and uses acoustic coupling to transmit data through a wide set of conventional or mobile phones.

3.2.2 Electronic Health Record

The Electronic Health Record application provides access (reading and/or writing) to the clinical data registered during the face to face encounters. Depending on their profiles, users are able to access to data for consultation and/or update tasks. The EHR application includes facilities to view the temporal variation of data, allowing for example to assess the weight or Hb1Ac evolution between visits or after the prescription of a new treatment.

Different reports may be automatically generated and printed out to be added to the paper-based health record or to be delivered to the patient (i.e. the definition of the insulin therapy and the diet). A preview function displays the report to check its content or to add further comments before printing.

3.2.3 Data analysis and decision making tools

A telemedicine system increases the amount of data and information used by doctors and modifies the therapeutic decision making procedures. Therefore, it must provide a set of data analysis tools for numeric and graphical visualisation data summaries, comparison of data monitoring for time periods associated with therapy changes and automatic report generation. These features can be supported by:

1. *Enhancement of data visualization in the electronic patient logbook*: Blood glucose values are colour coded to highlight whether they are within the target ranges, near to the limits or whether they classify as hypoglycaemia or hyperglycaemia according to the ranges selected by the physician for each individual patient.
2. *Graphical visualisation*: A range of data analysis possibilities are available for comparison of time spans with different therapies and for visualising data summaries to simplify and minimise time for decision making on therapy adjustments.
3. *Knowledge management tools*: The Knowledge Management (KM) tools can be classified into three types according to the way they are activated:

   •∈ Activated on users' demand: including mathematical simulation of blood concentrations for different insulin therapies;
   •∈ Triggered by data reception: including automatic data processing for alarm generation. A KM Agent is activated at the central server when new data are received and combines different methods to analyse blood glucose

monitoring data and insulin data, such as statistics, rule-based and model-based techniques, adapting its results to the data availability.

- ∈ Triggered by the system: including dynamic pre-programmed notification of events according to the preferences of the own users, both patients and physicians.

3.2.4 Cooperative electronic mail

The functionality of the cooperative electronic mail module is similar to an enhanced e-mail system where text messages can have attached non-textual information. This co-operative mail automatically integrates relevant information from the patient's electronic record, monitoring data, alarm messages, data summaries and acknowledgements of therapeutic changes reception. Health care professionals can check whether a text message or therapy adjustment created by any member of the care team has been delivered to the patient by consulting the output mailbox.

Patients can send messages to doctors including text and/or data collected by the electronic logbook. Doctors can also send text messages to the patients containing their comments or advice. In addition, any modification in the patient therapy automatically generates a message to be sent to the patient.

3.2.5 Care plan manager

This tool supports the definition of a care workflow and integrates a Care Plan View Line that shows information relevant to answer the so-called "W" questions in chronic care (who, what, when, where, why and how) (Pinelle et al., 2001). Using this service, users are able to view in a graphic interface all the activities, past or future, that make up the care plan of a certain patient, including visits, data sending, etc. This allows users to display the patient record and to schedule actions taking into account what other users are planning to do.

3.3. Users terminals

3.3.1 Web-based terminal

The main access method to reach the information in the multi-access system is the web graphical interface, offering the complete functionality of the telemedicine/shared care services with minimal human factors limitations.

The functional requirements for the graphical interface were, first, to provide different layouts adapted to professional and patient needs; second, to make possible access from a wide range of web platforms; and third, to minimize transfer delays. The usability requirements were also an initial commitment to obtain a user-friendly and easy to use system, with high learnability and an optimised number of interactions to perform the user's task.

Professionals use the Web application as their main working tool; therefore the professional interface has been designed to present at any moment the maximum relevant information, minimizing the need for secondary pages to perform a task. For this reason the professional application is optimised for a high-resolution display (1024 x 768) in order to present as much information as possible. Patients access the services using a heterogeneous range of possibilities (PCs, WebTV, Web phones, PDA navigators), usually more limited than the professional PC. For this reason, the Web application for patients is optimised for a medium resolution display (800 x 600) and scrolling is allowed when accessing with a lower resolution device.

Navigation is performed through a three level menu that defines the different scenarios and contains all the tasks performed by the application. In all the scenarios the screen appears divided into three sections: 1) The top of the screen shows the logo and the first and second menu levels that are the upper levels in the navigation hierarchy; 2) The left side of the screen shows the third menu level and additional information relevant to the scenario; 3) The central part of the screen is the working area and presents to the user the information related with the specific scenario.

3.3.2 PDA terminal

The PDA application is able to manage a local database, allowing the patient to work isolated from the telemedicine server without the need of a permanent active communication link. On demand by the user, the Patient Unit synchronizes the local database with the server database causing the automatic updating of all the new information in both directions.

The PDA application is developed in Java in order to be portable on multiple platforms. The portability of the application has been successfully tested in conventional PCs; in the PDA Psion NetBook™ with EPOC32 (Symbian) operating system; and in the Siemens SX-45 PDA with Pocket PC operating system and a GPRS mobile connection.

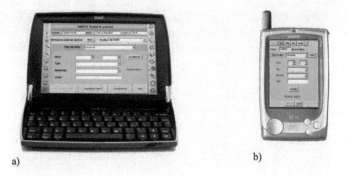

a)   b)

Figure 2: Mobile Patient Units. *a)* Psion netBook™ (resolution of 640x480 pixels)*; b) Siemens SX-45 PDA* (resolution of 320x240 pixels)

3.3.3 WAP application

The WAP application can be considered a web access where a very reduced terminal display is used. The constraints of the current WAP interfaces forced us to select a limited functionality of the telemedicine and shared care services. A preliminary prototype has been implemented: such a prototype gives information about the existence of new messages and allows the manual entering of self-monitored blood glucose data.

Figure 3: WAP application for mobile phones.

4. CONCLUSIONS

Telemedicine is providing innovative solutions in the effective treatment of chronic patients. In diabetes management, telecare services can provide people the tools they need to take better control of their illness. The benefits of telemedicine over conventional care methods are: 1) the increase of the quality and quantity of the information collected by patients affording a better decision-making process for doctors and patients; 2) the improvement on the number of therapy adjustments to be performed by doctors; 3) a better physician-patient communication procedure; and 4) the impact of the system to enhance the metabolic control of patients.

In spite of the extensive application of telemedicine in diabetes management, the evaluation of the clinical impact of telemedicine is an open issue. Clinical studies have been mainly focused on blood glucose monitoring of patients and telecare services experiences are still limited. Even more, few of them involve large-scale randomised controlled trials and the reported clinical benefits are in many cases not justified by adequate statistical significance (Hersh et al., 2001). The possibilities to perform the appropriate evaluation studies are limited by the organizational difficulties and the lack of resources required in the telemedicine evaluation plans as well as the high number of variables (Technical, Usability, Clinical, Quality of Life, Organizational, Economic) to be considered to carry out a thorough evaluation of a telemedicine system.

New paradigms in diabetes care, and in general in chronic diseases, are currently focusing in shared care. Telemedicine is playing an important role to facilitate the implementation of shared care programmes. The Electronic Health record is the most valuable and necessary piece to achieve such a shared care but the prompt and universal access to the clinical record is not sufficient to fulfil the team care group coordination and awareness needs. Wireless technologies, telemedicine and new mobile computer-supported cooperative work technologies are being integrated to create new mobile workspaces that forward and optimise the required collaboration between the care team, including the patient.

The implementation of these mobile workspaces requires the definition of multi-access architectures to provide all people involved, professionals and the patient, a common shared work area that can be accessed by any communication available means. The mobile workspace has to make possible the use of any service at any time and location and the shared care services and technologies have to be integrated with the information systems of the different health care levels containing relevant information to them. Strict requirements for security to guarantee patient privacy, data confidentiality and integrity and auditing of the clinical acts are mandatory. The use of middleware components eases flexibility, portability, system inter-operability and cost-effectiveness in the implementation of telemedicine shared care workspaces.

The combination of insulin pumps and continuous glucose monitoring systems seems to be the solution to achieve a good metabolic control for insulin-dependent diabetic people. Continuous glucose monitoring can detect glycaemia patterns that cannot be discovered only by means of few diary measurements (Bode et al., 1999). The possibility to obtain a complete blood glucose profile allows monitoring the suitability of an intensive therapy and can be used to fine-tune the therapy. The old concept of the artificial pancreas is starting to be a reality supported by the availability of these technologies and the integration of control systems able to close the loop modifying the pump parameters (Bellazzi et al., 2000; Pickup et al., 2000).

One of the latest efforts to achieve a close-loop control system based on a portable computer has been done within the EU funded research project ADICOL: "Advance Insulin Infusion using a Control Loop" (Vering and Reihl, 2002). The project developed a system, aiming to emulate an "artificial pancreas", to administer automatically a diabetic individual daily insulin dosage in a precise manner and at the required amount. The system has been evaluated in several trials but the difficulties to develop reliable and secure wearable systems and close-loop control algorithms are to be solved before investigating the patient acceptance of such systems.

The increased and wide acceptance by diabetic patients on the existing continuous glucose monitoring system and insulin pumps could extensively be improved by the integration of these new technologies and therapeutic procedures within a telemedicine system. One of the most advanced approaches to mobile telemedicine close-loop systems in diabetes management is currently under development within a new EU project named INCA:Intelligent Control Assistant for Diabetes (INCA Project, 2003). This project started in January 2003 and faces the challenge to create a mobile intelligent Personal Assistant for continuous self-monitoring glucose and subcutaneous insulin infusion integrated into a telemedicine diabetes management service. The personal assistant will provide patients with several close-loop control strategies that will offer an augmented information self-management environment to increase patient empowerment. Patient therapeutic decision making would be supervised by health care professional at any time by the use of a telemedicine central server.

The reliability and availability of continuous glucose sensors, insulin pumps, mobile computing and telemedicine is making a reality the development of new closed-loop glucose control systems as the best current solution to achieve a good metabolic control for insulin-dependent diabetes mellitus patients. The efforts made by undergoing research initiatives are moving in this direction to look for new frontiers to better the quality of care and quality of life of people with diabetes.

5. ACKNOWLEDGEMENTS

This work has been partially funded by the EU Project M^{2}DM (IST 1999-10315) and the Spanish MCYT Project "Middlecare" (TIC2002-02129). We give our thanks to all partners of the M2DM Project. The authors wish to acknowledge the valuable work of all researchers from the Grupo de Bioingeniería y Telemedicina, specifically to F.J. Perdices,

P. Alvarez and M. García. Our special gratitude to Prof. A. de Leiva, Dra. M. Rigla and E. Brugués from the Endocrinology Dept. of Hospital Sant Pau (Barcelona, Spain) for their fruitful help to this work.

6. REFERENCES

Ahring, KK, Joyce, C., Ahring, JPK, Farid, NR , 1992, Telephone modem access improves diabetes control in those with insulin-requiring diabetes, *Diabetes Care* **15**, 971-5.

American Diabetes Association, 2003, Implications of the Diabetes Control and Complications Trial, *Diabetes Care* **26** (Suppl. 1), S25-S27.

American Diabetes Association, 2003, Standards of Medical Care for Patients with Diabetes Mellitus *Diabetes Care* **26** (Suppl. 1), S33-S50

Andreassen, S., Gómez E.J. and Carson, E.R., 2002, Computers in Diabetes 2000, Comput. Meth. Prog. Biomed. **62**, 93-95.

Bellazzi, R. Larizza, C., Montani, S., Riva, A., Stefanelli, M., d´Annunzio, Lorini, R., Gómez, E.J., Hernando, E., Brugués, E., Cermeno, J., Corcoy, R. De Leiva, A., Cobelli, C., Nucci, G., Del Prato, S., Maran, A., Kilkki, E., Tuominen, J., 2002, A Telemedicine support for diabetes management: the T-IDDM project, *Comp,. Meth. and Prog. in Biomedicine* **69**, .147-162.

Bellazzi, R., Nucci G. and Cobelli, C., 2001, The Subcutaneous Route to Insulin-Dependent Diabetes Therapy, *IEEE Eng. Med. Biol. Magazine* **20** (1), 54-64.

Biermann, E., Dietrich, W., Rihl, J. and Standl E., 2002, Are there time and cost savings by using telemanagement for patients on intensified insulin therapy?: A randomised, controlled trial. *Comp,. Meth. and Prog. in Biomedicine* **69**(2).

Billiard, A., Rohmer, V., Roques, M.A., Joseph, M.G., Suraniti, S., Giraud, P., Limal, J.M., Fressinaud P. and Marre, M., 1991, Telematic transmission of computerised blood glucose profiles for IDDM patients. *Diabetes Care* **14**, 130-134.

Bode, B.W., Gross, T.M., Thornton, K.R., Mastrototaro, J. 1999, Continuous glucose monitoring used to adjust diabetes therapy improves glycosylated hemoglobin: a pilot study, *Diabetes Research and clinical Practice* **46**, 183-90.

Daniel J. Nigrin, Isaac S. Kohane, 2000, Glucoweb: A Case Study of Secure, Remote Biomonitoring and Communication. Proceedings of the AMIA 2000, Nov.4-8, 2000. Los Angeles, USA.

de Leiva, A., Lefèbvre P. and Nerup, J., 1995, European dimension of Diabetes Research, *Diabetologia* **39**, 5-11.

Del Pozo, F., Gómez E.J. and Arredondo, M.T., 1991, A Telemedicine approach to Diabetes Management, *Diab. Nutr. Metab.* **4** (Suppl. 1), 149-153.

Edmonds, M., Bauer, M., Osborn, S., Lutfiyya, H., Mahon, J., Doig, G., Grundy, P., Gittens, C. , Molenkamp, G,. Fenlon, D., 1998, Using the vista 350 telephone to communicate the results of home monitoring of diabetes mellitus to a central database and to provide feedback. *Int. Journal of Medical Informatics* **51**(2).

García-Olaya, A., Gómez, E.J., Hernando, M.E., Del Pozo, F., 2002, A Middleware CSCW Architecture for Diabetes Shared Care, IFMBE Proc. EMBEC'02, 3(2) , 1376-1377.

Gómez, E.J., Del Pozo, F. and Hernando, M.E., 1996, Telemedicine for Diabetes Care: the DIABTel approach towards Diabetes Telecare, *Med. Inform.* **21**(4), 283-295.

Gómez, E.J., Hernando, M.E., Garcia, A., del Pozo, F., Corcoy, R., Brugués, E., Cermeño, J. and de Leiva, A., 2002, Telemedicine as a tool for intensive management of diabetes: the DIABTel experience, *Comput. Meth. Prog. Biomed.* **69**, 163-177.

Hernando, M.E., Gomez, E.J., Garcia, A. and del Pozo, F., 1999, A multi-access server for the virtual management of diabetes, in Proc. ESEM 99 (Fifth conference of the European Society for Engineering and Medicine), 309-310.

Hersh, W.R. Helfand, M., Wallace, J., Kraemer, D., Patterson, P., Shapiro, S., Greenlick, M., 2001, Clinical outcomes resulting from telemedicine interventions: a systematic review, *BMC Medical Informatics and Decision Making* , 1-5.

INCA Project, 2003, http://www.ist-inca.org/.

Kindberg, T., Bryan-Kinns, N. and Makwana, R., 2000, Supporting the shared care of diabetes, http://www.dcs.qmul.ac.uk/research/distrib/Mushroom/index.html.

Lehmann E.D. (Ed.), 1996, Application of information technology in clinical diabetes care. Part 1 Databases, algorithms and decision support, *Med. Inform.* **21**(4), 255-374.

Marrero, D.G., Kronz, KK, Golden, MP, 1989, Clinical evaluation of a computer-assisted self-monitoring of the blood glucose, *Diabetes Care* **12**, 345-50.

Pickup, J. , 2000, Sensitive glucose sensing in diabetes, *The Lancet* **355** (9202), 427-8.

Pinelle, D., and Gutwin, C., 2001, Collaboration Requirements for Home Care, Research Report 01-01, Computer Science Department, University of Saskatchewan, http://hci.usask.ca/ publications

Tsang, M. W., Mok, M., Kam, G., Jung, M., Tang, A., Chan, U., Chu, C M., Li, I. and Chan, J. , 2001, Improvement in diabetes control with a monitoring system based on a hand-held, touch-screen electronic diary, *J. of Telemedicine and Telecare* **7**(1).

Vering, T. and Reihl, B., 2002, Concept and Realisation of a Control -Loop Application for the Automated Delivery of Insulin, 2nd Diabetes Technology Meeting, 31 Oct. - 2 Nov. 2002, Atlanta, USA.

Waldhäusl, W.K., 2001, Finally we have arrived in a new millennium, *Diabetologia* **44**, 1-1.

MOBILE MANAGEMENT AND PRESCRIPTION OF MEDICATION

Chris D. Nugent[*], Dewar D. Finlay, Norman D. Black, Tasos Falas, Constantinos Papadopoulos, and George A. Papadopoulos

1. INTRODUCTION TO MEDICATION MANAGEMENT

The major method of treating disease in the developed world is through the prescription of drugs. It is estimated that, on a global basis, 3 billion prescriptions are issued annually. The fundamental process of prescribing and issuing medication is common in most parts of the world. Typically a patient will visit a General Practitioner (GP) who will diagnose any health problems and issue a prescription accordingly. It is then the duty of a pharmacist to dispense the medication based on this prescription and advise the patient on how it should be taken. In practice this process is much more complex as GP's must not only diagnose and prescribe, but must also consider the patient's profile in terms of possible allergies to certain medications and how the new prescription will interact with other treatments that may already have been prescribed. In addition to this, the prescribe-to-intake process for medication usually incorporates more entities than just the GP, patient and pharmacist. In many cases health insurers, government agencies and even pharmaceutical wholesalers will form part of the chain. One of the major expenses in the management of medication is payroll-time spent on the administrative duties needed to service the requirements of various health plan providers. Although not all attributed to medication related administration, it is estimated that between \$0.25 and \$0.40 of every healthcare dollar in the US is spent on excessive administration costs (Lee et al., 2000).

As well as the inefficiencies introduced in the prescription/medication issue process, patient oriented issues are also prevalent. Medical compliance or adherence indicates the

---

[*] Chris D Nugent, University of Ulster, Jordanstown, Northern Ireland, BT37 0QB. Email: cd.nugent@ulster.ac.uk.

precision with which a patient follows a medical prescription. This is a relatively new research domain that is currently receiving much attention. Non-compliance has a negative impact on both the individual patient and the healthcare system and wider economy. The impact is clear: if a patient has been prescribed correctly following diagnosis, non-compliance becomes an important barrier in optimizing treatment. The individual costs of sub-optimal treatment can range from almost nil to critical when for example the effects on chronic hypertension lead to a stroke or when a transplanted organ is rejected (Aswad et al., 1993; Uquhart, 1993). It has been suggested that as high as 50% of patients fail to adhere to the terms of their prescriptions. In the United Kingdom it has been estimated that in the range of 6-10% of hospital admissions are due to problems associated with medications (DoH, 2000). In the United States of America, figures have been released to suggest that non-compliance in patients suffering from cardiovascular disease resulted in more than 125,000 deaths and several thousand hospitalisations a year. These represented 20 million lost working days, costing over $1.5 billion in lost earnings.

2. NON-COMPLIANCE

2.1. Introduction

The problem of non-compliance is a complex one, involving many stakeholders above and beyond the patient. Reasons for non-compliance may be identified as falling into one from a possible four categories: Drug-related issues, Patient related issues, Medical-practice-related issues and Follow-up-related issues. These are shown in more detail in Figure 1.

Non-compliance may be simple to explain for illnesses exhibiting weak symptoms, but non-compliance also occurs for severe diseases. It has been shown that patients with mild forms of diseases tend to comply less than patients with moderate to moderately severe diseases. In addition, patients with extremely severe or even terminal illnesses usually comply less than the moderate group.

It is not possible to identify a typical 'non-compliant' patient nor is it possible to suggest different population groups as being better or worse than others. Nevertheless, for the elderly population, it is possible to identify a number of features that may clearly aid to their non-compliance;

1. Elderly patients receive more prescriptions per head that any other group of the population.
2. Elderly patients are at more risk from complications arising from poor medication management. This is due to the physiological changes associated with ageing which affect the body's ability to handle drugs and to the potential adverse affects of multiple medication such as cognitive, physical and sensory impairment.
3. Demographic ageing may result in the overall level of non-compliance rising.

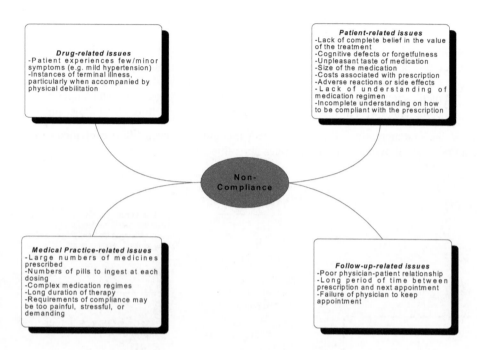

Figure 1. Issues relating to poor patient compliance.

The literature suggests that two related sets of factors can help explain non-compliance and that most often both are present:

- $\in$ The motivation, beliefs and capacities of the patient in relation to medicine taking.
- $\in$ The recognition of the importance of motivation by the doctor (or others) and their resulting actions.

Factors contributing to non-compliance may be patient-specific, some are systemic, some are the results of intentional actions and some are the unintentional consequences of other factors. Figure 2 shows a model of compliance which is based on an approach which emphasises perceptions and practicalities.

To assist with the issue of non-compliance a plethora of approaches have been devised, tested and employed. In general these may be segregated into two main approaches: direct and indirect.

- $\in$ Direct methods involve: Observation of patient taking medication; Measurement of the levels of a medicine in biological fluids (e.g. blood, urine);

Measurement of clinical attendance and count the number of missed and cancelled appointments; Screening of urine or blood samples for medicines prohibited by prescription (e.g. caffeine and nicotine).

•∈ Indirect methods involve: Questioning of a patient verbally or via a questionnaire during the treatment; Assessing patient compliance based on their clinical response; Undertaking pill counts; Use of electronic counters or medication monitors; Evaluation of patient diaries for completeness and compliance with instructions; Compliance assessment through family or nurse.

One of the commercially successful indirect methods has been the use of medication monitors. The following section elaborates upon this area.

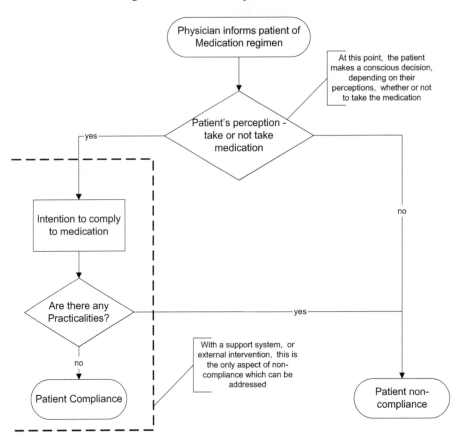

Figure 2. Model of compliance based on patient perceptions and practicalities.

2.2. Classical Compliance Aids

A medication compliance device can be defined as "a container or device which is normally designed to hold solid oral dosage forms for a defined period" (Walker, 1992). Although most common medication storage systems, including blister packs and conventional screw top medication bottles, meet these criteria, modern compliance aids tend to be something more elaborate. Commercially available solutions vary greatly in functionality; these devices can range from a partitioned container to a complex electromechanical device with telecommunication capabilities. Irrespective of the level of functionality the primary requirement of any compliance aid is to assist in self-medication management. Variations in the levels of functionality are purely as a result of manufacture's interpretation of constraints such as cost, technology and user requirements.

2.2.1. Types of Compliance Device

Typically compliance aids have been classified into three categories. This classification is based on functionality, and the categorization is as follows:

1. Pill holders
2. Alarm based aids
3. Monitoring devices

2.2.1a. Pill Holders.

This category includes devices that are primarily pill containers. Typically they consist of a container with several compartments each of which represents a dosage interval. This container can be carried with the patient and medication can be extracted as desired. The main variation in these devices between manufactures tends to be in terms of size, shape, number of pills they hold, and the number of medication intervals they support. A typical example of this type of device is illustrated in Figure 3a. It can be seen that this device has seven compartments; each compartment has a flip lid and is marked with a letter indicating a day of the week. Here the manufacturer has also included Braille markings to aid those who are visually impaired. In this particular implementation, each of the compartments is quite large and capable of holding large quantities of medication. This renders the device less portable and several manufactures have adopted a much smaller design with trade offs in the medication capacity. This type of system is capable of providing coarse guidance, as it does not cater for individual dosage intervals during the day.

Another popular approach has been the use of a plastic box with between 28-56 compartments (Corlett et al., 1996). This allows the medication for each day to be subdivided to represent different times during the day.

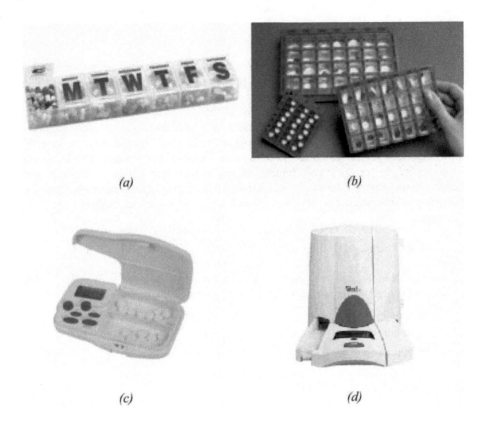

Figure 3. (a) 7 day pill holder (b) Dosset compliance aid (c) Alarmed pill holder (d) MD2 Compliance aid.

This is a realistic patient requirement. The Dosett system (www.dosett.com) shown in Figure 3b provides this functionality. This device is available in three sizes enabling the patient to choose which configuration suits best. Dosett is one of the most frequently studied compliance aids and has been shown to be beneficial to the elderly and to community based psychiatric patients (Rivers, 1992).

2.2.1b. Alarm Based Aids.

These devices are based around a timer and alarm mechanism and are usually accompanied by some sort of medication container similar to that mentioned above. Some (Szeto et al., 1997) have attempted to provide electronic reminders based on pill bottles. Generally these devices can memorize the medication schedule and provide different alarm types depending on the medication time. In some cases the device includes a reservoir of cartridges for a weekly supply of medication that can be inserted into the electronic reminder module.

These devices are sometimes referred to as active compliance aids, again the functionality and configuration vary among manufacturers. Figure 3c illustrates a typical system. Here the medication is stored in a two-compartment container; attached to the left of this is an electronic clock and alarm. The medication regimen is programmed into this timer using the buttons. When medication is to be taken an alarm sounds and the patient must open the compartment and remove the correct medication. This is a simple yet effective approach. Other solutions provide more compartments for medication, and several manufactures have designed systems were a weeks supply of medication can be stored in seven different containers, and the electronic module can be attached to each container on the appropriate day of the week.

2.2.1c. Monitoring Devices.

In most cases compliance monitoring devices offer the functionality of the aforementioned systems with additional recording and telecommunication capabilities. In some instances the dispensing process is also automated. The MD2 system, as shown in Figure 3d exhibits the characteristics of a compliance monitoring device. It is a home based unit and provides a comprehensive range of features. At pre-programmed intervals, MD2 will "announce" that it is time to take a medication dose. This announcement consists of multiple features: a voice message, a flashing light, and a textual message on an inbuilt display. To dispense the medication dose, the individual simply presses an exterior button, and the medication is dispensed in a small plastic cup. Instructions such as "Take with water" can be announced when the button is pressed. The remaining time until the next dose is displayed on the LCD screen. The device holds up to sixty of the plastic cups, which have to be manually preloaded with the correct dose of medication. The MD2 system also provides a means of telecommunication with a care model. In instances of non-compliance the MD2 system will automatically contact the call centre which subsequently places a call to either the user or a named relative to alert the instance of non-compliance.

2.3. Assessment of Current Compliance Aids

Acceptance of the aforementioned compliance aids is widespread. It is evident that the variance in functionality and complexity of the classical compliance solutions results in varied performance. The portability and low cost nature of the simple organisers and alarm based systems make them an attractive solution, however, the comprehensive functionality and in particular the telecommunications features of the larger home based monitoring systems render them as better solutions. Regardless of the functionality, most of the current compliance control systems are only effective if the patient is willing to take the medication but is impeded by practical issues, as illustrated in Figure 2.

There is a general feeling that in many cases critical care patients are apprehensive about leaving the home. However, it is believed that mobile solutions can alleviate these worries. Additionally, any attempt at further optimisation of any compliance control solution must address the broader issues not necessarily related to the patient. Enhancement of the interaction between stakeholders (Wertheimer and Santella, 2003) and in particular, real-time monitoring of compliance statistics, and optimisation of the prescribing process shall address some of the issues associated with non compliance. When these factors are taken into consideration it is evident that an optimal compliance solution must allow mobility and interaction between stakeholders. This is realizable through the use of mobile telecommunications technology.

3. ICT AND HEALTHCARE

3.1. Introduction

The application of Information and Communication Technologies (ICT) to healthcare can be described as e-health or telehealth. E-health, the most commonly used of these two terms, describes the practice of protecting and promoting health. E-health covers all aspects of ICT in healthcare and can incorporate all stakeholders in the healthcare arena.

3.2. Technology in Point of Care Applications

The acceptance and growth of Personal Computer (PC) based technology in point-of-care applications has been widespread. In 1999 64% of European GP's used a PC, see Figure 4 (InterCam, 1999). This is almost twice the number for 1996. In addition a staggering 90% to 100% of the 105,000 European pharmacies are currently equipped with a PC.

It is interesting to note, however, that the main use of PC's in GP's surgeries is secretarial or reception oriented, followed by management of the surgery. It is believed that the crucial indicator of acceptance of information technology in primary healthcare is the use of Electronic Patient Records (EPR).

EPR systems allow management of administrative, clinical and financial patient information. In recognition of the benefits of such an approach, many government and private healthcare administrations have encouraged its introduction. EPRs have almost been fully implemented in the UK, with 90% of doctors storing information on patients electronically. Closely following are the Netherlands and Denmark, with 80% and 75% respectively. The use of digital records in Germany and France is between 60% and 70% and in the USA 46% of GPs are using EPR.

Another critical factor in the uptake of ICT in healthcare is the application of internet technology. Internet based applications can be used to increase the quality of healthcare in remote areas, educate healthcare professionals and allow for sharing and clustering of healthcare resources and information. Statistics show that in the UK alone over 80% of GPs have access to the internet. However, similar to the case of the number of GP's using PC's, it is difficult to establish the nature of utilisation of this service.

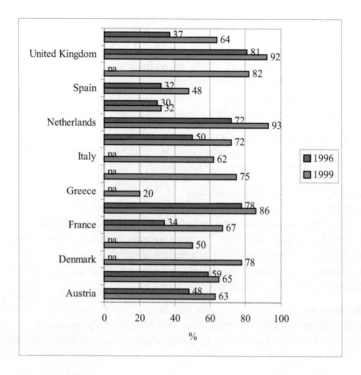

Figure 4. Proportions of GP's using PC's in European countries (Deloitte & Touche, 2000).

3.3. Electronic Prescribing

Much effort has been directed at streamlining the process of medication management from a medical professional's perspective. The area where most interest has been generated is that of electronic prescribing (e-prescriptions) or Electronic Transmission of Prescriptions (ETP). This is when computers and telecommunications technology are used to capture, store, print and in many cases transmit prescription information. Early E-prescription systems where little more than custom written software that enabled a GP to generate and print off a prescription which could be signed and physically passed on for further processing. Even such primitive systems had numerous advantages over conventional paper based techniques:

- ∈ Elimination of eligible prescriptions. This eliminates the need for a pharmacy to contact a GP to confirm what has been written and reduces the possibility of dispensing incorrect medication. Cases have been documented where physicians have been tried and found guilty of negligence for writing eligible prescriptions.

- ∈ Reduction in forged or stolen prescriptions. A common problem mainly attributed to substance abuse where individuals seek unauthorised quantities of medication by modifying or generating hand written prescriptions.
- ∈ Easy issue of repeat prescriptions. When a patient requires a repeat prescription, the issuer can easily regenerate the new prescription from a cached copy.

Modern e-prescription systems allow the prescription data to be transmitted securely, via the internet, to the next point of concern. As well as this, many of the previously manual and time consuming tasks are now automated, these include (Sternberg, 2003).

- ∈ Administrative tasks, such as billing.
- ∈ Drug interaction checking. This involves warning the GP of any adverse effects caused by introducing new medications to a patient's current regimen.
- ∈ Real time formulary checking. Particularly important in regions where managed care plans dictate what medication a patient is entitled to.
- ∈ Drug allergy screening. Based on a patient's profile the system can flag medications that are not suitable for the patient.

Although obstacles such as initial cost and lack of interoperability are a hindrance (Chin, 2003), healthcare agencies in most countries recognise the potential advantages of such techniques and many have already established or are conducting trials on e-prescription systems.

4. MOBILE TECHNOLOGY

4.1. Mobile Telephony

The de facto standard for mobile telephony in Europe is the Global System for Mobile communications (GSM). GSM first appeared in the early 90's and since then its acceptance has been wide spread. The goal initially was to provide a mechanism for good quality voice communications using a digital cellular architecture. This architecture also provided some data functionality including low bandwidth data transmission and most popular of all SMS (Short Message Service) (Beolchi, 2003). Although potentially nearing the end of its lifecycle, the growth in the application of GSM technology shows no sign of slowing. Since the late 90's there has been much anticipation over the introduction of GPRS (General Packet Radio Service) and third generation technologies, particularly UMTS (Universal Mobile Telecommunications Service). The intention is that these technologies will eventually replace GSM ultimately with UMTS, using GPRS during the transition. GPRS uses GSM infrastructure with additional hardware and software. The packet switched nature of the service enables an always on connection and it yields superior bandwidth to GSM. It was forecast that GPRS would be introduced in late 2000 and rapid acceptance and widespread penetration would soon follow. The reality is slightly different, as only now, at the time of writing, do most mobile operators offer GPRS services accessible through a moderate range of handsets. Transmission speeds are also lower than first expected and the cost of using the service remains quite high. The delay in the role out of GPRS has had an effect on the timescales associated with 3G technology. Boasting ultimate speeds of up to 2Mbps, the first commercial 3G services are only starting to emerge.

As intended, the main utilisation of GSM technology has been in voice communications and SMS. Other data services have also used GSM as a bearer technology for data communications. The most prevalent being the Wireless Application Protocol (WAP). WAP was first established in 1997 by a consortium of telecommunication solution providers who realised the need for a framework to enable portable devices to connect to the WWW. The protocol was developed paying particular attention to the constraints of handheld devices such as limited connectivity bandwidth, small screen size, and low power. WAP specifies an end to end framework that allows a mirco browser, installed on a mobile device to retrieve specially developed content from the WWW via a gateway or proxy server. Unlike standard web content which is composed using HTML this content is assembled using WML (Wireless Mark-Up Language). Most mobile phones are now supplied with a WAP micro-browser installed and most if not all mobile operators have WAP gateways.

4.2. Personal Digital Assistants

Although the promise of the wireless broadband internet has been slow to materialise, the rapid growth in the integration of handheld computers in the practice of medicine continues. The prominent genre of handheld computer used today is the Personal Digital Assistant (PDA). The PDA began life almost 20 years ago when Psion introduced the Psion 1. This device incorporated all the functionality of its ancestor, the pocket calculator, along with simple organiser functions and the ability to store a limited amount of text. Evolution of these early devices was restricted as manufactures were constrained by integrated circuit and display technology at that time. It was not until almost a decade later, when Apple entered the portable computing arena, that significant advancements were made. Their device, the MessagePad, represented a major milestone in the evolution of the PDA. Key features of the device included a touch sensitive screen and hand writing recognition.

Modern PDA's are not so much characterised by their manufacturer, but more by the operating system which they support. The two main rivals in the PDA operating system arena are Palm and Microsoft. The Palm software, PalmOS was first developed for their hardware, the Palm Pilot series, but it has since been incorporated by several manufacturers. Palm have claimed dominance for several years now, however significant ground has been gained by Microsoft powered devices, particularly since the introduction of their latest generation software 'PocketPC'. Evaluation of both technologies comparatively is difficult. From a user's perspective opinions are divided. Some favour the windows based PocketPC environment because of its ergonomic similarity to

Microsoft's desktop oriented products, others feel that the simplicity, flexibility and wide spread market acceptance of PalmOS make it a more appropriate choice. From an organisational or development perspective opinions are as equally divided. Integration of PalmOS based devices with existing infrastructure is relatively straight forward due to the open nature of the platform and the level of developer support attributed to the extent of Palm's existence in the marketplace. The arguments supporting the ease of integration of devices using PocketPC is equally as just, a particular advantage claimed is that due to the similarities between the PocketPC environment and full blown Windows, porting of desktop applications to mobile platforms requires only incremental costs. As opposed to the Palm platform, where software will require substantial or complete redesign during migration.

4.3. Smart Phones

Since mid 2000 a new genre of devices has arrived. Many observers assumed this inevitable. These devices commonly referred to as 'Smart Phones' incorporate both PDA and mobile telephone technology inside one unit. This gives the user access to organizer type functionality as well as voice and data features. Although many connectivity peripherals have been developed for the classical PDA (including modem cards, bluetooth units and GSM modules), this is the first series of devices that have incorporated both technologies at manufacture. The major challenge in developing these devices is realising a unit that is small and portable enough to compete in the modern mobile phone market, yet large enough (particularly with respect to the display size) to function as a full blown PDA.

5. STATE OF THE ART IN MEDICATION MANAGEMENT

Over the past decade an unprecedented number of information and communication technologies have promised to affect healthcare delivery (Jadad and Delemothe, 2003). From a physician's perspective, the e-health era has brought about significant changes in the ways in which medicine is practiced. Assistance to and automation of tasks such as patient management and administrative duties have greatly reduced the manual overheads that medical professionals encounter. This new approach to healthcare has led to increased efficiency and better quality of care for patients. Although significant, it is reckoned that this is only one of a number of milestones evident in the evolution of modern patient care. An ongoing development which is considered significant is the increased adoption of mobile computing technologies by medical professionals. Classically a clinician's interface to an e-heath system would have been desktop based; however this is restrictive as most medical professionals require mobility. Modern handheld devices have not only supplemented the classical fixed systems but in many cases, particularly with the evolution of wireless data transfer, these mobile platforms have become the successor. Indeed, for many doctors, the organiser functions of these devices alone justify the expenditure (Al-Ubaydli, 2004).

Several technologies have been introduced and discussed in previous sections of this article. It is evident that the favoured platform in mobile e-health delivery is the PDA. Although limited in terms of system resources these devices compare favourably over their closest contender and bigger brother the notebook computer. Superiorities include:

- ∈ Size and Weight. A typical PDA (a standard Palm Pilot or Pocket PC) can easily be carried in the user's pocket and weighs little more than a mobile phone.
- ∈ Battery Life. Mainly due to the absence of a moving media (hard disk drive) and small screen size, most PDA's can operate for considerable periods (10 hours plus) without the need for recharging.
- ∈ Instant on. PDA operating systems do not perform a boot sequence; this gives medical professionals instant access to records/services.
- ∈ Cost. Latest release PDA's are typically less than half the cost of low end notebook computers.

The issue of connectivity is also interesting as more and more applications are providing direct connectivity via a wireless network to central databases and servers, where previously connectivity was usually gained via synchronisation with a PC. It is worth noting that the adoption of WAP using mobile phone technology in healthcare has been slow, particularly from a medical professional's perspective. It is believed that this can mainly be attributed to the fact that applications can not be sufficiently represented on devices with small screens and limited input characteristics. Most PDA's are also capable of supporting WAP based applications however in instances where connectivity has been fundamental the trend has been more towards custom written communications applications using the internet as the underlying infrastructure. Other approaches are also common, including the use of cut down web browsers, and in some cases proprietary technology like the web clipping system developed by Palm.

A plethora of medical applications are currently available for all platforms and both Microsoft and Palm boast their superiority in the marketplace. Irrespective of platform, most of the high end commercial applications offer a full suite of functionality, including patient record access, e-prescribing and medical reference. The e-prescribing aspect of these systems allow physicians to utilises all the benefits of modern fixed e-prescribing systems while on the move, either on a hospital ward or in a patient's home. A typical example of such a system is Rx+. This e-prescribing tool is part of a suite of clinical automation tools called 'Touchworks' by 'Allscripts Healthcare Solutions' (http://www.allscripts.com). Allscripts Rx+ is designed to run on the PocketPC and is an ambulatory medication management and prescription communications tool featuring drug and allergy interaction checking and plan-specific formularies.

6. COMPLIANCE CONTROL IN THE FUTURE

Several types of compliance aids have been discussed in earlier sections of this article. Although many of these devices can be described as portable or mobile compliance aids, most of them that support connectivity are home based. Indeed many of the home based systems that do have communications functionality use proprietary protocols and do not utilise existing data communications infrastructures such as the Internet or wireless technology. All of the existing solutions are custom built and serve no other purpose other than managing medication. A slightly different approach to compliance controls and monitoring is the use of existing technology and devices. A typical example of such a system is the use of a pager or SMS to send an alert to a patient when it is time to take medication. This approach is particularly cost effective if the patient has a mobile telephone. Another system that has been adopted is the use of PDA's with customised software that alerts the patient at medication time. Some systems also allow the patient to flag that they have taken their medication. This is nothing more than an elaborate electronic diary system, however this is favoured over paper based records as when a patient flags that they have consumed the medication the event is time and date stamped. In paper based systems a patient can forge the results by amending the medication record at any time. This feature is particularly important when using compliance assistive techniques in clinical drug trials.

It is believed that the most efficient approach in modern compliance control is to utilise existing technologies and concepts in tailored devices. One example of this approach is the MEDICATE system. This system is being developed by a consortium of European entities and is the result of a research program investigating the issues surrounding the compliance problem. The MEDICATE solution consists of a home based compliance device that is capable of dispensing medication and recording patient compliance. A novel feature of this system is that it incorporates a detachable module that allows the patient to take their medication management out of the home. This detachable component can hold a day's supply of medication and incorporates the necessary electronics to remind the patient and record compliance. The detachable module can also be synchronised with the home based system when in close proximity (within the home) via a wireless link. The wireless link uses bluetooth technology, a standard designed to facilitate short range radio communications. As well as consisting of a patient oriented compliance device the MEDICATE system supports medication management from a medical professionals perspective through the implementation of a care model. This care model allows GP's, Pharmacists, care workers and any other entities in the supply to intake chain for medication to be interconnected. The home based compliance aid is also connected to this care model using an inbuilt modem and a conventional telephone line. The system uses internet technology and doctors can prescribe either from a PDA or a desktop platform. When a doctor enters a prescription it can be sent to the next operator concerned, usually a pharmacist. The pharmacist can then administer the medication appropriately. The prescription data is also sent directly to the unit in the patient's home, and when the actual medication is loaded into the device the patient can rely on the device for timely reminders of when medication should be taken, based on the up loaded data. When a patient fails to take their medication, the event will be recorded and if critical a carer can be contacted. A central server and database act as a hub for the system and through time a profile of patient's medication compliance patterns can be established. This profile can be used in further consultations and prescription issues as it will highlight instances of non-compliance which the GP can take action against.

It is clear that compliance is a significant issue from a patient and healthcare perspective. With advances in technology it is likely that complex care models supporting mobile medication management and prescribing will prevail.

7. ACKNOWLEDGEMENTS

This work was supported in part by a project IST-2000-27618 entitled 'The control, identification and delivery of prescribed medication' funded by the European Union.

8. REFERENCES

Al-Ubaydli, M., 2004, Clinical review: handheld computers, *British Medical Journal.* **328**:1181-1184.

Aswad, S., Devera-Sales, A., Zapanta, R., and Mendez, R., 1993, Costs for successful versus failed kidney transplantation: a two-year follow-up, *Transplantation Proceedings.* **25**: 3069-70.

Beolchi, L., 2003, *Telemedicine Glossary: Glossary of Concepts, Technologies, Standards and Users, 5TH Edition,* European Commission, Brussels.

Chin, T., 2003, 5 obstacles to e-prescribing, *American Medical News.* **46**(18):17-18.

Corlett, A. J., 1996, Aids to compliance with medication, *British Medical Journal.* **313**:926-929.

Department of Health, Prescriptions Dispensed in the Community Statistics for 1990 to 2000, England (August, 2000); http://www.doh.gov.uk/pdfs/sb0119.pdf

Deloitte & Touche, The Emerging European Health Telematics Industry, Market Analysis, February 2000.

Jadad, A. R., and Delamothe, T., 2003, From electronic gadgets to better health: where is the knowledge?, *British Medical Journal.* **327**(7410):300.

Lee, R. D., Conley, D. A., and Preikschat, A., eHealth 2000: Healthcare and the Internet in the New Millennium, WIT Capital industry report (2000); http://www.witcapital.com.

Rivers, P. H., 1992, Compliance aids – do they work?, *Drugs and Aging.* **2**:103-11.

Sternberg, D. J., 2003, Make life a little easier, *Marketing Health Services.* **23**(2):44.

Szeto, A. Y. J., and Giles, J. A., 1997, Improving oral medication compliance with an electronic aid, *Medicine & Biology Magazine.* **16**:48-58.

Uquhart, J., 1993, When outpatient drug treatment fails: identifying non-compliers as a cost-containment tool, *Medical Interface.* **6**:65-73.

Walker, R., 1992, Stability of medical products in compliance devices, *The Pharmaceutical Journal.* **248**:124-6.

Wertheimer, A. I., and Santella, T. M., 2003, Medication compliance research: still so far to go, *Journal of Applied Research in Clinical and Experimental Therapeutics.* **3**(3).

FUTURE CHALLENGES AND RECOMMENDATIONS

Dewar D. Finlay* and Chris D. Nugent

The preceding chapters have demonstrated how mobile and wireless technologies are a fundamental ingredient in the convergence of information systems, telecommunications and the practice of medicine. Addressing the provision of healthcare along the continuum that exists between the extremes of acute and chronic care, the presented scenarios and case studies have provided a useful insight into how real-life solutions can be realised and evaluated. In many cases in the past, similar proposed systems and services were speculative and were only promoted in theory. Evaluation of existing and more contemporary technologies has also played a significant part in several of the studies; trends in wireless communications, handheld devices and internet technology have all been reported.

From a purely technological perspective the main emphasis has been on the exchange of information in settings ranging from the distribution of medication to the management of large scale humanitarian efforts in the form of disaster relief. In the studies that are oriented toward the monitoring of chronically ill patients, the main objective has been to provide a communications platform that optimizes care and quality of life for the patient groups concerned. In the remaining studies the emphasis is on how acute care can be delivered in both naturally and politically hostile regions.

Although seemingly diverse in application, the assessments of the technologies that are available are similar. The main observation from the work performed is the trend away from the reliance on the underlying technological mediums. General Packet Radio Service (GPRS), Bluetooth and Wireless Local Area Networks (WLAN or Wireless LAN) are all examples of the contemporary technologies that are available, however, in most of the studies, the substantial innovations are apparent in the ways in which these technologies are utilised to provide a 'service'. A prime example of this emphasis on service arrangement is the need for guarantee of service in situations of disaster management. A slightly more subtle observation is the migration away from custom and

---

* Dewar D Finlay, University of Ulster, Jordanstown, Northern Ireland, BT37 0QB. Email: d.finlay@ulster.ac.uk.

∈

standalone devices. Architectures presented in several of the studies illustrate the model where common functionality, such as synchronisation and communication facilities, are located on a central device. This central device acts as a hub for satellite devices such as ECG monitors and medication dispensers.

The work presented is a comprehensive representation of the current state-of-the-art in patient and healthcare management. Nevertheless, it goes without saying that in terms of the evolution of such systems and services, we still have some ground to attain. Historically, the substantial challenges facing the developers of telemedicine and telecare systems originated from technological primacy. Now, however, apparent through the wide spread utilisation of Information Technology, we seem to be approaching a point of technical saturation. This convergence has helped observers realise and understand the new challenges and barriers that stand in the way of the ultimate telemedicine solution. The observation that success is no longer measured on the technical performance and validation of a solution, but rather on the uptake of the service that is being provided, leads us to what is probably the most profound challenge that lies ahead. Along with the focus on requirements at an organisational and multinational level, effective planning, training and integration with existing infrastructure are all factors that must be considered.

AUTHOR INDEX

SUBJECT INDEX

Lightning Source UK Ltd.
Milton Keynes UK
UKHW050836180822
407344UK00007B/10